Environmental Engineering: Technology and Implementation

Environmental Engineering: Technology and Implementation

Editor: Chloe Miller

STATES
ACADEMIC PRESS
www.statesacademicpress.com

States Academic Press,
109 South 5th Street,
Brooklyn, NY 11249, USA

Visit us on the World Wide Web at:
www.statesacademicpress.com

ISBN: 978-1-63989-181-8 (Hardback)

Cataloging-in-Publication Data

Environmental engineering : technology and implementation / edited by Chloe Miller.
p. cm.
Includes bibliographical references and index.
ISBN 978-1-63989-181-8
1. Environmental engineering. 2. Environmental protection--Technology transfer.
3. Green technology. I. Miller, Chloe.
TD145 .E58 2022
628--dc23

Table of Contents

Preface

Environmental engineering is a branch of engineering that integrates key principles of ecology, chemistry, biology, hydrology, geology, microbiology, hydraulics and mathematics. It aims to develop solutions for protecting and improving the health of organisms and quality of the environment. It is a sub-discipline of chemical and civil engineering. There is a need to understand the ramifications of our increased adoption of new technologies on the environment. Environmental issues such as global warming, acid rain and environmental pollution are a result of human activities. Environmental engineering seeks to address such local and worldwide phenomena, evaluate the severity of such events, and develop methods of treatment and containment. Its applications lie in the areas of wastewater management, water supply and water treatment, air pollution management, and environmental impact assessment and mitigation. This book is a valuable compilation of topics, ranging from the basic to the most complex advancements in the field of environmental engineering. The topics covered in this extensive book seek to address the emerging technologies developed using environmental engineering studies and their implementation. For someone with an interest and eye for detail, this book covers the most significant topics in this field.

This book unites the global concepts and researches in an organized manner for a comprehensive understanding of the subject. It is a ripe text for all researchers, students, scientists or anyone else who is interested in acquiring a better knowledge of this dynamic field.

I extend my sincere thanks to the contributors for such eloquent research chapters. Finally, I thank my family for being a source of support and help.

Editor

1

CHANGES IN HABITAT SUITABILITY FOR WATERBIRDS OF THE MOMOGE NATURE RESERVE OF CHINA DURING 1990–2014

Yue ZHANG[a, b, c], Zongming WANG[a], Chunying REN[a], Hao YU[a], Zhangyu DONG[a], Chunyan LU[a], Dehua MAO[a]

[a]*Key Laboratory of Wetland Ecology and Environment, Northeast Institute of Geography and Agroecology, Chinese Academy of Sciences, 130102 Changchun, China*

[b]*College of Resources and Environment, Jilin Agricultural University, 130118 Changchun, China*

[c]*University of Chinese Academy of Sciences, 100049 Beijing, China*

Abstract. There is increasing empirical evidence that changes in habitat quality play an important role in determining species distributions and biodiversity. However, most research has focused on habitat quality, and we still lack approaches for tracking habitat quality dynamics. In this paper, by establishing qualitative and quantitative relationships between waterbird populations and key habitat indicators such as water abundance, food, shelter conditions and disturbance, we developed an object-oriented classification method, in conjunction with a geographic information systems (GIS) based centroid moving method, to assess habitat suitability dynamics for waterbirds at the Momoge Nature Reserve, China. Our results showed that habitat suitability improved during 1990–2000 and declined during 2000–2014. Habitats with very good and good grades increased by 71.47 km^2 (4.88%) during 1990–2000 and decreased by 200.66 km^2 (13.78%) during 2000–2014. The habitat area with a good grade moved to the north, while the habitat area with a poor grade moved to the south during 1990–2014. This was mainly because the surrounding cropland area increased and shifted as oil and gas projects developed. These findings suggest that our object-oriented classification and centroid moving methods have great potential for use in biodiversity conservation and ecosystem management.

Keywords: habitat suitability, waterbirds, centroid, remote sensing, the Momoge Nature Reserve, China.

Introduction

Over the past 50 years, humans have caused rapid and extensive declines in global biodiversity (Millennium Ecosystem Assessment 2005) despite our increased focus on nature conservation and wildlife management. Globally, 85% of threatened bird species are at risk because of habitat loss and degradation. Waterbirds, among the most symbolic ecological assets in wetlands, are good indicators of environmental changes and are natural resources of great ecological value (Reid *et al.* 2013). Waterbirds show high mobility and aggregation in response to the fluctuations in resources and the demands of their different life requirements (Cumming *et al.* 2012). Natural and semi-natural habitats are exposed to growing pressure from the intensification of agriculture, forestry and climate change (Weiers *et al.* 2004). The changes of habitat quality would affect the distribution, diversity and abundance of waterbirds (Şekercioğlu *et al.* 2004). Wetlands support extensive food chains and biodiversity and are natural regulators of a number of environmental problems, such as flooding, drought and water quality (Williams 1990). Wetlands are critical habitats for a large variety of waterbirds (Ramsar 2012), and a major reason for decreasing bird populations is habitat degradation (Rönkä *et al.* 2008). Obtaining information on habitat distribution and monitoring habitat conservation is a prominent research area (Corbane *et al.* 2015) and requires both an evaluation of the current suitability of habitat and detection of habitat changes.

As the coupling of remote sensing and GIS technologies can provide quantitative and cost-effective assessment, these technologies are widely recognized for their applicability in wetland monitoring (Euliss *et al.* 2011) and ecological research. Remotely sensed data provide measurements and surrogates that are directly related to vegetation type and structure, biomass, and other ecosystem variables that collectively improve our understanding of habitat characteristics (Bradley *et al.* 2012). There are

Corresponding author: Zongming Wang
E-mail: zongmingwang@iga.ac.cn

numerous examples demonstrating that the addition of remotely sensed variables improves the accuracy of habitat suitability models. Waterbirds select habitats that meet their ecological requirements of providing food and shelter, as well as abundant water and minimum disturbance. Therefore, in many studies, remotely sensed variables serve as proxies for water, food, disturbance and other attributes of habitat quality (e.g., Dong *et al.* 2013; Tian *et al.* 2008). Human presence and human-induced environmental changes also affect the suitability of habitats, and thus, land cover change serves as a proxy for habitat loss (Wang *et al.* 2015). Accurate assessment of habitat quality depends on a profound understanding the habitat requirements of waterbirds, followed by effective modelling of habitat suitability.

In many cases, habitats can be very dynamic, affecting animal distributions and populations, so detecting habitat dynamics can improve our understanding of habitat-abundance targets (e.g., Gardner *et al.* 2007; Nakamura *et al.* 2016). Habitats can be dynamic in response to climate changes, conservation efforts, or land use changes (Anteau 2012). Measuring habitat status and understanding habitat dynamics have become important for waterbird conservation (Liu *et al.* 2010). However, to our knowledge, most of the debate to date has focused on habitat quantity, while changes in habitat quality have been overlooked (Hodgson *et al.* 2009). Accurately assessing habitat dynamics would clearly be useful for many aspects of research, conservation and management (Anteau *et al.* 2014).

The West Songnen Plain has the greatest concentration of marsh wetlands in China and is noteworthy for its biodiversity (Wang *et al.* 2009). This area also serves as an important stopover and wintering site for migratory waterbirds. However, wetland degradation, due to climate warming and anthropogenic pressures resulting from the reclamation of farmlands, construction of water conservation projects, and petroleum and gas developments (Pan *et al.* 2006), has become a serious environmental problem. The Momoge Nature Reserve (MNR), which was designated as a reserve of international importance, contains a large area of natural wetland in the Western Songnen Plain and has been severely impacted by intensive human activities. In recent years, waterbird populations in the Momoge Nature Reserve have declined, primarily in response to habitat loss, fragmentation and degradation (Xiao *et al.* 2014). Several of these waterbird species are red-listed (Ramsar 2012). Thus, there is an urgent need to protect and restore waterbird habitat. More attention should be given to quantitatively assessing habitat suitability for waterbirds and the dynamics of habitat quality in this region. The objects of this study were: (1) to select the key ecological indicators that have prominent impacts on waterbird habitat, (2) to quantitatively assess the habitat suitability grades in the Momoge Nature Reserve, and (3) to track and analyse dynamics and spatial characteristics of habitat suitability for waterbirds during 1990–2014. We hope that these results supply accurate information about habitat suitability changes and serve as a tool for biodiversity conservation.

1. Materials and methods

1.1. Study area

The MNR (45°42′33″–46°17′59″N, 123°27′09″–124°4′32″E) is located in the transition zone between deserts and grasslands in the northwestern part of Jilin Province, China (Fig. 1), at an altitude ranging from 128.0–160.7 m above sea level. The total area is

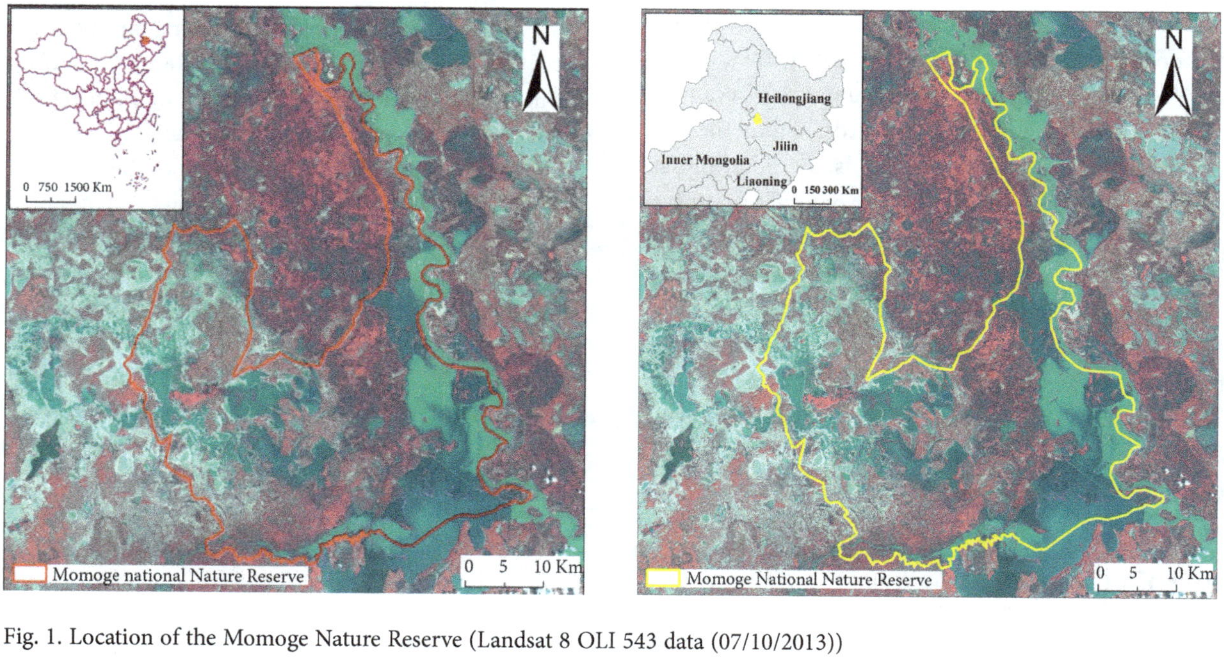

Fig. 1. Location of the Momoge Nature Reserve (Landsat 8 OLI 543 data (07/10/2013))

approximately 1440 km^2. The reserve has a temperate continental monsoon climate, with average annual precipitation of 391.8 mm and an average annual temperature of 4.2 °C (Yang *et al.* 2012). Winter is cold with heavy snow, while summer is hot with heavy rain. The MNR mainly draws off floodwater from the Nenjiang River, Tao'er River and Erlongtao River.

The MNR was established in 1981, listed in the Chinese National Nature Reserves in 1997 and designated as internationally important during the Ramsar Convention in 2013. The MNR plays a key role in regulating the regional climate, conserving biodiversity and protecting endangered waterbird species. Thus, the MNR receives great attention from international organizations, such as the WWF (World Wildlife Fund), GEF (Global Environment Facility) and ICF (International Crane Foundation) (Pan *et al.* 2006). It is an inland wetland ecosystem mainly composed of lakes and marshes, serving as an important habitat for a variety of threatened migratory birds, such as the Siberian crane, hooded crane, red-crowned crane and oriental stork. There are 193 recorded bird species in the MNR (Ramsar 2012). However, human activities, such as petroleum exploitation, have led to the degradation of the waterbird habitat in recent years.

1.2. Data sources

Satellite datasets from the Landsat Thematic Mapper (TM), Enhanced Thematic Mapper (ETM), Operational Land Imager (OLI), digital elevation model (DEM) and other ancillary data were used in this research. Landsat TM images were obtained from the United States Geological Survey (USGS) Landsat archive (http://glovis.usgs.gov/). The study area is located in the Path/Row 120/28 of the World Reference System 2 (WRS-2), and three images acquired from 25 June 1990, 6 July 2000, and 10 July 2014 were available for the study area. Satellite images from the same season were selected in order to decrease seasonal effects. Each Landsat TM image had a resolution of 30 m. DEM data were obtained from the Shuttle Radar Topography Mission (SRTM) (http://srtm.csi.cgiar.org/) with a resolution of 90 m and used to derive the slope factor. The ancillary data included road and river vector data, which were available at the AMUR-HEILONG RIVER BASIN Information Center (http://amur-heilong.net/). Prior to image classification for land cover mapping, all images were geo-rectified to topographic maps using ground control points (GCPs). The ENVI 4.8 software and ArcGIS 9.3 software were used for data processing.

1.3. Habitat suitability assessment approach

The object-oriented classification method was adopted for land cover maps, which were derived from Landsat images acquired in 1990, 2000 and 2014. To quantitatively assess habitat suitability, several indicators, including water availability, food, shelter conditions and disturbance, were selected to represent the habitat requirements of waterbirds. By numerically relating occurrence records for waterbirds to a suite of environmental variables from those locations, remote sensing-based indicators that were well correlated with field-based indicators of habitat status could be used as proxies (Strasser *et al.* 2014). In this paper, six indicators were identified in the analysis of habitat suitability for waterbirds, including shelter condition (land cover type and slope), food abundance (Normalized Difference

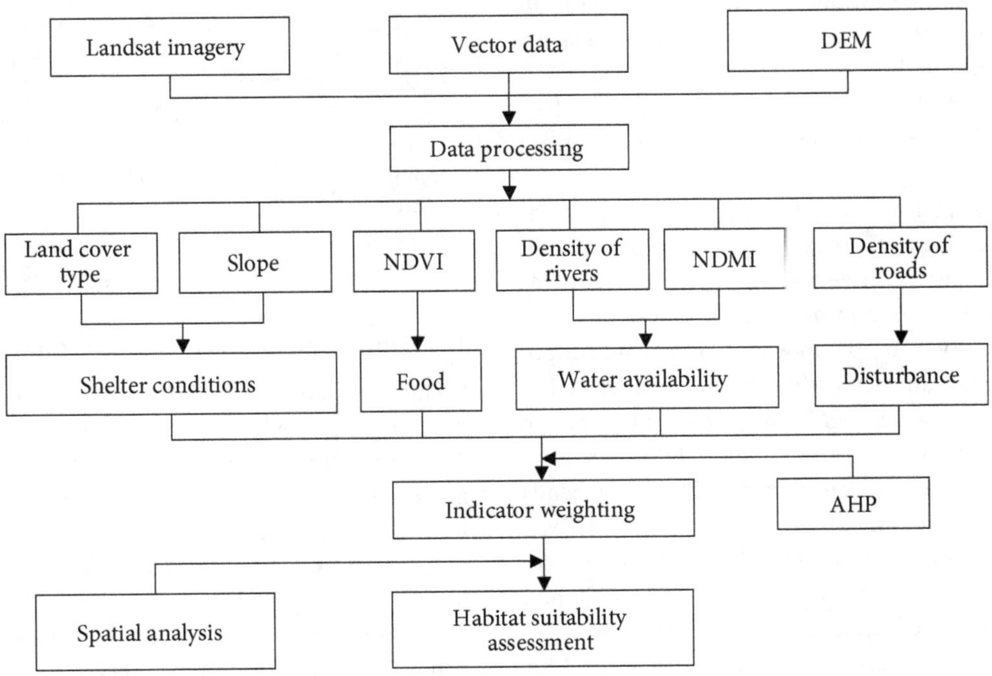

Fig. 2. Flow chart of the waterbird habitat suitability assessment in MNR

Vegetation Index, NDVI), water availability (the density of rivers and NDMI), and disturbance (the density of roads). Then, an analytic hierarchy process (AHP) method was applied to determine indicator weights. In this way, an assessment matrix of indicators and their weights was established. Based on the geo-referenced database, an assessment of habitat suitability could be performed. Figure 2 presents a flow chart showing how to obtain habitat quality information using Landsat imagery with object-oriented and AHP methods.

1.3.1. Identification of habitat indicators

Among the six indicators of habitat quality, land cover was the dominant proxy variable that informed habitat or supporting ecosystem services. Land cover data were obtained from Landsat images by object-based classification in conjunction with visual interpretation. Object-based segmentation is a method that allows for mapping complex, hierarchical habitat systems (Strasser, Lang 2015). The object-based classification was performed using eCognition 8.64 software (Kartikeyan *et al.* 1998). Based on natural vegetation succession and human land use types, land cover maps for 1990, 2000 and 2014 were classified into seven categories: wetland, cropland, woodland, grassland, water body, built-up land and barren land. The classification accuracy was assessed using historical data during 1990–2000 and through field investigations during 2013–2014. The total classification accuracies of the land cover maps for 1990, 2000 and 2014 were >90%, meeting the needs of this study. NDVI was significantly correlated with vegetation biomass, primary production and greenness and also accounted for intra- and inter-annual changes in the environment (Cord, Rödder 2011). NDVI could therefore be used as a potential measure of forage for waterbirds. In this study, NDVI was a normalized transformation of the Landsat near-infrared (TM4, ETM4, OLI5) to red (TM3, ETM3, OLI4) reflectance ratio, with values between −1–1. NDVI values for 1990, 2000 and 2014 were derived from Landsat imagery.

Water availability is another important aspect for waterbirds selecting a stopover and breeding site and depends on the density of rivers at the regional level. Otherwise, the plant community also depends on soil wetness and submersibility. Thus, the moisture of the land surface becomes an important factor at the microhabitat level. The density of rivers was derived from river vector data. The normalized difference moisture index (NDMI) (Wilson, Sader 2002) is the normalization of the difference in bands of moisture between Landsat near-infrared (TM4, ETM4, OLI5) and short wave-infrared (TM5, ETM5, OLI6) reflectance ratios. The NDMI values from 1990, 2000 and 2014 were also derived from Landsat imagery. Land cover and slope reflect the safety of shelter and are crucial for waterbirds selecting a place to rest (Brotons *et al.* 2004). The slope angle of the study area was derived from DEM data.

Finally, wild cranes are particularly wary and will abandon their nests if disturbed by human activity or predators (Zou, Wu 2009). The density of roads has a direct effect on waterbird habitat (Hirzel *et al.* 2001) and was derived from the road vector data using the spatial analysis function of ArcGIS 9.3.

1.3.2. Calculation of habitat suitability index

To calculate the Habitat Suitability Index, all habitat indicators were translated into numerical data, and all data were standardized using the spatial analysis tools of ArcGIS 9.3 software. Habitat suitability was graded as poor (0–25), fair (25–50), good (50–75) and very good (75–100).

As the impact of each indicator had a different effect on habitat suitability, reliable weights had to be determined for each indicator. An AHP method was applied for the indicator weighting. AHP can be used to create decision aids that simplify and expedite the shared decision making process, with effective and feasible results (Dolan 2008). This method can combine the relationships between habitat suitability indicators (Ludwig, Ianuzzi 2006). Table 1 presents the AHP results that were used to determine the weight of each indicator of habitat suitability. Then, a linear function was adopted to calculate an overall index of habitat suitability for waterbirds.

Table 1. The indicator weights for waterbird habitat suitability determined using an AHP method

Objective level	Criterion level		Index level	
	Indicators	Weights	Indicators	Weights
Waterbird Habitat Suitability Index in the Momoge Nature Reserve	Water abundance	0.45	Density of rivers	0.34
			NDMI	0.11
	Food	0.32	NDVI	–
	Shelter condition	0.15	Land cover type	0.12
			Slope	0.03
	Disturbance	0.08	Density of roads	–

1.4. Habitat suitability change detection

Changes in habitat suitability for waterbirds were derived from images acquired during the same season in 1990, 2000 and 2014. The spatio-temporal analysis of changes in habitat grades was performed using ArcGIS 9.3 software. Spatial analysis was carried out to describe patterns of grade changes and to measure the rate of changes that occurred during 1990–2014.

In this study, the centroid coordinates approach (Li *et al.* 2007) was used to obtain habitat suitability patches and to spatially track the dynamics of changes in habitat quality grades. Centroid coordinates were calculated as:

$$X_c = \sum_{i=1}^{n} C_i X_i / \sum_{i=1}^{n} C_i, Y_c = \sum_{i=1}^{n} C_i Y_i / \sum_{i=1}^{n} C_i , \quad (1)$$

where X_c and Y_c are the centroid coordinates of area-weighted suitability types, X_i and Y_i are the centroid coordinates of i th patch, C_i is the area of all patches in i th suitable grade, n is total numbers of all landscape patches.

To offer more details about movement process on centroids of different habitat suitability grades, the movement distance $d_{t_{i+1}-t_i}$ and angle $\theta_{t_{i+1}-t_i}$ were calculated as follows:

$$d_{t_{i+1}-t_i} = \sqrt{(X_{t_{i+1}} - X_{t_i})^2 + (Y_{t_{i+1}} - Y_{t_i})^2} ; \quad (2)$$

$$\theta_{t_{i+1}-t_i} = \arctan \frac{| Y_{t_{i+1}} - Y_{t_i} |}{| X_{t_{i+1}} - X_{t_i} |} \cdot 180° / \pi , \quad (3)$$

where X_{t_i}, Y_{t_i} and $X_{t_{i+1}}$, $Y_{t_{i+1}}$ are the centroid coordinates of a certain suitability type in t_i and t_{i+1} year, t_i and t_{i+1} are the beginning and terminal years.

2. Results

2.1. Indices of suitability indicators

Six suitability indicators were selected in this study. Some indicators were static, such as the density of roads and rivers, and the slope; and some indicators varied with time, such as land cover type, NDVI and NDMI, which were mapped in 1990, 2000 and 2014.

The density of roads and rivers ranged between 0–72.19 and 0–118.99, respectively (Fig. 3a, b). The distribution map of slope, ranging between 0–8.56, is presented in Figure 3c.

Land cover maps of the study area are presented in Figure 4a–c. As the maps show, the wetland area increased by 164.27 km² during 1990–2000 and decreased by 104.57 km² during 2000–2014. The area of cropland increased continuously during 1990–2014, while the area of water body declined continuously during 1990–2014. Meanwhile, the spatial distribution patterns of three main land cover types, namely, wetland, cropland and water body,

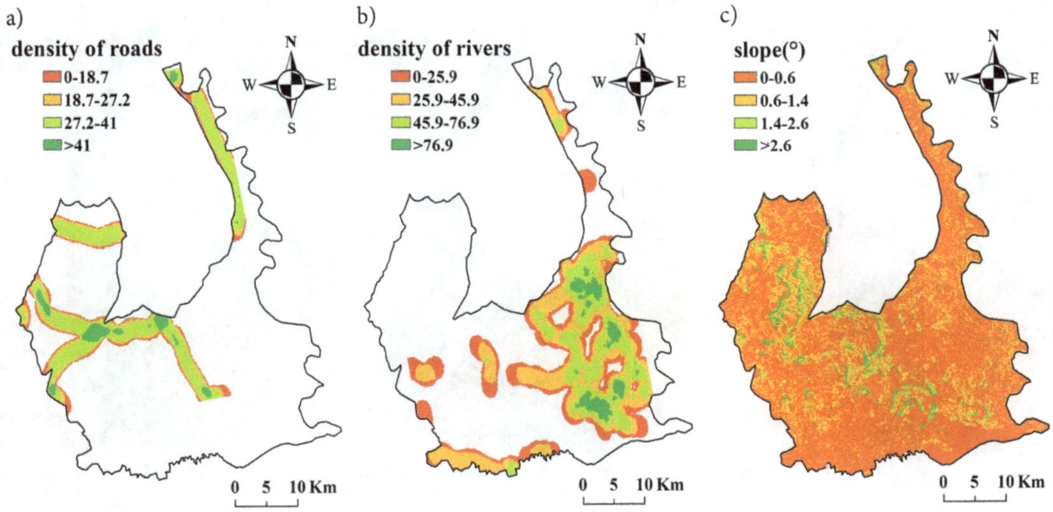

Fig. 3. Static waterbird habitat suitability indicators of the Momoge Nature Reserve: (a) density of roads, (b) density of rivers, and (c) slope

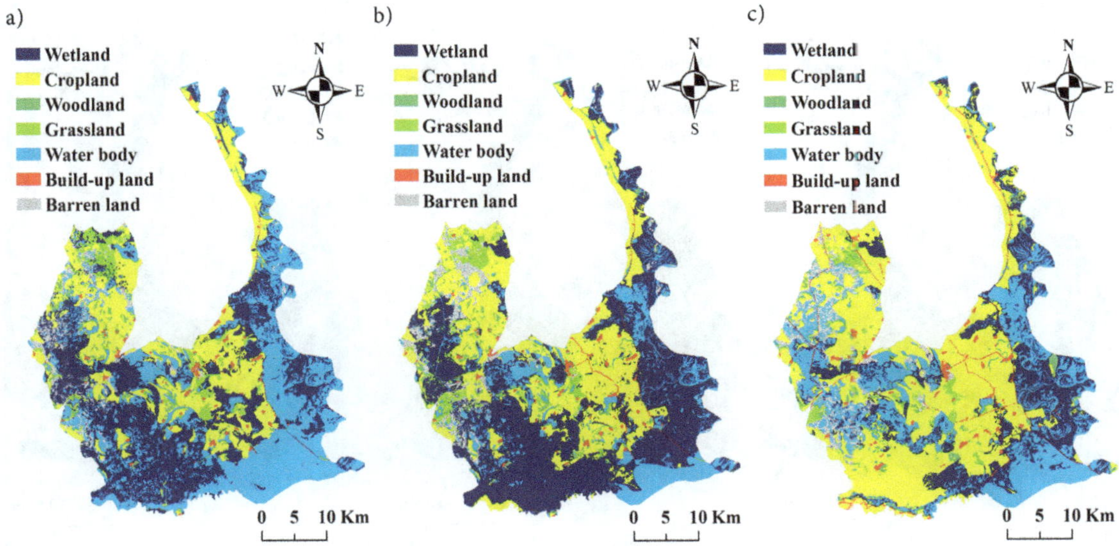

Fig. 4. Land cover maps for the Momoge Nature Reserve in 1990 (a), 2000 (b) and 2014 (c)

changed to different extents. The NDVI values extracted from the Landsat images in 1990, 2000 and 2014 ranged between −0.76–0.76, −0.49–0.74 and −0.48–0.75, respectively, as presented in Figure 5a–c. The area with a NDVI ranging between 0–0.2 increased, and the area with a NDVI of >0.5 declined. In the eastern part of the study area, NDVI values improved, and in the western part of the MNR, NDVI values became degraded. Wetness values were derived by NDMI transformation of Landsat TM images and ranged between −0.70–0.93, −0.53–0.93 and −0.36–0.50, as presented in Figure 6a–c. The area with a wetness value ranging between −80–0 declined, and the area with a wetness value <−80 increased. For the whole study area, the degree of wetness declined during the past 25 years.

2.2. Changes in the area of suitable waterbird habitat

The integrated index of waterbird habitat suitability was calculated according to the grades of very good, good, fair, and poor (Fig. 7). Most of the habitat patches of good and very good suitability were concentrated on both sides of the rivers and around the lakes or wetlands in the Nature Reserve. Habitat patches with a fair grade had the largest area, accounting for approximately 60–70% of the total study area. Habitat with a poor grade was the second-smallest in area, accounting for approximately 10–20% of the total study area. Most of the poor habitat patches were distributed around croplands or build-up lands.

The overall changes in suitability grades for waterbird habitat during 1990–2014 are displayed in Table 2. Our results showed that the least suitable habitat area, that with a grade of poor, decreased by 45.87 km^2 (3.13%), while the area with a fair grade increased by 176.43 km^2 (12.03%). The habitat areas with good and very good grades decreased by 129.19 km^2 (8.90%) during the study period. After the national reserve was established in 1997, local governments greatly improved management and water

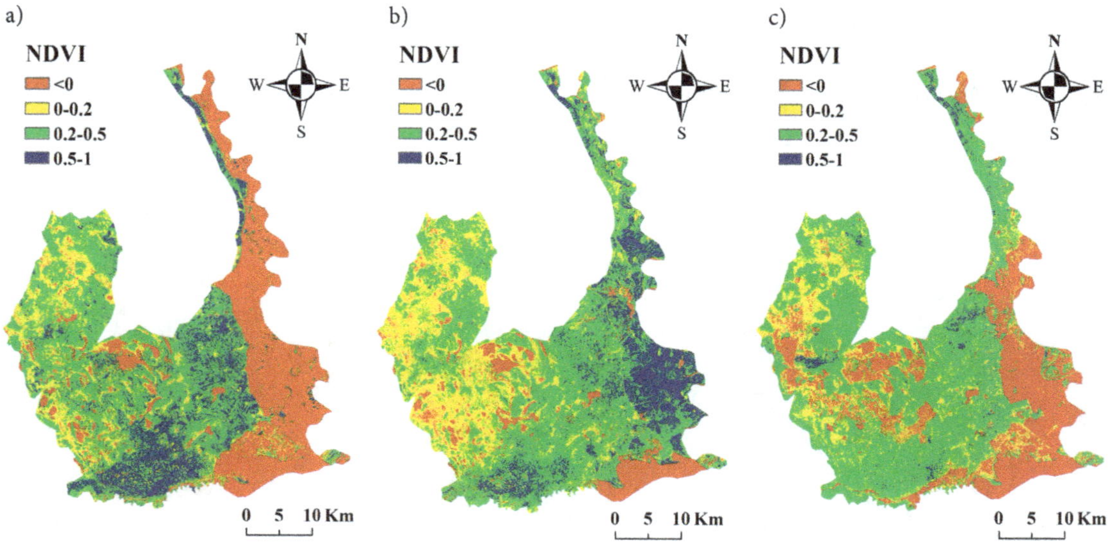

Fig. 5. NDVI maps for the Momoge Nature Reserve in 1990 (a), 2000 (b) and 2014 (c)

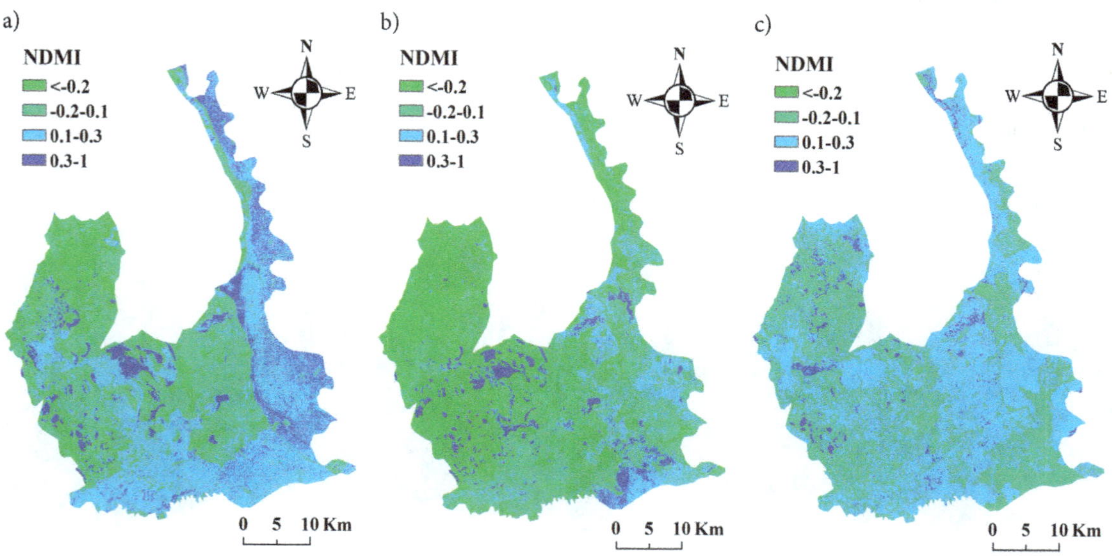

Fig. 6. NDMI maps for the Momoge Nature Reserve in 1990 (a), 2000 (b) and 2014 (c)

transfer projects for the wetlands, improving wetland habitat suitability to some extent. Habitats with very good and good grades increased by 71.47 km^2 (4.88%) during 1990–2000 and decreased by 200.66 km^2 (13.78%) during 2000–2014.

2.3. Trajectories of suitable grade centroids over time

The distribution of habitat suitability of each grade showed significant spatial differences. To determine the spatial habitat movements of each grade during 1990–2014, the moving centroid of each grade was tracked for different periods of time, as displayed in Figure 8. Our results showed that the centroid with a grade of fair shifted almost imperceptibly during the past 25 years. As the habitat with a fair grade always occupies the largest proportion of the study area, movement of the centroid was not obvious. For habitats with a poor grade, the centroid shifted by 10.57 km southwest during 1990–2000 and 6.59 km southeast during 2000–2014. For habitats with a good grade, the centroid shifted by 8.08 km northeast during 1990–2000 and 4.57 km northwest during 2000–2014. For habitats with very good grade, the centroid shifted by 3.72 km northeast during 1990–2000 and 4.44 km northwest during 2000–2014. In general, the habitat centroid shifted northward for habitats with a good grade and southward for habitats with a poor grade.

3. Discussion

3.1. Assessment of habitat suitability

The aim of our study was to determine the applicability of remotely sensed data and existing spatial data in the cost-effective assessment of habitat suitability changes for waterbirds. Most early studies applying species distribution models traditionally used spatial or climatic data only (Guisan, Zimmermann 2000). More recently, ecologists and conservation biologists have been increasingly utilizing remotely sensed indicators, such as NDVI and EVI (Enhanced Vegetation Index), in species habitat assessments. The habitat suitability assessment indicators should be selected carefully on the basis of the existing knowledge of the habitat requirements and characteristics of the species being modelled (Burnham, Anderson 2002). Waterbirds often have species-specific habitat requirements: Siberian Cranes exclusively use wetlands for nesting, feeding, and roosting; hooded cranes are found in shallow open wetlands and natural grasslands; and red-crowned cranes prefer to nest in marshes with relatively deep water and standing dead vegetation. Generally, wild cranes nest in shallow marshes with emergent herbaceous vegetation (Zou *et al.* 2003), and thus, areas distinguished as wetland are regarded as more suitable habitat. The six indicators selected in this study, land cover type, slope, NDVI, NDMI, and the density of rivers and roads,

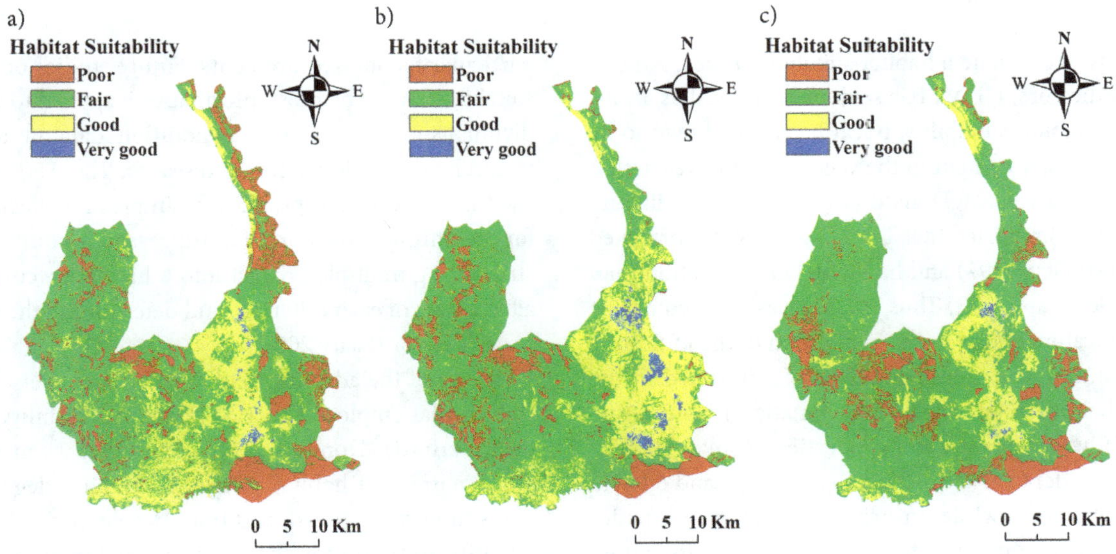

Fig. 7. Waterbird habitat suitability at the Momoge Nature Reserve in 1990 (a), 2000 (b) and 2014 (c)

Table 2. Changes in habitat area and proportion of habitat with different suitability grades

Grades	1990		2000		2014	
	Area (km^2)	Percentage (%)	Area (km^2)	Percentage (%)	Area (km^2)	Percentage (%)
Poor	279.06	19.04	186.85	12.75	233.19	15.91
Fair	853.49	58.22	874.24	59.63	1029.92	70.25
Good	326.27	22.26	379.78	25.91	198.64	13.55
Very good	6.78	0.48	24.74	1.71	5.22	0.29

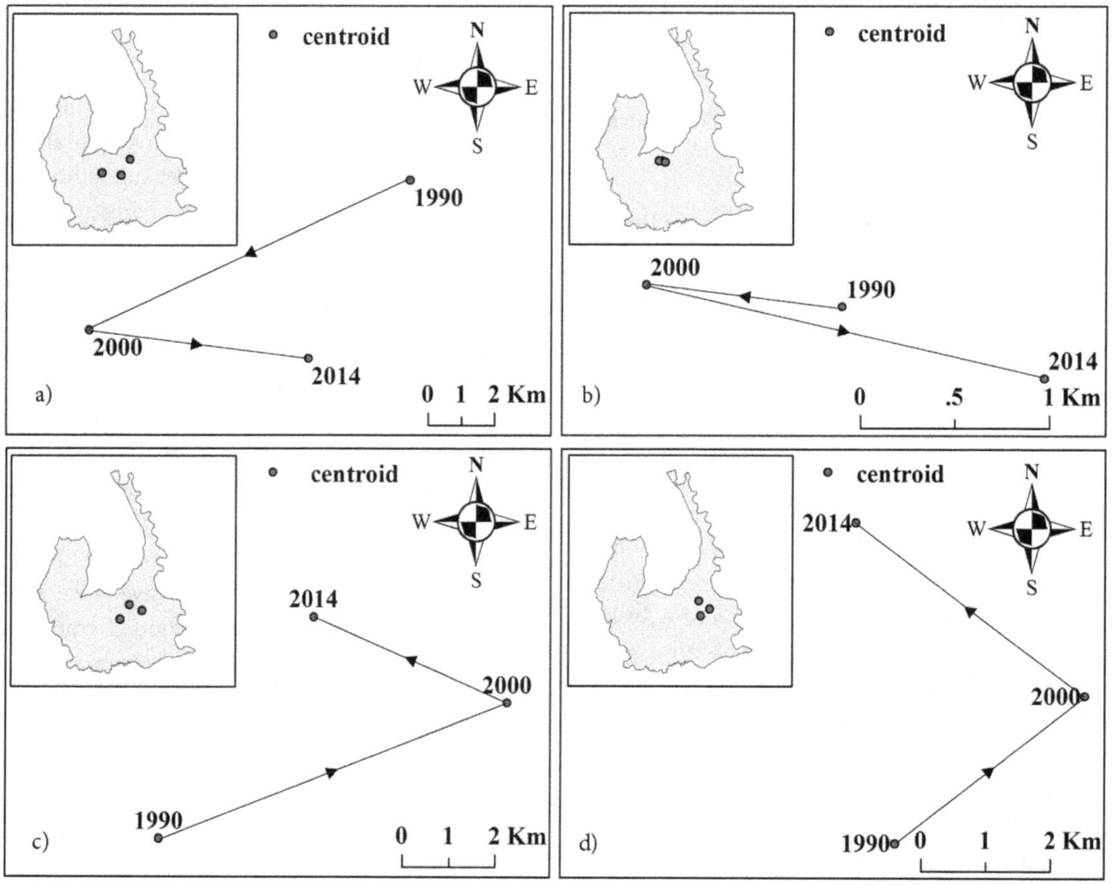

Fig. 8. The centroid trajectories of different habitat grades during 1990–2014: (a) poor suitability, (b) fair suitability, (c) good suitability, and (d) very good suitability

allowed us to calculate a habitat suitability index. Among the six indicators, NDMI has seldom been used as an indicator of habitat suitability, but it has been shown to be sensitive to water content in the vegetation and vegetation structure or cover (McDonald *et al.* 1998). Fiorella and Rippley (1993) reported that TM 4/5 was highly correlated with NDMI ($R^2 = 0.97$) and had a higher correlation than the Tasseled Cap Index. Thus, NDMI was selected as one of the indicators in this study. In addition to the indicators used here, other variables could be incorporated in the study of habitat suitability, such as climatic variables (Austin 2007), water quality, soil properties (Rönkä, Häkkilä 2005), water depth (Na *et al.* 2015), landscape and oil field distribution. Meanwhile, in order to quantify a broader range of vegetation characteristics, we could also extract summary statistics from the NDVI time series (e.g., start of season, length of growing season, and growth magnitude) (Wen *et al.* 2015). Moreover, the selection of all of the indicators could be applied at different spatial scales (e.g., sub-habitat level, patch level and landscape level), and multi-scale results for habitat suitability could be acquired. This multi-scale indicator selection method would make an analysis of habitat suitability more reasonable and feasible. However, although waterbirds co-exist and share common habitat elements, each species has its own particular habitat requirements. Future studies on habitat suitability would be most informative if applied to each individual species of interest. In addition, data on nest sites would be desirable for future research. The AHP method used in this research provided a simple and effective way for weighting indicators. Future research could involve structuring multiple criteria into a hierarchy, comparing alternatives of each criterion, and determining the weights of each factor (Saaty 2008).

One of the advantages of our research over previous studies that employed an HSI (Habitat Suitability Index) for waterbirds (Dong *et al.* 2013) was that we monitored the dynamics of habitat suitability. Habitat degradation tends to be widespread, and thus, developing methods to quantify and monitor changes in habitat quality or pressures are critical for adaptively managing protected areas (Nagendra *et al.* 2013). In this study, the direction and distance of changes were acquired by representing habitat shifts as vectors linking centroids from different time periods, describing details about the moving process of suitable habitats for waterbirds. Previous studies also gave more attention to habitat area, having no classification for habitat quality (Delgado, Marin 2013; Wen *et al.* 2015), while in this study, we made a detailed classification of the habitat area into very good, good, fair and poor grades.

In addition to applying unique methods satellite imagery from the same season was used to monitor temporal trends in habitat suitability. Multi-temporal characterizations of land cover, NDVI and wetness derived from imagery were good predictors of waterbird habitat and presented spatio-temporal characteristics in a form that was immediately usable in a geographic information system. This final result would add to evidence showing the importance of monitoring habitat dynamics. Possibly, environmental managers or researchers can use the methods described in this study to make habitat predictions as a substitute for field data when field surveys are not available. Silva *et al.* (2008) and Zomer *et al.* (2009) found that when dealing with wetland vegetation, Landsat images were useful for mapping macrophyte communities but not necessarily individual species. In the current research, high-spatial-resolution imagery (Belluco *et al.* 2006) and LiDAR (Zlinszky *et al.* 2012) have been used to map flooded and non-flooded vegetation types. Thus, further work could objectively consider using high-spatial-resolution imagery, aerial imagery and LiDAR data to perform genus-level wetland classification in order to improve the habitat suitability maps.

3.2. Changes in habitat suitability for waterbirds

Habitat loss, fragmentation and degradation are the greatest extinction threats to biodiversity (Fischer, Lindenmayer 2007). In the last few decades, over 50% of wetlands in the world have been lost, and the remaining wetlands have been degraded to different degrees because of human activities (Fraser, Keddy 2005). In that paper, it indicated that habitat recovery in the MNR occurred in the initial 10 years, but then habitat suitability became degraded during 2000–2014. In this study, we identified a 5.31% decline in the areas of very good and good habitat. The process of degradation involves gradual deterioration of habitat quality. In a degraded habitat, the number of species may decline or species may occur at lower density (Hazell *et al.* 2004) or have lower fitness. In the MNR, habitat degradation was associated with an increase in nearby cropland (62.34%) and a loss of water body (66.98%) during 1990–2014. Beginning in 1999, droughts lasting six years were another factor leading to the habitat degradation that occurred during 2000–2014 (Wang *et al.* 2009). Additionally, many hydraulic engineering projects that were built during the study period seriously damaged hydrological processes of the MNR wetlands.

Another concern for the MNR is the rich underground petroleum resources in the West Songnen Plain (Pan *et al.* 2006). The oilfield was listed as one of the four important oilfields that would be exploited more rapidly by the Sino-Petroleum Company in 2000. Crude oil output increased from 4 million tonnes to 260–300 million tonnes by the end of 2005 (Pan *et al.* 2006). Petroleum exploitation drives local economic development, but it leads to severe environmental impacts. Crude oil is toxic to water, soil and vegetation (Pan *et al.* 2006), with a great impact on land cover type. Petroleum exploitation led to wetland and water decline, causing habitat loss for waterbirds and imposing a great threat on the survival of waterbirds. This is consistent with the results that the habitat suitability of the study area declined from 2000–2014. To resolve the contradiction between a nature reserve and economic development, the boundaries of the core, buffer and transition ranges of the MNR were reshaped in July, 2002 (Wang *et al.* 2010). Meanwhile, the area of cropland surrounding the MNR has increased and shifted to the southwest in the past 25 years, which will also affect the spatial distribution of suitable habitat. As our results demonstrate, the centroid of habitat with a good grade shifted northward, while the centroid of habitat with a poor grade shifted southward. Generally, the moving process of suitable habitat reflected the impact of human activities.

In recent years, many researches have done much work on environmental protection (Ilgürel *et al.* 2016; Zhu *et al.* 2017; Li, Zhao 2017), and the results from this study could give some perspectives on environmental protection through assessing the present status of habitat quality and monitoring the changes in habitat quality for waterbirds. Considering the changing impact on waterbirds habitat, the ecological planning, design and construction of nature reserves is needed to meet the environmental requirement.

Conclusions

Landsat images and vector data, object-oriented segmentation, AHP and a calculation of centroid movement were used to assess the dynamics of habitat suitability for waterbirds in the MNR. These approaches showed that remote sensing and GIS can provide applicable information for spatial analysis methods, and thus be useful tools in assessing habitat quality and monitoring changes. Our results indicated that the habitat with a grade of poor decreased in area by 45.87 km^2 (3.13%), habitat with a fair grade increased in area by 176.43 km^2 (12.03%), and habitat with good and very good grades decreased in area by 129.19 km^2 (8.90%) from 1990–2014. The habitat suitability of the study area improved during 1990–2000, but habitat suitability declined during 2000–2014. As a whole, the centroid of habitat with a poor grade moved to the south and the centroid habitat with a good grade moved to the north during 1990–2014. For the habitat with a poor grade, the centroid shifted by 10.57 km to the southwest during 1990–2000 and 6.59 km to the southeast during 2000–2014. For the habitat with a good grade, the centroid shifted by 8.08 km to the northeast during 1990–2000 and 4.57 km to the northwest during 2000–2014. In conclusion, the approaches used here have great potential in modelling habitat suitability in highly heterogeneous and dynamic environments and may be used to help track habitat suitability dynamics in protected areas in other regions. Our findings provide up-to-date spatial information for biodiversity conservation in the MNR.

Acknowledgements

This study was jointly supported by the National Key Research and Development Project (No. 2016YFA0602301) and the National Natural Science Foundation of China (Nos. 41371403, 41401502 and 41671219).

Disclosure statement

We do not have any competing financial, professional, or personal interests from other parties.

References

Anteau, M. J. 2012. Do interactions of land use and climate affect productivity of waterbirds and prairie-pothole wetlands?, *Wetlands* 32: 1–9. https://doi.org/10.1007/s13157-011-0206-3

Anteau, M. J.; Wiltermuth, M. T.; Sherfy, M. H.; Shaffer, T. L. 2014. Measuring and predicting abundance and dynamics of habitat for piping plovers on a large reservoir, *Ecological Modelling* 272: 16–27. https://doi.org/10.1016/j.ecolmodel.2013.08.020

Austin, M. 2007. Species distribution models and ecological theory: a critical assessment and some possible new approaches, *Ecological Modelling* 200(1): 1–19. https://doi.org/10.1016/j.ecolmodel.2006.07.005

Belluco, E.; Camuffo, M.; Ferrari, S.; Modenese, L. 2006. Mapping salt-marsh vegetation by multispectral and hyperspectral remote sensing, *Remote Sensing of Environment* 105: 54–67. https://doi.org/10.1016/j.rse.2006.06.006

Bradley, B. A.; Olsson, A. D.; Wang, O.; Dickson, B. G. 2012. Species detection vs. habitat suitability: are we biasing habitat suitability models with remotely sensed data?, *Ecological Modelling* 244: 57–64. https://doi.org/10.1016/j.ecolmodel.2012.06.019

Brotons, L.; Thuiller, W.; Araújo, M. B.; Hirzel, A. H. 2004. Presence-absence versus presence-only modelling methods for predicting bird habitat suitability, *Ecography* 27: 437–448. https://doi.org/10.1111/j.0906-7590.2004.03764.x

Burnham, K. P.; Anderson, D. R. 2002. *Model selection and multimodel inference: a practical information-theoretic approach.* 2nd ed. New York: Springer-Verlag.

Corbane, C.; Lang, S.; Pipkins, K.; Alleaume, S. 2015. Remote sensing for mapping natural habitats and their conservation status-New opportunities and challenges, *International Journal of Applied Earth Observation and Geoinformation* 37: 7–16. https://doi.org/10.1016/j.jag.2014.11.005

Cord, A.; Rödder, D. 2011. Inclusion of habitat availability in species distribution models through multi-temporal remote-sensing data?, *Ecological Applications* 21: 3285–3298. https://doi.org/10.1890/11-0114.1

Cumming, G. S.; Paxton, M.; King, J.; Beuster, H. 2012. Foraging guild membership explains variation in waterbird responses to the hydrological regime of an arid-region flood-pulse river in Namibia, *Freshwater Biology* 57: 1202–1213. https://doi.org/10.1111/j.1365-2427.2012.02789.x

Delgado, L. E.; Marin, V. H. 2013. Interannual changes in the habitat area of the black-necked swan, cygnus melancoryphus, in the carlos anwandter sanctuary, southern Chile: a remote sensing approach, *Wetlands* 33: 91–99. https://doi.org/10.1007/s13157-012-0354-0

Dolan, J. G. 2008. Shared decision-making-transferring research into practice: the Analytic Hierarchy Process (AHP), *Patient Education Counseling* 73: 418–425. https://doi.org/10.1016/j.pec.2008.07.032

Dong, Z. Y.; Wang, Z. M.; Liu, D. W.; Li, L. 2013. Assessment of habitat suitability for waterbirds in the West Songnen Plain, China, using remote sensing and GIS, *Ecological Engineering* 55: 94–100. https://doi.org/10.1016/j.ecoleng.2013.02.006

Euliss, N. H.; Smith, L. M.; Liu, S.; Duffy, W. G. 2011. Integrating estimates of ecosystem services from conservation programs and practices into models for decision makers, *Ecological Applications* 21: S128–S134. https://doi.org/10.1890/09-0285.1

Fiorella, M.; Rippley, W. J. 1993. Analysis of conifer forest regeneration using Landsat thematic mapper data, *Photogrammetric Engineering and Remote Sensing* 59(9): 1383–1388.

Fischer, J.; Lindenmayer, D. B. 2007. Landscape modification and habitat fragmentation: a synthesis, *Global Ecology and Biogeography* 16: 265–280. https://doi.org/10.1111/j.1466-8238.2007.00287.x

Fraser, L. H.; Keddy, P. A. 2005. *The world's largest wetlands: ecology and conservation*. Cambridge: Cambridge University Press. https://doi.org/10.1017/CBO9780511542091

Gardner, T. A.; Barlow, J.; Peres, C. A. 2007. Paradox, presumption and pitfalls in conservation biology: the importance of habitat change for amphibians and reptiles, *Biological Conservation* I38: I66–I79. https://doi.org/10.1016/j.biocon.2007.04.017

Guisan, A.; Zimmermann, N. E. 2000. Predictive habitat distribution models in ecology, *Ecological Modelling* 135: 147–186. https://doi.org/10.1016/S0304-3800(00)00354-9

Hazell, D.; Hero, J. M.; Lindenmayer, D.; Cunningham, R. 2004. A comparison of constructed and natural habitat for frog conservation in an Australian agricultural landscape, *Biological Conservation* 119: 61–71. https://doi.org/10.1016/j.biocon.2003.10.022

Hirzel, A. H.; Helfer, V.; Metral, F. 2001. Assessing habitat-suitability models with a virtual species, *Ecological Modelling* 145: 111–121. https://doi.org/10.1016/S0304-3800(01)00396-9

Hodgson, J. A.; Moilanen, A.; Thomas, C. D. 2009. Metapopulation responses on patch connectivity and quality are masked by successional habitat dynamics, *Ecology* 90: 1608–1619. https://doi.org/10.1890/08-1227.1

Ilgürel, N.; Akdağ, N. Y.; Akdağ, A. 2016. Evaluation of noise exposure before and after noise barriers, a simulation study in Istanbul, *Journal of Environmental Engineering and Landscape Management* 24: 293–302. https://doi.org/10.3846/16486897.2016.1184671

Kartikeyan, B.; Sarkar, A.; Majumdar, K. L. 1998. A segmentation approach to classification of remote sensing imager, *International Journal of Remote Sensing* 19: 1695–1709. https://doi.org/10.1080/014311698215199

Li, C.; Zhao, J. 2017. Assessment of future urban growth impact on landscape pattern using cellular automata model: a case study of Xuzhou city, China, *Journal of Environmental*

Engineering and Landscape Management 25: 23–38. https://doi.org/10.3846/16486897.2016.1187620.

Li, X. Y.; Wang, Z. M.; Song, K. S.; Zhang, B. 2007. Assessment for salinized wasteland expansion and land use change using GIS and remote sensing in the west part of Northeast China, *Environment Monitoring and Assessment* 131: 421–437. https://doi.org/10.1007/s10661-006-9487-z

Liu, C. Y.; Sun, Y.; Liu, L. F.; Zhang, S. Q. 2010. 3S-based wetland habitat information management system in Songnen Plain, China, *UNEP/GEF Siberian Crane Wetland Project-Project Completion Workshop Proceedings*: 1–10.

Ludwig, D. F.; Ianuzzi, T. J. 2006. Habitat equivalency in urban estuaries: an analytical hierarchy process for planning ecological restoration, *Urban Ecosystems* 9: 265–290. https://doi.org/10.1007/s11252-006-0007-2

McDonald, A. J.; Genmell, F. M.; Lewis, P. E. 1998. Investigation of the utility of spectral indices for determining information on coniferous forests, *Remote Sensing of the Environment* 66: 250–272. https://doi.org/10.1016/S0034-4257(98)00057-1

Millennium Ecosystem Assessment. 2005. *Ecosystems and Human Well-Being: Synthesis*. Berlin: Springer.

Na, X. D.; Zang, S.; Zhang, Y. H.; Li, W. 2015. Assessing breeding habitat suitability for the endangered red-crowned (*Grus Japonensis*) based on multi-source remote sensing data, *Wetlands* 35: 955–967. https://doi.org/10.1007/s13157-015-0686-7

Nagendra, H.; Lucas, R.; Honrado, J. P.; Jongman, R. H. G. 2013. Remote sensing for conservation monitoring: Assessing protected areas, habitat extent, habitat condition, species diversity, and threats, *Ecological Indicators* 33: 45–59. https://doi.org/10.1016/j.ecolind.2012.09.014

Nakamura, M.; Masuda, R.; Tsukamoto, K.; Otake, T. 2016. Narrowed temperature adaptability in non-natal osmotic environments of two euryhaline wanderers, dace and black porgy: implications for seasonal habitat changes, *Fisheries Science* 82: 261–268. https://doi.org/10.1007/s12562-015-0956-6

Pan, X. L.; Zhang, D. Y.; Quan, L. 2006. Interactive factors leading to dying-off *Carex tato* in Momoge wetland polluted by crude oil, Western Jilin, China, *Chemosphere* 65: 1772–1777. https://doi.org/10.1016/j.chemosphere.2006.04.063

Ramsar. 2012. *The Ramsar Convention on Wetlands* [online], [cited 8 July 2012]. Available from Internet: http://www.ramsar.org/

Reid, J. R. W.; Colloff, M. J.; Arthur, A. D.; McGinness, H. M. 2013. Influence of catchment condition and water resource development on waterbird assemblages in the Murray-Darling Basin, Australia, *Biological Conservation* 165: 25–34. https://doi.org/10.1016/j.biocon.2013.05.009

Rönkä, M. T. H.; Häkkilä, K. 2005. Environmental changes and population trends of breeding waterfowl in northern Baltic Sea, *Annales Zoologici Fennici* 42: 587–602.

Rönkä, M.; Tolvanen, H.; Lehikoinen, E.; Numers, M. V. 2008. Breeding habitat preferences of 15 bird species on southwestern Finnish archipelago coast: Applicability of digital spatial data archives to habitat assessment, *Biological Conservation* 141: 402–416. https://doi.org/10.1016/j.biocon.2007.10.010

Saaty, T. L. 2008. Decision making with the analytic hierarchy process, *International Journal of Services Sciences* 1: 83–98. https://doi.org/10.1504/IJSSCI.2008.017590

Şekercioğlu, Ç. H.; Daily, G. C.; Ehrlich, P. R. 2004. Ecosystem consequences of bird declines, *Proceedings of the National Academy of Sciences of the United States of America* 101: 18042–18047. https://doi.org/10.1073/pnas.0408049101

Silva, T. S. F.; Costa, M. P. F.; Melack, J. M.; Novo, E. M. L. M. 2008. Remote sensing of aquatic vegetation: theory and applications, *Environmental Monitoring and Assessment* 140: 131–145. https://doi.org/10.1007/s10661-007-9855-3

Strasser, T.; Lang, S. 2015. Object-based class modelling for multi-scale riparian forest habitat mapping, *International Journal of Applied Earth Observation and Geoinformation* 37: 29–37. https://doi.org/10.1016/j.jag.2014.10.002

Strasser, T.; Lang, S.; Riedler, B.; Pernkopf, L. 2014. Multiscale object feature library for habitat quality monitoring in riparian forests, *IEEE Geoscience and Remote Sensing Letters* 11: 559–563. https://doi.org/10.1109/LGRS.2013.2278335

Tian, B.; Zhou, Y. X.; Zhang, L. Q.; Yuan, L. 2008. Analyzing the habitat suitability for migratory birds at the Chongming Dongtan Nature Reserve in Shanghai, China, *Estuarine, Coastal and Shelf Science* 80: 296–302. https://doi.org/10.1016/j.ecss.2008.08.014

Wang, X. Y.; Feng, J.; Zhao, J. 2010. Effects of crude oil residuals on soil chemical properties in oil sites, Momoge Wetland, China, *Environmental Monitoring and Assessment* 161: 271–280. https://doi.org/10.1007/s10661-008-0744-1

Wang, Z. M.; Mao, D. H.; Li, L.; Jia, M. M. 2015. Quantifying changes in multiple ecosystem services during 1992–2012 in the Sanjiang Plain of China, *Science of the Total Environment* 514: 119–130. https://doi.org/10.1016/j.scitotenv.2015.01.007

Wang, Z. M.; Song, K. S.; Zhang, B.; Liu, D. W. 2009. Shrinkage and fragmentation of grasslands in the West Songnen Plain, China, *Agriculture, Ecosystems & Environment* 12: 315–324. https://doi.org/10.1016/j.agee.2008.10.009

Weiers, S.; Bock, M.; Wissen, M.; Rossner, G. 2004. Mapping and indicator approaches for the assessment of habitats at different scales using remote sensing and GIS methods, *Landscape and Urban Planning* 67: 43–65. https://doi.org/10.1016/S0169-2046(03)00028-8

Wen, L.; Saintilan, N.; Yang, X. H.; Hunter, S. 2015. MODIS NDVI based metrics improve habitat suitability modelling in fragmented patchy floodplains, *Remote Sensing Applications: Society and Environment* 1: 85–97. https://doi.org/10.1016/j.rsase.2015.08.001

Williams M. (Ed.). 1990. *In Wetlands: a threatened landscape*. Oxford: Blackwell Press.

Wilson, E. H.; Sader, S. A. 2002. Detection of forest harvest type using multiple dates of Landsat TM imagery, *Remote Sensing of Environment* 80: 385–396. https://doi.org/10.1016/S0034-4257(01)00318-2

Xiao, H. Y.; Zhang, M. X.; Xiao, R. 2014. Dynamic evaluation of main ecological service functions in Momoge wetlands, *Wetland Science* 4: 451–458.

Yang, F. Y.; Lu, X. G.; Lou, Y. J.; Lou, X. N. 2012. Fish diversity in Momoge National Nature Reserve, *Wetland Science* 2: 214–222.

Zhu, C. Y.; Ji, P.; Li, S. H. 2017. Effects of urban green belts on the air temperature, humidity and air quality, *Journal of Environmental Engineering and Landscape Management* 25: 39–55. https://doi.org/10.3846/16486897.2016.1194276

Zlinszky, A.; Mücke, W.; Lehner, H.; Briese, C. 2012. Categorizing wetland vegetation by airborne laser scanning on Lake Balaton and Kis-Balaton, Hungary, *Remote Sensing* 4: 1617–1650. https://doi.org/10.3390/rs4061617

Zomer, R. J.; Trabucco, A.; Usin, S. L. 2009. Building spectral libraries for wetlands land cover classification and hyperspectral remote sensing, *Journal of Environmental Management* 90: 2170–2177. https://doi org/10 1016/j jenvman 2007 06 028

Zou, H. F.; Wu, Q. M. 2009. Internal distribution pattern of the nests and home ranges of red-crowned cranes in Zhalong Nature Reserve, *Acta Ecologica Sinica* 29: 1710–1718.

Zou, H. F.; Wu, Q. M.; Ma, J. Z. 2003. The nest-site selection of red-crowned crane in Zhalong Nature Reserve after burning and irrigating, *Journal of Northeast Normal University* 35: 54–59.

Yue ZHANG. PhD student, lecturer of College of Resources and Environment, Jilin Agricultural University. Doctor of Cartography and geographic information system. Publications: author of over 10 scientific publications, participant of five international conferences. Research interests: wetland vegetation ecology.

Zongming WANG. PhD, Professor of Northeast Institute of Geography and Agroecology, Chinese Academy of Sciences. Publications: author of over 30 scientific publications. Research interests: remote sensing and ecosystem services assessment of wetlands, and regional environmental changes.

Chunying REN. PhD, Associate professor of Northeast Institute of Geography and Agroecology, Chinese Academy of Sciences. More than 20 scientific publications. Research interests: GIS technology, wetland landscape change analysis, remote sensing of resource and environment.

Hao YU. PhD student of Northeast Institute of Geography Agroecology, Chinese Academy of Sciences. Research interests: Resource and environment remote sensing, and land degradation and restoration.

Zhangyu DONG. PhD, lecturer of Hefei University of Technology. Doctor of Cartography and geographic information system. More than 30 scientific publications. Research interests: wetland restoration modelling.

Chunyan LU. PhD, lecturer of Computer and Information Sciences College, Fujian Agriculture and Forestry University. Doctor of Cartography and geographic information system. More than 20 scientific publications. Research interests: GIS technology, wetland landscape change analysis, remote sensing information processing.

Dehua MAO. PhD, Assistant Professor of Northeast Institute of Geography and Agroecology, Chinese Academy of Sciences. Publications: author of over 20 scientific publications. Research interests: remote sensing of wetland, landscape pattern and ecosystem services, vegetation net primary productivity and modelling.

2

PULVERIZED FUEL ASH SOLIDIFICATION/STABILIZATION OF WASTE: COMPARISON BETWEEN BENEFICIAL REUSE OF CONTAMINATED MARINE MUD AND SEDIMENT

Ivan Y. TANG[a], Dickson Y. S. YAN[b], Irene M. C. LO[a] and Tongzhou LIU[c]

[a]*Department of Civil and Environmental Engineering, The Hong Kong University of Science and Technology, Hong Kong, China*

[b]*Faculty of Science and Technology, Technological and Higher Education Institute of Hong Kong, Hong Kong, China*

[c]*Harbin Institute of Technology Shenzhen Graduate School, Shenzhen Key Laboratory of Water Resource Utilization and Environmental Pollution Control, Shenzhen 518055, China*

Abstract. This study aimed to maximize the utilization of contaminated marine mud and sediment for beneficial reuse by solidification/stabilization (S/S) treatment with cement and pulverized fuel ash (PFA). For the purposes of waste maximization and enhancing the mechanical property of the S/S mixtures, mixing 75% by mass of either contaminated marine mud or sediment with 20% and 5% of cement and PFA, respectively, was found to be the optimal mix design. Their unconfined compressive strengths reached up to 8.32 MPa and 4.47 MPa, respectively. Apart from the mechanical property, according to the U.S.EPA, the TCLP results show that all regulated heavy metals were immobilized to a safe level and are available for engineering application such as fill material. The results of XRD reveal that the formation of CSH gel in the S/S treated mud and sediment is responsible for the strength development and heavy metal immobilization.

Keywords: cement, compressive strength, fill material, heavy metal immobilization, TCLP.

Introduction

Solidification/stabilization (S/S) is a widely used technology to treat contaminated soils in many countries, so as to minimize the level of contaminant leaching prior to landfill disposal. Cement has been intensively utilized as the binder for S/S treatment due to its low material cost and abundance in resources (Ubbrìaco, Calabrese 1998; Conner 1999; Gong, Bishop 2003). To further reduce the treatment cost and promote the concept of green remediation, partial replacement of cement with pulverized fuel ash (PFA), which is the residue resulting from the coal combustion process, has been applied to S/S technology recently for the purpose of engineering applications such as road bases (Lav *et al.* 2006; Ghosh, Dey 2009) and concrete products (Siddique 2004; Durán-Herrera *et al.* 2011). Apart from the waste utilization and reduction of treatment cost, the use of PFA has been proven to enhance the strength of cement mixed soil (Kogbara *et al.* 2013) and to reduce the extent of contaminant leaching (Perera, Al-Tabbaa 2005; Moon *et al.* 2010). The silica and alumina contents of PFA were considered to play important roles in improving the quality of the S/S treated soil for beneficial reuse, as both silica and alumina react with the calcium hydroxide in cement to form calcium silicate hydrates or calcium aluminium silicate hydrates, which are the main products of cement hydration to provide strength and contaminant binding in cement-based soil (Bouzalakos *et al.* 2008; Li *et al.* 2012).

Comparatively, the development of S/S technology in relation to contaminated marine mud and sediment for beneficial reuse is not as comprehensive as that of soil. The high plasticity index of marine mud and sediment was reported to be inversely related to the strength of cement mixture (Saadeldin, Siddiqua 2013) thus limiting most of

Corresponding authors: Dickson Yan, Irene Lo
E-mails: dicksonyan@vtc.edu.hk; irene.lo@ust.hk

their engineering applications. In addition, the contamination levels of sediment and marine mud are relatively higher than that of coarse soil due to their large surface areas and abundant clay content (Sparks 2003). Two common ways to handle dredged marine mud and sediment are ocean dumping and landfill disposal. With consideration of the adverse environmental impacts, the US Federal regulations – marine protection, research and sanctuaries act of 1972 (MPRSA) have restricted the ocean dumping of dredged sediment from harbours such as those of New York and Newark, and thus left the local port authorities no choice, but an unsustainable option to dispose of the dredged sediment in landfills.

In view of the recent developments of S/S technology, dredged marine mud and sediment indeed can be potentially reused for various construction purposes such as pavement bases (Yan *et al.* 2014) and fill material (Zhang *et al.* 2011). Due to the successful strength enhancement of cement mixed soil after the addition of PFA, PFA was recently studied regarding S/S processes in mud and sediment. It was found that PFA can successfully overcome the low strength development of cement mixed sediment and red mud. By mixing 15% by mass of red mud and 15% PFA with 53% cement, the compressive strength of the cement mixture reached up to 35 MPa after 28 days of the curing process (Yao *et al.* 2013). However, from the sustainability point of view, the percentage of red mud utilization is considered rather low to promote the beneficial reuse of waste. On the contrary, the waste utilization was maximized by 94% and 3% by mass of sediment and PFA, respectively, for S/S treatment (Zentar *et al.* 2012). The unconfined compressive strength of the resulting cement-based material at the 28th day was about 0.6 MPa which may not be enough for most engineering applications. With all the aforementioned concerns, striking a balance between waste maximization and broadened engineering application is believed to be one of the potential perspectives in the sustainable development of S/S treatment for sediment and marine mud.

This study aimed at the maximization of using contaminated sediment and marine mud for beneficial reuse as sustainable fill material by S/S treatment with cement and PFA. The physical and mechanical properties and the contaminant leaching behavior of the S/S treated sediment and marine mud was evaluated. In order to provide an in-depth understanding of S/S treated contaminated sediment and marine mud, their performances were compared with the consideration of feasible engineering applications. This study enables further improvements of the S/S technology in relation to contaminated sediment and marine mud, and provides an alternative for sustainable management of dredged sediment and marine mud.

1. Experimental

1.1. Materials

The contaminated marine mud and sediment were obtained from a contaminated site in Hong Kong. The contaminated marine mud was collected from 10–12 cm below the ground surface and the contaminated sediment was dredged from 10 to 100 cm below the seawater/sediment interface. The samples were then sealed in plastic bags, placed in covered buckets and stored in a cold room at 4 °C to avoid any possible deterioration. Both the contaminated marine mud and sediment were air dried for 2 weeks prior to S/S treatment and then sieved through a 5 mm size sieve to remove oversized debris. The particle size distribution of the contaminated marine mud and sediment are shown in Figure 1, whereas their physical properties are given in Table 1. The particle size of the marine mud was smaller than that of sediment, with about 20%, 50% and 30% of sand, silt and clay for marine mud, and 45%, 50% and 5% of sand, silt and clay for sediment, respectively. According to the Unified Soil Classification System (ASTM D 2487), the marine mud (with plasticity index of 30.1) and sediment (with plasticity index of 23.2) were classified as CH (fat clay with sand) and MH (sandy elastic silt), respectively. Both marine mud and sediment were contaminated and their corresponding heavy metal contents are shown in Table 1. The marine mud was mainly contaminated by Ba and Mn, with 323.3 and 481.0 mg/kg, respectively. On the other hand, high concentration of Ba, Cr, Pb, Zn, Mn and especially Cu were found in the sediment, registering 224.7, 232.3, 152.0, 638.3, 573.3 and 2171.9 mg/kg, respectively.

The cement used was Hong Kong commercially available ordinary Portland cement (OPC) which is equivalent to the American Society Testing and Material (ASTM) type I cement. The pulverized fuel ash used was obtained from a power plant in Hong Kong and is equivalent to the ASTM Class F fly ash. The chemical compositions of OPC

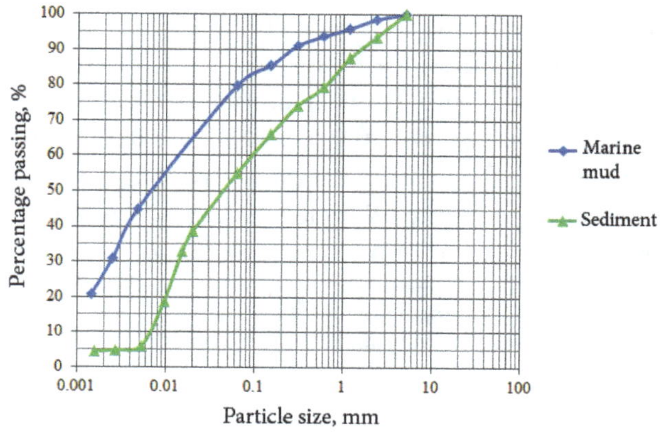

Fig. 1. Particle size distribution of contaminated marine mud and sediment

Table 1. Physical properties and heavy metal contents of contaminated marine mud and sediment

	Mud	Sediment
Physical properties		
Sand (> 63 μm) (% by mass)	20.4±1.0	45.0±0.6
Silt (63 ~ 2 μm) (% by mass)	49.6±0.6	50.4±0.6
Clay (< 2 μm) (% by mass)	30.0±1.0	4.6±0.1
Moisture Content (%)	28.3±0.92	56.6±1.27
pH	8.39±0.04	8.54±0.03
Liquid Limit	50.9	52.4
Plastic Limit	20.8	29.2
Plasticity Index	30.1	23.2
Group Symbol	CH[1]	MH[2]
Heavy metal content (mg/kg by dry weight)		
Sb	45.7±3.6	49.1±5.7
As	36.3±0.9	27.0±1.9
Ba	323.3±56.8	224.7±25.2
Cd	6.0±3.0	3.8±0.6
Cr	55.1±2.4	232.3±27.5
Pb	51.5±4.9	152.0±17.4
Hg	0.9±0.2	2.3±0.4
Ni	31.8±3.6	41.1±3.1
Zn	98.0±8.1	638.3±58.9
Co	5.3±4.4	2.8±6.4
Cu	18.4±5.0	2171.9±117.1
Mn	481.0±9.2	573.3±37.5
Mo	11.7±1.8	18.4±1.8
Sn	5.3±0.2	8.8±5.5

[1] CH refers to clay with a high liquid limit.
[2] MH refers to silt with a high liquid limit.

Table 2. Chemical compositions of ordinary Portland cement and pulverized fuel ash

Chemical Composition	Cement (% by dry mass)	PFA[1] (% by dry mass)
CaO	65.4	13.9
SiO_2	19.5	44.0
Al_2O_3	3.8	17.9
MgO	1.5	4.7
Fe_2O_3	3.2	13.6
SO_4	5.3	2.5
Loss on ignition	1.3	3.4

[1] PFA refers to pulverized fuel ash.

and PFA are shown in Table 2. The OPC comprises CaO, SiO_2, Al_2O_3 and Fe_2O_3 (i.e. 65.4%, 19.5%, 3.8% and 3.2%, respectively), while PFA comprises less CaO (13.9%) but more SiO_2 (44.4%) and fair amounts of Al_2O_3 and Fe_2O_3 (i.e. 17.9% and 13.6%, respectively). Potable water was used for the S/S treatment in this study.

1.2. Preparation of S/S treated contaminated marine mud and sediment

The design mixes of the S/S treatment for contaminated marine mud and sediment are shown in Table 3. Symbols are used to represent their corresponding materials (i.e. M: mud, S: sediment, C: OPC, F: PFA). For instance, the sample M70C15F15 refers to 70% mud mixed with 15% OPC and 15% PFA. The dry contaminated marine mud and sediment were mixed with OPC and PFA in a mechanical mixer for 2 min to ensure homogenous mixing. Portable water was then added and mixed with the dry components for another 2 min. Afterwards, the maximum dry density and optimum moisture content of the S/S treated contaminated marine mud and sediment were determined according to the standard Proctor compaction test (BS 1377-4: 1990) as shown in Figures 2a and 2b, respectively, and summarized in Table 3.

Table 3. Mix proportions of S/S treated marine mud and sediment

Sample ID	Mass Ratio (%)			Water/ Binder Mass Ratio[1]	Maximum Dry Density[2] (kg/m³)
	Mud	Cement	PFA		
M70C15F15	70	15	15	0.64	1724
M70C10F20	70	10	20	0.68	1683
M70C5F25	70	5	25	0.71	1675
M70F30	70	0	30	0.73	1660
M75C20F5	75	20	5	0.64	1700
M75C15F10	75	15	10	0.66	1694
M80C15F5	80	15	5	0.63	1672
Sample ID	Mass Ratio (%)			Water/ Binder Mass Ratio[1]	Maximum Dry Density[2] (kg/m³)
	Sediment	Cement	PFA		
S70C15F15	70	15	15	0.73	1435
S70C10F20	70	10	20	0.80	1421
S70C5F25	70	5	25	0.83	1402
S70F30	70	0	30	0.84	1398
S75C20F5	75	20	5	0.83	1380
S75C15F10	75	15	10	0.86	1359
S80C15F5	80	15	5	0.83	1372

[1] Binder refers to cement and pulverized fuel ash.
[2] Water/Binder ratio is obtained based on the optimum moisture content and maximum dry density of the S/S samples.

The compacted S/S treated contaminated marine mud and sediment samples were then cast into cubic stainless steel moulds (with a dimension of 50×50×50 mm) with the same effort as the maximum dry density obtained previously in the standard Proctor compaction according to BS EN 12390-2:2009. All S/S samples were prepared in duplicate to show the reproducibility and reliability of the experiments. All samples were demoulded after 1 day and placed in a 100% relative humidity curing room at room temperature prior to testing.

1.3. Testing procedures

The unconfined compressive strength (UCS) of the hardened S/S treated contaminated marine mud and sediment, after curing for 7, 14 and 28 days, was determined according to ASTM D 1633-00, 2000. The UCS of the S/S samples was measured by an automatic compressive testing machine (MTS 815 Test System) at a constant loading rate of 0.1 mm/min. The experimental data points were plotted with error bars and are reported in terms of mean ± standard deviation. To provide a better understanding of the strength development of S/S treated contaminated marine mud and sediment, the micro-structural characterization was also studied through X-ray diffraction (XRD) analysis. The samples after curing for 28 days were crushed into powder for investigation of their crystalline information. The XRD analysis was carried out by a Powder X-ray Diffraction System (Philips PW 1830) with an X-ray source scanning the sample between 2θ angles from 5° to 80° at a speed of 0.02° per minute.

To assess the potential environmental impacts of the S/S treated contaminated marine mud and sediment, the heavy metal leachability of the S/S mixtures at the 28th day were studied through toxicity characteristic leaching procedure (TCLP U.S.EPA method 1311). The samples were crushed to a size of less than 1 cm in diameter and mixed with glacial acetic acid (pH 4.93±0.05) at a solid to solution ratio of 1:20 by maintaining an end-over-end rotation of 30 rpm for 18 h. The crushed S/S samples were then separated from the TCLP leachates with 0.45 um cellulose acetate membranes. The heavy metal concentrations of the TCLP leachates were determined by inductively coupled plasma-atomic emission spectroscopy (ICP-AES, Varian 725-ES). All TCLP tests were performed in triplicate and the average concentrations of the leachates were reported with corresponding standard deviations.

2. Results and discussion

2.1. Compaction characteristics of S/S treated marine mud and sediment

The correlation between maximum dry density and optimum water content of various mixes of S/S treated marine mud and sediment are given in Figures 2a and 2b, respectively. It was found that the optimal water content of S/S treated marine mud and sediment decreased from 21% to 18.5% and 24% to 21.5%, respectively, when their cement content increased from 5% to 15% at a fixed amount of marine mud/sediment (i.e. 70% by mass). Indeed, a similar reduction of optimal water content was also found in a previous study (Zentar *et al.* 2012) and the reason for such is related to the small specific area of cement. Since the ability of water absorption is proportionally related to the specific area of particles, increasing the amount of cement, which has a smaller specific area compared with that of sediment and marine mud, possibly reduces the optimal water content of the S/S mixtures. Nevertheless, by increasing the amount of cement from 5% to 15%, the maximum dry densities of the S/S mixtures increase. The increase of the maximum dry density with increasing cement content is attributed to the higher specific gravity of the cement (i.e. 3.15 g/cm^3) which may benefit the strength

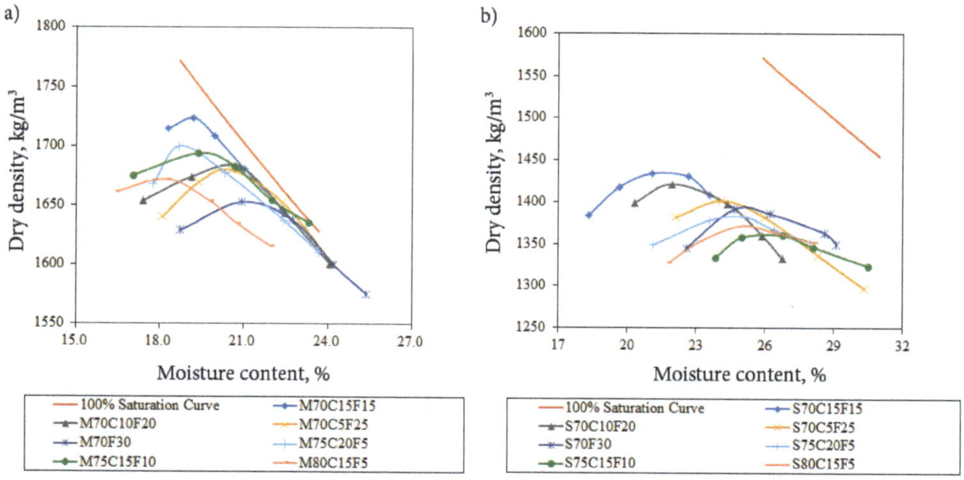

Fig. 2. Compaction curves of (a) S/S treated marine mud and (b) S/S treated sediment

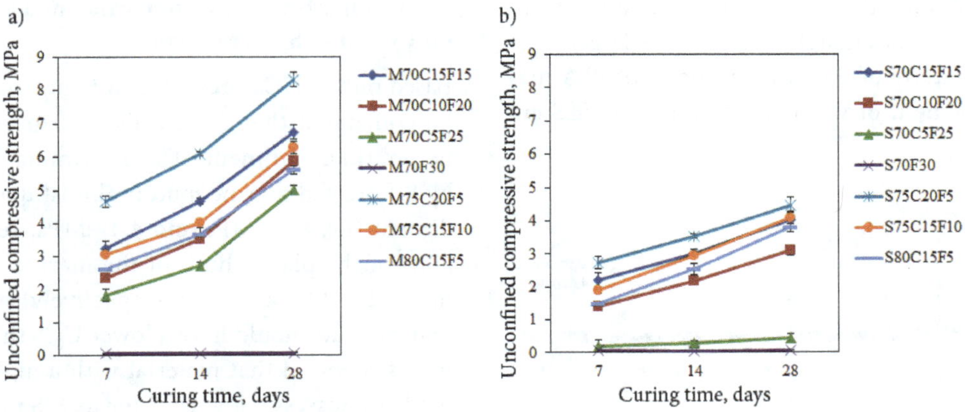

Fig. 3. Unconfined compressive strength of (a) S/S treated marine mud and (b) S/S treated sediment

development of S/S treated marine mud and sediment. However, instead of further increase, the maximum dry densities of both S/S treated marine mud and sediment decreased when the cement content increased to 20% by mass. Since the purpose of this study was to maximize the utilization of marine mud and sediment for S/S treatment so as to enable the reuse of waste for engineering application, the amount of marine mud and sediment increased to 75% by mass whereas the percentage of cement increased to 20%. It is suspected that the reduction in maximum dry densities of the corresponding S/S mixtures is due to the relatively low specific gravity of marine mud and sediment (i.e. 2.65 g/cm^3).

2.2. Unconfined compressive strength of S/S treated marine mud and sediment

The unconfined compressive strengths of the S/S treated contaminated marine mud and sediment samples at the curing time of 7, 14 and 28 days are given in Figures 3a and 3b, respectively, and were found to increase gradually from the 7th to 28th day. In the absence of cement, neither the UCS of PFA-marine mud mixture nor the UCS of PFA-sediment mixture built up with time, even with 30% by mass of PFA. In addition, by utilizing 70% by mass of marine mud or sediment, the UCS of the S/S mixtures decreased with increasing amount of PFA (i.e. from 15% to 25% by mass). PFA is considered as an alternative of cement due to its pozzolanic components. However, it was found that there was no enhancement in the compressive strength of the S/S mixtures after applying PFA. By comparing the chemical composition between cement and PFA, it was found that the PFA used in this study contained less calcium oxide, but higher silica and alumina contents (Table 2). In view of the essential elements for the formation of C-S-H or C-A-H gel (i.e. CaO, SiO_2 and Al_2O_3), apparently CaO is the limiting factor of the PFA used in this study. By increasing the utilization rate of marine mud and sediment from 70% to 80% by mass (with 15% of cement), the UCS at the 28th day reduced from 6.74 to 5.62 MPa and 4.02 to 3.82 MPa, for S/S treated marine mud and S/S treated sediment, respectively. As previously discussed, the role of cement in strength development of the S/S mixtures is much more significant than that of PFA due to its limited CaO content. In order to maintain the UCS of the S/S mixtures with increased amounts of marine mud and sediment, more cement is required instead of PFA. In consideration of waste and UCS maximization, 20% by mass of cement was applied on 75% marine mud and sediment for S/S treatment, respectively. The corresponding UCS at the 28th day was 8.32 MPa and 4.47 MPa, respectively. In view of the various engineering applications, on-site reuse of S/S treated waste as fill material is one of the most common ways to beneficially reuse hazardous waste. According to the U.S.EPA Handbook for Solidification/Stabilization of Hazardous Wastes (U.S.EPA 540/2-86/001), the cylinder strength of S/S samples should be at least 1 MPa. Assuming that the cube strength of S/S samples is 0.8 times higher than that of the cylinder, the maximum UCS of the designed S/S treated marine mud and sediment samples (i.e. 20%, 75% and 5% of cement, marine mud/sediment and PFA by mass) satisfied the mechanical requirement for on-site reuse as fill material. To provide a better understanding of the strength development of S/S treated contaminated marine mud and sediment, XRD analysis was performed after 28 days of sample curing and the results are given in Figure 4. For both S/S treated contaminated marine mud and sediment, C-S-H gel dominated the cement hydration product, while a small amount of ettringite was found in the S/S treated contaminated sediment. The result indicates that the strength development of the S/S mixtures was mainly contributed by C-S-H gel, but not C-A-H gel. The cement used consists of 3.8% of Al_2O_3, while 17.9% for PFA (Table 2). Since only 5%

by mass of PFA was added, the content of Al_2O_3 in the S/S mixtures was not enough to form C-A-H gel. Therefore, further study of the different types of PFA may enable improvement of waste utilization and UCS enhancement.

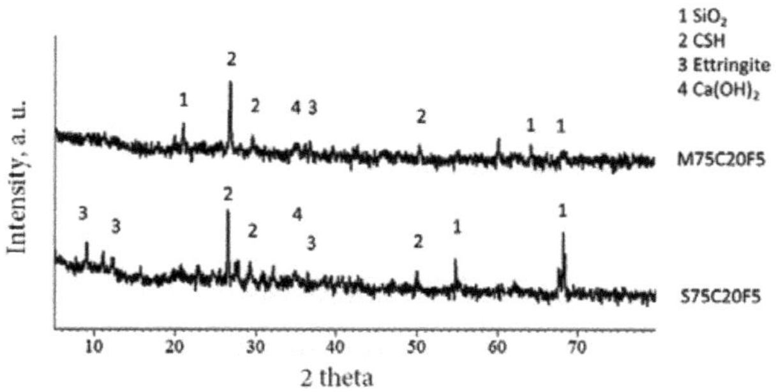

Fig. 4. XRD patterns of S/S treated marine mud and S/S treated sediment after 28 days of curing.

2.3. Comparison between marine mud and sediment on UCS after S/S treatment

Based on the UCS presented in Figure 3a and 3b, it can be concluded that S/S treated marine mud performs better than sediment. The maximum UCS obtained by S/S treated marine mud is almost double that of sediment (i.e. 8.32 MPa and 4.47 MPa, respectively). In view of the plastic limit of the marine mud and sediment (i.e. 30.1 and 23.2, respectively), the S/S treated marine mud should have a lower UCS than sediment as it was reported that material with a high plastic index tends to possess a low UCS for its S/S mixture (Saadeldin, Siddiqua 2013). It is believed that the high UCS of S/S treated marine mud is caused by its high maximum dry density (Table 2) and small particle size (Fig. 1). According to the result of standard Proctor compaction, the average maximum dry density achieved by

Table 4. TCLP results of S/S treated marine mud after 28 days of curing

Metals	TCLP limits of UTS[2] (mg/L)	TCLP leachate of marine mud and PFA mixture (mg/L)	M70C15F15 (mg/L)	M70C10F20 (mg/L)	M70C5F25 (mg/L)	M70F30 (mg/L)	M75C20F5 (mg/L)	M75C15F10 (mg/L)	M80C15F5 (mg/L)
Sb (mg/L)	1.15	0.146	N.D.[1]	N.D.	N.D.	N.D.	N.D.	N.D.	N.D.
As (mg/L)	5	0.025	N.D.	N.D.	N.D.	N.D.	N.D.	N.D.	N.D.
Ba (mg/L)	21	0.172	0.89±0.07	0.86±0.04	0.97±0.08	0.65±0.01	0.31±0.01	0.70±0.03	0.64±0.06
Cd (mg/L)	0.11	0.018	N.D.	N.D.	N.D.	N.D.	N.D.	N.D.	N.D.
Cr (mg/L)	0.6	0.037	N.D.	N.D.	N.D.	N.D.	N.D.	N.D.	N.D.
Pb (mg/L)	0.75	1.266	0.74±0.04	0.90±0.01	1.08±0.12	1.23±0.09	0.71±0.09	0.63±0.06	0.70±0.03
Hg (mg/L)	0.025	0.022	N.D.	N.D.	N.D.	N.D.	N.D.	N.D.	N.D.
Ni (mg/L)	11	0.057	0.06±0.00	0.10±0.00	0.12±0.00	0.23±0.00	0.02±0.00	0.05±0.00	0.07±0.00
Zn (mg/L)	4.3	0.169	0.05±0.00	0.07±0.00	0.06±0.00	0.28±0.00	0.01±0.00	0.03±0.00	0.08±0.00

Metals	TCLP requirement of UTS (%)	TCLP leachate of marine mud and PFA mixture (mg/L)	M70C15F15 (%)	M70C10F20 (%)	M70C5F25 (%)	M70F30 (%)	M75C20F5 (%)	M75C15F10 (%)	M80C15F5 (%)
Co (%)[3]	90%	0.063	N.D.	N.D.	N.D.	N.D.	N.D.	N.D.	N.D.
Cu (%)[3]	90%	0.010	N.D.	N.D.	N.D.	N.D.	N.D.	N.D.	N.D.
Mn (%)[3]	90%	2.961	95.5±1.6	90.8±3.0	90.2±9.1	93.0±0.1	99.8±0.1	88.4±1.9	55.3±7.1
Mo (%)[3]	90%	0.038	N.D.	N.D.	N.D.	N.D.	N.D.	N.D.	N.D.
Sn (%)[3]	90%	0.004	N.D.	N.D.	N.D.	N.D.	N.D.	N.D.	N.D.

[1] N.D. refers to non-detectable.

[2] UTS refers to Universal Treatment Standards (UTS 40 CFR 268.48).

[3] For Co, Cu, Mn, Mo and Sn, at least 90 percent reduction in mobility is required after Stabilization/Solidification treatment according to Universal Treatment Standards.

S/S treated marine mud is higher than that of sediment (i.e. 1711 and 1395 kg/m^3 for S/S treated marine mud and sediment with 20% by mass of cement, respectively). Since marine mud contains more clay portion and it is generally finer than sediment which can easily fill up the void spaces in cementitious material, its strength is thereby increased (Neville 1995).

Apart from their mechanical property, in order to beneficial reuse of contaminated marine mud and sediment, the S/S mixtures have to be minimized as suggested by the U.S.EPA. The TCLP results of S/S treated marine mud and sediment at the 28th day are given in Tables 4 and 5, respectively. The addition of cement to the contaminated marine mud and sediment resulting in a S/S matrix shows positive effects on decreasing the mobility of heavy metals. Usage of more than 15% cement in 70% and 75% contaminated marine mud and sediment presents reliable S/S effectiveness such that these TCLP results all met the TCLP regulatory standard. Usage of less than 15% cement in both 70% and 75% contaminated marine mud and sediment for S/S treatment were unable to effectively immobilize heavy metals, especially Pb and Mn. Before the S/S treatment, over hundreds mg/kg of Ba and Mn was found in the marine mud. The extent of contamination in the sediment was found to be more serious, with over hundreds 100 mg/kg of Ba, Cr, Pb, Zn and Mn, and even over two thousands mg/kg of Cu. However, with slightly more cement addition (i.e. 20% cement in both 75% contaminated marine mud and sediment), the extent of heavy metal leaching was well controlled within the universal treatment standard, especially for Cu in the sediment with 93% reduction in its leachability, not posing any adverse health and environmental impacts. Overall of the regulated heavy metals were immobilized.

Table 5. TCLP results of S/S treated sediment after 28 days of curing

Metals	TCLP limits of UTS[2] (mg/L)	TCLP leachate of sediment and PFA mixture (mg/L)	S70C15F15 (mg/L)	S70C10F20 (mg/L)	S70C5F25 (mg/L)	S70F30 (mg/L)	S75C20F5 (mg/L)	S75C15F10 (mg/L)	S80C15F5 (mg/L)
Sb (mg/L)	1.15	0.113	N.D.[1]	N.D.	N.D.	N.D.	N.D.	N.D.	N.D.
As (mg/L)	5	0.031	N.D.	N.D.	N.D.	N.D.	N.D.	N.D.	N.D.
Ba (mg/L)	21	0.461	0.41±0.07	0.67±0.03	0.66±0.02	0.32±0.00	0.27±0.01	0.41±0.03	0.32±0.01
Cd (mg/L)	0.11	0.043	N.D.	N.D.	N.D.	N.D.	N.D.	N.D.	N.D.
Cr (mg/L)	0.6	0.123	N.D.	N.D.	N.D.	N.D.	N.D.	N.D.	N.D.
Pb (mg/L)	0.75	0.984	0.69±0.08	0.92±0.09	0.95±0.00	0.99±0.01	0.57±0.03	0.92±0.09	0.97±0.01
Hg (mg/L)	0.025	0.033	N.D.	N.D.	N.D.	N.D.	N.D.	N.D.	N.D.
Ni (mg/L)	11	1.388	0.14±0.00	0.25±0.02	0.63±0.06	0.91±0.00	0.11±0.00	0.13±0.01	0.09±0.00
Zn (mg/L)	4.3	13.650	0.07±0.01	0.19±0.01	3.73±0.42	8.98±0.93	0.04±0.01	0.08±0.00	0.04±0.00
Metals	**TCLP requirement of UTS (%)**	**TCLP leachate of marine mud and PFA mixture (mg/L)**	**S70C15F15 (%)**	**S70C10F20 (%)**	**S70C5F25 (%)**	**S70F30 (%)**	**S75C20F5 (%)**	**S75C15F10 (%)**	**S80C15F5 (%)**
Co (%)[3]	90%	0.085	N.D.	N.D.	N.D.	N.D.	N.D.	N.D.	N.D.
Cu (%)[3]	90%	13.619	92.7±1.0	87.7±1.0	85.7±0.5	86.1±2.1	92.9±3.3	90.6±4.1	80.2±3.1
Mn (%)[3]	90%	7.954	95.3±5.0	85.3±3.3	57.3±3.5	43.5±1.4	99.9±0.0	92.8±4.0	96.6±1.2
Mo (%)[3]	90%	0.048	N.D.	N.D.	N.D.	N.D.	N.D.	N.D.	N.D.
Sn (%)[3]	90%	0.012	N.D.	N.D.	N.D.	N.D.	N.D.	N.D.	N.D.

[1] N.D. refers to non-detectable.

[2] UTS refers to Universal Treatment Standards (UTS 40 CFR 268.48).

[3] For Co, Cu, Mn, Mo and Sn, at least 90 percent reduction in mobility is required after Stabilization/Solidification treatment according to Universal Treatment Standards.

Conclusions

The present study investigated the sustainable and economic management of contaminated marine mud and sediment by mixing them with abandoned PFA in S/S treatment. Based on the result in this study, conclusions are drawn as follows:

1. By increasing the amount of cement from 5% to 15% by mass, the optimal water content of the corresponding S/S treated mixture (with 70% of marine mud and sediment, individually) decreases, but the maximum dry densities increase. However, the increases of either marine mud or sediment in the S/S mixtures decrease their maximum dry densities.
2. The UCS of the S/S treated marine mud and sediment using cement and PFA as binders reached up to 8.32 MPa and 4.47 MPa, respectively, at the curing time of 28 days. The UCS achieved enables engineering applications with a high strength requirement. As the most common way to reuse waste, the S/S treated marine mud and sediment qualify as fill material which generally require 1 MPa of UCS.
3. With 15% by mass of cement, most of the heavy metals were immobilized to acceptable levels, except Pb and Mn. However, by increasing the amount of cement to 20% with 5% of PFA, according to the U.S.EPA, all regulated heavy metals, including Sb, As, Ba, Cd, Cr, Pb, Hg, Ni, Zn, Co, Cu, Mn, Mo and Sn, were successfully immobilized.
4. The results of XRD proved that the formation of CSH gel in the S/S treated mud and sediment is responsible for the strength development and heavy metal immobilization.
5. For the purposes of waste maximization and enhancing the mechanical property of the S/S mixtures, mixing 75% by mass of either contaminated marine mud or sediment with 20% and 5% of cement and PFA, respectively, were found to be the optimal mix designs to provide a sustainable green remediation alternative in the future.

Acknowledgements

The authors would like to thank National Natural Science Foundation of China (Grant No. 51208145) for its partial support.

References

Bouzalakos, S.; Dudeney, A. W. L.; Cheeseman, C. R. 2008. Controlled low-strength materials containing waste precipitates from mineral processing, *Mineral Engineering* 21: 252–263. http://dx.doi.org/10.1016/j.mineng.2007.09.006

Conner, J. R. 1999. *Chemical fixation and solidification of hazardous wastes.* New York: Van Nostrand Reihold.

Durán-Herrera, A.; Juárez, C. A.; Valdez, P.; Bentz, D. P. 2011. Evaluation of sustainable high-volume fly ash concretes, *Cement and Concrete Composites* 33: 39–45. http://dx.doi.org/10.1016/j.cemconcomp.2010.09.020

Ghosh, A.; Dey, U. 2009. Bearing ratio of reinforced fly ash overlying soft soil and deformation modulus of fly ash, *Geotext Geomembranes* 27: 313–320. http://dx.doi.org/10.1016/j.geotexmem.2008.12.002

Gong, P.; Bishop, P. L. 2003. Evaluation of organics leaching from solidified/stabilized hazardous wastes using a powder reactivated carbon additive, *Environmental Technology* 24: 445–455. http://dx.doi.org/10.1080/09593330309385579

Kogbara, R. B.; Al-Tabbaa, A.; Yi, Y.; Stegemann, J. A. 2013. Cement-fly ash stabilisation/solidification of contaminated soil: Performance properties and initiation of operating envelopes, *Applied Geochemistry* 33: 64–75. http://dx.doi.org/10.1016/j.apgeochem.2013.02.001

Lav, A. H.; Lav, M. A.; Goktepe, A. B. 2006. Analysis and design of a stabilized fly ash as pavement base material, *Fuel* 85: 2359–2370. http://dx.doi.org/10.1016/j.fuel.2006.05.017

Perera, R.; Al-Tabbaa, A. 2005. The role of accelerated carbonation in the accelerated ageing of stabilised/solidified waste forms, in *The International Conference of Stabilisation/Solidification Treatment and Remediation,* 12–13 April 2005, London, Cambridge, UK, 181–191. http://dx.doi.org/10.1201/9781439833933.ch23

Siddique, R. 2004. Performance characteristics of high-volume Class F fly ash concrete, *Cement and Concrete Research* 34: 487–493. http://dx.doi.org/10.1016/j.cemconres.2003.09.002

Ubbrìaco, P.; Calabrese, D. 1998. Solidification and stabilization of cement paste containing fly ash from municipal solid waste, *Thermochimica Acta* 321: 143–150. http://dx.doi.org/10.1016/S0040-6031(98)00453-5

Li, X. G.; Lv, Y.; Ma, B. G.; Chen, Q. B.; Yin, X. B.; Jian, S. W. 2012. Utilization of municipal solid waste incineration bottom ash in blended cement, *Journal of Cleaner Production* 32: 96–100. http://dx.doi.org/10.1016/j.jclepro.2012.03.038

Moon, D. H.; Lee, J. R.; Grubb, D. G.; Park, J. H. 2010. An assessment of Portland cement, cement kiln dust and Class C fly ash for the immobilization of Zn in contaminated soils, *Environmental Earth Sciences* 61: 1745–1750. http://dx.doi.org/10.1007/s12665-010-0596-1

Neville, A. M. 1995. *Properties of concrete.* 4th ed. UK: Addison Wesley, Longman Limited.

Saadeldin, R.; Siddiqua, S. 2013. Geotechnical characterization of a clay-cement mix, *Bulletin of Engineering Geology and the Environment* 72: 601–608. http://dx.doi.org/10.1007/s10064-013-0531-2

Sparks, D. L. 2003. *Environmental soil chemistry.* Amsterdam: Academic Press.

Yan, D. Y. S.; Tang, I. Y.; Lo, I. M. C. 2014. Development of controlled low-strength material derived from beneficial reuse of bottom ash and sediment for green construction, *Construction and Building Materials* 64: 201–207. http://dx.doi.org/10.1016/j.conbuildmat.2014.04.087

Yao, Y.; Li, Y.; Liu, X.; Jiang, S.; Feng, C.; Rafanan, E. 2013. Characterization on a cementitious material composed of red mud and coal industry byproducts, *Construction and Building Materials* 47: 496–501. http://dx.doi.org/10.1016/j.conbuildmat.2013.05.030

Zentar, R.; Wang, D.; Abriak, N. E.; Benzerzour, M.; Chen, W. 2012. Utilization of siliceous-aluminous fly ash and cement for solidification of marine sediments, *Construction and Building Materials* 35: 856–863. http://dx.doi.org/10.1016/j.conbuildmat.2012.04.024

Zhang, C. L.; Jin, W.; Wu, X. Y.; Wang, S. C. Properties of solidified dredged sediment used as subgrade construction material, in *International Conference on Electric Technology and Civil Engineering*, 22–24 April 2011, Lushan, 5774642: 2468–2470.

Standards and Regulations

ASTM D1633-00: 2000. Standard Test Method for Compressive Strength of Molded Soil-cement Cylinders.

ASTM D2487: 2011. Standard Practice for Classification of Soils for Engineering Purposes [Unified soil classification system].

BS EN 12390-2: 2009. Testing hardened concrete. Making and curing specimens for strength tests.

BS 1377-4: 1990. Methods of test for soils for civil engineering purposes. Compaction-related tests.

MPRSA. Marine Protection, Research and Sanctuaries Act of 1972.

U.S.EPA 540/2-86/001: 1986. Handbook for stabilization/solidification of hazardous wastes.

UTS 40 CFR 268.48. Universal treatment standards.

Ivan Y. TANG graduated in Environmental Engineering from Sun Yat-Sen University in 2012. He obtained his Master of Science in 2013 from The Hong Kong University of Science and Technology and is now studying MPhil program at the Department of Civil and Environmental Engineering in the same University. Research interests: solid waste treatment and management; sustainable construction materials; and environmental management.

Dickson Y. S. YAN, Teaching Fellow at the Faculty of Science and Technology, the Technological and Higher Education Institute of Hong Kong. He obtained his PhD in Environmental Engineering from The Hong Kong University of Science and Technology in 2013 and has published about 10 scientific papers. Research interests: carbon footprint assessment on remediation technology; life cycle assessment of products and production lines; waste material recycling and reuse; nanotechnology for environmental application; waste disposal and treatment; and remediation of contaminated soils, sediments and groundwater.

Irene M. C. LO, Professor at the Department of Civil and Environmental Engineering, Hong Kong University of Science and Technology. She is currently Academician of European Academy of Sciences and Arts and Chairman of Environmental Division, the Hong Kong Institution of Engineers. Research interests: remediation of contaminated soils, sediments and groundwater; waste material recycling and reuse; nanoparticle transport and fate; nanotechnology for environmental application; waste disposal and treatment; carbon footprint assessment on remediation technology; construction carbon management and reduction measures; and life cycle assessment of products and production lines.

Tongzhou LIU, Associate Professor at Harbin Institute of Technology Shenzhen Graduate School. He received his PhD degree in Environmental Engineering from The Hong Kong University of Science and Technology in 2009 and has published more than 10 scientific papers. Research interests include fate and transport of contaminants in subsurface environment; contaminants control in sediment and dredge materials; remediation of contaminated site and river; and physicochemical processes for recalcitrant contaminants control.

3

TRACTOR'S ENGINE EFFICIENCY AND EXHAUST EMISSIONS' RESEARCH IN DRILLING WORK

Antanas Juostas, Algirdas Janulevičius

Aleksandras Stulginskis University, Studentų g. 15, 53361 Akademija, Kauno r., Lithuania

Abstract. This paper provides an overview of possibilities for determining tractor's engine load, fuel consumption and exhaust emissions in real operating conditions. The use of accumulated database in tractor's electronic control modules for the analysis of engine load, fuel consumption and exhaust emissions is analysed. The methodology for analysis of engine power, speed and exhaust emissions' dependencies, also for analysis of engine exhaust emissions is presented. This paper presents testing results of the unit combined of tractor "Massey Ferguson MF 6499" and drilling machine "Vaderstad Rapid" by engine load, fuel consumption and exhaust emissions. Drilling process time, engine load, fuel consumption and exhaust emission components' distribution are presented in different engine speed and cyclic fuel injection modes. Test results are analysed separately for technological drilling and work processes at the headland. In the technological process of drilling, if the tractor engine speed and, correspondingly, the transmission gear ratio were reduced to get the set working speed, fuel consumption decreased, CO and CO_2 emissions varied slightly, but the NO_x increased significantly. Significant part of exhaust emissions occurred at headlands. The conclusion is that the fuel consumption and exhaust emissions, including harmful components, can be reduced only by complex optimization of technological processes and tractor operating modes.

Keywords: tractor, engine load, fuel consumption, engine speed, exhaust emissions, NO_x, CO, CO_2, environmental impact assessment, environment monitoring, environmental processes modelling.

Reference to this paper should be made as follows: Juostas, A.; Janulevičius, A. 2014. Tractor's engine efficiency and exhaust emissions' research in drilling work, *Journal of Environmental Engineering and Landscape Management* 22(02): 141–150. http://dx.doi:10.3846/16486897.2013.852556

Introduction

Tractors are the main power machines in agriculture whose energy comes from the fuel burned in their engine. As total amount of powerful tractors, cars and other agricultural machinery is increasing, more and more various pollutants are emitted into the environment (Baltrėnas *et al.* 2008, 2004; Pérez-Martínez 2012; Szendro *et al.* 2012; Tretjakovas *et al.* 2012; Vaitiekūnas, Banaitytė 2007; Worldwide Emissions Standards 2011/2012). Fuel consumption and emissions during specific operations depend on engine speed and load characteristics. The engine load can be affected by various operation techniques, as well as by the transmission design and utilization of implements of various performance (Ashrafur Rahman *et al.* 2013; Negoižescu, Tokar 2013; Lindgren *et al.* 2010, 2011; Müllerová *et al.* 2011; Sendžikienė *et al.* 2012; Sirvydas *et al.* 2013). Farmers frequently have a goal, especially in the heavier work (ploughing, cultivation), to achieve maximum productivity, reduce fuel consumption, rather than make an adverse effect on the environment (toxic emissions).

Many studies were focused on environmental impact assessment and improvements in fuel consumption and efficiency of fuel utilization for specific tractors, vehicles and other machines (Grisso, Pitman 2010; Juostas, Janulevičius 2009; Pérez-Martínez 2012; Rakopoulos, Giakoumis 2009). Also, many studies were focused on engine exhaust emissions environment monitoring and measures for reduction of harmfulness of gases, i.e. reduction of emission of components that are harmful to the environment (Ashrafur Rahman *et al.* 2013; Lebedevas *et al.* 2010; Li *et al.* 2006; Lindgren, Hansson 2004; Lindgren *et al.* 2010, 2011). Most of the data for fuel

Corresponding author: Algirdas Janulevičius
E-mail: algirdas.janulevicius@asu.lt

consumption and environmental impact assessment today are obtained according to test data when the mode of engine is steady. During various field applications, engine speed and load characteristics are constantly changing. Hansson *et al.* (2003) and Lindgren *et al.* (2011) have found that fuel efficiency increases proportionally with increasing the engine load. Lindgren *et al.* (2010) studies have shown that emissions, when working with a mobile loader, significantly increased compared to the steady-mode test results. Speed selection in field work has a great influence on the efficiency of work and engine load, as well as fuel consumption and exhaust emissions. In the works of tillage cultivation, tractor operation principle "Gear up, speed down" gives significantly reduced fuel consumption (Grisso, Pitman 2010; Grisso *et al.* 2008). On the other hand, emissions of CO, NO_x and PM exhaust gases increased (Negoižescu, Tokar 2013; Li *et al.* 2006).

The result may be that using different management strategies and environmental processes modelling for machines and different machines' operating modes that are optimized for specific conditions, emissions may be affected to a large extent without affecting operation time or fuel consumption. This means that there is a potential for solving emissions problems. However, today we still lack of information about efficiency of tractor engines, fuel consumption efficiency and toxicity of exhaust emissions in specific field works of a particular tractor operation.

Purpose of the first part of the research is to develop methodologies and tools for data collection to determine the amount and harmfulness of engine exhaust emissions when tractor is working under natural conditions. The purpose of the second part of the tests is to investigate working performance and exhaust emissions for engine of the tractor in working conditions.

1. Methodology and methods

The main gas emission components of engine exhaust are as follows: NO_x *(nitrogen oxides)*, PM *(particulate matter)*, HC *(hydrocarbons)*, CO *(carbon monoxide)* and CO_2 *(carbon dioxide)*. In all cases, the CO_2 emission is directly dependent on fuel consumption. CO_2 emissions gradually increase with increasing of fuel consumption (An *et al.* 2012; Ashrafur Rahman *et al.* 2013). From ecological point of view, CO_2 gas is also dangerous because it traps heat on the surface of earth like a sort of shell. NO comprises about 90% of all nitrogen oxides in exhaust gases. The latter is formed where gas temperature is very high and there is sufficient amount of oxygen. When the combustion process is complete, the cycle temperature is higher, and then more NO is in the exhaust gases. When this process is not complete, more CO and HC are in exhaust gases. Harmful substances generated by diesel engines depend on the load and speed of the engine (McLaughlin, Layer 2003a; Lindgren *et al.* 2011; Lebedevas *et al.* 2012).

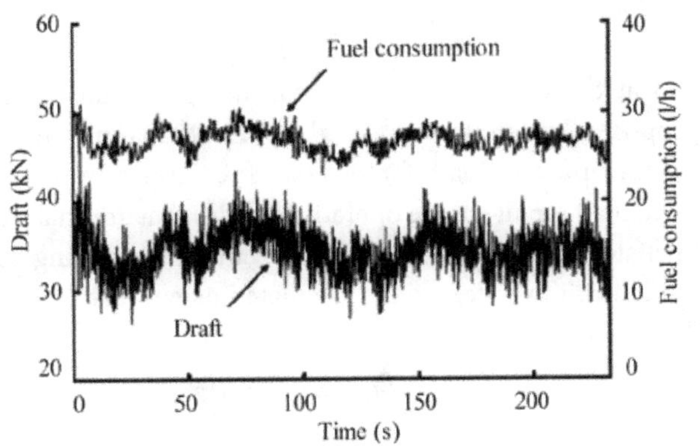

Fig. 1. Dependencies of traction force and fuel consumption and their interrelation in real-time soil preparation work (Li *et al.* 2006)

Environment monitoring of engine emissions in field conditions is an expensive and time consuming process, especially if operational strategies are studied. In addition, in field conditions it is almost impossible to directly measure and control all the variables affecting engine emissions (McLaughlin, Layer 2003a). Computer simulations and calculations are suitable for such studies. This requires a methodology, environmental processes modelling and data, where specific conditions are taken into account.

Analysis of data obtained in the field work tests (Li *et al.* 2006; McLaughlin, Layer 2003a) has showed that the fuel consumption curve for the tractor is very similar to the traction force curve (Fig. 1). In the results of a field work test by McLaughlin and Layer (2003b) hourly fuel consumption is very well reflected in the engine power curve.

Engine power is characterized by the product of tractor's traction force and speed. Fuel consumption can be expressed in mega-joules of traction power energy per litre of fuel (MJ l^{-1}) and dependence on time. Thus, the fuel efficiency, expressed in such a way, i.e. as a sum of efficiencies for the engine, transmission and traction power, is a measure of pure energy per unit of the fuel, when the tractor performs intended pulling activities under field conditions. If we compare the fuel consumption, fuel efficiency relatively depends on the traction force in conventional field work (Li *et al.* 2006).

Lebedevas *et al.* (2011) presented development of an algorithm for calculating energy characteristics of combustion products. The relation between fuel consumption and emissions is highly dependent on engine load (the acting torque) (Lindgren *et al.* 2011). Therefore, the reduction in fuel consumption can also reduce emissions. NO_x formation is closely related with the engine torque and combustion temperature which generally increases with the increase of engine speed and engine load (An *et al.* 2012; Asprion *et al.* 2013; Li *et al.* 2006). An *et al.* (2012) study shows that the engine idle speed of 800 rpm may also have a significant affect on the NO_x emission formation processes. Results of Li *et al.* (2006) and Bostic *et al.* (2009) tests for NO_x emissions very well reflect the curve of the traction force, but it is with increased emissions in the area of grater traction force and, correspondingly, at the higher engine loads (Fig. 2). The increase of NO_x emissions is treated as a result of increased need for fuel consumption and higher combustion temperatures when higher traction force is needed.

In test results of Li *et al.* (2006) and Bostic *et al.* (2009) exhaust gas temperature reflected the curve of traction force very well, but initially the temperature were low and temperature transition process lacked behind from the transition process of traction force (Fig. 3).

Bostic *et al.* (2009), Lebedevas *et al.* (2011), Junevičius *et al.* (2011) emulate engine performance from empirical relation between the engine speed, engine torque and fuel consumption, i.e. having at least two variables, the third can be derived. Tests of Hansson *et al.* (2003), Lindgren *et al.* (2010) and Bostic *et al.* (2009) showed that in the real field work fuel consumption and emissions cannot be accurately calculated if actual engine load in the field work is not taken into account. Tests of Lindgren *et al.* (2011, 2010) and Fathollahzadeh *et al.* (2010) showed a very high-frequency fluctuations in traction force and fuel consumption at different typical works. Such high-frequency fluctuations are normal for in-field use and depend on the soil surface roughness, which influences the tractor and the implement performance disparity and uneven depth of tillage. Therefore, using mathematical models, the load of the engine, fuel consumption, emissions, etc. can also be calculated with good accuracy only by possessing the data collected at high frequency.

Engines in older tractors and other machines have mechanically controlled fuel injection system. However, technical progress has led to wide introduction of new, electronically controlled engines with fuel injection systems. Electronic systems dynamically control engine parameters and manage its work (Mousazadeh 2013). Depending on the configuration, such motors can make rapid changes in engine speed and load conditions. Review of factors, affecting the fast changes in engine load and speed, were conducted by Rakopoulos, Giakoumis (2009). Sudden changes in engine load or speed are particularly relevant to tractors.

To test engine performance, exhaust harmfulness and emissions, tractor model "*Massey Ferguson MF 6499*" was selected. The histogram "ECU Load Profile" in microprocessors of tractors "Massey Ferguson" stores the operating parameters: working hours at particular

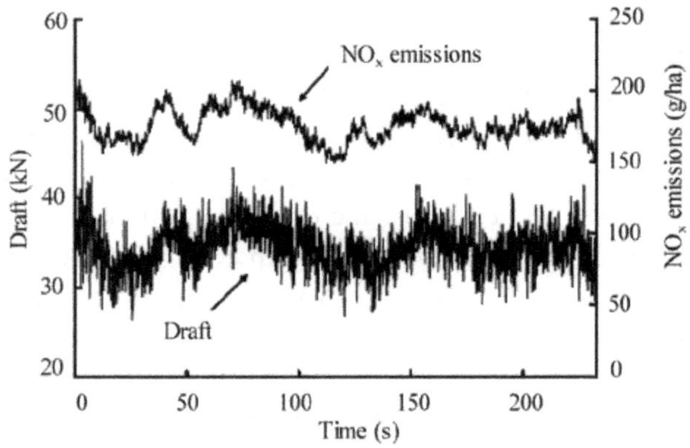

Fig. 2. Dependencies of traction force and NOx emissions and their interrelation in real-time soil preparation work (Li *et al.* 2006)

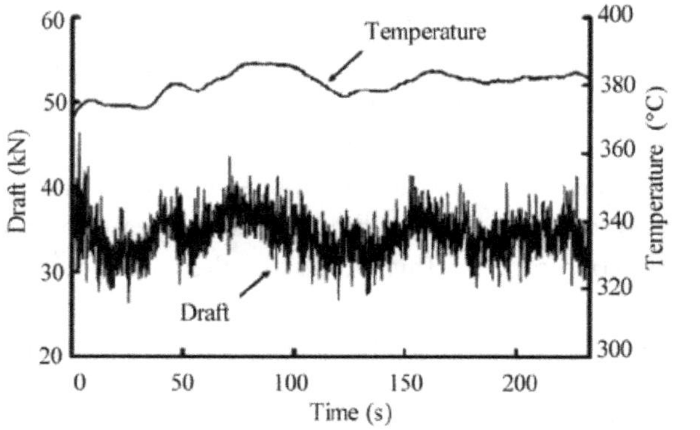

Fig. 3. Dependencies of traction force and combustion gas temperatures and their interrelation in real-time soil preparation work (Li *et al.* 2006)

engine speed and cyclic injection modes (Janulevičius *et al.* 2013, 2010a, b). Histogram and its table show the total tractor's operational periods, in seconds, in various engine modes. Modes of operation are divided by engine speed and the cyclic fuel injection. Histogram and the table show how long the engine of the tractor were operating at 700–900, 900–1100, 1100–1300, etc. rpm speed. It also shows how long the engine of the tractor were operating at the cyclic injection modes of 0–10, 10–20, 20–30, etc. mg.

From these histograms, recorded at the beginning and at the end of tested drilling, using methodologies proposed by Janulevičius *et al.* (2013, 2010a, b), engine torque, power, load factor and fuel consumption is calculated for all engine operating modes that are analysed in histogram. CO_2, CO and NO_x emissions during the tested drilling process were calculated using the following dependencies:

$$E_{(x)} = \frac{\varepsilon\, E_{(x,\,PPM)} \cdot \sigma_E \cdot \rho_E}{P}, \qquad (1)$$

where: $E_{(x)}$ is the emission of the engine exhaust gas component (x, i.e. CO_2, CO and NO_x), $gkW^{-1}h^{-1}$; $E_{(x,\,PPM)}$

is the emission of the engine exhaust gas component (x, i.e. CO_2, CO and NO_x), ppm; ε is the coefficient that evaluates the changes in engine load *(transient)*; σ_E is the gas flow, m^3h^{-1}; ρ_E is the gas density, kg $kmol^{-1}$; *P* is the engine power at various operating mode, kW.

$$\sigma_E = \frac{T_E + 460}{540}\sigma_O, \tag{2}$$

where: σ_O is the flow of supplied fresh air, m^3h^{-1}; T_E is the temperature of exhaust gases, (F); Bostic *et al.* (2009) has proposed mathematical model for calculation of exhaust gas temperature.

$$\sigma_O = \frac{\upsilon_e\, n}{210{,}86}\eta_{\sigma_O}, \tag{3}$$

where: υ_e is the engine displacement in cm^3; *n* is the engine speed, min^{-1}; η_{σ_O} is the volumetric efficiency coefficient of supplied air flow.

We can write the final equation for calculation of emissions of engine exhaust gas components (CO_2, CO and NO_x):

$$E0_{(x)} = \frac{\varepsilon E_{(x,\,PPM)}\, n\, \upsilon_e\, \eta_{\sigma_O}\, \rho_E\, A}{P}\left(\frac{T_E + 540}{1{,}139\cdot 10^5}\right), \tag{4}$$

where *A* is the performed engine work, kJ.

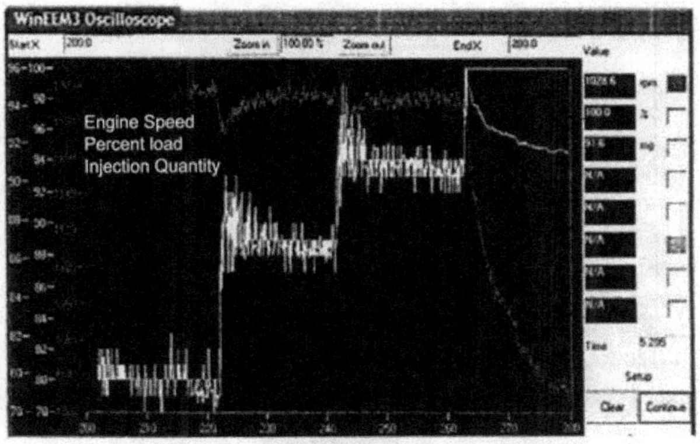

Fig. 4. Oscillogram of engine mode indicators (sample)

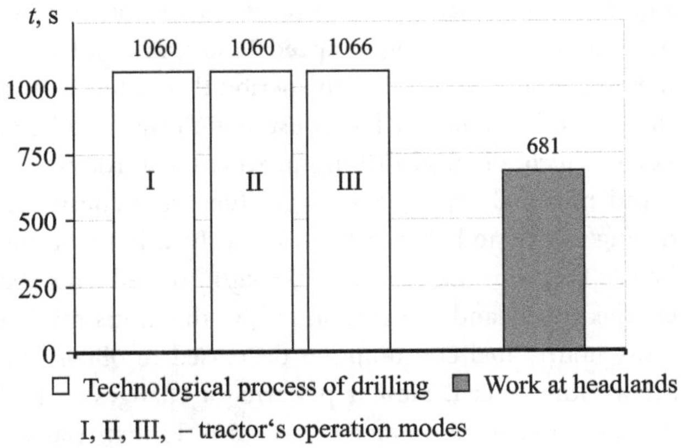

Fig. 5. Distribution of drilling process times between operating modes

$$A = Pt, \tag{5}$$

where *t* is the working hours, h.

Emissions for separate components (CO_2, CO, NO_x, etc.) at specific engine operating modes can be determined from the "ECU Load Profile" histograms and the engine characteristics *M*, B_d, b_c, NO_x, CO, CO_2 = *f*(*n*). These characteristics for the tractor "*MF 6499*" were compiled using a stand for the tractor's load via PTO shaft (tractor dynamometer (AW 400) on PTO test results) and standard test methodology (OECD Standard Code 2. 2011). In our case, the difference was that more parameters were measured and in much more points. In addition, oscillograms were recorded during each test (Fig. 4) to determine engine parameters, which could not be measured using the stand AW 400.

The variation of the engine speed (red curve), engine load (white curve) and fuel cyclic injection quantity (yellow curve) oscillograms (Fig. 4) during the test time were taken.

2. Results of the research

Engine performance and exhaust emissions tests were conducted during the drilling operations using the combination of tractor "Massey Ferguson MF-6499" and the drill "Vaderstad Rapid". For the test, a smooth, nearly even structures, loamy wheat-stubble field with a length of 403±7 m were selected. Before the test, the field were ploughed at 21±1 cm depth and cultivated at 10±1.5 cm depth. Soil moisture at a depth of 15 cm were 17.7±0.12%, soil hardness 0.58±0.11 MPa. During the studies, tractor slippage did not exceed 11% and theoretical speeds of 3.5±0.14 m s^{-1} were maintained. The selected speed were realized in three versions: I – gear 2b at engine speed of 1600 rpm; II – gear 2a at engine speed of 1800 rpm; III – gear 1d at engine speed of 2000 rpm. Figure 5 shows the distribution between the technological and auxiliary (work at headlands) working times in the analysed drilling process. Efficiency of working time for the unit τ = 0.824, productivity *W* – averagely about 4 ha h^{-1}.

Figure 6 shows the distribution of test time (3867 s) for the unit in the modes of engine speed and cyclic fuel injection. In this figure we can see the separation between the modes and their durations when the unit were making technological work and work at the headlands (in turns). In headland operating modes, large proportion of operational time apparently is concentrated in the mode of small engine speed (900–100 min^{-1}) and small quantities of cyclic fuel injection (10–20 mg). This mode took about 35% of full operational time at headlands, which accounted for 17.6% of total test time for drilling process. During the technological process of drilling, tractor engine were running in the modes when cyclic

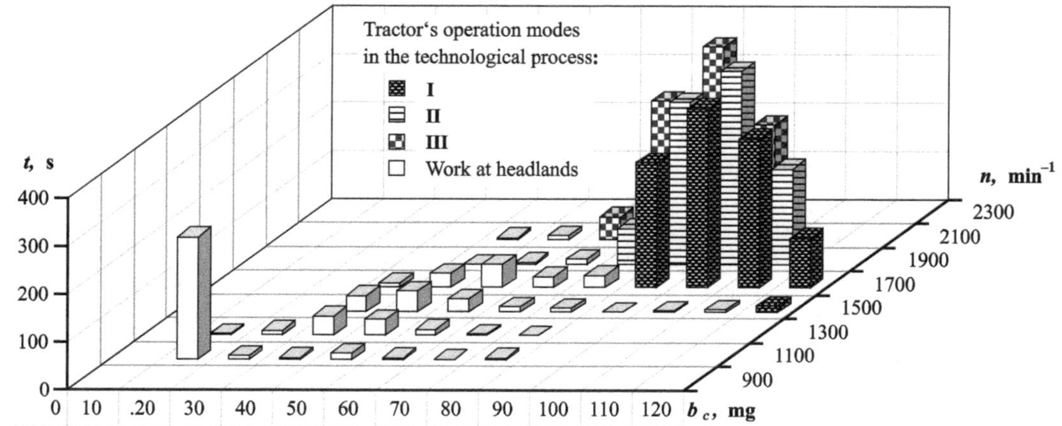

900												
1100		255	7	3	13	3	1	3				
1300		3	10	40	33	13	3	1				
1500				33	43	26	10	7	0	1	4	13
1700				10	30	50	23	26	260	372	310	104
1900					0	3	10	74	335	453	198	
2100					5	9	47	287	494	238		
2300												

Fig. 6. Distribution of unit's working time period (3867 s) in the modes of engine speed (n) and cyclic fuel injection quantity (b_c)

fuel injection quantity were in the range of 60–120 mg. By reducing the engine speed and, correspondingly, the transmission ratio, cyclic injection quantities increased slightly to obtain the set working speed. The tested tractor's working mode III (gear 1d, engine speed 2000 rpm) were related with cyclic injection quantities in the range of 60–100 mg. Fuel injection quantity of 80–90 mg prevailed. Mode II (gear 2a, engine speed 1800 rpm) were related with cyclic injection quantities in the range of 70–100 mg. Fuel injection quantity of 90–100 mg prevailed. Mode I (gear 2b, engine speed 1600 rpm) were related with cyclic injection quantities in the range of 80–120 mg. Fuel injection quantity of 90–110 mg prevailed. The average time period for technological drilling process in all three tested modes was 1060 s.

Figure 7 shows fuel consumption during the test period of drilling. During the test period of drilling (3867 s) 28.1 kg of fuel were consumed. Part of that fuel, 26.05 kg (92.7%) were used for technological drilling process, and 2.05 kg (7.3%) were used for work at headlands.

Figure 8 shows the distribution of fuel consumption during the test time (3867 s) for the unit in the modes of engine speed and cyclic fuel injection. Fuel consumption, in relative terms, reflects the power of the engine and working hours in that mode (Janulevičius *et al.* 2010a; Li *et al.* 2006). In tractor mode I (gear 2b, engine speed 1600 rpm), 8.27 kg of fuel were consumed during the technological process of drilling that lasted 1060 s. In this mode the hourly fuel consumption was 28.1 kg/h. For the

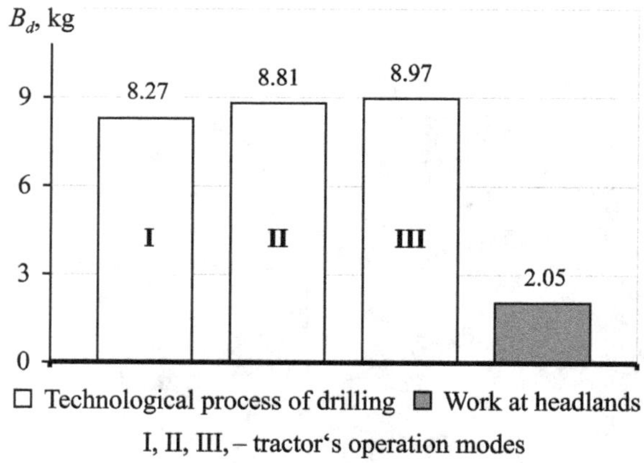

Fig. 7. Distribution of fuel consumption during the process of drilling test (3867 s) by the operating modes

similar technological drilling process, when tractor were in mode II (gear 2a, engine speed 1800 rpm), 8.81 kg of fuel were consumed during the period that lasted 1060 s. In this mode the hourly fuel consumption was 29.9 kg/h. For the technological drilling process (gear 1d, engine speed min 2000 rpm), 8.97 kg of fuel were consumed during the period that lasted 1066 s. In this mode the hourly fuel consumption was 30.3 kg/h. By reducing the engine speed and, correspondingly, the transmission ratio, hourly fuel consumption decreased to obtain the set working speed. Similar patterns of changes in fuel consumption were obtained by Grisso, Pitman (2010), Grisso *et al.* (2008), Lindgren *et al.* (2010, 2011). For the work at headlands 2.05 kg of fuel were consumed during

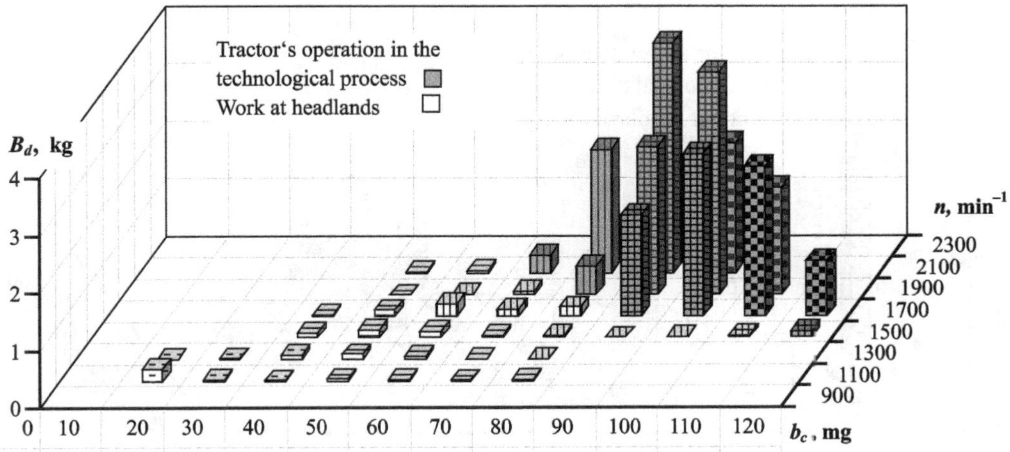

n	10	20	30	40	50	60	70	80	90	100	110	120
900												
1100		0.20	0.01	0.01	0.03	0.01	0.00	0.01				
1300		0.00	0.01	0.08	0.09	0.04	0.01	0.00				
1500				0.08	0.14	0.10	0.05	0.03	0.00	0.01	0.03	0.11
1700				0.03	0.11	0.22	0.12	0.16	1.77	2.83	2.61	0.96
1900					0.00	0.02	0.06	0.50	2.56	3.87	1.88	
2100					0.02	0.05	0.31	2.15	4.20	2.26		
2300												

Engine's load factor
- <0.2
- 0.2–0.4
- 0.4–0.6
- 0.6–0.8
- >0.8

Fig. 8. Distribution of unit's fuel consumption during the test period (3867 s) in the modes of engine speed (n) and cyclic fuel injection quantity (b_c)

the process that lasted 681 s. In the headlands' mode the hourly fuel consumption was 10.84 kg/h.

In Figure 8 the modes of engine speed and cyclic fuel injection are marked by the size of engine load. Engine load is set by the extent of torque usage. From Figure 8 we can see that for the investigated drilling technological process, when engine speed were 2000 rpm and gear 1d were selected, the recommended engine load (load factor >0.8) lasted 22% of the total duration of that mode, while 78% of the time the engine load factor were in the range from 0.6 to 0.8. In mode II of the tractor, when engine speed were 1800 rpm and a gear 2a were selected, the recommended engine load (load factor >0.8) lasted 19% of the total duration of that mode, while 81% of the time the engine load factor were in the range from 0.6 to 0.8. And when engine speed were 1600 rpm and gear 1d were selected, the recommended engine load (load factor >0.8) lasted 39% of the total duration of that mode, while 61% of the time the engine load factor were in the range from 0.6 to 0.8. The lowest fuel consumption and the maximum engine load were while operating at 1600 rpm engine speed and when gear 2b was selected.

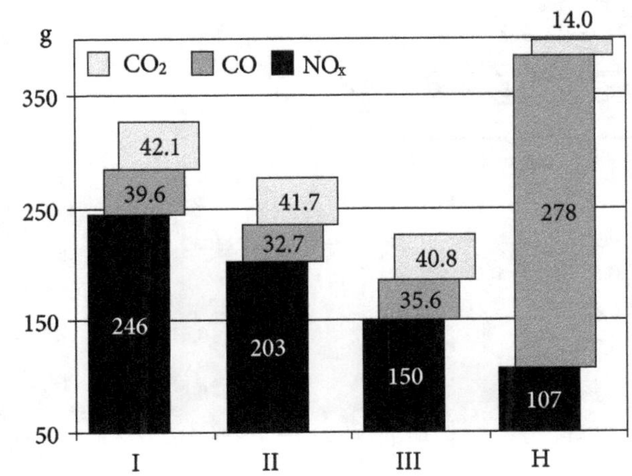

Fig. 9. Distribution of emitted combustion components CO, CO_2 and NO_x during the test period (3867 s) in the drilling process by operating modes; I, II, III – tractor's operation modes, H – mode at headland

While working at headlands, engine load factor were less than 0.4. About half of working time at headlands it were between 0.2 and 0.4, and the other half of the working time it were less than 0.2.

Figures 9, 10, 11 and 12 show the CO, CO_2 and NO_x emissions exhaust during the test period of drilling. During the drilling test period (3867 s), CO emission were approximately 386 g, CO_2 emission were 103 g and NO_x emission were 706 g. Part of these were for the technological process of drilling, namely: CO emissions – 108 g (28%), CO_2 – 124.6 10^3 g (90%) and NO_x – 600 g (85%). When working at headlands, CO emissions were 278 g (72%), CO_2 – 14 10^3 g (10%) and NO_x – 107 g (15%).

In the technological process of drilling, if the tractor engine speed and, correspondingly, the transmission gear ratio were reduced to get the set working speed, CO and CO_2 emissions varied slightly, but the NO_x emission increased significantly. In the technological process of drilling, when tractor were in mode III (gear 1d, engine

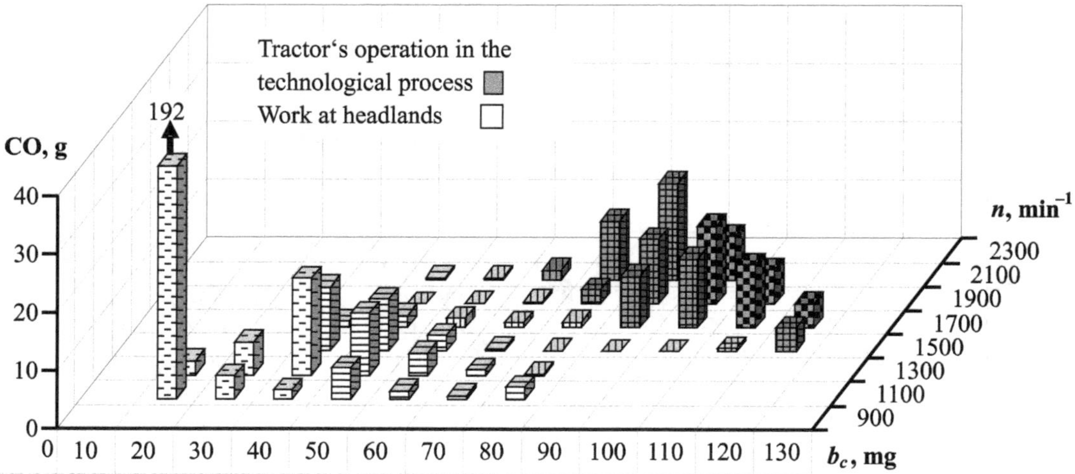

n \ b_c	10	20	30	40	50	60	70	80	90	100	110	120	130
900													
1100		192.1	3.9	1.5	5.3	1.4	0.5	2.1					
1300		2.4	5.6	16.6	10.6	3.7	0.9	0.3					
1500				10.9	9.0	2.8	0.4	0.2	0.0	0.1	0.6	4.1	
1700				1.1	2.1	1.6	0.8	0.9	8.8	11.8	10.8	4.2	
1900					0.0	0.1	0.3	2.3	11.1	13.4	5.9		
2100					0.3	0.3	1.7	10.0	16.5	7.4			
2300													

Fig. 10. Distribution of CO emissions during the test period of drilling (3867 s) in the modes of engine speed (n) and cyclic fuel injection quantity (b_c)

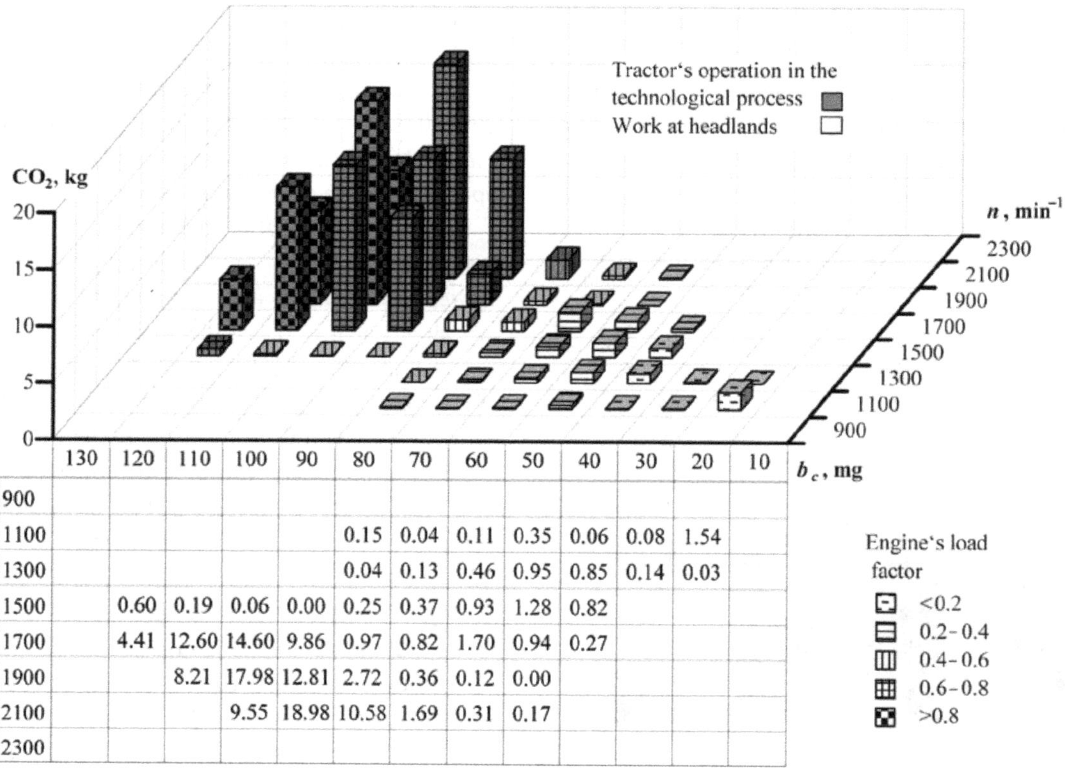

n \ b_c	130	120	110	100	90	80	70	60	50	40	30	20	10
900													
1100						0.15	0.04	0.11	0.35	0.06	0.08	1.54	
1300						0.04	0.13	0.46	0.95	0.85	0.14	0.03	
1500		0.60	0.19	0.06	0.00	0.25	0.37	0.93	1.28	0.82			
1700		4.41	12.60	14.60	9.86	0.97	0.82	1.70	0.94	0.27			
1900			8.21	17.98	12.81	2.72	0.36	0.12	0.00				
2100			9.55	18.98	10.58	1.69	0.31	0.17					
2300													

Fig. 11. Distribution of CO_2 emissions during the test period of drilling (3867 s) in the modes of engine speed (n) and cyclic fuel injection quantity (b_c)

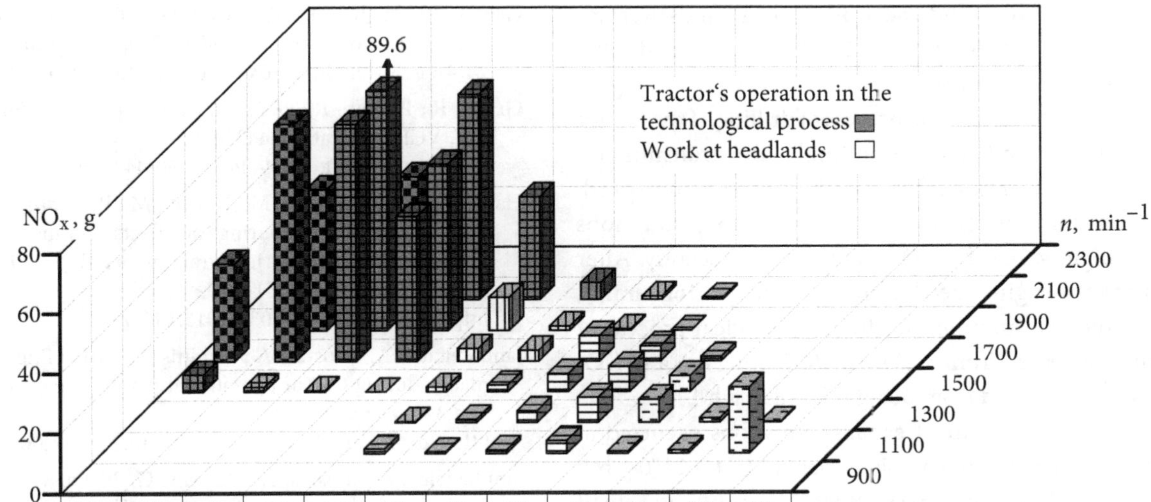

	120	110	100	90	80	70	60	50	40	30	20	10
900												
1100					1.6	0.4	1.2	4.3	0.8	1.1	22.0	
1300					0.3	1.1	3.9	8.3	7.7	1.4	0.3	
1500	5.8	1.6	0.4	0.0	1.6	2.3	6.0	8.5	5.3			
1700	32.6	79.4	79.3	48.4	4.2	3.6	8.5	5.1	1.4			
1900		47.2	89.6	55.6	11.0	1.4	0.5	0.0				
2100			41.2	68.8	34.4	5.7	1.1	0.7				
2300												

Fig. 12. Distribution of NO_x emissions during the test period of drilling (3867 s) in the modes of engine speed (n) and cyclic fuel injection quantity (b_c)

speed 2000 min^{-1}), during the period of 1066 s NO_x emission were 150 g. Hourly NO_x emissions for this mode were 506 g h^{-1}. For the similar technological drilling process, when tractor were in mode II (gear 2a, engine speed 1800 rpm), NO_x emissions were 203 g during the period that lasted 1060 s. Hourly NO_x emissions for this mode were 680 g h^{-1}. During the technological process of drilling (gear 2b, engine speed 1600 rpm) NO_x emissions were 246 g during the period that lasted 1060 s. Hourly NO_x emissions for this mode were 836 g h^{-1}. Hourly NO_x emissions increased when engine speed and, correspondingly, transmission ratio were reduced to obtain the set working speed. When engine speeds were reduced from 2000 min^{-1} to 1600 min^{-1}, NO_x emissions increased from 506 g h^{-1} to 836 g h^{-1}. An hourly NO_x emission, when working at headlands, was approximately 566 g h^{-1}.

In the technological process of drilling, hourly CO emissions varied in the range from 110 to 134 g h^{-1}, and CO_2 emissions – from 138 to 143 kg h^{-1}. The highest CO emissions were in the work processes at headlands. The hourly CO emissions here reached 1470 g h^{-1}. In the diagram zone of low speed and low cyclic injection, the tractor most of the time (255 s) operated in the mode of 900–1000 min^{-1} and 10–20 mg of fuel injection. This mode corresponds the engine idling. From Figures 2 and 6 we can determine that in the process of drilling, when engine speeds were in the mode from 900 to 1100 min^{-1} and cyclic fuel injection from 10 to 20 mg, average hourly emissions of CO were 2710 g h^{-1}. Similar patterns of fuel consumption and amount of CO_2, CO, NO_x emission variations are also obtained by other scientists in their research works (An *et al.* 2012; Asprion *et al.* 2013; Li *et al.* 2006; Lindgren, Hansson 2004; Lindgren *et al.* 2010, 2011).

Conclusions

1. Tractor engine performance, fuel consumption and exhaust emission components can be evaluated by using information stored in the engine control microprocessor, where engine speed and fuel cyclic injection modes and periods are registered in detail. Such database in tractors "Massey Ferguson MF 6499" is called "ECU Load Profile."

2. For the tested drilling process, technological drilling work accounted for 82% of total working time, and work at the headlands accounted for 18% of total working time. In the drilling process fuel consumption was 26.2 kg h^{-1}. Of these, 92.7% accounted for the technological work, and 7.3% – for the work at headlands. Distribution of exhaust emissions' components in the

technological work and the work at headlands were as follows: NO_x – 84.8% and 15.2%; CO – 28% and 72%; CO_2 – 89.9% and 10.1%.

3. In the technological process of drilling, if the tractor engine speed and, correspondingly, the transmission gear ratio were reduced to get the set working speed, fuel consumption decreased, CO and CO_2 emissions varied slightly, but the NO_x increased significantly. After reducing the engine speed from 2000 min^{-1} to 1600 min^{-1}, hourly fuel consumption in the technological process of drilling decreased from 30.3 kg h^{-1} to 28.1 kg h^{-1}, and NO_x emissions increased from 506 g h^{-1} to 836 g h^{-1}.

4. Significant part of exhaust emissions occurred at headlands. The highest emissions were of CO – 1470 g h^{-1}. The mode of low engine speed (900–1100 min^{-1}) and small cyclic fuel injection quantity (10–20 mg) distinguished here; hourly emissions of CO in this mode were 2710 g h^{-1}, and this mode took 37.4% of working time at the headlands or 6.6% of total working time.

5. Fuel consumption and exhaust emissions, including harmful components, can be reduced only by complex optimization of technological processes and tractor operating modes.

References

An, H.; Yang, W. M.; Chou, S. K.; Chua, K. J. 2012. Combustion and emissions characteristics of diesel engine fueled by biodiesel at partial load conditions, *Applied Energy* 99: 363–371. http://dx.doi.org/10.1016/j.apenergy.2012.05.049

Ashrafur Rahman, S. M.; Masjuki, H. H.; Kalam, M. A.; Abedin, M. J.; Sanjid, A.; Sajjad, H. 2013. Impact of idling on fuel consumption and exhaust emissions and available idle-reduction technologies for diesel vehicles – a review, *Energy Conversion and Mamagement* 74: 171–182. http://dx.doi.org/10.1016/j.enconman.2013.05.019

Asprion, J.; Chinellato, O.; Guzzella, L. 2013. A fast and accurate physics-based model for the NOx emissions of diesel engines, *Applied Energy* 103: 221–233. http://dx.doi.org/10.1016/j.apenergy.2012.09.038

Baltrėnas, P.; Vaitiekūnas, P.; Vasarevičius, S.; Jordaneh, S. 2008. Modelling of motor transport exhaust gas influence on the atmosphere, *Journal of Environmental Engineering and Landscape Management* 16(2): 65–75. http://dx.doi.org/10.3846/1648-6897.2008.16.65-75

Baltrėnas, P.; Vaitiekūnas, P.; Mincevič, I. 2004. Investigation on the impact of transport exhaust emissions on the air, *Journal of Environmental Engineering and Landscape Management* 12(1): 3–11. http://dx.doi.org/10.1080/16486897.2004.9636809

Bostic, D. R.; Watson, D. G.; Harrison, T. V. 2009. Modeling no emissions of an off-road diesel engine based on emission tests, *Agricultural Engineering International: The CIGR Journal* Vol. XI. 19 p.

Fathollahzadeh, H.; Mobli, H.; Rajabipour, A.; Minaee, S.; Jafari, A.; Tabatabaie, S. M. H. 2010. Average and instantaneous fuel consumption of Iranian conventional tractor with moldboard plow in tillage, *ARPN Journal of Engineering and Applied Sciences* 5(2): 30–35.

Grisso, R.; Vaughan, D.; Roberson, G. 2008. Fuel prediction for specific tractor models, *Applied Engineering in Agriculture* 24(4): 423–428. http://dx.doi.org/10.13031/2013.25139

Grisso, R.; Pitman, R. 2010. *Gear up and throttle down: saving fuel*, VCE publication: 442–450. http://pubs.ext.vt.edu/442/442-450/442-450

Hansson, P.-A.; Lindgren, M.; Nordin, M.; Petersson, O. 2003. A methodology for measuring the effects of transient loads on the fuel efficiency of agricultural tractors, *Applied Engineering in Agriculture* 19(3): 251–257. http://dx.doi.org/10.13031/2013.13657

Janulevičius, A.; Juostas, A.; Pupinis, G. 2013. Engine performance during tractor operational period, *Energy Conversion and Management* 68: 11–19. http://dx.doi.org/10.1016/j.enconman.2013.01.001

Janulevičius, A.; Juostas, A.; Pupinis, G. 2010a. Tractor engine load and fuel consumption in road construction works, *Transport* 25(4): 409–416. http://dx.doi.org/10.3846/transport.2010.50

Janulevičius, A.; Juostas, A.; Pupinis, G. 2010b. Engine working modes during tractors operational period, *Mechanika* 3(83): 409–416.

Junevičius, R.; Bogdevičius, M.; Török, A. 2011. Modifying mathematical models for calculating operational chatacteristics of diesel engines burning RME biofuels, *Transport* 26(1): 50–60. http://dx.doi.org/10.3846/16484142.2011.561528

Juostas, A.; Janulevičius, A. 2009. Evaluating working quality of tractors by their harmful impact on the environment, *Journal of Environmental Engineering and Landscape Management* 17(2): 106–113. http://dx.doi.org/10.3846/1648-6897.2009.17.106-113

Lebedevas, S.; Lebedeva, G.; Makarevičiene, V.; Kazanceva, I.; Kazancev, K. 2010. Analysis of ecological parameters of the diesel engine powered with biodiesel fuel containing methyl esters from Camelina sativa oil, *Transport* 25(1): 22–28. http://dx.doi.org/10.3846/transport.2010.04

Lebedevas, S.; Lebedeva, G.; Bereišienė, K. 2011. Modelling of internal combustion engines' emission through the use of traffic flow mathematical models, *Transport* 26(3): 271–278. http://dx.doi.org/10.3846/16484142.2011.621978

Lebedevas, S.; Pukalskas, S.; Žaglinskis, J.; Matijošius, J. 2012. Modelling of internal combustion engines' emission through the use of traffic flow mathematical models, *Transport* 27(2): 171–177. http://dx.doi.org/10.3846/16484142.2012.694078

Li, Y. X.; McLaughlin, N. B.; Patterson, B. S.; Burtt, S. D. 2006. Fuel efficiency and exhaust emissions for biodiesel blends in an agricultural tractor, *Canadian Biosystems Engineering/Le Génie des Biosystèmes au Canada* 48: 2.15–2.22.

Lindgren, M.; Arrhenius, K.; Larsson, G.; Bäfver, L.; Arvidsson, H.; Wetterberg, C.; Hansson, P.-A.; Rosell, L. 2011. Analysis of unregulated emissions from an off-road diesel engine during realistic work operations, *Atmospheric Environment* 45(30): 5394–5398. http://dx.doi.org/10.1016/j.atmosenv.2011.06.046

Lindgren, M.; Hansson, P.-A. 2004. Effects of transient conditions on exhaust emissions from two non-road diesel engines, *Biosystems Engineering* 87(1): 57–66. http://dx.doi.org/10.1016/j.biosystemseng.2003.10.001

Lindgren, M.; Larsson, G.; Hansson, P.-A. 2010. Evaluation of factors influencing emissions from tractors and construction equipment during realistic work operations using diesel fuel

and bio-fuels as substitute, *Biosystems Engineering* 107(2): 123–130.
http://dx.doi.org/10.1016/j.biosystemseng.2010.07.010

McLaughlin, N. B.; Layer, M. 2003a. *Instrumentation for field measurement of tractor exhaust gas emissions.* CSAE/SCGR Paper 03-224 [online], [cited 08 December 2005]. Winnipeg, MB. Available from Internet: http://www.engr.usask.ca/societies/csae/PapersCSAE2003/CSAE03-224.pdf

McLaughlin, N.; Layer, M. 2003b. *Field measurement of agricultural tractor exhaust gas emissions,* Agriculture and Agri-Food Canada, Environment Canada [online], [cited 11 March 2012]. Available from Internet: http://www.epa.gov/ttnchie1/conference/ei12/poster/mclaughlin.pdf

Mousazadeh, H. 2013. A technical review on navigation systems of agricultural autonomous off-road vehicles, *Journal of Terramechanics* 50(3): 211–232.
http://dx.doi.org/10.1016/j.jterra.2013.03.004

Müllerová, D.; Landis, M.; Schiess, I.; Jablonický, J.; Prístavka, M. 2011. Operating parameters and emission evaluation of tractors running on diesel oil and biofuel, *Research in Agricultural Engineering* 57(Special Issue): S35–S42.

Negoižescu, A.; Tokar, A. 2013. Investigations on car emissions under the urban traffic conditions with the influence on Timişoara air quality, *Transport* 28(1): 38–45.
http://dx.doi.org/10.3846/16484142.2013.781060

OECD standard code for the official testing of agricultural and forestry tractor performance. 2011. CODE 2 – February 2011.

Pérez-Martínez, P. J. 2012. Energy consumption and emissions from the road transport in Spain: a conceptual approach, *Transport* 27(4): 383–396.
http://dx.doi.org/10.3846/16484142.2012.751051

Rakopoulos, C. D.; Giakoumis, E. G. 2009. *Diesel engine transient operation: principles of operation and simulation analysis.* London: Springer-Verlag. 390 p.

Sendžikienė, E.; Makarevičienė, V.; Kalenska, S. 2012. Exhaust emissions from the engine running on multi-component fuel, *Transport* 27(2): 111–117.
http://dx.doi.org/10.3846/16484142.2012.690138

Sirvydas, A.; Nadzeikienė, J.; Kerpauskas, P.; Ūksas, T. 2013. The principle of heat conversion into mechanical work, *Mechanika* 19(3): 358–362.

Szendro, G.; Csete, M.; Torok, A. 2012. Unbridgeable gap between transport policy and practice in Hungary, *Journal of Environmental Engineering and Landscape Management* 20(2): 104–109.
http://dx.doi.org/10.3846/16486897.2012.660881

Tretjakovas, S.; Girgždys, A.; Masilevičius, R. 2012. The total ozone amount chantes over Lithuania and neighbouring countries, *Journal of Environmental Engineering and Landscape Management* 20(2): 147–154.
http://dx.doi.org/10.3846/16486897.2011.641721

Vaitiekūnas, P.; Banaitytė, R. 2007. Modeling of motor transport exhaust pollutant dispersion, *Journal of Environmental Engineering and Landscape Management* 15(1): 39–46.
http://dx.doi.org/10.1080/16486897.2007.9636906

Worldwide Emissions Standards. Heavy Duty and Off-Road Vehicles, DELPHI, Innovation for the real world. 2011/2012.

Antanas JUOSTAS. Dr, Power and Transport Machinery Engineering Institute, Aleksandras Stulginskis University (ASU). Doctor of Science, ASU, 2012; Master of Science (Mechanical Engineering), LŽŪU, 2007. Publications: author of 8 scientific publications. Research interests: tractors parameters from dynamic, ecological and economical point of view.

Algirdas JANULEVIČIUS. Dr, Assoc. Prof., Power and Transport Machinery Engineering Institute, Aleksandras Stulginskis University (ASU). Doctor of Science, 1993. Employment: Associate Professor (1997). Publications: author or co-author of more than 100 scientific publications, 39 inventions. Research interests: transport and power machinery parameters from dynamic, economical and ecological point of view.

4

SOME TECHNICAL ASPECTS OF THE RHEOLOGICAL PROPERTIES OF HIGH CONCENTRATION FINE SUSPENSIONS TO AVOID ENVIRONMENTAL DISASTERS

József FAITLI, Imre GOMBKÖTŐ

Institute of Raw Material Preparation and Environmental Processing, University of Miskolc, Miskolc – Egyetemváros H-3515, Hungary

Abstract. The behavior of slurries and suspensions made by mixing solid particles and liquids is very important for various industrial applications. The latest accidental failure events at tailings facilities (Kolontár, Baia Mare) have focused public interest into this field. Nowadays, environmental practice is turning to use dry deposition techniques or at least as high concentration slurries or pastes as possible, to avoid large spills in case of an accidental failure of an embankment. High concentration slurries are becoming widely accepted in many environmentally related operations such as underground backfilling or simple tailings deposition. However, the hydraulic transport of pastes or high density slurries requires higher energy input via pumps, and, in addition, the energy requirement or pressure loss calculation methods are also different because the rheology of pastes differs from that of dilute slurries. At the University of Miskolc, Institute of Raw Materials Preparation and Environmental Processing, Miskolc, Hungary, this topic has been studied for many decades. The fine suspension – coarse mixture flow model was introduced, and it has been determined that the flow behavior of fine suspensions made of solid particles smaller than a limit particle size can be measured and interpreted in almost the same way as for single phase clear liquids. Based on these measured rheological parameters of fine suspensions, the frictional energy loss can be calculated. The aim of this paper is to give a summary and data base about the rheological behavior of different industrial materials based on pilot scale hydraulic loop measurements. An Anton-Paar rotational viscometer and a tube viscometer with three measuring pipe sections, – developed by our institute – were used for testing. The results of measurements of various granular materials, such as sands, fly ashes, perlite, tailings and red mud are presented in connection with environmental applications. Based on these results, empirical relationships are presented, where the parameters are determined by simple function fitting into the data of measurements carried out at discrete concentration values. The rheology of fine suspensions of any concentration up to the measured maximum can be calculated by these relationships.

Keywords: fine suspension rheology, dense slurries, paste technology, environmental sciences.

Introduction

The flow behavior of single phase or clear liquids is described by the constitutive equation (Govier, Aziz 1972 [2004]). The constitutive equation is the relation between the velocity of deformation (shear rate tensor) and the shear stress (stress tensor). Generally, it is a tensor equation. One dimensional constitutive equations (the relation between the shear rate and shear stress is analyzed only in one specific direction) are used for engineering applications. The constitutive equation of Newtonian fluids was determined by Newton, using flows between parallel plains – one plain was steady, one was moving at a constant velocity – and found that the flow velocity profile was linear, thus shear rate and shear stress are proportional and the proportionality factor between them is the absolute viscosity. Newtonian fluids, such as air or water flow according to the Newtonian constitutive equation. There is a group of fluids in which flow behavior changes with shear time or the shear itself. At a constant shear rate, the resistance of the internal friction of some fluids decreases, and this is called thixotropy. Others have increasing resistance of internal friction over time, and these fluids are rheopectic. Viscoelastic liquids show elastic features, as they regain some of their original shape after the deformation forces cease; however they possess viscous properties as well. Usually, suspensions of fine solids in the practice of mechanical processing do not show elastic behavior, and their flow properties are generally constant over time; however in higher concentrations they behave as non-Newtonian fluids.

Corresponding author: József Faitli
E-mail: ejtfaitj@uni-miskolc.hu

Non-Newtonian fluids are viscous fluids, at constant shear rate the shear stress is also constant, but their flow cannot be described by the Newtonian constitutive equation. The following (Table 1) engineering flow models (nonphysical models) were applicable for the test materials used in our investigations.

Table 1. Engineering flow models have been used for all tests described in this paper

Flow model	Constitutive equation	Rheological parameters
Newtonian	$\tau = \mu\left(\frac{du}{dr}\right)$	μ – absolute viscosity
Bingham plastic	$\tau = \tau_0 + \eta\left(\frac{du}{dr}\right)$	τ_0 – yield-stress η – coefficient of rigidity
Power law $n > 1$ – pseudoplastic, $n < 1$ – dilatant	$\tau = K\left(\frac{du}{dr}\right)^n$	K – consistency index n – flow behavior index

The three most important devices for measuring rheology of single phase liquids are the falling sphere, capillary, and the rotational viscometer.

Mechanical processing deals with coarse disperse systems. The disperse part contains particles, bubbles and drops larger than 0.1 μm dispersed in a solid, liquid or gas phase. If solid particles are mixed with liquid (dispersed, for example, in water), the description of the flow behavior of these mixtures is more difficult than the one for single phase fluids described earlier. What is flowing? The solid particle itself is not deforming and there is no shear stress inside it. Is only the liquid flowing? In resting liquids, dense particles larger than 0.1 μm are settling down. Multiphase flow was early investigated by several authors (Newitt *et al.* 1955; Condolios, Chapus 1963; Wasp *et al.* 1977; Hill *et al.* 1986; Hanks 1980). Recently, scientific interest has been focused on suspension rheology related to wet grinding (He *et al.* 2004, 2006; Johnson *et al.* 2000; Matijasic *et al.* 2008). In the case of ultra-fine grinding, both of the two different approaches should be applied: colloid chemistry (<0.1μm), where physical-chemical surface phenomena are dominant and mechanical processing (>0.1μm) where mechanical forces are the most important. In the mechanical processing industry, the most important application is the flow of solid – liquid mixtures in pipelines, – namely hydraulic transport – and this basic phenomenon can definitely be described by the rheological behavior of the flowing materials. In Hungary, the biggest pipelines transport coal power plant fly-ashes, and this is our primary interest. According to the Tarján–Faitli fine suspension – coarse mixture flow model (Tarján, Faitli 1998; Bőhm *et al.* 2007), a limit particle size can be determined for every solid material. Solid – liquid mixtures made by particles smaller than the limit can be transported as fine suspension flow in pipelines. For example, the limit particle size is 160 μm for fly ashes and 50 μm for sands. Such small particles are able to penetrate into the boundary layer near the pipe wall. The integral of the shear stress on the particle surface is a take-off the wall force and it is negligible for small particles; therefore the particle is able to alter the flow behavior of the boundary layer and indirectly the rheology of the fine suspension also changes. The fine suspension flow can be considered as single phase liquid flow with its own density and rheology. Flow behavior of such mixtures can be measured with viscometers, and on the basis of the measured rheological parameters the pressure loss can be calculated for any flow rate. The described Tarján–Faitli fine suspension flow model is consistent with the Sengun and Probstein (Sengun, Probstein 1989a, 1989b) bimodal model. Based on the bimodal model the fine fraction can be subdivided into a submicron – colloidal – and a micron range fraction. The colloidal size range is responsible for non-Newtonian behaviour and the relatively course – some 10 microns-size particles contribute to the viscosity rise through hydrodynamic dissipation.

If the transported particles are coarser than the particle size limit, particles do not fit into the boundary layer, the surface integral of the shear stress results in a considerable take-off force for higher flow rates. This model explains the often measured fact, that at high flow rates the pressure drop of coarse sand – water mixtures flow is almost the same as just the water flow alone, but at lower flow rates coarse particles are sliding along the pipe wall, and the pressure drop is high, – much higher than the water's – because of the mechanical, not flow friction between the particles and the pipe wall. In this coarse mixture flow case, the fluid is the only phase which flows and the particle movement and particle friction on the pipe wall are purely the consequence of mechanical forces. The force of mechanical friction does not depend on the velocity of the moving bodies, but on the normal force and the coefficient of friction. Contrary to mechanical friction, flow frictional loss highly depends on the flow velocity of the fluid. According to the model, the rheology of the coarse mixture cannot be interpreted, only the rheology of the liquid. In extreme conditions – when a river is rolling huge rocks – it is expedient that only the fluid is flowing while the movement of the solid rocks is the result of various mechanical forces due to the flow. This paper does not deal with this coarse mixture flow.

The most important equipment for measuring the rheology of fine suspensions is the tube viscometer. The principal of the measurement is that in straight horizontal pipes, stable laminar flow has to be established and the pressure loss as a function of the flow rate has to be measured. The tube viscometer developed by our Institute (Fig. 1) has three pipe sections, each six meters long, while their diameters vary at 16; 21 and 27 mm. The mixed suspension (120 liter) is circulated by a rigid characteristic screw pump. Another device for measuring the rheology of fine suspensions is the rotational viscometer. There is an Anton-Paar rotational viscometer at our institute (cylinder-cylinder and cone-plate test systems, 50 cm^3 sample volume, 0.007–7 Pas viscosity and 40–1200 1/s shear rate measuring range). Measuring the rheology of fine suspensions, the cylinder – cylinder geometry is used with a 0.5 mm gap in the Couette ring. The tube viscometer is a large laboratory instrument, but at the same time it is a small pilot scale industrial hydraulic transport system and therefore, its test data is reliable for industrial size design. Contrary, test data of rotational viscometer measurements can be used for verifying only; however the results of the two devices generally do correspond. Measurement temperatures were 22 °C room temperature for both viscometers. The rotational viscometer is equipped with a thermostat. The tube viscometer – however – is equipped with a casing pipe section where cooling water was circulated to stabilize the temperature. There are vertical up and down pipe sections built into the tube viscometer to mix the transported suspension. In addition there are separate sampling points in the main tube where samples can be taken from the upper half, lower half and full cross sections of the pipe. In addition, any accidental sedimentation can be monitored based on the measured pressure loss. In such case the revolution number of the screw pump must be switched into maximum. Both viscometers were calibrated by calibrating the sensors. In the case of the rotational viscometer the revolution number and the torque sensors, in the case of the tube viscometer the pressure sensors were calibrated. The flow rate measuring technique – filling a tank and measuring the time – is absolute technique, no additional calibration was necessary.

In the following the results of some rheological tests with different materials are presented. Based on the Tarján – Faitli fine suspension – coarse mixture flow model (Tarján, Faitli 1998; Bőhm *et al.* 2007) – the rheological properties of only the fine fraction were measured in the tube viscometer. The full granular materials were tested in a hydraulic test loop – 400 liter sample. There are many working coal power plant fly-ash deposition system installations worldwide (Jacksonville Power Plant - USA, Craiova 2, Isalnita, Rovinari, Turceni Power Plants – Romania, Mátra Power Plant – Hungary, etc…) which were designed on the basis of the model and measurements presented.

1. Materials and methodology

1.1. Flow behavior of fine suspensions, measurement results

Before starting a rheological test the physical properties of the materials had to be measured. The particle densities of the solid materials were measured by laboratory picnometers. Particle size distributions of all the tested bulks had been determined by laboratory sieving using 200 mm laboratory sieve and micro sieve series. The measured particle size distribution of most of the tested materials can be seen on Figure 2. The rheological tests

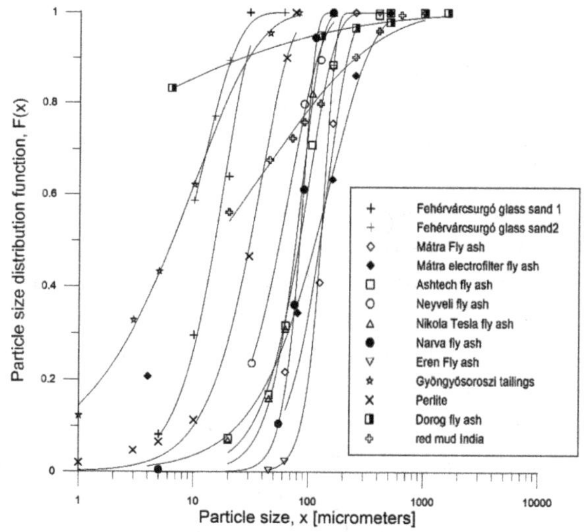

Fig. 2. Measured particle size distributions of most of the tested bulks

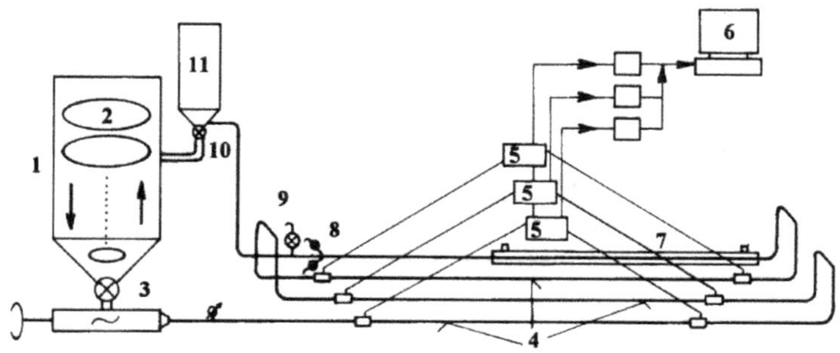

1. Mixing tank
2. Mixing hole discs
3. Screw pump
4. Measuring pipes
5. Differential pressure transducers
6. Data acquisition system
7. Casing pipe for cooling water
8. Sampling points
9. Sampling point
10. Valve
11. Flow rate measuring tank

Fig. 1. Tube viscometer with three measuring pipes

presented here were carried out mainly with fine suspensions. Coarse fractions were removed by sieving at the limit particle size before the tests.

The tube viscometer measurements were started using pure water firstly, and then the required masses of the solid for different concentrations were added into the system step by step. By this method discrete concentration suspensions were tested gradually. The device is equipped by a propeller mixer. After the feeding of a solid portion some time was waited while the suspension was continuously circulated at high speed by the pump until pressure loss equilibrium.

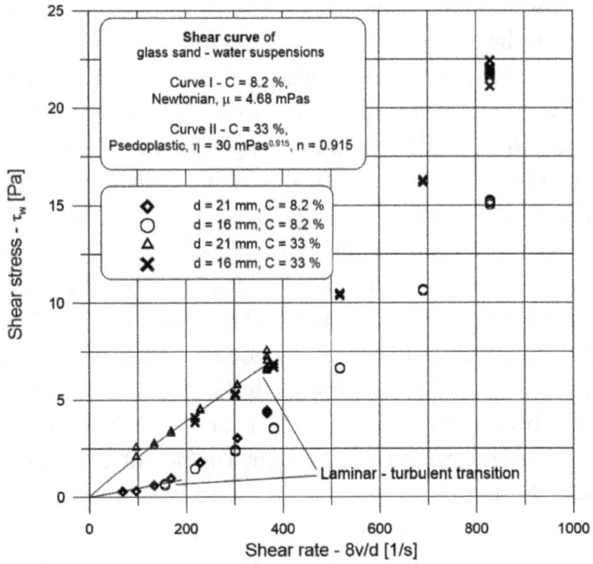

Fig. 3. Flow curves of sand glass from Fehérvárcsurgó

This paper reports about many rheological tests, it summarizes the results of almost a hundred shear curves. It is not possible to show them here, just two shear curves of tube viscometer tests are shown in Figure 3 as an example.

Sand glass is mined at Fehérvárcsurgó as a basic raw material for glass production. During mineral processing, the raw material is also classified, thus fine, narrow particle size classes can be achieved. These narrow particle size fractions are really good test materials for testing flow behavior. Sand glass with a particle size less than 50 μm can be hydraulically transported as a fine suspension flow. Fine sand glass has been frequently tested (Tables 2 and 3) and it behaves as a Newtonian fluid below 30% volumetric concentration, and generally becomes pseudoplastic at higher concentrations.

The most important industrial application of hydraulic transport in Hungary, as previously mentioned, is slag and fly ash transport from coal fired power plants to tailings areas. Fly ash particles with a size less than 160 μm can be transported as fine suspensions in normal size pipelines. During the last decade flow properties of several fly ash samples were analyzed from all over the world. These tests were carried out with particles finer than the limit particle size of fly ashes – 160 μm, and they were sieved if it was necessary. It was determined that all the high concentration fly ash fine suspension samples behaved as a Bingham – plastic fluid (Tables 4–8).

Some of the tested fly ash samples (Tables 4–8) had moderate friction resistance with increasing concentration and they became Bingham plastic over 20% volumetric concentration and the coefficient of rigidity remains under 100 mPas up to 30% volumetric concentration.

Table 2. Flow properties of sand glass from Fehérvárcsúrgó (5 ... 30 μm, ρ_s = 2644 kg/m³)

Volumetric concentration	μ absolute viscosity [mPas]		Flow behavior
0.1	2.99		Newtonian
0.2	4.63		Newtonian
0.3	8.47		Newtonian
	K Consistency index	m flow behavior index	
0.35	19.3 mPas$^{0.916}$	0.916	pseudoplastic
0.4	75.5 mPas$^{0.713}$	0.713	pseudoplastic

Table 3. Flow properties of sand glass from Fehérvárcsúrgó (x_{90} = 20.4 μm, ρ_s = 2734 kg/m³)

C Volumetric concentration	μ absolute viscosity [mPas]		Flow behavior
0.082	4.68		Newtonian
0.196	5		Newtonian
0.26	8		Newtonian
	K Consistency index	m flow behavior index	
0.33	30 mPas$^{0.915}$	0.915	pseudoplastic

Table 4. Flow properties of Mátra Power Plant, Fly Ash: (x_{95} = 160 μm, ρ_s = 1900 kg/m³)

C Volumetric concentration	μ absolute viscosity [mPas]		Flow behavior
0.13	5		Newtonian
0.16	5.6		Newtonian
0.2	7		Newtonian
	η coefficient of rigidity [mPas]	τ_0 Yield Stress [Pa]	
0.23	9	0.8	Bingham plastic
0.27	17	1.1	Bingham plastic
0.33	51	1.5	Bingham plastic
0.35	102	2.3	Bingham plastic

Table 5. Flow properties of Mátra Power Plant, Fly Ash from electrofilter: (x_{90} = 200 μm, ρ_s = 1984 kg/m^3)

C Volumetric concentration [-]	η coefficient of rigidity [mPas]	τ_0 Yield Stress [Pa]	Flow behavior
0.16	3.4	1.24	Bingham plastic
0.25	11.3	1.89	Bingham plastic
0.31	60	3.63	Bingham plastic

Table 6. Flow properties of Ashtech – India, Fly Ash (x_{80} = 110 μm, ρ_s = 1974 kg/m^3)

C Volumetric concentration	μ absolute viscosity [mPas]		Flow behavior
0.1	6		Newtonian
0.19	8		Newtonian
0.26	10		Newtonian
0.3	13		Newtonian
0.36	16		Newtonian
C Volumetric concentration	η coefficient of rigidity [mPas]	τ_0 Yield Stress [Pa]	
0.4	42	1	Bingham plastic
0.5	80	4	Bingham plastic

Table 7. Flow properties of Neyveli – India, Fly Ash: (x_{95} = 160 μm, ρ_s = 2492 kg/m^3)

C Volumetric concent-ration	μ absolute viscosity [mPas]		Flow behavior
0.1	9		Newtonian
C Volumetric concent-ration	η coefficient of rigidity [mPas]	τ_0 Yield Stress [Pa]	
0.25	12	0.75	Bingham plastic
0.3	15	1.05	Bingham plastic
0.36	22	4.5	Bingham plastic
0.38	86	15	Bingham plastic
0.43	140	29	Bingham plastic

Table 8. Flow properties of Nikola Tesla B Power Plant – Serbia, Fly Ash: (x_{95} = 160 μm, ρ_s = 1630 kg/m^3)

C Volumetric concentration	μ absolute viscosity [mPas]		Flow behavior
0.1	11		Newtonian
C Volumetric concentration	η coefficient of rigidity [mPas]	τ_0 Yield Stress [Pa]	
0.195	15	0.86	Bingham plastic
0.336	24	1.9	Bingham plastic
0.448	99	6	Bingham plastic

The fly ash samples (Tables 9 and 10) were originally generated by another type of combustion method. The water mixing of these fly ashes was much more difficult and they had higher friction resistance. The measuring range of our tube viscometer is 1 to 250 mPas; this is determined by the differential pressure transducers and the drive system of the screw pump. These types of fly ashes reached a level of approximately 250 mPas, around 35% volumetric concentration; therefore, there is no measured point indicated in the tables over this concentration.

The base metal mine near Gyöngyösoroszi, Hungary was decommissioned a long time ago; however final closure work is still continuing. One of the problems was to handle the former froth flotation tailings deposited on the tailing facility. The planned method of handling was to utilize these tailings for abandoned underground mine backfilling. This tailing material has a very fine particle size range and contains a high amount of clay minerals. These material features are good to make paste state mixtures. Using this tailing for tests, it was determined that it becomes Bingham plastic over 18% of volumetric concentration with extreme high coefficient of rigidity and yield stress corresponding to increasing concentration (Table 11). For testing this material, a pilot scale hydraulic transport test loop was built applying a high capacity

Table 9. Flow properties of Narva Pow. Plant N°8. Balti CFB Furnace, Fly Ash: (x_{95} = 160 μm, ρ_s = 2701 kg/m^3)

C Volumetric concentration	η coefficient of rigidity [mPas]	τ_0 Yield Stress [Pa]	Flow behavior
0.1	9	0.4	Bingham plastic
0.22	12	0.6	Bingham plastic
0.3	44	7.5	Bingham plastic
0.34	190	21	Bingham plastic
0.37	230	31	Bingham plastic

Table 10. Flow properties of Eren Pow. Plant – Turkey, Fly Ash: (x_{95} = 160 μm, ρ_s = 2148 kg/m^3)

C Volumetric concentration	μ absolute viscosity [mPas]		Flow behavior
0.1	8.5		Newtonian
0.15	22		Newtonian
0.2	20		Newtonian
0.22	26		Newtonian
C Volumetric concentration	η coefficient of rigidity [mPas]	τ_0 Yield Stress [Pa]	
0.28	70	6	Bingham plastic
0.3	130	10	Bingham plastic
0.34	125	16	Bingham plastic
0.35	220	18	Bingham plastic

Mono type screw pump able to produce a high pressure head and able to transport paste like suspensions over the test loop. Pressure transducers were changed as well to be able to measure extreme pressures in the system. The test material was considered to be paste by a standard Slump Cone Test (Phasias *et al.* 1996; Gombkötő, Faitli 2008), over 48% of volumetric concentration.

Tests were carried out for perlite – water mixtures as well (Table 12). It was determined that perlite suspensions remain Newtonian fluids up to 40% of volumetric concentration.

The flow behavior of all of the previously presented granular materials was time independent. This means that during the tube viscometer tests, the pressure loss of a given concentration fine suspension flow, at a set constant flow rate, remained the same for over an hour, while the suspension was circulated in the closed loop.

The following two samples showed rheopectic behavior. These tests were carried out in the rotational viscometer, where the rheology of the suspensions was determined during the first minute of the test by changing the shear rate. The initial results of the rheology tests are presented. The time dependency of resistance against shear was analyzed using the same revolution for the rotor of the viscometer over a long time interval. The torque of the rotor became two – three times higher than the initial value over a few minutes, and thus the apparent viscosity become higher as well. Fly ash originating from the hazardous waste treatment facility in Dorog (Table 13) behaved as Bingham plastic over 10% of volumetric concentration and it was a time dependent rheopectic fluid.

Table 11. Flow properties of base metal tailing, Gyöngyösoroszi: (x_{95} = 50 μm, ρ_s = 2701 kg/m^3)

C Volumetric concentration	η coefficient of rigidity [mPas]	τ_0 Yield Stress [Pa]	Flow behavior
0.18	30	26	Bingham plastic
0.26	60	32	Bingham plastic
0.4	95	122	Bingham plastic
0.485	285	308	Bingham plastic
0.525	490	638	Bingham plastic

Table 12. Flow properties of perlite – water mixtures

C Volumetric concentration	μ absolute viscosity [mPas]	Flow behaviour
0.1	8	Newtonian
0.2	15	Newtonian
0.3	30	Newtonian
0.4	59	Newtonian

The other sample that had rheopectic behavior was red mud from an Indian bauxite operation (Table 14). Even though the Dorog fly ash was Bingham plastic, Indian red mud was determined to be Newtonian over a short time period; however its shear resistance was increased over time as well.

Table 13. Flow properties of fly ash, Dorog hazardous waste treatment facility (x_{80} = 60 μm, ρ_s = 2429 kg/m^3), 1st minute

C Volumetric concentration	η coefficient of rigidity [mPas]	τ_0 Yield Stress [Pa]	Flow behaviour
0.1	3.8	0.21	Bingham plastic
0.2	6.5	0.46	Bingham plastic
0.3	150	7.5	Bingham plastic

Table 14. Flow properties of red Mud, India: (x_{80} = 1250 μm, ρ_s = 3680 kg/m^3), 1st minute

C Volumetric concentration	μ absolute viscosity [mPa]	Flow behaviour
0.35	5	Newtonian
0.4	13	Newtonian
0.45	40	Newtonian
0.5	63	Newtonian
0.55	145	Newtonian

2. Results and discussion

It can be seen that in the cases of the presented solid – liquid fine suspensions, viscosity is slightly increasing in a nearly linear way over concentrations from $C = 0$ (clear water) of μ_0 = 1 mPas. Generally, these suspensions become non-Newtonian fluids over 20% of volumetric concentration and the viscosity like parameter (absolute viscosity – Newtonian, coefficient of rigidity – Bingham plastic, consistency index – Pseudoplastic) gets exponentially higher. The highest tested concentration was 55% by volume. Parameters of rheology can be calculated by function fitting over the measured domain. Viscous flow of suspensions is possible up to the theoretical maximum concentration C_{max} – solid particles are placed with the greatest amount of space filled in. Experimental determination of C_{max} is difficult. In the case of fly ash tests, samples were put into settling columns – Batch settling test – in order to determine the final concentration the settled solid can reach over time. It was experienced, that higher final concentration was achieved using higher initial concentration.

The following well known relation between concentration and viscosity is taken from the literature:

$$\mu = \mu_0 \left[1 + \frac{2.5C}{2\left(1 - \frac{C}{C_{max}}\right)} \right]. \quad (1)$$

The Einstein equation can be recognized in it, $\mu = \mu_0(1+2.5C)$ which is valid only at low concentrations (C < 2% by vol.), and viscosity becomes infinite at C_{max}. However, this equation does not describe our experiences! At low concentrations viscosity slightly increases almost linearly, but at higher concentrations it starts to increase exponentially. To better describe the phenomenon, the following equation was introduced. It consists of a linear and an exponential part.

$$\eta = \mu_0\left(1 + K_1C + K_2C^{K_3}\right), \qquad (2)$$

where μ_0 – viscosity of the carrier liquid.

Viscosity should be written on the left side of the equation in the case of perlite – water mixtures, because they are Newtonian fluids at even high concentrations too. The coefficient of rigidity (η) and the consistency index (K) should be involved in case of Bingham plastic – fly ashes – and pseudoplastic – sands – fluids. Most of the tested materials behaved as Bingham – plastic fluids. For yield stress the following equation can be used:

$$\tau_0[Pa] = K_4 e^{K_5 C}. \qquad (3)$$

Equation 3 is not a physical relation. If the concentration is substituted as a non-dimensional value, the yield stress can be calculated in Pa units. The rheological measurements should be carried out at some discrete concentrations of a given material, as was shown earlier. Then, the K_1–K_5 constants in Equations 2 and 3 can be determined by curve fitting, and rheological parameters can be calculated for any concentration – within the measured range. Figure 4 summarizes the measured rheological parameters and the fitted curves (Eqs 2 and 3) of all the presented tests.

The constants for each tested material can be seen in Table 15. Using these constants and equations 2 and 3,

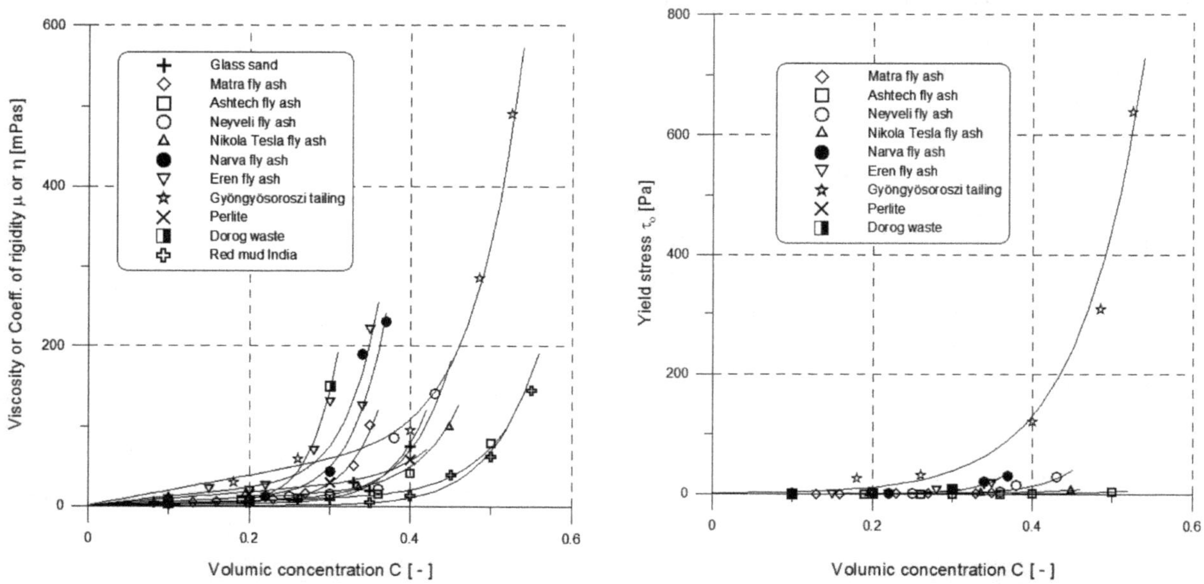

Fig. 4. Corresponding measurement results (symbols) and calculated functions (continuous lines) according to Eqs 2 and 3

Table 15. Collected constants comparison of all the tested material

Material	Measured max. Conc.	K_1	K_2	K_3	R-squared	K_4	K_5	R-squared
Fehérvárcsurgó, sand glass	0.4	12	280,000	9	0.988	–	–	–
Mátra, fly ash	0.35	20	1,100,000	9	0.986	0.002	21	0.866
Ashtech, fly ash	0.5	30	30,000	9	0.930	0.0002	19.8	0.980
Neyveli, fly ash	0.43	30	220,000	9	0.917	0.002	22	0.955
Nikola, fly ash	0.45	50	110,000	9	0.999	0.04	11.4	0.991
Narva, fly ash	0.37	50	1,700,000	9	0.903	0.002	26.5	0.951
Eren, fly ash	0.35	110	2,100,000	9	0.875	0.002	26.5	0.925
Gyöngyösoroszi, tailing	0.53	190	120,000	9	0.998	1	12.2	0.979
Perlite	0.4	85	85,000	9	0.995	–	–	–
Dorog, fly ash (1st minute)	0.3	20	7,000,000	9	0.999	0.009	22	0.996
Red mud (1st minute)	0.55	1	35,000	9	0.989	–	–	–

rheological parameters can be calculated for any concentration. This is a simple engineering method, but it is also a very powerful method to design industrial sized applications. In Table 15 the coefficient of determination – R-squared – values of curve fittings are also shown, and they indicate good fit. The underground backfilling possibilities of the abandoned base metal mine near Gyöngyösoroszi were also investigated. For this purpose mixed solids were chosen to be used. The tested mixtures were made of Mátra fly ash (x_{57} = 160 μm, ρ_s = 2355 kg/m^3), sand (x_{57} = 160 μm, ρ_s = 2720 kg/m^3), and Mátraszele bentonite (x_{41} = 160 μm, ρ_s = 2574 kg/m^3). For these materials, detailed flow parameters are not presented, but the K_1 – K_5 parameters can be found in Table 16. The fine suspensions of these mixed materials (bentonite was added to fly ash and sand to alter the hydraulic conductivity of the backfill material) were Bingham plastic fluids, even sand – bentonite mixtures too.

Table 16. K_1–K_5 Constants of Mátra fly ash (x_{57} = 160 μm, ρ_s = 2355 kg/m^3), sand: (x_{57} = 160 μm, ρ_s = 2720 kg/m^3), and bentonite (Mátraszele): (x_{41} = 160 μm, ρ_s = 2574 kg/m^3)

Material	Measured max. Conc.	K_1	K_2	K_3	K_4	K_5
fly ash – bentonite (5%) – water	0.35	30	19,500	5	0.01	26
fly ash – bentonite (10%) – water	0.35	30	39,000	5	0.01	26
sand – bentonite (5%) – water	0.35	50	14,000	5	0.01	26
sand – bentonite (10%) – water	0.35	50	17,000	5	0.01	26

Conclusions

Systematic slurry testing has been done in the last decades on the basis of the Tarján–Faitli fine suspension – coarse mixture flows model. This paper summarizes the results of the rheological testing of the separated fine fractions.

The empirical equations of Equation 2 and 3 were introduced based on the presented results of many different particulate materials from many places of the world. These equations better characterize the presented test results than the well-known Equation 1.

The rheological parameters of any feasible concentration and any kind of solid/liquid mixtures can be determined by using these equations. However, these equations are empirical ones and the K_1 – K_5 parameters in them should be determined by curve fitting. They are really powerful for the engineering design of slurry handling, and hydraulic transport systems. This knowledge and practice are also important for fields where environmental awareness is required.

The described method, using the Tarján-Faitli model to determine pressure loss of hydraulic transport pipelines using measured rheology of given concentration fine suspensions is a practical approach. The application of this method necessitates a concentration – viscosity function. The most important advantage of the introduced method against theoretical concentration – viscosity equations in the literature is that equation 2 aptly describes the first, almost linear, part (0 – ~20%) and also the second exponential part (~20% – C_{max}).

References

Bőhm, J.; Debreczeni, Á.; Faitli, J.; Gombkötő, I.; Meggyes, T. 2007. High-concentration hydraulic transport of tailings, *Land Contamination and Reclamation* 15(2): 195–217. http://dx.doi.org/10.2462/09670513.853

Condolios, E.; Chapus, E. E. 1963. Designing solids-handling pipelines, *Chemical Engineering* 73: 131–138.

Gombkötő, I.; Faitli, J. 2008. Application of paste technology for tailings handling, *Proceedings of XXIV International Mineral Processing Congress*, 24–28 September 2008, Beijing, China, 3522–3529.

Govier, G. W.; Aziz, K. 1972 [2004]. *The flow of complex mixtures in pipes*. Van Nostrand Reinhold.

Hanks, R. W. 1980. The influence of non-Newtonian rheology in mixed homogeneous–heterogeneous slurry flow, *Proceedings of the 5th International Technical Conference on Slurry Transportation*, 8–10 March 1980, Lake Tahoe, USA. 251 p.

He, M.; Wang, Y.; Forssberg, E. 2004. Slurry rheology in wet ultrafine grinding of industrial minerals: a review, *Powder Technology* 147(1–3): 94–112. http://dx.doi.org/10.1016/j.powtec.2004.09.032

He, M.; Wang, Y.; Forssberg, E. 2006. Parameter studies on the rheology of limestone slurries, *International Journal of Mineral Processing* 78(2): 63–77. http://dx.doi.org/10.1016/j.minpro.2005.07.006

Hill, R. A.; Snock, P. E.; Gandhi, R. L. 1986. Hydraulic transport of solids, in I. J.Karassik, J. P. Messina, P. Cooper, C. C. Heald (Eds.). *The pump handbook*. 2nd ed. New York: McGraw-Hill.

Johnson, S. B.; Franks, G. V.; Scales, P. J.; Boger, D. J.; Healy, T. W. 2000. Surface chemistry – rheology relationships in concentrated mineral suspensions, *International Journal of Mineral Processing* 58(1–4): 267–304. http://dx.doi.org/10.1016/S0301-7516(99)00041-1

Matijasic, G.; Zizek, K.; Glasnovic, A. 2008. Suspension rheology during wet comminution in planetary ball mill, *Chemical Engineering Research and Design* 86(4): 384–389. http://dx.doi.org/10.1016/j.cherd.2007.11.013

Newitt, D. M.; Richardson, J. F.; Abbott, M.; Turtle, R. B. 1955. Hydraulic conveying of solids in horizontal pipes, *Transactions Institute Chemical Engineering* 33: 93–113.

Phasias, N.; Boger, D. V.; Summers, J.; Glenister, D. J. 1996. A fifty cent rheometer for yield stress measurement, *Journal of Rheology* 40(6): 1179–1189. http://dx.doi.org/10.1122/1.550780

Sengun, M. Z.; Probstein, R. F. 1989a. Bimodal model of slurry viscosity with application to coal slurries, Part I, *Rheologica Acta* 28(5): 382–393. http://dx.doi.org/10.1007/BF01336805

Sengun, M. Z.; Probstein, R. F. 1989b. Bimodal model of slurry viscosity with application to coal slurries, Part II, *Rheologica Acta* 28(5): 394–401. http://dx.doi.org/10.1007/BF01336806

Tarján, I.; Faitli, J. 1998. The distinction of the fine suspension flow from the coarse mixture flow by measuring of the pressure loss on a horizontal pipe, *Mineral Economy Journal – Gospodarka Surowcani Mineralnymi* 14(3): 61–71.

Wasp, E. L.; Kenny, J. P.; Gandhi, R. L. 1977. *Solid Liquid Flow, Slurry Pipeline Transportation*. 1st ed. Germany, Clausthal: Trans-Tech Publications.

József FAITLI. Associate Professor, since 2012, Head of the Institute of Raw Materials Preparation and Environmental Processing at the University of Miskolc. Graduated in 1989 as mining mechanical and electrical engineer. Since graduation he has been working for this Institute in different positions. PhD in 1998, in mechanical processing. Number of scientific publications is 119 so far. Research interest include: operations with solid – liquid disperse systems, hydraulic transport, particle motion in different media, control and simulation of mechanical processing operations, processing of secondary raw materials.

Imre GOMBKÖTŐ. PhD, graduated as mineral process engineer in 2000 at the Faculty of Earth Science and Engineering at the University of Miskolc. Currently he is academic teacher at the Institute of Raw Material Preparation and Environmental Processing at the University of Miskolc. He was responsible for the Innovation Research and Education Centre for Earth and Environmental Sciences operated by the Faculty of Earth Science and Engineering at the University of Miskolc, currently he is Head of the Department of Mechanical Processing. PhD since 2008 where topic of work was flow of disperse systems, determination of pressure loss in special conditions and paste technology. Author and co-author of many publications about mechanical raw material preparation and separation, waste utilisation, underground mine backfilling, paste technology and permeable reactive barriers as remediation tools printed in journals and presented on Hungarian and international conferences. He is actively working for private companies as well for implementing new techniques into processing technologies on demand.

5

RESPONSES OF ABOVE- AND BELOW-GROUND CARBON STOCKS TO ENVIRONMENTAL DRIVERS IN TIBETAN ALPINE GRASSLANDS

Hongsheng LIU[a, b], Zewei MIAO[c]

[a]*Institute of Geography Sciences and Natural Resource Research, Chinese Academy of Sciences, 11A Datun Road, Beijing 100101, China*

[b]*Department of Biology, University of Texas at Arlington, 701 Nedderman Dr. Arlington, Texas, 76013, USA*

[c]*Nicholas School of the Environment, Duke University, 308 Research Drive, Durham, NC 27708, USA*

Abstract. Paucity in the knowledge of responses of grassland carbon dynamics to environmental variables constrains our ability to predict future ecosystem productivity. The aim of this study was to investigate differential responses of above- and below-ground carbon stocks to environmental drivers in Tibetan alpine Plateau at both regional and local scales. Variance partitioning and non-linear regression between carbon stocks and environmental driving variables suggested that both above- and below-ground carbon stocks showed a significant negative relationship with temperature and a positive relationship with soil moisture. Annual accumulated temperature constrained above-ground carbon at regional scale ($r^2 = 0.50$, $P < 0.0001$), while soil moisture controlled below-ground carbon at local scale ($r^2 = 0.48$, $P < 0.0001$). Scale-specific responses of above- and belowground carbon storage to temperature and soil moisture complicated the influences of abiotic environmental variables on ecosystem productivity. Soil carbon had significant unimodal ($r^2 = 0.11$, $P = 0.0073$) and linear ($r^2 = 0.37$, $P < 0.0001$) relationships with mean annual temperature and soil moisture, respectively. Since the driving factors of aboveground and soil carbon content are specific to spatial scales, the relationships of grassland carbon storage and environmental factors at small scales are not applicable to a large spatial scale.

Keywords: biomass, spatial scales, soil moisture, annual precipitation, annual mean temperature, variation partition.

Introduction

Grasslands play an important role in global terrestrial carbon cycle due to their large area and high carbon density. Grassland stores up to 10% of terrestrial biomass carbon, 10–30% of global soil organic carbon, and likely provide an annual sink of ~0.5 Pg C (Scurlock *et al.* 2002). Identifying the responses of above-ground and below-ground carbon dynamics to environmental variables is critical to investigate the effect of climate change on grassland ecosystem functions.

The relationship between carbon dynamics, especially belowground carbon and environmental variables, is not certain in Tibetan grassland yet. Majority of previous literature only includes measurements of above-ground carbon or biomass storage (Sala *et al.* 1988; Swemmer *et al.* 2007; Yang *et al.* 2008; Sun *et al.* 2013). Although belowground carbon has been estimated to account for more than 50% of the net carbon stocks of grassland ecosystems, fewer studies focused on below-ground carbon in Tibetan plateau (Hendricks *et al.* 1993; Wu *et al.* 2011). Relative to the aboveground biomass, determinations of root biomass distribution and belowground processes are much more challenging due to the intensive labour and time consuming. In addition, environmental variables including climate, soil physical, chemical and biological characteristics, and plant properties all affect the distribution of root carbon (Gong *et al.* 2015; Xu *et al.* 2015). Therefore, it is important for Tibetan grassland ecosystem management to investigate the effects of environmental factors on belowground carbon distribution in various regions.

The driving mechanism of environmental factors on above-ground carbon and below-ground carbon are different. Above-ground carbon turnover rates increase as moisture availability increases (Chabot, Hicks 1982; Reich *et al.* 1992; Piao *et al.* 2006; Bai *et al.* 2008; Sistla *et al.* 2013). This has been generalized to roots, and it has been

Corresponding author: Zewei Miao
E-mail: miaozewei@gmail.com

assumed that root and shoot biomass ratio follow roughly the same extent regardless of environmental conditions (Hendricks *et al.* 1993; Ostertag 2001; Hui, Jackson 2005). It is specific to environmental conditions for grass to capture nutrients, water, and light, and hence maximize their growth rate and to allocate to carbon among various organs (Chapin *et al.* 1986; Enquist, Niklas 2002). Thus, environmental drivers and driving mechanism of roots and foliage carbon may be different in allocating carbohydrates between above-ground and below-ground tissues (McConnaughay, Coleman 1999). For example, carbon is primarily allocated to tissues where resources are most limiting (Chapin *et al.* 1990; Ostertag 2001). Therefore, when environmental variables including moisture are positively related to the above-ground carbon and negatively related to below-ground carbon, the negative influences of environmental variables on soil carbon (SOC) could be offset by high growth rate.

Fewer studies of carbon stocks in Tibetan plateau grasslands have been carried out at a large spatial scale. The influence of environmental factors on plant biomass and soil carbon stock varied greatly from scale to scale. Generally, at a small scale, the above-ground carbon and SOC were controlled by nutrient level and moisture availability, whereas temperature played a much more important role than moisture at a large scale. For example, at local scale, above-ground carbon storage, cycling and their responses to disturbance have been well described by soil properties, soil moisture, grazing (Patton *et al.* 2007), land use change (Gibbon *et al.* 2010), nutrient availability (De Deyn *et al.* 2009), and/or topography (Avohou, Sinsin 2009). At large spatial scales, grass primary productivity had good relationships with regional factors, including annual precipitation (Sala *et al.* 1988; Bai *et al.* 2008; Medvigy *et al.* 2010), variation in annual precipitation (Knapp *et al.* 2002; Bai *et al.* 2004), and the interaction of temperature and precipitation (Swemmer *et al.* 2007). Therefore, environmental drivers of carbon sequestrations and SOC content are sensitive to spatial scales in grassland ecosystems. The cause-effect relationship of ecosystem carbon stock and driving factors discovered at local scales may be invalid at regional and/or global scales. Climatic patterns may be too broad to account for the observed variations in carbon productivity at a small geographic scale. As a result, determining environmental drivers and driving mechanism of carbon stock is of great importance at different scales.

The objectives of this research focus on (a) the driving forces of above-, below-ground, and soil organic carbon content at local and regional scales, (b) partitioning the influences of resources availability on the components of grassland carbon content at local and regional scales, and (c) predicting the influence of climate change on plant and soil carbon content in the Tibetan Plateau. The hypothesis of this research is that the limiting factors of above-, below-ground, and soil organic carbon content are different and scale-dependent. To test the hypothesis, we collected plants and soil samples across the Tibetan Plateau, which is one of the largest alpine grasslands in the world with extending about 2.40×10^6 km^2 in western China. The spatial gradients of temperature and moisture, limited grass species and sparse ecosystem types together with little human disturbance made the Tibetan plateau an ideal study area for investigating response of plant and soil carbon stock to various resources availability.

1. Methods and materials

1.1. Study sites

Plant and soil samples were collected at 45 sites and 96 quadrates on the Tibetan Plateau grasslands by cutting or grazing during the summer (July to August) of 2010. The climate on the plateau is cold and relatively humid, with mean annual temperature of 1.61°C and precipitation of 413.6 mm (Yang *et al.* 2008). The altitude of the Plateau ranges from 200 to 8840 m with mean value of 4000 to 5000 m above sea level. Information obtained from the vegetation atlas of China with a scale of 1:1000,000 showed that vegetation is mainly dominated by the alpine steppe and meadow, with a total cover of >60% area of the plateau (Chinese Academy... 2001). The spatial patterns of alpine grasslands are closely related to the precipitation gradient across the plateau. Alpine steppe occurs in cold and arid regions and consists of xerophytic grasses such as *Carex moorcroftii Falc. ex Boott, Stipa purpurea Griseb,* and *Polygonum viviparum L.*. Alpine meadow dominated by *Kobresia pygmaea C. B. Clarke* is distributed in humid and relatively warm areas. Soil types relate to alpine steppe and meadow are cold calcic and felty soils, respectively, which both have been defined as cambisols in the Food and Agriculture Organization (FAO) soil taxonomy system (Wu *et al.* 2003).

1.2. Sampling and measurements

To achieve systematic sampling, the spatial interval between sampling sites was set at about 100 km. Because of inaccessibility of traffic and limitations of topography, some samplings were conducted along the major roads, while others were more widely spaced and the average distance was >100 km, especially in areas with altitude higher than 4000 a.s.l. The geographical locations of sampling sites were determined by a Global Positioning Satellite device together with records of altitude, soil condition, grassland type, and total vegetation cover. At each site (50×50 m), all plants in plots (1×1 m) were harvested to ground level to measure above-ground biomass. Below-ground biomass was sampled by taking 2–10 soil cores of 12 cm diameter within each quadrate at depths of 0–10,

10–20, 20–40, 40–60, 60–80, and 80–100 cm. Roots and other belowground plant biomass, e.g., rhizomes contained within the excavated soil were separated with water through a 0.3-mm-mesh sieve. Biomass samples were oven-dried at 65°C to constant mass. A correction factor of 0.45 was applied to convert biomass to carbon. In situ air (Temp-air) and soil (Temp-soil) temperature was measured by putting mercury thermometer either 1 m above the ground or 15 cm below the ground.

Soil samples at each quadrate were obtained at depths of 0–10 cm and then handpicked to remove plant detritus and rock. All soil samples were sieved through a 2 mm mesh and grounded with a ball mill for carbon and nutrient analysis. Soil organic carbon content was analysed using wet oxidation method. Carbon contents (%) of soil samples were converted to soil organic carbon (SOC) content (kg Cm^{-2}) for a single layer:

$$SOC = C \times h \times \rho \times A, \quad (1)$$

where C represents carbon contents of soil samples (%); h represents soil layer thickness (m); ρ is the mean bulk density (g cm^{-3}); A is the top surface area of soil core (104 cm^{-2} is this case).

Soil moisture was measured gravimetrically after 24 h desiccation at 105 °C. Bulk density was calculated as the ratio of the oven-dry soil mass to the volume (g cm^{-3}) of soil cylinder. Soil texture was determined by particle size analyzer (Mastersizer 2000, Malvern, UK) after removal of organic matter and calcium carbonates. Mean annual temperature (MAT), mean annual precipitation (MAP), and accumulated annual temperature (the sum of temperature higher than zero degree, AAT) used in our analysis were derived from the climate database of the Tibetan Plateau at 0.1×0.1 degree resolution during 2001–2009. Because there was a quadratic relation between soil organic matter and MAT, a squared MAT was used to improve the fitness of regression analysis. The climate database was generated from records of 77 total climatic stations across the plateau, 45 of which located above 3000 m in elevation.

1.3. Data analysis and statistic

All data were tested for normality and made log-transformations when necessary. Ordinary least squares (OLS) regression analyses were conducted to evaluate the relationships between carbon stocks and local (including *in situ* measured air and soil temperature, soil moisture, soil bulk density and texture) and regional environmental variables including MAT, MAP, AAT, latitude, longitude, altitude. A stepwise regression was used to assess the impact of the "pure" local and regional factors and the interaction of them on carbon stocks. All analyses were performed using SAS 9.3 and figures were generated by using Sigmaplot 10.00 for Windows (SYSTAT Inc. Chicago, Illinois, USA).

Table 1. Plant and soil carbon contents of alpine steppe and alpine meadow grasslands in Tibetan Plateau

Variables	Alpine Steppe	Alpine Meadow	Mean
Above-ground carbon content (kg m^{-2})	0.029±0.024	0.050±0.045	0.039±0.036
Belowground carbon content (kg m^{-2})	0.101±0.096	0.335±0.296	0.207±0.241
Plant carbon content (kg m^{-2})	0.129±0.103	0.387±0.335	0.249±0.244
Soil carbon content (kg m^{-2})	2.1±1.9	4.2±4.1	3.1±3.0
Total carbon content (kg m^{-2})	2.3±2.0	4.3±4.1	3.2±3.0

Note: soil carbon content is sampled to depth of 10 cm. The same as below.

2. Results

2.1. Spatial distributions of grassland carbon stocks

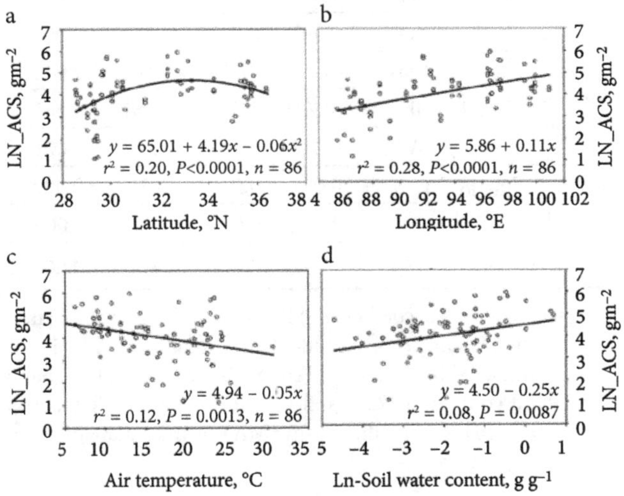

Fig. 1. Relationships between log-transformed above-ground carbon stock (ACS) and environmental variables: (a) latitude; (b) longitude; (c) air temperature; (d) soil moisture

2.2. Environmental drivers of above-ground carbon (AGC) stocks

Above-ground carbon content was mainly influenced by regional factors (Table 2). AGC had significant ($P < 0.001$) unimodal and linear relationship, respectively, with latitude and longitude (Figs 1a, 2b). Although both air temperature and soil moisture had significantly ($P < 0.01$) correlated with AGC, they could only explain about 10% of the variation (Figs 1c, 2d). Step-wise regression results indicated that regional variables (latitude, longitude and mean annual temperature) explained 49.8% of the variation of AGC, while local factors (i.e. air temperature, soil water content, and sand content) explained less than one third of the variation (Table 3). Of the variables examined by regional factors, longitude explained the largest proportion (88.4%) of the variation, followed by latitude (68.5%) and annual accumulated temperature (51.4%). At local scale, the AGC was mainly influenced by air temperature (44.3%) and then followed by soil water content (31.9%) and sand content (27.9%). We then applied

Table 2. Summary of the results obtained from multiple linear regressions, showing the effects of regional and local factors on three carbon pools in alpine grassland of Tibetan Plateau

			CO	SE	SC	*P* value
Above-ground carbon content (g m^{-2})	Regional	Lat	−0.305	0.057	−0.685	<0.0001
		Lon	0.237	0.034	0.884	<0.0001
		AAT	−0.001	0.000	−0.514	<0.0001
	Local	Tem-air	−0.091	0.021	−0.443	<0.0001
		Sand	0.02	0.008	0.279	0.012
		SWC	0.359	0.121	0.319	0.004
Below-ground carbon content (g m^{-2})	Regional	Lat	0.085	0.039	0.21	0.032
		MAT	−0.057	0.013	−0.441	<0.0001
	Local	Clay	−0.645	0.315	−0.164	0.044
		BD	−0.783	0.411	−0.261	0.060
		SWC	0.468	0.143	0.451	0.001
Soil carbon content (Kg m^{-2})	Regional	Lon	0.091	0.016	0.5	<0.0001
		MAT	−0.023	0.008	−0.25	0.005
	Local	SWC	0.497	0.066	0.629	<0.0001

Abbreviations: $^{+}$CO is coefficient; SE is standard error; SC is standardized coefficient; Lat is latitude; Lon is longitude; AAT is annual accumulated temperature; Tem-air is air temperature; Sand is soil sand content; SWC is soil moisture content; Clay is soil clay content; BD is soil bulk density; MAT^2 is the square of mean annual temperature.

Table 3. Summary of the ANOVA results, shows the different influence of regional and local factors on three carbon pools in alpine grassland of Tibetan Plateau

		SS^{+}	MS	*R* square	*F* ratio	*P* value
Above-ground carbon content (gm^{-2})	Regional	70.943	23.652	0.498	28.124	<0.0001
	Local	41.239	13.746	0.296	11.762	<0.0001
Below-ground carbon content (gm^{-2})	Regional	24.294	12.147	0.204	11.403	<0.0001
	Local	55.972	18.641	0.482	26.392	<0.0001
Soil carbon content (Kgm^{-2})	Regional	21.657	10.892	0.331	21.731	<0.0001
	Local	26.398	13.199	0.407	29.112	<0.0001

variance partitioning method to estimate the contribution of local and regional factors to AGC, and found that only 3.1% of the variation of AGC was attributed to "pure" local factors, while 23.3% of the variation can be explained by "pure" regional factors (Fig. 2a). Overall results implied that the spatial distribution of AGC is mainly limited by regional factors, especially temperature.

2.3. Environmental drivers of belowground carbon (BGC) stocks

Unlike AGC, BGC was weakly related to regional factors (Table 2). Rather, significant relationships ($P < 0.0001$) were observed between BGC and soil water content ($r^2 =$ 42%), and between BGC and soil bulk density ($r^2 = 46\%$), though the overall trends were opposite (Fig. 3a, b). Notable, BGC showed a moderately significant negative relationship with mean annual temperature ($r^2 = 0.07$, $P =$ 0.011) (Fig. 3c). Compared with regional factors, BGC was affected more strongly by local (20.4% vs. 48.2%) factors (Table 3). The result of variance partitioning showed that only 9.0% of the BGC variation was attributed to "pure" regional factors, while 36.8% was explained by "pure" local factors (Fig. 2b). Of the variables examined by local factors, BGC was mainly controlled by soil water content (45.1%), followed by bulk density (26.1%), and clay content (16.4%) (Table 2). In general, BGC is highly controlled by local factors, in particular soil moisture.

2.4. Environmental drivers of soil organic carbon (SOC) stocks

Both regional and local factors had relatively similar yet significant influence on SOC (Table 2). SOC had significant linear relationships with soil water content and longitude (Fig. 3d, e), but had a unimodal relationship with mean annual temperature (Fig. 3f). At regional scale, SOC was mainly constrained by longitude (50.0%) and the square of mean annual precipitation (25.0%), while limited by soil water content (62.9%) at local scale. Stepwise regression suggested that the effects of regional and

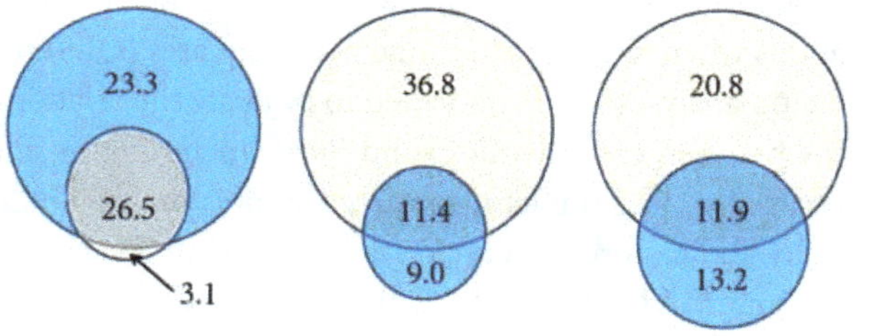

Fig. 2. Partitioning of the variances in carbon stock into percentages explained by regional properties (blue cycle), local factors (orange cycle), and their covariance (intersection). Numbers in the respective circles outside the intersection represent the variance explained by pure regional properties and pure local factors

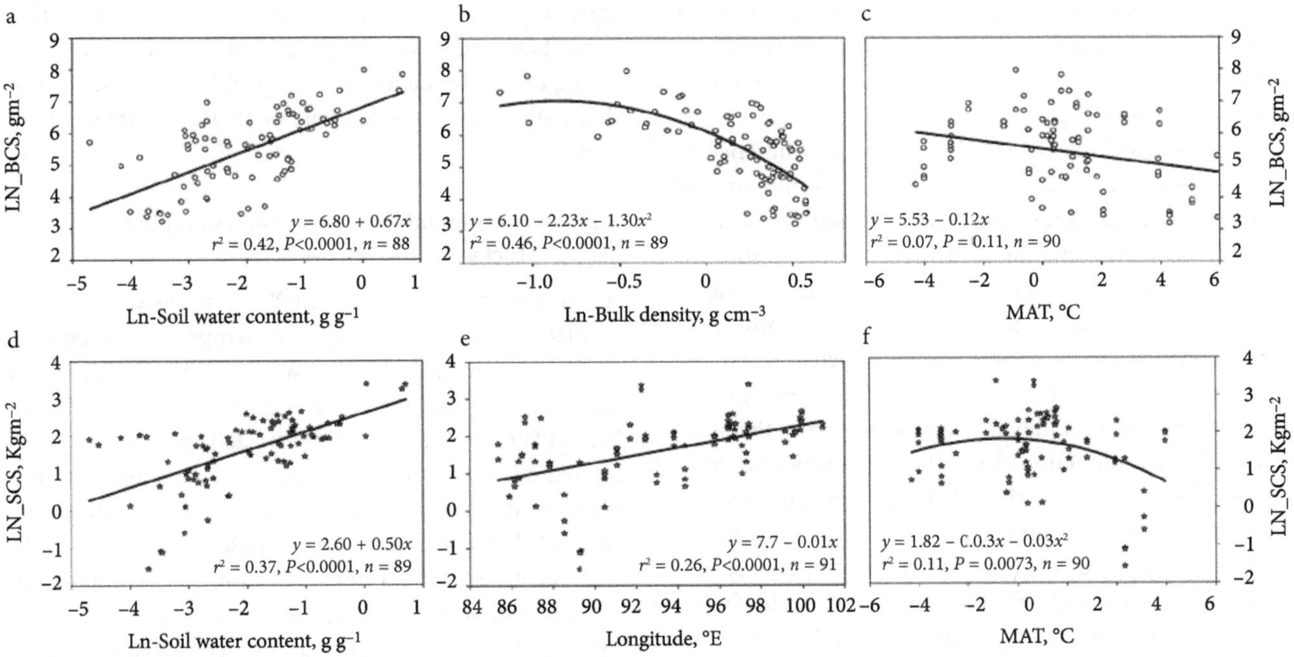

Fig. 3. Relationships of log-transformed belowground and soil carbon stock to environmental variables: (a) soil moisture; (b) soil bulk density; (c) mean annual temperature; (d) soil moisture; (e) longitude; (f) mean annual temperature. BCS is belowground carbon stock; SCS is soil carbon stock; MAT is mean annual temperature

local factors on soil organic carbon stock are relatively similar, with explaining 33.1 and 40.7% of the variation, respectively. Variance partitioning showed similar result, i.e. "pure" regional and "pure" local factors were compatible (13.2% vs 20.8%) in explaining the variation of SOC.

3. Discussion

3.1. Scale-specific responses of grassland carbon stocks to environmental drivers

Carbon dynamics in Tibetan high-altitude grasslands are of special interest because of its high belowground carbon proportion and fragile feedbacks to climate change (Mack *et al.* 2004; Zimov *et al.* 2006; Davidson, Janssens 2006; Freschet *et al.* 2015). Responses of carbon dynamics to environmental factors remain uncertain in Tibetan plateau, mainly due to paucity of field belowground carbon data and high spatial heterogeneity (Jobbagy, Jackson 2000; Sun *et al.* 2013). Field belowground carbon survey and scaling-up approaches were thus combined to estimate ecosystem carbon content of grassland in those remote and high-altitude areas. There is an explicit fundamental assumption associated with scaling-up and/or scaling-down approach: the site-scale models should be applied to scale up to regional scales, i.e., the relationships between independent environmental predictors and carbon stocks dependent variables are applicable across various scales. Our study showed that the driving forces of aboveground carbon (AGC), belowground carbon (BGC), and SOC in the Tibetan Plateau were not constant, but strongly scale dependent (Table 2 and 3). The results of variance partitioning showed that the variation of AGC explained by "pure" regional factors was 7.8 (23.3% vs 3.1%) times higher than "pure" local factors, while the variation of BGC was explained 4.1 (36.8% vs 9.0%) times more by "pure" local factors than by "pure" regional factors. Compared with the significant but opposite patterns of regional and local factors on AGC and BGC, both local and regional factors had relatively similar (20.8% vs. 13.2%) influence on SOC. Overall, temperature at large scales play an important role in AGC, whereas soil moisture availability constrains BGC, both of them determining SOC.

Such climatic variables controlling AGC at great geological scale and edaphic factors limiting BGC and SOC at fine-scale is caused by a couple of reasons. The first reason is sampling efforts. Ecosystem carbon stocks have been approximated by mean annual temperature and precipitation at large geological scale (Amundson 2001; Gibbon *et al.* 2010; Jiang *et al.* 2015). At small local scale, however, ecosystem carbon content vary with respect to simple, easily quantified soil variables such as soil nutrient availability, texture, and moisture (Wynn *et al.* 2006; Miao, Li 2007; Miao *et al.* 2011; Qin *et al.* 2015). Leith and Whittaker (1975) pointed out that these conflicting results regarding dissimilar driving factors of ecosystem processes at different spatial scales may be related to sampling efforts. They thought that climatic factors generally limit grassland production at a continental scale, but typically too broad to account for the observed variations in productivity at smaller geographic scales. The scale-dependent limiting factors of carbon stocks have important implications for continental to global scale carbon modeling efforts: the interpolating and/or extrapolating between small and large scales seem infeasible.

The second reason is the functional difference of plant organs. The functional difference between shoot and root, as well as the heterogeneous distribution of resources, such as solar radiation and moisture determine the response of AGC and BGC to environmental conditions, and hence SOC and the terrestrial ecosystem carbon cycling. Compared with edaphic variables such as soil moisture and nutrients, the distribution of solar energy and air temperature are highly homogeneous across the landscape. The main function of shoot is to absorb solar energy and transform it into bioavailable energy. Consequently, a positive relationship between AGC and temperature at large scales can be expected, since air temperature and solar irradiation are closely related (Piao *et al.* 2006; Sun *et al.* 2013). By contrast, root support life cycle of plants by providing water and nutrients, which are heterogeneously distributed either temperately or spatially. Keeping large root system will benefit to plants water budget by absorbing more water through increased absorption area. Tissues that are expensive to build because of a scarcity of mineral nutrients or slow acquisition of carbon should be long-lived because it will take longer for them to return the resources spent in constructing them to the plant. Consequently, plants growing on water limited sites often have long-lived leaves with low carbon-gaining capacities. Conversely, plants on water limited sites often have leaf tissues that require a shorter time period before the resources spent in building them are returned to the plant. Water addition in arid and semi-arid regions could stimulate root system growth by allocating more carbohydrate into root system (Chapin *et al.* 1986). As a result, the local factors play a determinative role in root growth, such as soil moisture and nutrient availability. The magnitude of SOC fundamentally depends on the rate of input (litter fall and find root) and output (decomposition and transport) (Amundson 2001). Despite this simplicity, measurement of SOC at regional scales is problematic, due to the heterogeneous distribution of SOC across the landscape and non-linear relationships between SOC, temperature and moisture (Davidson, Janssens 2006). Compared with the significant but opposite patterns of regional and local factors on the above- and below-ground carbon stock, both local and regional factors had relatively similar (20.8% vs. 13.2%) influence on SOC. It seems that the offsetting responses of AGC and BGC to air temperature and soil moisture seem to have negated any overwhelm influence of either resource on SOC. Our result generally in line with others that precipitation could stimulate plant production and thus contribute to the accumulation of soil organic matter in a water-limiting area (Jobbagy, Jackson 2000; Wynn *et al.* 2006; Miao *et al.* 2011). On the other hand, SOC tends to be high in cold and/or wet regions, owing to a combination of biological factors related to either high biomass productivity or slow decomposition, or both.

3.2. Environmental drivers of Tibetan alpine carbon stocks

As one of the largest alpine grasslands in the world, Tibetan Plateau is characterized by low temperature, long term snow coverage, and short growing season. It is generally assumed that projected global warming will benefit the Tibetan Plateau grassland carbon stock by prolonging growing season, promoting photosynthesis, and enhancing nutrient availability (Piao *et al.* 2006). However temperature showed linear and unimodal relationship, respectively, with plant and soil carbon stocks, implying complex impacts of global warming on the carbon stocks of Tibetan Plateau. The positive correlations between temperature and primary productivity are widely reported across various terrestrial ecosystems (Hui *et al.* 2005; Sun *et al.* 2013; Miao, Li 2010; Miao *et al.* 2011). The responses can also be related to the particular soil resource that is most limiting. For example increased temperature could also inhibit plant growth in arid regions due to worse soil moisture (Niu *et al.* 2008). In addition, increase in temperature could reduce nutrient supply as metabolic enhanced microbial will compete with plants for available resources (Rustad *et al.* 2001; King *et al.* 2008). In this study, the opposite trends of plant biomass and temperature indicated that AGC and BGC are not limited by temperature in Tibetan Plateau. By contrast, the overall unimodal relationships of SOC with MAT (mean annual temperature) revealed that SOC increased with MAT when MAT was lower than –0.50 °C, while SOC decreased with an increasing of temperatures when temperature is above –0.50 °C. The decrease of SOC under moderate temperature (–0.50 °C of MAT) is likely due to the decrease of soil moisture with an increase of temperature.

Alpine regions are believed to be exposed to a higher rate of warming compared to any other parts of the globe and are most sensitive and vulnerable to global change (King *et al.* 2008). The strong significant ($r^2 = 0.42$, $P < 0.0001$) linear relationship between BGC and soil moisture, and between SOC and soil moisture ($r^2 = 0.37$, $P < 0.0001$) indicated that BGC and SOC were mainly limited by soil moisture. The weak relationships between BGC and MAT ($r^2 = 0.07$, $P = 0.01$) and between SOC and MAT ($r^2 = 0.11$, $P = 0.01$) suggested that temperature played a minor role in determining BGC and SOC. Our results were consistent with many other reports (Kardol *et al.* 2010; Wu *et al.* 2011). The scale-specific responses of AGC, BGC and SOC to temperature and moisture were helpful to predict the responses of alpine grassland community to warming-induced decrease of soil moisture in alpine grassland community. Climatic warming increases soil temperature and hence accelerate soil water transpiration by stimulating evapotranspiration (Rustad *et al.* 2001; Fontaine *et al.* 2004; Niu *et al.* 2008). Soil moisture aggravation will significant negatively impact on the above- and below-ground properties related to climate change. As a result, soil moisture variation is an important possible consequence of global warming. Although soil moisture was measured only one time, the significant correlation ($r^2 = 0.86$, $P < 0.0001$) between soil moisture and soil texture suggested a general trend of soil moisture distribution in the Plateau. Understanding interannual variation in carbon sequestration and allocation will require more detailed analysis of the influences of soil resources on the components of ecosystem carbon stocks.

Conclusions

Overall results indicated that the responses of above- and below-ground carbon stocks to environmental drivers are different in Tibetan alpine Plateau at both regional and local scales.

1. Aboveground carbon storage was mainly constrained by regional factors, especially air temperature.
2. Belowground carbon stocks was limited by local factors, especially soil moisture, whereas both regional and local factors had similar influence on soil carbon.
3. Global warming would reduce carbon stocks and complicate alpine grassland carbons mosaics, mainly caused by increasing air temperature and decreasing soil moisture.

Acknowledgements

We thank Dr. Laura Gough and Dr. Majie Fan for their valuable suggestions and reviewing early version. The authors are also grateful to Mei Hang, Zhengjia Liu, Xiujing Yang, Lei Li for assistance in field data collection, and Guiduo Shang, Lei Li, and Zhaosheng Wang for laboratory assistance. This work was supported by the National Natural Science Foundation of China. This research has been financially supported by the National Sciences Foundation of China (No. 30500067, 31070393, and 40975845).

References

Amundson, R. 2001. The carbon budget in soils, *Annual Review of Earth Planet Sciences* 29: 535–562. http://dx.doi.org/10.1146/annurev.earth.29.1.535

Avohou, H. T.; Sinsin, B. 2009. The effects of topographic factors on aboveground biomass production of grasslands in the Atacora Mountains in North-western Benin, *Mountain Research and Development* 29(3): 250–254. http://dx.doi.org/10.1659/mrd.00028

Bai, Y. F.; Wu, J. G.; Xing, Q.; Pan, Q.; Huang, J.; Yang, D.; Han, X. 2008. Primary production and rain use efficiency across a precipitation gradient on the Mongolia plateau, *Ecology* 89(8): 2140–2153. http://dx.doi.org/10.1890/07-0992.1

Bai, Y. F.; Han, X.; Wu, J.; Chen, Z.; Li, L. 2004. Ecosystem stability and compensatory effects in the Inner Mongolia grassland, *Nature* 431(7005): 181–184. http://dx.doi.org/10.1038/nature02850

Chabot, B. F.; Hicks, D. J. 1982. The ecology of leaf life spans, *Annual Review of Ecology, Evolution, and Systematics* 13: 229–259. http://dx.doi.org/10.1146/annurev.es.13.110182.001305

Chapin, F. S.; Schulze, E. D.; Mooney, H. A. 1990. The ecology and economics of storage in plants, *Annual Review of Ecology, Evolution, and Systematics* 21: 423–47. http://dx.doi.org/10.1146/annurev.es.21.110190.002231

Chapin, F. S. III; Vitousek, P. M.; Cleve, K. V. 1986. The nature of nutrient limitation in plant communities, *The American Naturalist* 127(1): 48–58. http://dx.doi.org/10.1086/284466

Chinese Academy of Sciences. 2001. *Vegetation atlas of China.* Beijing: Science Press.

Davidson, E. A.; Janssens, I. A. 2006. Temperature sensitivity of soil carbon decomposition and feedbacks to climate change, *Nature* 440: 165–173. http://dx.doi.org/10.1038/nature04514

De Deyn, G. B.; Quirk, H.; Yi, Z.; Oakley, S.; Ostle, N. J.; Bardgett, R. D. 2009. Vegetation composition promotes carbon and nitrogen storage in model grassland communities of contrasting fertility, *Journal of Ecology* 97: 864–875. http://dx.doi.org/10.1111/j.1365-2745.2009.01536.x

Enquist, B. J.; Niklas, K. J. 2002. Global allocation rules for biomass partitioning in seed plants, *Science* 295: 1517–1520. http://dx.doi.org/10.1126/science.1066360

Freschet, G. T.; Kichenin, E.; Wardle, D. A. 2015. Explaining within-community variation in plant biomass allocation: a balance between organ biomass and morphology above vs below ground?, *Journal of Vegetation Science* 26(3): 431–440. http://dx.doi.org/10.1111/jvs.12259

Fontaine, S.; Bardoux, G.; Abbadie, L.; Mariotti, A. 2004. Carbon input to soil may decrease soil carbon content, *Ecology Letter* 7(4): 314–320.

Gibbon, A.; Silman, M.; Malhi, Y.; Fisher, J. B.; Meir, P.; Zimmermann, M.; Dargie, G. C.; Farfan, W. R.; Garcia, K. C. 2010. Ecosystem carbon storage across the grassland-forest transition in the high Andes of Manu National Park, Peru, *Ecosystems* 13: 1097–1111. http://dx.doi.org/10.1007/s10021-010-9376-8

Gong, X. Y.; Fanselow, N.; Dittert, K.; Taube, F.; Lin, S. 2015. Response of primary production and biomass allocation to nitrogen and water supplementation along a grazing intensity gradient in semiarid grassland, *European Journal Agronomy* 63: 27–35. http://dx.doi.org/10.1016/j.eja.2014.11.004

Hendricks, J. J.; Nadelhoffer, K. J.; Aber, J. D. 1993. Assessing the role of fine roots in carbon and nutrient cycling, *Trends in Ecology & Evolution* 8(5): 174–178. http://dx.doi.org/10.1016/0169-5347(93)90143-D

Hui, D.; Jackson, R. B. 2005. Geographic and interannual variability in biomass partitioning in grassland ecosystems: a synthesis of field data, *New Phytologist* 169(1): 85–93. http://dx.doi.org/10.1111/j.1469-8137.2005.01569.x

Jiang, Y.; Tao, J.; Huang, Y.; Zhu, J.; Tian, L.; Zhang, Y. 2015. The spatial pattern of grassland aboveground biomass on Xizang

Plateau and its climatic controls, *Journal Plant Ecology* 8(1): 30–40. http://dx.doi.org/10.1093/jpe/rtu002

Jobbagy, E. G.; Jackson, R. B. 2000. The vertical distribution of soil organic carbon and its relation to climate and vegetation, *Ecological Application* 10(2): 423–436. http://dx.doi.org/10.1890/1051-0761(2000)010[0423:TVDOSO]2.0.CO;2

Kardol, P.; Campany, C. E.; Souza, L.; Norby, R. J.; Weltzin, J. F.; Classen, A. T. 2010. Climate change effects on plant biomass alter dominance patterns and community evenness in an experimental old-field ecosystem, *Global Change Biology* 16(12): 2676–2687. http://dx.doi.org/10.1111/j.1365-2486.2010.02162.x

King, A. J.; Meyer, A. F.; Schmidt, S. K. 2008. High levels of microbial biomass and activity in unvegetated tropical and temperate alpine soils, *Soil Biology & Biochemistry* 40: 2605–2610. http://dx.doi.org/10.1016/j.soilbio.2008.06.026

Knapp, A. K.; Fay, P. A.; Blair, J. M.; Collins, S. L.; Smith, M. D.; Carlisle, J. D.; Harper, C. W.; Danner, B. T.; Lett, M. S.; McCarron, J. K. 2002. Rainfall variability, carbon cycling, and plant species diversity in a mosaic grassland, *Science* 298: 2202–2205. http://dx.doi.org/10.1126/science.1076347

Leith, H.; Whittaker, R. (Eds.). 1975. *Primary productivity of the biosphere.* Berlin, Heidelberg, New York: Springer. http://dx.doi.org/10.1007/978-3-642-80913-2

Mack, M. C.; Schuur, E. A. G.; Bret-Harte, M. S.; Shaver, G. R.; Chapin F. S. III. 2004. Ecosystem carbon storage in arctic tundra reduced by long-term nutrient fertilization, *Nature* 431: 440–443. http://dx.doi.org/10.1038/nature02887

McConnaughay, K. D. M.; Coleman, J. S. 1999. Biomass allocation in plants: ontogeny or optimality? A test along three resource gradients, *Ecology* 80(8): 2581–2593. http://dx.doi.org/10.1890/0012-9658(1999)080[2581:BAIPOO]2.0.CO;2

Medvigy, D.; Wofsy, S. C.; Munger, J. W; Moorcroft, P. 2010. Responses of terrestrial ecosystems and carbon budgets to current and future environmental variability, *Proceedings of the National Academy of Sciences* 107(18): 8275–8280. http://dx.doi.org/10.1073/pnas.0912032107

Miao, Z.; Lathrop, R. G.; La Puma, I. P.; Clark, K. L.; Hom, J.; Skowronski, N.; Van Tuyl, S. 2011. Simulation and sensitivity analysis of carbon storage and fluxes in the New Jersey Pinelands, *Environmental Modelling and Software* 26(9): 1112–1122. http://dx.doi.org/10.1016/j.envsoft.2011.03.004

Miao, Z.; Li, C. 2007. *Biomass estimates for major boreal forest species in west-central Canada.* Canadian Wood Fibre Centre, Canadian Forest Service, Information Report FI-X-002, 1–37.

Miao, Z.; Li, C. 2010. Predicting tree growth dynamics of boreal forest in response to climate change, in C. Li, R. Lafortezza and J. Chen (Eds.). *Landscape Ecology in Forest Management and Conservation.* Springer-HEP Publisher, 176–205.

Niu, S. L.; Wu, M. Y.; Han, Y., Xia, J.; Li, L.; Wan, S. 2008. Water-mediated responses of ecosystem carbon fluxes to climatic change in a temperate steppe, *New Phytologist* 177(1): 209–219.

Ostertag, R. 2001. Effects of nitrogen and phosphorus availability on fine-root dynamics in Hawaiian montane forests, *Ecology* 82: 485–499. http://dx.doi.org/10.1890/0012-9658(2001)082[0485:EONAPA]2.0.CO;2

Patton, B. D.; Dong, X.; Nyren, P. E.; Nyren, A. 2007. Effects of grazing intensity, precipitation, and temperature on forage production, *Rangeland and Ecological Management* 60: 656–665. http://dx.doi.org/10.2111/07-008R2.1

Piao, S. L.; Mohammat, A.; Fang, J. Y.; Cai, Q.; Feng, J. 2006. NDVI based increase in growth of temperate grasslands and its responses to climate changes in China, *Global Environmental Change* 16(4): 340–348. http://dx.doi.org/10.1016/j.gloenvcha.2006.02.002

Qin, Y.; Yi, S.; Chen, J.; Ren, S.; Ding, Y. 2015. Effects of gravel on soil and vegetation properties of alpine grassland on the Qinghai-Tibetan plateau, *Ecological Engineering* 74: 351–355. http://dx.doi.org/10.1016/j.ecoleng.2014.10.008

Reich, P. B.; Walters, M. B.; Ellsworth, D. S. 1992. Leaf life-span in relation to leaf, plant, and stand characteristics among diverse ecosystems, *Ecological Monograph* 62(3): 365–392. http://dx.doi.org/10.2307/2937116

Rustad, L. E.; Campbell, J. L.; Marion, G. M.; Norby, R. J.; Mitchell, M. J.; Hartley, A. E.; Cornelissen, J. H. C.; Gurevitch, J. 2001. A meta-analysis of the response of soil respiration, net nitrogen mineralization, and aboveground plant growth to experimental ecosystem warming, *Oecologia* 126(4): 543–562. http://dx.doi.org/10.1007/s004420000544

Sala, O. E.; Parton, W. J.; Joyce, L. A.; Lauenroth, W. K. 1988. Primary production of the central grassland region of the United States, *Ecology* 69: 40–45. http://dx.doi.org/10.2307/1943158

Scurlock, J. M. O.; Johnson, K.; Olson, R. J. 2002. Estimating net primary productivity from grassland biomass dynamics measurements, *Global Change Biology* 8(8):736–753. http://dx.doi.org/10.1046/j.1365-2486.2002.00512.x

Sistla, S. A.; Moore, J. C.; Simpson, R. T.; Gough, L.; Shaver, G. R.; Schimel, J. P. 2013. Long-term warming restructures Arctic tundra without changing net soil carbon storage, *Nature* 497: 615–618. http://dx.doi.org/10.1038/nature12129

Sun, J.; Cheng, G. W.; Li, W. P. 2013. Meta-analysis of relationships between environmental factors and aboveground biomass in the alpine grassland on the Tibetan Plateau, *Biogeosciences* 10: 1707–1715. http://dx.doi.org/10.5194/bg-10-1707-2013

Swemmer, A. M.; Knapp, A.; Snyman, H. A. 2007. Intra-seasonal precipitation patterns and aboveground productivity in three perennial grasslands, *Journal of Ecology* 95(4): 780–788. http://dx.doi.org/10.1111/j.1365-2745.2007.01237.x

Wu, G. L.; Zhao, L. P.; Shi, Z. H.; Shang, Z. P. 2011. Above- and below-ground response to soil moisture change on an alpine wetland ecosystem in the Qinghai-Tibetan Plateau China, *Biogeosciences Discussion* 8(4): 7141–7164. http://dx.doi.org/10.5194/bgd-8-7141-2011

Wu, H; Guo, Z; Peng, C. 2003. Distribution and storage of soil organic carbon in China, *Global Biogeochemical Cycles* 17(2): 1048. http://dx.doi.org/10.1029/2001GB001844

Wynn, J. G.; Bird, M. I.; Vallen, L.; Grand-Clement, E.; Carter, J.; Berry, S. L. 2006. Continental-scale measurement of the soil organic carbon pool with climatic, edaphic, and biotic controls, *Global Biogeochemical Cycles* 20(1): GB1007. http://dx.doi.org/10.1029/2005GB002576

Xu, M.; Peng, F.; You, Q.; Guo, J.; Tian, X.; Xue, X.; Liu, M. 2015. Year-round warming and autumnal clipping lead to downward transport of root biomass, carbon and total nitrogen in soil of an alpine meadow, *Environmental and Experimental Botany* 109: 54–62. http://dx.doi.org/10.1016/j.envexpbot.2014.07.012

Yang, Y. H.; Fang, J. Y.; Ma, W. H.; Wang, W. 2008. Relationship between variability in aboveground net primary production

and precipitation in global grasslands, *Geophysical Research Letters* 35(23): L23710.
http://dx.doi.org/10.1029/2008GL035408

Zimov, S. A.; Schuur, E. A. G.; Chapin, F. S. III. 2006. Permafrost and the global carbon budget, *Science* 312: 1612–1613.
http://dx.doi.org/10.1126/science.1128908

Hongsheng LIU, PhD, research assistant, Institute of Geography Sciences and Natural Resource Research, Chinese Academy of Sciences, and Department of Biology, University of Texas at Arlington, USA. He is an author of about 30 publications. The major research area is ecosystem ecology.

Zewei MIAO, PhD, Research scientist, Nicholas School of the Environment, Duke University, USA. He is an author of about 50 publications. His major research area is eco-environmental modelling and biomass feedstock mechanical preprocessing for bioenergy production. Currently, he is an agronomic data scientist of Monsanto Co.

6

INTUITIONISTIC FUZZY EDAS METHOD: AN APPLICATION TO SOLID WASTE DISPOSAL SITE SELECTION

Cengiz KAHRAMAN[a], Mehdi KESHAVARZ GHORABAEE[b], Edmundas Kazimieras ZAVADSKAS[c], Sezi CEVIK ONAR[a], Morteza YAZDANI[d], Basar OZTAYSI[a]

[a] *Department of Industrial Engineering, Istanbul Technical University, 34367 Macka, Istanbul, Turkey*
[b] *Department of Industrial Management, Faculty of Management and Accounting, Allameh Tabataba'i University, Tehran, Iran*
[c] *Department of Construction Technology and Management, Faculty of Civil Engineering, Vilnius Gediminas Technical University, Lithuania*
[d] *Department of Management Science, Universidad Europea de Madrid, Madrid, Spain*

Abstract. Evaluation based on Distance from Average Solution (EDAS) is a new multicriteria decision making (MCDM) method, which is based on the distances of alternatives from the average scores of attributes. Classical EDAS has been already extended by using ordinary fuzzy sets in case of vague and incomplete data. In this paper, we propose an interval-valued intuitionistic fuzzy EDAS method, which is based on the data belonging to membership, nonmembership, and hesitance degrees. A sensitivity analysis is also given to show how robust decisions are obtained through the proposed intuitionistic fuzzy EDAS. The proposed intuitionistic fuzzy EDAS method is applied to the evaluation of solid waste disposal site selection alternatives. The comparative and sensitivity analyses are also included.

Keywords: EDAS, ordinary fuzzy sets, intuitionistic fuzzy sets, solid waste, site selection, average solution, multicriteria decision making (MCDM).

Introduction

EDAS method uses average solution for appraising the alternatives. It was developed by Keshavarz Ghorabaee *et al.* (2015). In this method, two measures which is called PDA (positive distance from average) and NDA (negative distance from average) are considered for the appraisal. This method is very useful when we have some conflicting criteria. In the compromise MCDM methods such as VIKOR and TOPSIS, the best alternative is obtained by calculating the distance from positive and negative ideal solutions. The best alternative has least distance from positive ideal solution (PIS) and largest distance from negative ideal solution (NIS) in these MCDM methods. In EDAS method, the best alternative is selected with respect to the distances from average solution: The positive distance from average (PDA) and the negative distance from average (NDA). An alternative having higher values of PDA and lower values of NDA is better (Keshavarz Ghorabaee *et al.* 2015).

The fuzzy set theory is a powerful tool in order to overcome ambiguous information. It was developed by Zadeh (1965) and has been used in various areas including multicriteria decision making (Liu *et al.* 2012), aggregation operations (Liu *et al.* 2016; Liu, Jin 2012; Liu, P. D., Liu, Y. 2014), definition of uncertain linguistic variables (Liu *et al.* 2011; Liu, Yu 2014), etc. EDAS has been extended to ordinary fuzzy EDAS by Keshavarz Ghorabaee *et al.* (2016) to handle the MCDM problems under fuzzy environment. Fuzzy sets have a history starting from ordinary fuzzy sets and extending to other types of fuzzy sets as illustrated in Figure 1 (Kahraman *et al.* 2015, 2016).

Intuitionistic fuzzy sets (IFS) have been proposed by Atanassov (1986). IFSs are very effective to deal with uncertainty and vagueness. Therefore, the decision makers can utilize intuitionistic fuzzy sets, especially the interval-valued intuitionistic fuzzy sets (IVIFS) introduced by Atanassov and Gargov (1989) to better express the information of the candidates under incomplete and

Corresponding author: Cengiz Kahraman
E-mail: kahramanc@itu.edu.tr

uncertain information environment. IFS can consider the membership and nonmembership degrees of the elements of a set at the same time, which their sum is not necessarily equal to one. Multicriteria decision making with interval-valued intuitionistic fuzzy sets has received a great deal of attention from researchers recently (Cevik Onar *et al.* 2015; Oztaysi *et al.* 2015). Later, intuitionistic fuzzy sets have been considered as a special subset of neutrosophic sets by Smarandache (1998) and have been extended by some researchers (Liu, Shi 2015; Liu, Wang 2014).

It is evident from the previous studies in the literature, there has been no work extending EDAS to IVIF EDAS model to solve a multicriteria problem under fuzzy environment. The proposed method can be used to compare electronic waste or solid waste disposal technologies, disposal site alternatives, and hazardous material removal technologies for environmental protection.

To the best of our knowledge, this is the first paper extending EDAS method to IVIF EDAS. The performance of the proposed IVIF EDAS is illustrated through a solid waste disposal site selection problem. Solid waste is the unwanted or useless solid materials generated from combined residential, industrial and commercial activities in a given area. Management of solid waste reduces or eliminates adverse impacts on the environment and human health and supports economic development and improved quality of life. A number of processes are involved in effectively managing waste for a municipality. These include monitoring, collection, transport, processing, recycling and disposal.

The rest of this paper is organized as follows. Section 1 includes a literature review on EDAS method and gives the steps of classical EDAS. Section 2 presents the preliminaries of ordinary fuzzy sets and intuitionistic fuzzy sets together with the steps of ordinary fuzzy EDAS and intuitionistic fuzzy EDAS. Section 3 presents the applications of classical (crisp), ordinary fuzzy, and intuitionistic fuzzy EDAS for the solution of a solid waste disposal site selection problem. The last Section concludes the paper.

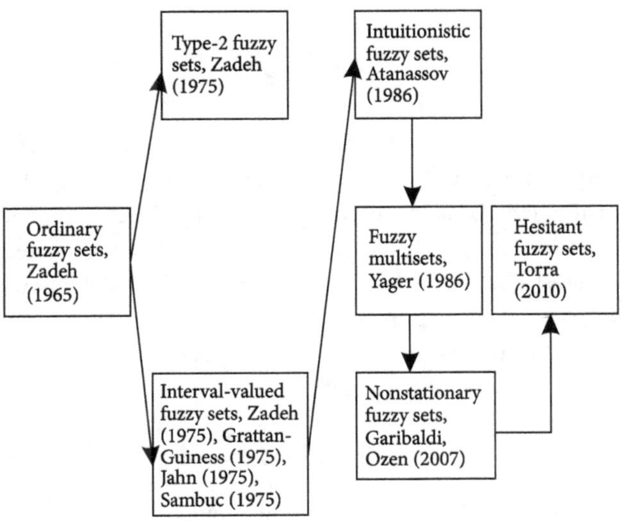

Fig. 1. Extensions of fuzzy sets

1. EDAS method

Curiosity of researchers to inventing new MCDM methods is getting competitive. Trends of MCDM methods development since initial days showed enormous interest among researchers in this area to build robust structures in order to handle complex decisions. Their precise effort was concentrated on realizing requirements of decision system, then implementing variables and parameters, and interconnecting those variables in a logical manner to establish specific tool. Many MCDM methods have been created and then extended. Analytical hierarchy process (Deng *et al.* 2014), ANP (Van Horenbeek, Pintelon 2014), TOPSIS (Büyüközkan, Çifçi 2012; Seçme *et al.* 2009), VIKOR (Mardani *et al.* 2016), ARAS (Turskis *et al.* 2013), COPRAS (Turanoglu Bekar *et al.* 2016), SAW (Hashemkhani Zolfani *et al.* 2012), ELECTRE (Figueira *et al.* 2013), PROMETHEE (Corrente *et al.* 2013), DEMATEL (Yeh, Huang 2014), MOORA (Stanujkic *et al.* 2014; Yazdani 2015), WASPAS (Chakraborty, Zavadskas 2014), SWARA (Kouchaksaraei *et al.* 2015), and so on are examples of MCDM methods which are applied in different situations. Sort of methods have been invented, developed and expanded conjoining with fuzzy theories or in any other scale of integration as hybrid MCDM models. The method *Evaluation Based on Distance from Average Solution* (EDAS) has been developed by Keshavarz Ghorabaee *et al.* (2015). TOPSIS and VIKOR are among top and most popular MCDM methods (Behzadian *et al.* 2012; Yazdani, Payam 2015). The logical function of these methods considers the optimal solution based on maximum distance from negative solution simultaneously minimum distance from best or positive ideal solution. However, the best alternative in the EDAS method is corresponded to the distance from average solution (AV). In this work we first present the crisp and ordinary fuzzy EDAS methods and then extend to its intuitionistic fuzzy version.

In EDAS method, first two measures are delivered as the positive distance from average (PDA), and the negative distance from average (NDA). These measures can show the difference between each solution (alternative) and the average solution. Therefore higher values of PDA and lower values of NDA will indicate optimal solution. In fact, the higher values of PDA and/or lower values of NDA represent that the solution (alternative) is better than average solution. Classical algorithm of EDAS can be followed using the following steps:

Step 1. Choose the most relevant attributes which describe decision alternatives for specific decision problem.

Step 2. If x_{ij} is the performance rating of ith alternative $A_1, A_2, ..., A_n$, $(i = 1,2,....,n)$ respecting to the jth criterion $C_1, C_2, ..., C_m$ $(j = 1,2,...,m)$. So, to form the interval decision matrix X and weight of each criterion, following table and variables should be considered:

$$X = [x_{ij}]_{n\times m} = \begin{bmatrix} x_{11} & x_{12} & \cdots & x_{1m} \\ x_{21} & x_{22} & \cdots & x_{2m} \\ \cdot & \cdot & \cdot & \cdot \\ \cdot & \cdot & \cdot & \cdot \\ \cdot & \cdot & \cdot & \cdot \\ x_{n1} & x_{n2} & \cdots & x_{nm} \end{bmatrix}. \quad (1)$$

$$W = [w_1, w_2, ..., w_m]$$

for $(i = 1,2,....,n)$ and $(j = 1,2,...,m)$, where w_j is the weight of criterion j^{th}.

Step 3. According to the definition of EDAS method, the average solution with respect to all criteria must be determined as shown following formulas:

$$AV_j = \frac{\sum_{i=1}^{n} x_{ij}}{n}. \quad (2)$$

Step 4. The positive distance from average (*PDA*) and the negative distance from average (*NDA*) matrixes need to be calculated in this step according to lower and upper values of matrix as shown:

$$PDA_{ij} = \frac{\max(0,(x_{ij} - AV_j))}{AV_j}, \quad (3)$$

and

$$NDA_{ij} = \frac{\max(0,(AV_j - x_{ij}))}{AV_j}. \quad (4)$$

In this way PDA_{ij} and NDA_{ij} represent the positive and negative distance of ith alternative from average solution in terms of jth criterion for the lower level of decision matrix, respectively.

Step 5. Obtain weighted summation of the positive distance and negative distances from average matric:

$$SP_i = \sum_{j=1}^{m} w_j PDA_{ij}, \quad (5)$$

and

$$SN_i = \sum_{j=1}^{m} w_j NDA_{ij}. \quad (6)$$

Step 6. Identify the normalized values of SP_i and SN_i for all alternatives, shown as follows:

$$NSP_i = \frac{SP_i}{\text{Max}_i(SP_i)}, \quad (7)$$

and

$$NSN_i = 1 - \frac{SN_i}{\text{Max}_i(SN_i)}. \quad (8)$$

Step 7. Detect the appraisal score *AS* for all alternatives, shown as follows:

$$AS_i = \frac{1}{2}(NSP_i + NSN_i), \quad (9)$$

where $0 \le AS_i \le 1$.

Step 8. Rank the alternatives according to the decreasing values of appraisal score (AS_i). The alternative with the highest AS_i is the best choice among the candidate alternatives.

2. Fuzzy extension of EDAS

Fuzzy decision making models alongside development of MCDM methods are growing rapidly. The crucial importance of fuzzy decision models is evidence among research groups. Different MCDM methods with different instruments transformed to a fuzzy model. Fuzzy TOPSIS, fuzzy VIKOR, fuzzy COPRAS and etc. are examples in this area. In the following, ordinary fuzzy EDAS method developed by Keshavarz Ghorabaee *et al.* (2016) will be introduced. Later Intuitionistic fuzzy EDAS method is proposed. We first give the preliminaries for the ordinary fuzzy EDAS (Keshavarz Ghorabaee *et al.* 2016).

Definition 1. Let $\tilde{A} = (a_1, a_2, a_3, a_4)$ be a trapezoidal fuzzy number. The defuzzified value of $\tilde{A}$ can be obtained by Eq. 10:

$$\kappa(\tilde{A}) = \frac{1}{3}\left(a_1 + a_2 + a_3 + a_4 - \frac{a_3 a_4 {}_- a_1 a_2}{a_3 + a_4 - a_1 - a_2}\right). \quad (10)$$

Definition 2. Suppose that $\tilde{A} = (a_1, a_2, a_3, a_4)$ is a trapezoidal fuzzy number. The function ψ represents the maximum between $\tilde{A}$ and zero:

$$\psi(\tilde{A}) = \begin{cases} \tilde{A}, \ if\ \kappa(\tilde{A}) > 0 \\ 0, \ if\ \kappa(\tilde{A}) \le 0 \end{cases}. \quad (11)$$

2.1. Ordinary fuzzy EDAS (Keshavarz Ghorabaee *et al.* 2016)

Let us assume the fuzzy decision matrix as this:

$$\tilde{X} = [\tilde{x}_{ij}]_{n\times m} = \begin{bmatrix} \tilde{x}_{11} & \tilde{x}_{12} & \cdots & \tilde{x}_{1m} \\ \tilde{x}_{21} & \tilde{x}_{22} & \cdots & \tilde{x}_{2m} \\ \cdot & \cdot & \cdot & \cdot \\ \cdot & \cdot & \cdot & \cdot \\ \cdot & \cdot & \cdot & \cdot \\ \tilde{x}_{n1} & \tilde{x}_{n2} & \cdots & \tilde{x}_{nm} \end{bmatrix}. \quad (12)$$

Consider i^{th} alternative $A_1, A_2, ..., A_n$, $(i = 1,2,....,n)$ respecting to the jth criterion, so each member of decision matrix is denoted by triangular fuzzy number $\tilde{x}_{ij} = (x_{ij1}, x_{ij2}, x_{ij3})$. For each criterion weights are stated as $\tilde{w} = (w_1, w_2, w_3)$. So, to solve fuzzy EDAS model following steps must be pursued:

Step 1. Determine the fuzzy average decision matrix regarding to all of the criteria:

$$\widetilde{AV}_j = \frac{\sum_{i=1}^{n} \tilde{x}_{ij}}{n}. \quad (13)$$

Step 2. The optimal solution should have maximum distance from negative feasible solutions while as the same time minimum distance by best and ideal solutions, so, identify the fuzzy positive distance from average and the fuzzy negative distance from average:

$$PDA = \left[\widetilde{pda}_{ij}\right]_{n\times m}, \tag{14}$$

and

$$NDA = \left[\widetilde{nda}_{ij}\right]_{n\times m}, \tag{15}$$

and

$$\widetilde{pda}_{ij} = \begin{cases} \dfrac{\psi\left(\tilde{x}_{ij} \ominus \widetilde{av}_j\right)}{\kappa\left(\widetilde{av}_j\right)}, & if\ j \in B \\ \dfrac{\psi\left(\widetilde{av}_j - \tilde{x}_{ij}\right)}{\kappa\left(\widetilde{av}_j\right)}, & if\ j \in C \end{cases}, \tag{16}$$

and

$$\widetilde{nda}_{ij} = \begin{cases} \dfrac{\psi\left(\widetilde{av}_j - \tilde{x}_{ij}\right)}{\kappa\left(\widetilde{av}_j\right)}, & if\ j \in B \\ \dfrac{\psi\left(\tilde{x}_{ij} \ominus \widetilde{av}_j\right)}{\kappa\left(\widetilde{av}_j\right)}, & if\ j \in C \end{cases}. \tag{17}$$

Step 3. Compute the fuzzy weighted summation of the positive distance and negative distances from average matric:

$$\widetilde{sp}_i = \oplus_{j=1}^{m}\left(\tilde{w}_j \oplus \widetilde{pda}_{ij}\right), \tag{18}$$

and

$$\widetilde{sn}_i = \oplus_{j=1}^{m}\left(\tilde{w}_j \oplus \widetilde{nda}_{ij}\right). \tag{19}$$

Step 4. Determine the fuzzy normalized values of $\widetilde{SN}_i$ and $\widetilde{SP}_i$.

$$\widetilde{nsp}_i = \frac{\widetilde{sp}_i}{\max_i\left(\kappa\left(\widetilde{sp}_i\right)\right)}, \tag{20}$$

and

$$\widetilde{nsn}_i = 1 - \frac{\widetilde{sn}_i}{\max_i\left(\kappa\left(\widetilde{sn}_i\right)\right)}. \tag{21}$$

Step 5. Calculate the fuzzy appraisal score for all alternatives. Then this score must be deffuzified and finally highest appraisal score is the best choice among the candidate alternatives:

$$\widetilde{as}_i = \frac{\widetilde{nsp}_i \oplus \widetilde{nsn}_i}{2}. \tag{22}$$

Step 6. Rank the alternatives according to the decreasing values of defuzzified appraisal scores. In other words, the alternative with the highest appraisal score is the best choice among the candidate alternatives.

2.2. The proposed IVIF EDAS

In the fuzzy set theory, the membership of an element to a fuzzy set is a single value between zero and one. However, the degree of non-membership of an element "$v_{\tilde{A}}(x)$" in a fuzzy set may not be equal to "$1-\mu_{\tilde{A}}(x)$" since there may be some hesitation degree. Therefore, a generalization of fuzzy sets was proposed by Atanassov (1986) as intuitionistic fuzzy sets (IFS) which incorporate the degree of hesitation, which is defined as "$1-\mu_{\tilde{A}}(x)-v_{\tilde{A}}(x)$".

Let $X \neq \emptyset$ be a given set. An intuitionistic fuzzy set in X is an object A given by

$$\tilde{A} = \left\{x, \mu_{\tilde{A}}(x), v_{\tilde{A}}(x); x \in X\right\}, \tag{23}$$

where $\mu_{\tilde{A}} : X \to [0,1]$ and $v_{\tilde{A}} : X \to [0,1]$ satisfy the condition:

$$0 \le \mu_{\tilde{A}}(x) + v_{\tilde{A}}(x) \le 1, \tag{24}$$

for every $x \in X$. Hesitancy is equal to "$1-\left(\mu_{\tilde{A}}(x)+v_{\tilde{A}}(x)\right)$".

The definition of interval-valued intuitionistic fuzzy sets (IVIFS) is given as follows. Let $D \subseteq [0,1]$ be the set of all closed subintervals of the interval and X be a universe of discourse. An interval-valued intuitionistic fuzzy set in $\tilde{A}$ over X is an object having the form:

$$\tilde{A} = \left\{< x, \tilde{\mu}_A(x), \tilde{v}_A(x) > \middle| x \in X\right\}, \tag{25}$$

where $\tilde{\mu}_{\tilde{A}} \to D \subseteq [0,1]$, $\tilde{v}_{\tilde{A}}(x) \to D \subseteq [0,1]$ with the condition $0 \le \sup\tilde{\mu}_{\tilde{A}}(x) + \sup\tilde{v}_{\tilde{A}}(x) \le 1, \quad \forall x \in X.$

The intervals $\tilde{\mu}_{\tilde{A}}(x)$ and $\tilde{v}_{\tilde{A}}(x)$ denote the membership function and the non-membership function of the element x to the set A, respectively. Thus, for each $x \in X$, $\tilde{\mu}_{\tilde{A}}(x)$ and $\tilde{v}_{\tilde{A}}(x)$ are closed intervals and their lower and upper end points are denoted by $\tilde{\mu}_{\widetilde{AL}}(x)$, $\tilde{\mu}_{\widetilde{AU}}(x)$, $\tilde{v}_{\widetilde{AL}}(x)$, and $\tilde{v}_{\widetilde{AU}}(x)$, respectively. Interval-valued intuitionistic fuzzy set $\tilde{A}$ is then denoted by

$$\tilde{A} = \left\{< x, \left[\tilde{\mu}_{\widetilde{AL}}(x), \tilde{\mu}_{\widetilde{AU}}(x)\right], [\tilde{v}_{\widetilde{AL}}(x), \tilde{v}_{\widetilde{AU}}(x)] > \middle| x \in X\right\}, \tag{26}$$

where

$$0 \le \tilde{\mu}_{\widetilde{AU}}(x) + \tilde{v}_{\widetilde{AU}}(x) \le 1, \tilde{\mu}_{\widetilde{AL}}(x) \ge 0, \tilde{v}_{\widetilde{AL}}(x) \ge 0.$$

For each element x, we can compute the unknown degree (hesitancy degree) of an interval-valued intuitionistic fuzzy interval of $x \in X$ in $\tilde{A}$ defined as follows:

$$\pi_{\tilde{A}(x)} = 1 - \tilde{\mu}_{\tilde{A}}(x) - \tilde{v}_{\tilde{A}}(x) = \left(\left[1-\tilde{\mu}_{\widetilde{AU}}(x)-\tilde{v}_{\widetilde{AU}}(x)\right], \left[1-\tilde{\mu}_{\widetilde{AL}}(x)-\tilde{v}_{\widetilde{AL}}(x)\right]\right). \tag{27}$$

For convenience, let

$$\tilde{\mu}_{\tilde{A}}(x) = \left[\mu^-, \mu^+\right], \tilde{v}_{\tilde{A}}(x) = \left[v^-, v^+\right],$$

so $\tilde{A} = \left(\left[\mu^-, \mu^+\right], \left[v^-, v^+\right]\right)$. Figure 2 illustrates an interval-valued intuitionistic fuzzy set.

Some arithmetic operations with interval-valued intuitionistic fuzzy sets and $\lambda \ge 0$ are given in the following: Let $\tilde{I}_1 = \left(\left[\mu_1^-, \mu_1^+\right], \left[v_1^-, v_1^+\right]\right)$ and $\tilde{I}_2 = \left(\left[\mu_2^-, \mu_2^+\right], \left[v_2^-, v_2^+\right]\right)$ be two interval-valued intuitionistic fuzzy sets. Then,

$$\tilde{I}_1 \oplus \tilde{I}_2 = \left(\left[\mu_1^- + \mu_2^- - \mu_1^-\mu_2^-, \mu_1^+ + \mu_2^+ - \mu_1^+\mu_2^+\right], \left[v_1^-v_2^-, v_1^+v_2^+\right]\right); \tag{28}$$

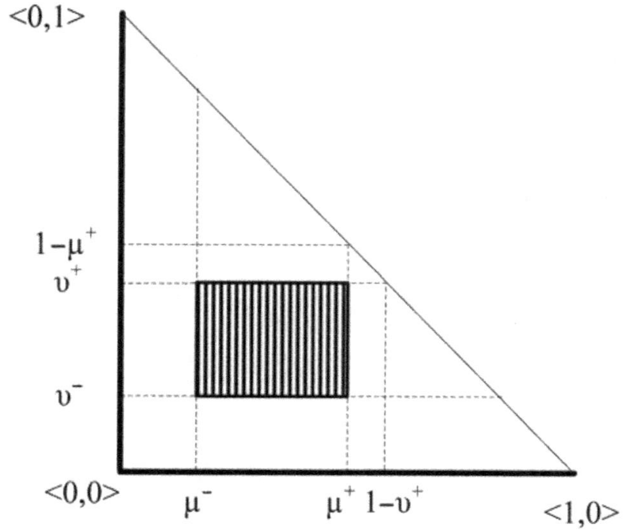

Fig. 2. Interval-valued intuitionistic fuzzy set

$$\tilde{I}_1 \otimes \tilde{I}_2 = \left(\left[\mu_1^-\mu_2^-, \mu_1^+\mu_2^+\right], \left[v_1^- + v_2^- - v_1^-v_2^-, v_1^+ + v_2^+ - v_1^+v_2^+\right]\right); \quad (29)$$

$$\tilde{I}_1 \ominus \tilde{I}_2 = \begin{cases} \left(\left[\dfrac{\mu_1^- - \mu_2^-}{1-\mu_2^-}, \dfrac{\mu_1^+ - \mu_2^+}{1-\mu_2^+}\right], \left[\dfrac{v_1^-}{v_2^-}, \dfrac{v_1^+}{v_2^+}\right]\right) \\ \left(\left[0,0\right],\left[1,1\right]\right), \text{ otherwise} \end{cases} \quad (30)$$

Eq. (28) is valid if $\mu_1^- \geq \mu_2^-$, $\mu_1^+ \geq \mu_2^+, v_1^- \leq v_2^-$, $v_1^+ \leq v_2^+$ and $v_2^- > 0, v_2^+ > 0$ and $v_1^-\left(1-\mu_2^-\right) \leq v_2^-\left(1-\mu_1^-\right)$, $v_1^+\left(1-\mu_2^+\right) \leq v_2^+\left(1-\mu_1^+\right)$.

$$\tilde{I}_1 \oslash \tilde{I}_2 = \begin{cases} \left(\left[\dfrac{\mu_1^-}{\mu_2^-}, \dfrac{\mu_1^+}{\mu_2^+}\right], \left[\dfrac{v_1^- - v_2^-}{1-v_2^-}, \dfrac{v_1^+ - v_2^+}{1-v_2^+}\right]\right) \\ \left(\left[0,0\right],\left[1,1\right]\right), \text{ otherwise} \end{cases} \quad (31)$$

Eq. (29) is valid if $\mu_1^- \leq \mu_2^-$, $\mu_1^+ \leq \mu_2^+, v_1^- \geq v_2^-$, $v_1^+ \geq v_2^+$ and $\mu_2^- > 0, \mu_2^+ > 0$ and $\mu_1^-\left(1-v_2^-\right) \leq \mu_2^-\left(1-v_1^-\right)$, $\mu_1^+\left(1-v_2^+\right) \leq \mu_2^+\left(1-v_1^+\right)$.

Multiplication by a fixed value, λ ≥ 0, is given by Eq. (32).

$$\lambda \tilde{I}_1 = \left(\left[1-\left(1-\mu_1^-\right)^\lambda, 1-\left(1-\mu_1^+\right)^\lambda\right], \left[\left(v_1^-\right)^\lambda, \left(v_1^+\right)^\lambda\right]\right). \quad (32)$$

The necessary definitions used in the proposed methodology are stated as follows:

Definition 3. Let $\tilde{\alpha}_j = \left(\left[\mu_j^-, \mu_j^+\right], \left[v_j^-, v_j^+\right]\right)$ $(j=1,2,\ldots,n)$ be a collection of interval-valued intuitionistic fuzzy numbers and let IIFWA: $Q^n \to Q$, if

$$IIFWA_w\left(\tilde{\alpha}_1, \tilde{\alpha}_2, \ldots, \tilde{\alpha}_n\right) = w_1\tilde{\alpha}_1 \oplus w_2\tilde{\alpha}_2 \oplus \ldots \oplus w_n\tilde{\alpha}_n. \quad (33)$$

Then, IIFWA is called an interval-valued intuitionistic fuzzy weighted averaging (IIFWA) operator, where Q is the set of all IVIFNs, $w = \left(w_1, w_2, \ldots, w_n\right)$ is the weight vector of the IVIFNs $\tilde{\alpha}_j\left(j=1,2,\ldots,n\right)$, and $w_j > 0$, $\sum_{j=1}^n w_j = 1$. The IIFWA operator can be further transformed in to the following form:

$$IIFWA_w\left(\tilde{\alpha}_1, \tilde{\alpha}_2, \ldots, \tilde{\alpha}_n\right) = \left(\left[1-\prod_{j=1}^n\left(1-\mu_j^-\right)^{w_i}, 1-\prod_{j=1}^n\left(1-\mu_j^+\right)^{w_i}\right], \left[\prod_{j=1}^n\left(v_j^-\right)^{w_i}, \prod_{j=1}^n\left(v_j^+\right)^{w_i}\right]\right). \quad (34)$$

Especially if $w = \left(\frac{1}{n}, \frac{1}{n}, \ldots, \frac{1}{n}\right)$, then the IIFWA operator reduces to an interval-valued intuitionistic fuzzy averaging (IIFA) operator, where

$$IIFA\left(\tilde{\alpha}_1, \tilde{\alpha}_2, \ldots, \tilde{\alpha}_n\right) = \frac{1}{n}\left(\tilde{\alpha}_1 \oplus \tilde{\alpha}_2 \oplus \ldots \oplus \tilde{\alpha}_n\right) =$$
$$\left(\left[1-\prod_{j=1}^n\left(1-\mu_j^-\right)^{1/n}, 1-\prod_{j=1}^n\left(1-\mu_j^+\right)^{1/n}\right], \left[\prod_{j=1}^n\left(v_j^-\right)^{1/n}, \prod_{j=1}^n\left(v_j^+\right)^{1/n}\right]\right). \quad (35)$$

In the following, the steps of the proposed interval-valued intuitionistic fuzzy EDAS method are given:

Step 1. Choose the most relevant attributes which describe decision alternatives for specific decision problem.

Step 2. If x_{ij} is the performance rating of ith alternative $A_1, A_2, \ldots, A_n$, $(i=1,2,\ldots,n)$ respecting to the jth criterion $C_1, C_2, \ldots, C_m$ $(j=1,2,\ldots,m)$, So, to form the interval decision matrix X and weight of each criterion, following table and variables should be considered;

$$X = [x_{ij}]_{n \times m} = \begin{bmatrix} \tilde{x}_{11} & \tilde{x}_{12} & \cdots & \tilde{x}_{1m} \\ \tilde{x}_{21} & \tilde{x}_{22} & \cdots & \tilde{x}_{2m} \\ \vdots & \vdots & & \vdots \\ \tilde{x}_{n1} & \tilde{x}_{n2} & \cdots & \tilde{x}_{nm} \end{bmatrix} =$$
$$\begin{bmatrix} \left(\left[\mu_{11}^-, \mu_{11}^+\right], \left[v_{11}^-, v_{11}^+\right]\right) & \left(\left[\mu_{12}^-, \mu_{12}^+\right], \left[v_{12}^-, v_{12}^+\right]\right) \ldots & \left(\left[\mu_{1m}^-, \mu_{1m}^+\right], \left[v_{1m}^-, v_{1m}^+\right]\right) \\ \left(\left[\mu_{21}^-, \mu_{21}^+\right], \left[v_{21}^-, v_{21}^+\right]\right) & \left(\left[\mu_{22}^-, \mu_{22}^+\right], \left[v_{22}^-, v_{22}^+\right]\right) \ldots & \left(\left[\mu_{2m}^-, \mu_{2m}^+\right], \left[v_{2m}^-, v_{2m}^+\right]\right) \\ \vdots & \vdots & \vdots \\ \left(\left[\mu_{n1}^-, \mu_{n1}^+\right], \left[v_{n1}^-, v_{n1}^+\right]\right) & \left(\left[\mu_{n2}^-, \mu_{n2}^+\right], \left[v_{n2}^-, v_{n2}^+\right]\right) \ldots & \left(\left[\mu_{nm}^-, \mu_{nm}^+\right], \left[v_{nm}^-, v_{nm}^+\right]\right) \end{bmatrix}, \quad (36)$$

where

$$\tilde{W}=(\tilde{w}_1,\tilde{w}_2,....,\tilde{w}_m)=\Big(\big([\mu_{w1}^-,\mu_{w1}^+],[v_{w1}^-,v_{w1}^+]\big),\big([\mu_{w2}^-,\mu_{w2}^+],[v_{w2}^-,v_{w2}^+]\big),...,\big([\mu_{wm}^-,\mu_{wm}^+],[v_{wm}^-,v_{wm}^+]\big)\Big)$$

and $\tilde{w}_j$ is the weight of criterion j; $(i=1,2,....,n)$ and $(j=1,2,...,m)$.

Step 3. According to the definition of EDAS method, the average solution with respect to all criteria must be determined as shown in Eq. (37):

$$\tilde{AV}_j=\frac{\left(\left[\sum_{i=1}^{n}\mu_{ij}^- - \sum_{\substack{i=1\\ i\neq k}}^{n}\mu_{ij}^-\times\mu_{kj}^-,\sum_{i=1}^{n}\mu_{ij}^+ - \sum_{\substack{i=1\\ i\neq k}}^{n}\mu_{ij}^+\times\mu_{kj}^+\right],\left[\prod_{i=1}^{n}v_{ij}^-,\prod_{i=1}^{n}v_{ij}^+\right]\right)}{n},$$

$$j=1,2,\ldots,m, \tag{37}$$

Or

$$\tilde{AV}_j=\left(\left[\left(1-\sum_{i=1}^{n}\mu_{ij}^- + \sum_{\substack{i=1\\ i\neq k}}^{n}\mu_{ij}^-\times\mu_{kj}^-\right)^{1/n},\left(1-\sum_{i=1}^{n}\mu_{ij}^+ + \sum_{\substack{i=1\\ i\neq k}}^{n}\mu_{ij}^+\times\mu_{kj}^+\right)^{1/n}\right],\left[\left(\prod_{i=1}^{n}v_{ij}^-\right)^{1/n},\left(\prod_{i=1}^{n}v_{ij}^+\right)^{1/n}\right]\right). \tag{38}$$

Step 4. The positive distance from average (*PDA*) and the negative distance from average (*NDA*) matrixes need to be calculated in this step according to lower and upper values of matrix as below:

$$\tilde{PDA}_{ij}=\frac{\max(0,\left([\mu_{ij}^-,\mu_{ij}^+],[v_{ij}^-,v_{ij}^+]\right)-\tilde{AV}_j)}{\kappa\left(\tilde{AV}_j\right)}, \tag{39}$$

and

$$\tilde{NDA}_{ij}=\frac{\max(0,\tilde{AV}_j-\left([\mu_{ij}^-,\mu_{ij}^+],[v_{ij}^-,v_{ij}^+]\right))}{\kappa\left(\tilde{AV}_j\right)}. \tag{40}$$

In this way $\tilde{PDA}_{ij}$ and $\tilde{NDA}_{ij}$ represent the positive and negative distance of ith alternative from average solution in terms of jth criterion for the lower level of decision matrix, respectively.

We propose the following ranking method for determining the maximum term in Eqs (39) and (40). Let $\tilde{\alpha}=([a,b],[c,d])$ be an interval-valued intutionistic fuzzy number. The following score function is proposed for defuzzifying $\tilde{\alpha}$:

$$I(\tilde{\alpha})=\frac{a+b+(1-c)+(1-d)+a\times b-\sqrt{(1-c)\times(1-d)}}{4}. \tag{41}$$

In Eq. (41), the terms (1–c) and (1–d) convert non-membership degrees to membership degrees while the term $\sqrt{(1-c)\times(1-d)}$ decreases the defuzzified value.

Step 5. Obtain weighted summation of the positive and negative distances from average matrix:

$$\tilde{SP}_i=\left(\left[\sum_{j=1}^{m}\mu_{w_j}^-\times\mu_{PDA_{ij}}^-,\sum_{j=1}^{m}\mu_{w_j}^+\times\mu_{PDA_{ij}}^+\right],\left[\sum_{j=1}^{m}v_{w_j}^- - \sum_{\substack{j=1\\ j\neq k}}^{m}v_{w_j}^-\times v_{PDA_{kj}}^-,\sum_{j=1}^{m}v_{w_j}^+ - \sum_{\substack{j=1\\ j\neq k}}^{m}v_{w_j}^+\times v_{PDA_{kj}}^+\right]\right),\ i=1,2,\ldots,n, \tag{42}$$

and

$$\tilde{SN}_i=\left(\left[\sum_{j=1}^{m}\mu_{w_j}^-\times\mu_{NDA_{ij}}^-,\sum_{j=1}^{m}\mu_{w_j}^+\times\mu_{NDA_{ij}}^+\right],\left[\sum_{j=1}^{m}v_{w_j}^- - \sum_{\substack{j=1\\ j\neq k}}^{m}v_{w_j}^-\times v_{NDA_{kj}}^-,\sum_{j=1}^{m}v_{w_j}^+ - \sum_{\substack{j=1\\ j\neq k}}^{m}v_{w_j}^+\times v_{NDA_{kj}}^+\right]\right),$$

$$i=1,2,\ldots,n. \tag{43}$$

Step 6. Identify the normalized values of $\tilde{SP}_i$ and $\tilde{SN}_i$ for all alternatives. A defuzzification process is needed to select the maximum $\tilde{SP}_i$ and $\tilde{SN}_i$. The defuzzification equation given in Step 4 is reused for this aim. Based on the division operation between IVIFNs, intuitionistic $\tilde{NSP}_i^{intu}$ and $\tilde{NSN}_i^{intu}$ are obtained as follows:

$$\tilde{NSP}_i^{intu}=\left(\left[\frac{\sum_{j=1}^{m}\mu_{w_j}^-\times\mu_{PDA_{ij}}^-}{\mathrm{Max}\left(\sum_{j=1}^{m}\mu_{w_j}^+\times\mu_{PDA_{ij}}^+\right)},\frac{\sum_{j=1}^{m}\mu_{w_j}^+\times\mu_{PDA_{ij}}^+}{\mathrm{Max}\left(\sum_{j=1}^{m}\mu_{w_j}^+\times\mu_{PDA_{ij}}^+\right)}\right],\left[\frac{\sum_{j=1}^{m}v_{w_j}^- - \sum_{\substack{j=1\\ j\neq k}}^{m}v_{w_j}^-\times v_{PDA_{kj}}^- - \mathrm{Max}\left(\sum_{j=1}^{m}v_{w_j}^- - \sum_{\substack{j=1\\ j\neq k}}^{m}v_{w_j}^-\times v_{PDA_{kj}}^-\right)}{1-\mathrm{Max}\left(\sum_{j=1}^{m}v_{w_j}^- - \sum_{\substack{j=1\\ j\neq k}}^{m}v_{w_j}^-\times v_{PDA_{kj}}^-\right)},\frac{\sum_{j=1}^{m}v_{w_j}^+ - \sum_{\substack{j=1\\ j\neq k}}^{m}v_{w_j}^+\times v_{PDA_{kj}}^+ - \mathrm{Max}\left(\sum_{j=1}^{m}v_{w_j}^+ - \sum_{\substack{j=1\\ j\neq k}}^{m}v_{w_j}^+\times v_{PDA_{kj}}^+\right)}{1-\mathrm{Max}\left(\sum_{j=1}^{m}v_{w_j}^+ - \sum_{\substack{j=1\\ j\neq k}}^{m}v_{w_j}^+\times v_{PDA_{kj}}^+\right)}\right]\right) \tag{44}$$

and

$$\tilde{NSN}_i^{intu} = 1 - \left(\begin{bmatrix} \dfrac{\sum_{j=1}^{m} \mu_{w_j}^{-} \times \mu_{NDA_{ij}}^{-}}{\text{Max}\left(\sum_{j=1}^{m} \mu_{w_j}^{+} \times \mu_{NDA_{ij}}^{+} \right)}, \dfrac{\sum_{j=1}^{m} \mu_{w_j}^{+} \times \mu_{NDA_{ij}}^{+}}{\text{Max}\left(\sum_{j=1}^{m} \mu_{w_j}^{+} \times \mu_{NDA_{ij}}^{+} \right)} \end{bmatrix}, \begin{bmatrix} \dfrac{\sum_{j=1}^{m} v_{w_j}^{-} - \sum_{\substack{j=1 \\ j \neq k}}^{m} v_{w_j}^{-} \times v_{NDA_{kj}}^{-} - \text{Max}\left(\sum_{j=1}^{m} v_{w_j}^{-} - \sum_{\substack{j=1 \\ j \neq k}}^{m} v_{w_j}^{-} \times v_{NDA_{kj}}^{-} \right)}{1 - \text{Max}\left(\sum_{j=1}^{m} v_{w_j}^{-} - \sum_{\substack{j=1 \\ j \neq k}}^{m} v_{w_j}^{-} \times v_{NDA_{kj}}^{-} \right)}, \\ \dfrac{\sum_{j=1}^{m} v_{w_j}^{+} - \sum_{\substack{j=1 \\ j \neq k}}^{m} v_{w_j}^{+} \times v_{NDA_{kj}}^{+} - \text{Max}\left(\sum_{j=1}^{m} v_{w_j}^{+} - \sum_{\substack{j=1 \\ j \neq k}}^{m} v_{w_j}^{+} \times v_{NDA_{kj}}^{+} \right)}{1 - \text{Max}\left(\sum_{j=1}^{m} v_{w_j}^{+} - \sum_{\substack{j=1 \\ j \neq k}}^{m} v_{w_j}^{+} \times v_{NDA_{kj}}^{+} \right)} \end{bmatrix} \right) \quad (45)$$

Step 7. Detect the intuitionistic appraisal score $\tilde{AS}^{intu}$ for all alternatives, shown as follows:

$$\tilde{AS}^{intu} = \frac{1}{2}(\tilde{NSP}_i^{intu} + \tilde{NSN}_i^{intu}), \quad (46)$$

where $0 \leq \tilde{AS}^{intu} \leq 1$.

Step 8. Rank the alternatives according to the decreasing values of appraisal score (AS^{intu}). The alternative with the highest AS^{intu} is the best choice among the candidate alternatives.

3. Application

Various environmental protection problems have been solved by using multi criteria decision making techniques (Turskis *et al.* 2012; Zavadskas *et al.* 2009; Khan, Samadder 2015). Solid waste disposal site selection is also an important environmental problem that we have to use many tangible, intangible, and conflicting criteria in its solution.

In a multicriteria solid waste disposal site selection problem, there are three alternatives and three criteria which are water pollution (W), distance to residential areas (D), and slope (S). The evaluation scales for crisp case, ordinary fuzzy case, and intuitionistic fuzzy case are given in Table 1.

3.1. Application of Crisp EDAS

The crisp decision matrix is given as in Table 2. All criteria are equally weighted.

Table 1. Evaluation scale

Linguistic term	Crisp scale	Ordinary fuzzy scale	Intuitionistic fuzzy scale
AH	100	(0.80, 0.90, 1, 1)	([0.85, 1.00], [0, 0])
VH	85	(0.70, 0.80, 0.80, 1)	([0.75, 0.95], [0, 0.05])
H	70	(0.50, 0.60, 0.70, 0.80)	([0.60, 0.80], [0.05, 0.20])
M	50	(0.40, 0.50, 0.50, 0.60)	([0.40, 0.60], [0.15, 0.40])
L	30	(0.20, 0.30, 0.40, 0.50)	([0.05, 0.20], [0.60, 0.80])
VL	20	(0.10, 0.20, 0.20, 0.30)	([0, 0.05], [0.75, 0.95])
AL	10	(0, 0, 0.10, 0.20)	([0, 0], [0.85, 1.00])

Table 2. Crisp Decision Matrix

Alternatives	W	D	S
SWDS-1	95	65	77
SWDS-2	80	85	74
SWDS-3	86	72	80
Average	87	74	77

Using Eqs (3) and (4), crisp PDA and NDA values are obtained as in Table 3.

Table 3. PDA and NDA values

PDA $j = 1$	PDA $j = 2$	PDA $j = 3$	NDA $j = 1$	NDA $j = 2$	NDA $j = 3$
0.092	0.000	0.000	0.000	0.000	0.000
0.000	0.149	0.000	0.080	0.002	0.039
0.000	0.000	0.039	0.011	0.000	0.000

Using Eqs (5), (6), (7), and (8), SP_i and NSP_i values are obtained as in Table 4.

Table 4. SP_i and NSP_i values

SP_i	SN_i	NSP_i	NSN_i
0.031	0.000	0.619	1.000
0.050	0.040	0.999	0.000
0.013	0.004	0.262	0.910

Based on the values in Table 4, ranking of the alternatives are obtained as in Table 5.

Table 5. ASi values

Alternative	$AS_i = (NSP_i + NSN_i) / 2$	Rank
SWDS-1	0.809	1
SWDS-2	0.499	3
SWDS-3	0.584	2

SWDS-1 is the best of all.

3.2. Application of Ordinary Fuzzy EDAS

In this case, the decision matrix in Table 6 is filled in by using the linguistic scale given in Table 1.

Table 6. Linguistic Decision Matrix

Alternatives	W	D	S
SWDS -1	AH	H	VH
SWDS -2	VH	VH	H
SWDS -3	VH	H	VH

The fuzzy average for each criterion is calculated as follows:

$$\widetilde{AV}_W = (0.69, 0.85, 0.85, 0.97);$$

$$\widetilde{AV}_D = (0.61, 0.79, 0.79, 0.91);$$

$$\widetilde{AV}_S = (0.62, 0.79, 0.79, 0.94).$$

The fuzzy positive and negative distances from average are found as follows:

For SWDS-1,

$\widetilde{PDA}_{1W} = (-0.20947,\ 0.07031, 0.21092,\ 0.536983)$, $\widetilde{PDA}_{1D} = (0,\ 0,\ 0,\ 0)$, $\widetilde{PDA}_{1S} = (-0.32594,\ 0.016159, 0.016159, 0.78243)$,

For SWDS-2,

$\widetilde{PDA}_{2W} = (-0.33269,\ -0.07031, -0.07031, 0.536983)$, $\widetilde{PDA}_{2D} = (0,\ 0,\ 0,\ 0)$, $\widetilde{PDA}_{2S} = (-0.32594,\ 0.016159, 0.016159,\ 0.78243)$,

For SWDS-3,

$\widetilde{PDA}_{3W} = (-0.33269,\ -0.07031, -0.07031, 0.536983)$, $\widetilde{PDA}_{3D} = (0,\ 0,\ 0,\ 0)$, $\widetilde{PDA}_{3S} = (-0.32594,\ 0.016159, 0.016159, 0.78243)$,

For SWDS-1,

$\widetilde{NDA}_{1W} = (0,0,0,0)$, $\widetilde{NDA}_{1D} = (-0.24675,\ 0.116883, 0.246753, 0.532468)$, $\widetilde{NDA}_{1S} = (0,0,0,0)$,

For SWDS-2,

$\widetilde{NDA}_{2W} = (-0.37052,\ 0.059761, 0.059761, 0.322709)$, $\widetilde{PDA}_{2D} = (-0.24675,\ 0.116883,\ 0.246753,\ 0.532468)$, $\widetilde{PDA}_{2S} = (0,\ 0,\ 0,\ 0)$,

For SWDS-3,

$\widetilde{NDA}_{3W} = (-0.37052,\ 0.059761, 0.059761, 0.322709)$, $\widetilde{PDA}_{3D} = (-0.24675,\ 0.116883,\ 0.246753,\ 0.532468)$, $\widetilde{PDA}_{3S} = (0,\ 0,\ 0,\ 0)$,

The fuzzy weighted summations of the positive distance and negative distances from average matrix using equal criteria weights are as follows:

$$\widetilde{SP}_1 = (-0.17669, 0.028534, 0.074937, 0.435406);$$

$$\widetilde{SN}_1 = (-0.08143, 0.038571, 0.081429, 0.175714);$$

$$\widetilde{SP}_2 = (-0.21735, -0.01787, -0.01787, 0.435406);$$

$$\widetilde{SN}_2 = (-0.2037, 0.058293, 0.10115, 0.282208);$$

$$\widetilde{SP}_3 = (-0.21735, -0.01787, -0.01787, 0.435406);$$

$$\widetilde{SN}_3 = (-0.2037, 0.058293, 0.10115, 0.282208).$$

The fuzzy normalized values of $\widetilde{SN}_i$ and $\widetilde{SP}_i$ are calculated as follows:

$$\widetilde{NSP}_1 = (-3.39836, 0.548819, 1.441324, 8.374566);$$

$$\widetilde{NSN}_1 = (-2.37647, -0.56471, 0.258824, 2.564706);$$

$$\widetilde{NSP}_2 = (-4.18045, -0.34369, -0.34369, 8.374566);$$

$$\widetilde{NSN}_2 = (-4.42283, -0.94366, -0.12013, 4.914225);$$

$$\widetilde{NSP}_3 = (-4.18045, -0.34369, -0.34369, 8.374566);$$

$$\widetilde{NSN}_3 = (-4.42283, -0.94366, -0.12013, 4.914225).$$

The defuzzified appraisal score (as) for each alternative is calculated as follows:

$$defuzz(\widetilde{as}_1) = 0.974093;$$

$$defuzz(\widetilde{as}_2) = 0.615543;$$

$$defuzz(\widetilde{as}_3) = 0.615543.$$

The ranking of alternatives is as follows: 1 > 2 = 3. The best alternative is the same as when the crisp EDAS was used. However, the other alternatives are equivalent when ordinary fuzzy EDAS is used. This is caused because of the vagueness in the linguistic evaluations.

3.3. Application of Interval-Valued Intuitionistic Fuzzy EDAS

In this case, the decision matrix is filled in by using the linguistic scale given in Table 1 as in Table 7.

Table 7. Intuitionistic Decision Matrix

Alternatives	*W*	*D*	*S*
SWDS-1	([0.85,1.00], [0, 0])	([0.60, 0.80], [0.05, 0.20])	([0.75, 0.95], [0, 0.05])
SWDS-2	([0.75, 0.95], [0, 0.05])	([0.75, 0.95], [0, 0.05])	([0.60, 0.80], [0.05, 0.20])
SWDS-3	([0.75, 0.95], [0, 0.05])	([0.60, 0.80], [0.05, 0.20])	([0.75, 0.95], [0, 0.05])

The average values with respect to all criteria are determined as in Table 8:

Table 8. Average values with respect to criteria

W	([0.787703, 0.966383],[0, 0])
D	([0.67679, 0.848093], [0, 0.125992])
S	([0.713022, 0.897317], [0, 0.07937])

PDA and NDA values of Alternatives 1, 2, and 3 are calculated as in Table 9.

Table 9. PDA and NDA values of alternatives

$\widetilde{PDA}_{1W}$ = ([−0.13248, 0.242425],[0, 0])	$\widetilde{NDA}_{1W}$ = ([−0.24243, 0.132479], [0, 0])
$\widetilde{PDA}_{1D}$ = ([−0.32665, 0.162226],[−0.10006, 0.263332])	$\widetilde{NDA}_{1D}$ = ([−0.16223, 0.326654], [−0.26333, 0.100056])
$\widetilde{PDA}_{1S}$ = ([−0.1835, 0.295182], [−0.09886, 0.06228])	$\widetilde{NDA}_{1S}$ = ([−0.29518, 0.1835], [−0.06228, 0.098864])
$\widetilde{PDA}_{2W}$ = ([−0.24631, 0.18551],[0, 0.056915])	$\widetilde{NDA}_{2W}$ = ([−0.18551, 0.24631], [−0.05692, 0])
$\widetilde{PDA}_{2D}$ = ([−0.12915, 0.359725],[−0.16589, 0.065833])	$\widetilde{NDA}_{2D}$ = ([−0.35973, 0.129155], [−0.06583, 0.165889])
$\widetilde{PDA}_{2S}$ = ([−0.37034, 0.108341], [−0.03658, 0.249122])	$\widetilde{NDA}_{2S}$ = ([−0.10834, 0.370341], [−0.24912, 0.036584])
$\widetilde{PDA}_{3W}$ = ([−0.24631, 0.18551],[0, 0.056915])	$\widetilde{NDA}_{3W}$ = ([−0.18551, 0.24631], [−0.05692, 0])
$\widetilde{PDA}_{3D}$ = ([−0.32665, 0.162226],[−0.10006, 0.263332])	$\widetilde{NDA}_{3D}$ = ([−0.16223, 0.326654], [−0.26333, 0.100056])
$\widetilde{PDA}_{3S}$ = ([−0.1835, 0.295182], [−0.09886, 0.06228])	$\widetilde{NDA}_{3S}$ = ([−0.29518, 0.1835], [−0.06228, 0.098864])

SP, NSP, and the defuzzified $\widetilde{as}$ values are given below.

$$\widetilde{SP}_1 = \left(\left[-0.21421,\ 0.233278\right],\left[-0.06631, 0.108538\right]\right);$$

$$\widetilde{SN}_1 = \left(\left[-0.23328, 0.214211\right],\left[-0.10854, 0.066307\right]\right);$$

$$\widetilde{NSP}_1 = \left(\left[-0.82574, 0.899241\right],\left[-0.2556,\ 0.418392\right]\right);$$

$$\widetilde{NSN}_1 = \left(\left[-0.86566, 0.794904\right],\left[-0.40277, 0.246054\right]\right);$$

$$defuzz\left(\widetilde{as}_1\right) = 0.493021;$$

$$\widetilde{SP}_2 = \left(\left[-0.2486,\ 0.217859\right],\left[-0.06749, 0.123957\right]\right);$$

$$\widetilde{SN}_2 = \left(\left[-0.21786, 0.248602\right],\left[-0.12396, 0.067491\right]\right);$$

$$\widetilde{NSP}_2 = \left(\left[-0.91305, 0.800139\right],\left[-0.24788,\ 0.455261\right]\right);$$

$$\widetilde{NSN}_2 = \left(\left[-0.77849, 0.888345\right],\left[-0.44294, 0.241169\right]\right);$$

$$defuzz\left(\widetilde{as}_2\right) = 0.443874;$$

$$\widetilde{SP}_3 = \left(\left[-0.25215,\ 0.214306\right],\left[-0.06631, 0.127509\right]\right);$$

$$\widetilde{SN}_3 = \left(\left[-0.21431, 0.252155\right],\left[-0.12751, 0.066307\right]\right);$$

$$\widetilde{NSP}_3 = \left(\left[-0.97201, 0.826109\right],\left[-0.2556,\ 0.491524\right]\right);$$

$$\widetilde{NSN}_3 = \left(\left[-0.79526, 0.935707\right],\left[-0.47317, 0.246054\right]\right);$$

$$defuzz\left(\widetilde{as}_3\right) = 0.430828.$$

According to the obtained results, the ranking of the alternatives is 1 > 2 > 3. The best alternative is again SWDS-1. However, the second best alternative is SWDS-2 with intuitionistic fuzzy EDAS where as it is 3 with crisp EDAS. Alternatives SWDS-2 and SWDS-3 are equivalent with ordinary fuzzy EDAS. These differences come from the structures of the data, which are exact in crisp case, trapezoidal in ordinary fuzzy case, and interval-valued in intuitionistic fuzzy case. A decision maker should select the proper approach with respect to the data he/she has.

3.4. Sensitivity analysis

The weights of the alternatives have been significantly changed as it can be seen in Table 10. However, the ranking has not changed. SWDS-1 is the best alternative in all cases. Alternative 2 is the second best alternative in all cases.

Table 10. Sensitivity analysis by changing criteria weights

Weights	Rankings of alternatives
$w_w = 0.3,\ w_D = 0.35,\ w_S = 0.35$	1 > 2 > 3 (AS_1 = 0.495, AS_2 = 0.449, AS_3 = 0.438)
$w_w = 0.4,\ w_D = 0.4,\ w_S = 0.2$	1 > 2 > 3 (AS_1 = 0.489, AS_2 = 0.433, AS_3 = 0.416)
$w_w = 0.1,\ w_D = 0.3,\ w_S = 0.6$	1 > 2 > 3 (AS_1 = 0.499, AS_2 = 0.483, AS_3 = 0.480)
$w_w = 0.6,\ w_D = 0.2,\ w_S = 0.2$	1 > 2 > 3 (AS_1 = 0.472, AS_2 = 0.400, AS_3 = 0.371)
$w_w = 0.1,\ w_D = 0.8,\ w_S = 0.1$	1 > 2 > 3 (AS_1 = 0.499, AS_2 = 0.483, AS_3 = 0.480)
$w_w = 0.4,\ w_D = 0.2,\ w_S = 0.4$	1 > 2 > 3 (AS_1 = 0.489, AS_2 = 0.433, AS_3 = 0.416)

Table 10 indicates that the ranking result obtained by IVIF EDAS is very robust. From slight changes to extreme changes in criteria weights, SWDS-1 is selected without any doubt. Smaller weights of the criterion "water pollution (W)" cause the scores of alternatives 2 and 3 get closer.

We also checked the effect of a change in linguistic evaluations to the final decision. For this aim, the linguistic evaluations of SWDS-1 have been changed from AH, H, VH to AH, H, H. This change produces the following defuzzified appraisal scores (as): $defuzz\left(\widetilde{as}_1\right) = 0.356$, $defuzz\left(\widetilde{as}_1\right) = 0.444$, and $defuzz\left(\widetilde{as}_1\right) = 0.432$. SWDS-2 becomes the best alternative in this case.

Conclusions

EDAS method needs fewer computations with respect to most of the other multiattribute decision-making methods while it can produce the same ranking of alternatives. The evaluation of alternatives in this method is based on distances of each alternative from the average solution with respect to each criterion. Ordinary fuzzy extension of crisp EDAS has been developed based on trapezoidal fuzzy numbers. We have proposed the interval-valued intuitionistic fuzzy EDAS (IVIF EDAS) method in this paper. The linguistic evaluations have been represented by interval-valued intuitionistic fuzzy sets and their arithmetic operations have been applied. Since the vagueness included by trapezoidal fuzzy sets and interval-valued intuitionistic fuzzy sets are basically different, the obtained rankings may change from ordinary fuzzy EDAS (OF EDAS) to intuitionistic fuzzy EDAS. The calculations of IVIF EDAS are more tedious with respect to ordinary fuzzy EDAS. Different ranking equations must be used in OF EDAS and IVIF EDAS.

For further research, the other extensions of fuzzy sets such as type-2 fuzzy sets, hesitant fuzzy sets, or neutrosophic sets may be used to develop the other versions of fuzzy EDAS.

References

Atanassov, K. T. 1986. Intuitionistic fuzzy sets, *Fuzzy Sets and Systems* 20: 87–96. https://doi.org/10.1016/S0165-0114(86)80034-3

Atanassov, K. T.; Gargov, G. 1989. Interval valued intuitionistic fuzzy sets, *Fuzzy sets and systems* 31(3): 343–349. https://doi.org/10.1016/0165-0114(89)90205-4

Behzadian, M.; Otaghsara, S. K.; Yazdani, M.; Ignatius, J. 2012. A state-of the-art survey of TOPSIS applications, *Expert Systems with Applications* 39(17): 13051–13069. https://doi.org/10.1016/j.eswa.2012.05.056

Büyüközkan, G.; Çifçi, G. 2012. A novel hybrid MCDM approach based on fuzzy DEMATEL, fuzzy ANP and fuzzy TOPSIS to evaluate green suppliers, *Expert Systems with Applications* 39(3): 3000–3011. https://doi.org/10.1016/j.eswa.2011.08.162

Cevik Onar, S.; Oztaysi, B.; Otay, İ; Kahraman, C. 2015. Multi-expert wind energy technology selection using interval-valued intuitionistic fuzzy sets, *Energy* 90: 274–285. https://doi.org/10.1016/j.energy.2015.06.086

Chakraborty, S.; Zavadskas, E. K. 2014. Applications of WASPAS method in manufacturing decision making, *Informatica* 25(1): 1–20. https://doi.org/10.15388/Informatica.2014.01

Corrente, S.; Greco, S.; Słowiński, R. 2013. Multiple criteria hierarchy process with ELECTRE and PROMETHEE, *Omega* 41(5): 820–846. https://doi.org/10.1016/j.omega.2012.10.009

Deng, X.; Hu, Y.; Deng, Y.; Mahadevan, S. 2014. Supplier selection using AHP methodology extended by D numbers, *Expert Systems with Applications* 41(1): 156–167. https://doi.org/10.1016/j.eswa.2013.07.018

Figueira, J. R.; Greco, S.; Roy, B.; Słowiński, R. 2013. An overview of ELECTRE methods and their recent extensions, *Journal of Multi-Criteria Decision Analysis* 20(1–2): 61–85. https://doi.org/10.1002/mcda.1482

Garibaldi, J. M.; Ozen, T. 2007. Uncertain fuzzy reasoning: a case study in modelling expert decision making, *IEEE Transactions on Fuzzy Systems* 15: 16–30. https://doi.org/10.1109/TFUZZ.2006.889755

Grattan-Guiness, I. 1975. Fuzzy membership mapped onto interval and many-valued quantities, *Zeitschrift fuer Mathematische Logik und Grundlagen der Mathematik*, 22.

Hashemkhani Zolfani, S. H.; Sedaghat, M.; Zavadskas, E. K. 2012. Performance evaluating of rural ICT centers (telecenters), applying fuzzy AHP, SAW-G and TOPSIS Grey, a case study in Iran, *Technological and Economic Development of Economy* 18(2): 364–387. https://doi.org/10.3846/20294913.2012.685110

Jahn, K. U. 1975. Intervall-wertige Mengen, *Mathematische Nachrichten* 68: 115–132. https://doi.org/10.1002/mana.19750680109

Kahraman, C.; Cevik Onar, S.; Oztaysi, B. 2015. Fuzzy multicriteria decision-making: a literature review, *International Journal of Computational Intelligence Systems* 8(4): 637–666. https://doi.org/10.1080/18756891.2015.1046325

Kahraman, C.; Oztaysi, B.; Cevik Onar, S. 2016. A comprehensive literature review of 50 years of fuzzy set theory, *International Journal of Computational Intelligence Systems* 9(1): 3–24. https://doi.org/10.1080/18756891.2016.1180817

Keshavarz Ghorabaee, M.; Zavadskas, E. K.; Olfat, L.; Turskis, Z. 2015. Multi-criteria inventory classification using a new method of Evaluation Based on Distance from Average Solution (EDAS), *Informatica* 26(3): 435–451. https://doi.org/10.15388/Informatica.2015.57

Keshavarz Ghorabaee, K.; Zavadskas, E. K.; Amiri, M.; Turskis, Z. 2016. Extended EDAS method for fuzzy multi-criteria decision-making: an application to supplier selection, *International Journal of Computers Communications & Control* 11(3): 358–371. https://doi.org/10.15837/ijccc.2016.3.2557

Khan, D.; Samadder, S. R. 2015. A simplified multi-criteria evaluation model for landfill site ranking and selection based on AHP and GIS, *Journal of Environmental Engineering and Landscape* 23(4): 267–278. https://doi.org/10.3846/16486897.2015.1056741

Kouchaksaraei, R. H.; Hashemkhani Zolfani, S. H.; Golabchi, M. 2015. Glasshouse locating based on SWARA-COPRAS approach, *International Journal of Strategic Property Management* 19(2): 111–122. https://doi.org/10.3846/1648715X.2015.1004565

Liu, P. D.; Jin, F.; Zhang, X.; Su, Y.; Wang, M. 2011. Research on the multi-attribute decision-making under risk with interval probability based on prospect theory and the uncertain linguistic variables, *Knowledge-Based Systems* 24(4): 554–561. https://doi.org/10.1016/j.knosys.2011.01.010

Liu, P. D.; He, L.; Yu, X. C. 2016. Generalized hybrid aggregation operators based on the 2-dimension uncertain linguistic information for multiple attribute group decision making, *Group Decision and Negotiation* 25(1): 103–126. https://doi.org/10.1007/s10726-015-9434-x

Liu, P. D.; Jin, F. 2012. A multi-attribute group decision-making method based on weighted geometric aggregation operators of interval-valued trapezoidal fuzzy numbers, *Applied Mathematical Modelling* 36(6): 2498–2509. https://doi.org/10.1016/j.apm.2011.09.006

Liu, P. D.; Liu, Y. 2014. An approach to multiple attribute group decision making based on intuitionistic trapezoidal fuzzy power generalized aggregation operator, *International Journal of Computational Intelligence Systems* 7(2): 291–304. https://doi.org/10.1080/18756891.2013.862357

Liu, P. D.; Shi, L. L. 2015. The generalized hybrid weighted average operator based on interval neutrosophic hesitant set and its application to multiple attribute decision making, *Neural Computing and Applications* 26(2): 457–471. https://doi.org/10.1007/s00521-014-1736-4

Liu, P. D.; Wang, Y. M. 2014. Multiple attribute decision-making method based on single valued neutrosophic normalized weighted Bonferroni mean, *Neural Computing and Applications* 25(7–8): 2001–2010. https://doi.org/10.1007/s00521-014-1688-8

Liu, P. D.; Yu, X. C. 2014. 2-dimension uncertain linguistic power generalized weighted aggregation operator and its application for multiple attribute group decision making, *Knowledge-Based Systems* 57(1): 69–80. https://doi.org/10.1016/j.knosys.2013.12.009

Liu, P. D.; Zhang, X.; Jin, F. 2012. A multi-attribute group decision-making method based on interval-valued trapezoidal fuzzy numbers hybrid harmonic averaging operators, *Journal of Intelligent & Fuzzy Systems* 23(5):159–168.

Mardani, A.; Zavadskas, E. K.; Govindan, K.; Amat Senin, A.; Jusoh, A. 2016. VIKOR technique: a systematic review of the state of the art literature on methodologies and applications, *Sustainability* 8(1): 37. https://doi.org/10.3390/su8010037

Oztaysi, B.; Cevik Onar, S.; Goztepe, K.; Kahraman, C. 2015. Evaluation of research proposals for grant funding using interval-valued intuitionistic fuzzy sets, *Soft Computing* 1–16. https://doi.org/10.1007/s00500-015-1853-8

Sambuc, R. 1975. *Fonctions ϕ-floues. Application l'aide au diagnostic en pathologie thyroidienne*: PhD Thesis. Université Marseille, France.

Seçme, N. Y.; Bayrakdaroğlu, A.; Kahraman, C. 2009. Fuzzy performance evaluation in Turkish banking sector using analytic hierarchy process and TOPSIS, *Expert Systems with Applications* 36(9): 11699–11709. https://doi.org/10.1016/j.eswa.2009.03.013

Smarandache, F. 1998. *Neutrosophy: neutrosophic probability, set, and logic.* Rehoboth, USA: American Research Press.

Stanujkic, D.; Magdalinovic, N.; Milanovic, D.; Magdalinovic, S.; Popovic, G. 2014. An efficient and simple multiple criteria model for a grinding circuit selection based on moora method, *Informatica* 25(1): 73–93. https://doi.org/10.15388/Informatica.2014.05

Torra, V. 2010. Hesitant fuzzy sets, *International Journal of Intelligent Systems* 25: 529–539. https://doi.org/10.1002/int.20418

Turanoglu Bekar, E.; Cakmakci, M.; Kahraman, C. 2016. Fuzzy COPRAS method for performance measurement in total productive maintenance: a comparative analysis, *Journal of Business Economics and Management* 17(5): 663–684. https://doi.org/10.3846/16111699.2016.1202314

Turskis, Z.; Lazauskas, M.; Zavadskas, E. K. 2012. Fuzzy multiple criteria assessment of construction site alternatives for non-hazardous waste incineration plant in Vilnius city, applying ARAS-F and AHP methods, *Journal of Environmental Engineering and Landscape Management* 20(2): 110–120. https://doi.org/10.3846/16486897.2011.645827

Turskis, Z.; Zavadskas, E. K.; Kutut, V. 2013. A model based on ARAS-G and AHP methods for multiple criteria prioritizing of heritage value, *International Journal of Information Technology & Decision Making* 12(1): 45–73. https://doi.org/10.1142/S021962201350003X

Van Horenbeek, A.; Pintelon, L. 2014. Development of a maintenance performance measurement framework – using the analytic network process (ANP) for maintenance performance indicator selection, *Omega* 42(1): 33–46. https://doi.org/10.1016/j.omega.2013.02.006

Yager, R. R. 1986. On the theory of bags, *International Journal of General Systems* 13: 23–37. https://doi.org/10.1080/03081078608934952

Yazdani, M. 2015. New intuitionistic fuzzy approach with multi-objective optimisation on the basis of ratio analysis method, *International Journal of Business and Systems Research* 9(4): 355–374. https://doi.org/10.1504/IJBSR.2015.072584

Yazdani, M.; Payam, A. F. 2015. A comparative study on material selection of microelectromechanical systems electrostatic actuators using Ashby, VIKOR and TOPSIS, *Materials & Design* 65: 328–334. https://doi.org/10.1016/j.matdes.2014.09.004

Yeh, T. M.; Huang, Y. L. 2014. Factors in determining wind farm location: Integrating GQM, fuzzy DEMATEL, and ANP, *Renewable Energy* 66: 159–169. https://doi.org/10.1016/j.renene.2013.12.003

Zadeh, L. A. 1965. Fuzzy sets, *Information and Control* 8: 338–353. https://doi.org/10.1016/S0019-9958(65)90241-X

Zadeh, L. A. 1975. The concept of a linguistic variable and its application to approximate reasoning – 1, *Information Sciences* 8: 199–249. https://doi.org/10.1016/0020-0255(75)90036-5

Zavadskas, E. K.; Kaklauskas, A.; Turskis, Z.; Kalibatas, D. 2009. An approach to multi-attribute assessment of indoor environment before and after refurbishment of dwellings, *Journal of Environmental Engineering and Landscape* 17(1): 5–11. https://doi.org/10.3846/1648-6897.2009.17.5-11

Cengiz KAHRAMAN is a full Professor at Industrial Engineering Department of Istanbul Technical University (ITU). His research areas are engineering economics, quality management, statistical decision making, multiple criteria decision making, and fuzzy decision making. He published about 200 journal papers, about 160 conference papers and 70 book chapters. He is in the editorial boards of 20 international journals and edited 10 international books from Springer. He guest-edited many special issues of international journals and organized international conferences. He was the vice dean of ITU Management Faculty between the years 2004–2007 and the head of ITU Industrial Engineering Department between the years 2010–2013.

Mehdi KESHAVARZ GHORABAEE received the BS degree in electrical engineering from the University of Guilan, Rasht, Iran in 2010 and the MS degree in production management from the Allame Tabataba'i University, Tehran, Iran in 2013. He is currently working toward the PhD degree in management (Operations Research) at Allame Tabataba'i University. He has published some papers in leading international journals. His research interests include multi-criteria decision making (MCDM), multi-objective evolutionary algorithms, genetic algorithm, fuzzy MCDM, inventory control, supply chain management, scheduling and reliability engineering.

Edmundas Kazimieras ZAVADSKAS is a full professor and the Head of the Department of Construction Technology and Management of Vilnius Gediminas Technical University, Lithuania. Senior Research Fellow at the Research Institute of Smart Building Technologies. PhD in Building Structures (1973). Dr Sc. (1987) in Building Technology and Management. A member of Lithuanian and several foreign Academies of Sciences. Doctore Honoris Causa from Poznan, Saint Petersburg and Kiev universities. The Honorary International Chair Professor in the National Taipei University of Technology. A member of international organizations; a member of steering and programme committees at many international conferences; a member of the editorial boards of several research journals; the author and co-author of more than 400 papers and a number of monographs in Lithuanian, English, German and Russian. Editor-in-chief of journals *Technological and Economic Development of Economy* and *Journal of Civil Engineering and Management*. Research interests: building technology and management, decision-making theory, automation in design and decision support systems.

Sezi CEVIK ONAR is a full time Associate Professor at Industrial Engineering Department of Istanbul Technical University (ITU). Her research interests include multiple criteria decision making, supply chain management and fuzzy systems. She has published 15 articles, 35 conference papers and 8 book chapters. She has edited a book on Engineering Management which is published by Springer. Dr Cevik Onar teaches courses on human resources management, business planning, management and organization.

Morteza YAZDANI is a PhD student in business and economics in Faculty of Business and Communication at Universidad Europea de Madrid, Spain. His main research area is the application of multi criteria decision making in supply chain management, sustainability and strategic planning. He has published papers in some international journals such as *International Journal of Logistics Research*, *International Journal of Strategic Decision Science*, *Expert Systems with Applications*, *Engineering Economics*, *Journal of Civil Engineering and Management*, *Technological and Economic Development of Economy* and *Materials & Design*.

Basar OZTAYSI is a full time Associate Professor at Industrial Engineering Department of Istanbul Technical University (ITU). His research interests include multiple criteria decision making, data mining and intelligent systems. He has published 25 articles, 32 conference papers and 10 book chapters. He has edited a book on Supply Chain Management which is published by Springer. Başar Öztayşi teaches courses on data management, information systems management and business intelligence and decision support systems.

STRUCTURAL FORMATION AND LEACHING BEHAVIOR OF MECHANICALLY ACTIVATED LIGNITE FLY ASH BASED GEOPOLYMER

Gábor MUCSI[a], Ágnes SZENCZI[a], Zoltán MOLNÁR[a], János LAKATOS[b]

[a]*Institute of Raw Material Preparation and Environmental Processing, University of Miskolc, 3515 Miskolc-Egyetemváros, Hungary*

[b]*Institute of Chemistry, University of Miskolc, 3515 Miskolc-Egyetemváros, Hungary*

Abstract. Geopolymers are inorganic polymers which can be produced by the reaction of aluminium silicate oxides and alkaline silicates in an alkaline medium. Nowadays, silicate bearing wastes, such as fly ash or slag and other by-products of power stations are very common raw materials for geopolymers.

In our research a lignite type fly ash was mechanically activated in a high energy density mill (HEM) for different retention time, and geopolymer specimens were produced from these materials. After determining the optimal fly ash fineness, the concentration and composition of the alkaline activator solution were also investigated. The highest compressive strength was reached using fly ash with 2 m^2/g specific surface area, activated with a mixture of Na-K silicate (water glass) and NaOH solution. The total dosage of alkaline activators to the fly ash was 40 m/m%, the NaOH activator of 12 M solution represented 25 m/m%, while Na-K silicate 75 m/m%. Then, the leaching properties of the main components and the mobility of toxic elements of the produced geopolymers were tested in distilled water, 1M acetic acid and 1M hydrochloric acid as well. The lowest mobility of the elements was obtained in most cases by the activator containing 100m/m% Na-K silicate solution (water glass) using the mechanically activated fly ash.

Keywords: fly ash, mechanical activation, geopolymer, leaching.

Introduction

Coal firing power plants generate fly ash and slag in huge amount worldwide, about 780 million tons/year (Heidrich *et al.* 2013). Due to the relatively low utilization rate, enormous quantities of these wastes are landfilled in ponds. The fly ash is the finest grained particle fraction of the solid ash which is transported with flue gas, and removed by electrostatic or bag filters. The fly ash has pozzolanic properties and pozzolanic activity, the latter itself does not react with water, but in the presence of calcium hydroxide solution hardens and solidifies, producing a practically insoluble in water reaction product (Opoczky 2001). The most important property of the fly ash regarding geopolymerization is its reactivity, which depends on the chemical and phase composition, the dispersity (specific surface area, particle size distribution), and the morphology of the fly ash particles, and determined by the physical properties and chemical composition of the fly ash.

Geopolymers concerning their chemical structure are polysialates, where the sialate (silicon-oxo-aluminate) network is built up from the bonds of Si-oxide and Al-oxide by sharing oxygen atoms (Davidovits 1988).

The main disadvantage of the fly ash in production of geopolymers is its low reactivity. One way of improving its reactivity is the mechanical activation (MA). According to Pietersen *et al.* (1989) and Brouwers and van Eijk (2002), the inner surface of the fly ash is more active than the outer shell, therefore, not only a larger surface area is obtained during the mechanical activation, but the inner surface with higher reactivity is liberated and will be available for the reaction, so the geopolymerization can be more efficient and faster.

The geopolymerization contains the following steps according to Hardjito and Rangan (2005):

- dissolving the solid aluminium-, silicon oxides and silicates in the alkaline activator solution (MOH wherein M: alkali metal, Na or K);
- diffusion or transport of dissolved Si and Al complexes from the particle surface to the inter-particle reaction space;
- finally, the polymerization of dissolved particle parts and added silicate solution providing the hardening of silicate structure.

Corresponding author: Gábor Mucsi
E-mail: ejtmucsi@uni-miskolc.hu

Arioz *et al.* (2012) activated an F-type fly ash with 12M NaOH solution and water glass in their experiments. The specimens were cured at 40, 80 and 120 °C for 6, 15 and 24 hours residence time. The broken specimen at 28 days was ground and the sample was tested by leaching using ICP-OES for the leached element analysis. According to the US EPA 1311 method the pH of the acetic acid used as a solvent was set to 4.93±0.05 before the leaching test. The ratio of liquid-to-solid material was 20:1. After 18 hours contact time the slurry was filtered and the liquid phase was measured by ICP-OES. It was concluded, that the As and Hg could be immobilized in the structure, while the other examined metal ions (Zn, Cr, Pb and Cd) were not.

Cheng *et al.* (2012) examined the leaching of different heavy metals (Pb^{2+}, Cu^{2+}, Cr^{3+} and Cd^{2+}) from the geopolymers. The geopolymer was produced as a mixture of metakaolin and alkaline solution, and then it was crushed down to the prescribed grain size for the leaching tests. It was established that the above mentioned heavy metals can be all immobilized in the structure of the geopolymer.

Izquierdo *et al.* (2009) compared the solubility of elements in geopolymer and in raw fly ash. They found that the fly ash based geopolymer is suitable for the immobilization of many trace elements, such as Be, Bi, Cd, Co, Cr, Cu, Nb, Ni, Pb, Sn, Th, U, Y, Zr and the rare earth elements. However, the leached amount of some elements was higher in geopolymers in aqueous solution as compared with raw fly ash. As the effect of strong alkaline condition after geopolymerisation caused by the unreacted alkali content, the elements due to their mobility can form As, B, Mo, Se, V, W oxianions. Their concentration was found to be increased. This means that the ash composition, an optimal composition of the activator solution, the ratio of activator solution and the circumstances of production are very important in order to achieve a stable final product exhibiting a low mobility of the elements.

It is well known that the elements mobility can depend on its oxidation state. According to Füleky (2011) the toxic lead in the water and in the soil normally can be found in the form of Pb^{2+} ions. In the case of oxidizing circumstances it can turn into insoluble Pb (IV)-oxide. It can occur as adsorbed layer on the surface of tiny particles or in hydroxide or basic carbonate form as a precipitated material due to its small ionic radius and cation adsorption capacity of soils. Therefore, the adsorbable amount of Pb by plants is low. Although Bokányi *et al.* (2011) experienced the outstanding sequestering capacity towards lead by many micro- and macro algae biosorbents. Its mobility is negligible in weak acidic circumstances. The opposite can be found in case of Cr(III)-Cr(VI) system where the higher oxidation state form is the mobile one (Lakatos 2006).

Palacios and Palomo (2004) and Palomo and Palacios (2003) studied the immobilization of lead in fly ash based geopolymer systems activated by sodium-hydroxide. It was established that the main reason of the effective encapsulation of lead is the formation of insoluble Pb_3SiO_5. The immobilization was more efficient if the activation state of the fly ash was higher.

According to Fernández-Jiménez *et al.* (2004, 2005) the As dissolution became higher in the case of geopolymer compared to the raw fly ash. Furthermore, authors stated that As occurs in Fe-rich regions of fly ash, which is in agreement with the previous observations, which claimed that arsenic could adsorb on the surface of Fe-hydroxide.

Chromium is in aqueous phase or as adsorbed on the surface of soil particles precipitated in the Cr^{3+} oxidation state. Only under highly acidic circumstances and in the presence of complex forming agents Cr(III) can become mobile. It cannot be leached by natural way from soil-groundwater system. Under strong oxidative circumstances chrome exists as Cr(VI) oxidation state, which is very toxic. In this form the chromium can be adsorbed on the surface of soil particles only in a small amount and could not precipitate, therefore the mobility of the Cr(VI) is much higher. Chromate ion is more oxidative under acidic circumstances, so easily oxidize organic materials, while the chromate transforms to Cr(III) state which is not toxic and has lower mobility, so it is more advantageous for the environment (Füleky 2011).

Deja (2002) examined the immobilization of Cr(VI) form in alkali activated slag based binders. It was concluded that in this structure the Cr(VI) is immobilized and the compressive strength of the product increased in case of carbonate and silicate activated systems. Reductive circumstances (by addition of free S^{2-} ions) effected positively the immobilization reaction due to the reduction of the Cr(VI) to Cr(III).

The fitoremediation of Cd is affected by the soil pH and the organic material content. Under acidic circumstances the adsorption of Cd in plants increased. In cadmium contaminated soils the lower pH resulted in the multiplying of Cd concentration in plants. Under alkaline or almost neutral conditions Cd can be adsorbed on soil particles, therefore with the addition of lime the absorption of Cd in plants will decrease. In case of high organic content the Cd mobility also decreases due to the low amount of the available Cd, since it has been fixed to the large organic molecules (e.g. humic materials) (Barati 2002).

Cadmium can be easily immobilized in Portland cement based binder materials due to the replacement of Ca, Ca/Cd forming silicate gel. Immobilization can also be achieved in alkaline silicate or carbonate activated slag systems, due to similar replacement mechanism (Deja 2002).

Zinc toxicity can appear only in acidic and organic material-rich soils in the case of high Zn concentration.

Sandy soils contain less mobile zinc, while solonetz soils contain more. The mobility of the zinc in the soil depends on the pH of the soil. Decreasing pH can result in the increase of the mobile zinc content (Barati 2002).

Copper is an essential element for the organisms of plants, animals and humans. Plants can extract copper from soil in Cu(II) ionic form. However, at higher concentration copper become toxic for living organism. Cu(II) ions exist in soil mainly in adsorbed state on soil particles or as precipitated hydroxide or basic carbonate form due to its relatively small ionic radius and high cation exchange capacity of soils. The mobility of the copper increases due to the acidification of the soils (Füleky 2011).

Temuujin *et al.* (2011) investigated the leachability of calcinated geopolymers. Calcination of the fly ash based geopolymers at 600°C resulted in a decrease of amorphous components from 63.4 to 61.6 wt.%. However, the solubility of the Al, Si and Fe ions in 14 M NaOH and 18% HCl after 5 days immersion decreased from 1.3 to 16-fold in comparison to as prepared geopolymer samples. Calcination of the geopolymer also resulted in a 30% reduction in compressive strength.

Several authors revealed that the mechanical activation of the geopolymer raw material (fly ash, granulated blast furnace slag) affected positively the properties of the final product (Kumar *et al.* 2007b; Temuujin *et al.* 2009; Mucsi *et al.* 2014).

Investigations of leachability show that mobility of a given element strongly depends on the structure which immobilizes it. Immobilization can be carried out in the structure with covalent bonds, or the fixed ions can be cohered only to dislocations or functional groups binding them by weaker forces. It is also important from the point of view of mobility of the fixed elements that the ionic forms of them are generated in accordance with the acid-base character of the leaching medium.

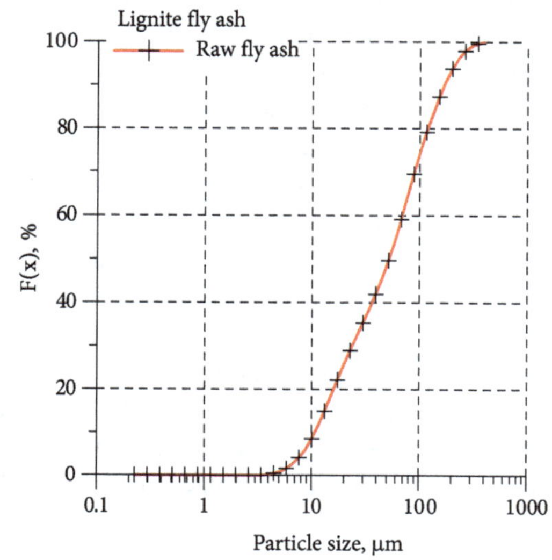

Fig. 1. Particle size distribution of the raw fly ash

Concerning the geopolymer production and application, the mobility of the main components of geopolymers (like Si, Al and Fe), as well as the toxic materials, is important.

The goal of the present work is to monitor the leaching characteristics of a mechanically activated fly ash based geopolymer under different conditions focusing on the main constituents and toxic elements. The main components mobility is essential concerning the network formation during geopolymerisation. Based on the toxic elements leaching the environmental compatibility of the obtained geopolymer could be evaluated. Further task of the research was to determine the optimal grinding fineness of the fly ash carried out by the mechanical activation from the point of view of compressive strength and specimen density.

1. Materials and methods

1.1. Materials

A lignite type fly ash was chosen for the investigation originated from Mátra Power Station, Visonta, Hungary. The determination of the particle density of the raw fly ash was carried out by pycnometer method in alcohol medium, the result of the raw fly ash was 1.88 kg/dm^3. The initial moisture content of the fly ash was determined by drying at 105 °C in a drying cabinet until constant mass reached, resulted in 4.28%.

The particle size distribution was measured by Horiba LA 950 V2 type laser particle size analyzer and specific surface area was calculated from these data with shape factor 1.

Particle size distribution of the raw fly ash can be seen in Figure 1. Characteristic particle sizes and specific surface area of the raw fly ash are as follows: x_{50} = 52.04 μm, x_{80} = 119.10 μm, SSA (cm^2/g) = 1500 cm^2/g.

In order to determine the chemical (oxidic) composition of the fly ash XRF analysis was carried out, the results are given in Table 1.

Main components presents in 95.2%, while other oxides in 4.8%. The SiO_2 content was relatively low 45.9%, while the Al_2O_3 and Fe_2O_3 contents were average, the lime (CaO) content was higher than a typical in case of the F type fly ash. Additionally, the important SO_3 and L.O.I. content were 3.8% and 2.3% respectively.

The main mineral phases of fly ash are quartz, anhydrite and albite. The detailed quantitative phase composition of the fly ash can be seen in Table 2.

1.2. Methods

1.2.1. Mechanical activation

The mechanical activation of the raw fly ash was carried out in a ceramic lined stirred media mill developed

Table 1. Chemical composition of the fly ash from (XRF analysis)

Main components	m/m%
SiO_2	45.9
Al_2O_3	16.8
CaO	13.0
MgO	2.9
Na_2O	0.50
K_2O	1.8
Fe_2O_3	12.1
L.O.I.	2.3
Other oxides	
SO_3	3.4
P_2O_5	0.28
TiO_2	0.48
SrO	0.05
ZrO_2	0.02
MnO	0.18
As_2O_3	0.03
CuO	0.02
ZnO	0.03

Table 2. Phase composition of raw fly ash

Mineral composition, m/m%	
Quartz	20.34
Maghemite	4.22
Hematite	3.91
Anhidrite	7.08
Albite	4.71
Albite K0.16	5.58
Lime	1.61
Amorphous	52.5

Table 3. Operating parameters of the stirred media mill

Parameter	Value
Diameter of the grinding balls, mm	1.10–1.20
Diameter of the rotor disc, mm	153
Inner diameter of the grinding chamber, mm	174
Length of the grinding chamber, mm	250
Volume of the grinding chamber, dm^3	4.00
Material density of the grinding balls, kg/m^3	3591
Bulk density of the grinding balls, kg/m^3	2160
Bulk volume of the grinding media, dm^3	2.40
Porosity of the grinding media, %	39.80
Bulk density of the fly ash, kg/m^3	850
Power of the motor, kW	5.50
Velocity of the rotor, m/s	5.00

by the Institute of Raw Material Preparation and Environmental Processing. The material of the liners and the seven stirrer discs were made of high wear resistant Al_2O_3 ceramic. Grinding media was ceramic grinding beads with the diameter of 1.1–1.2 mm. Table 3 contains the main operating parameters of the mill.

The residence time of mechanical activation in the stirred media mill was 1, 3, 5, 7 and 10 minutes in dry state. Prior to the grinding the raw fly ash was dried in a drying cabinet during overnight to decrease moisture content.

1.2.2. Particle size distribution

The particle size distribution and the calculated specific surface area (SSA) of the raw fly ash and its ground products were determined using a HORIBA LA-950V2 type laser particle size analyzer. From the measured data the computer calculated the particle size distribution according to the Mie-theory which takes into consideration the refractive index of the material. The measurement range of the analyzer is between 10 nm and 3 mm.

1.2.3. Uniaxial compressive strength

The uniaxial compressive strength test was performed by a hydraulic strength testing machine (250 kN force).

1.2.4. Fourier Transform Infrared Spectroscopy

The structural changes due to the geopolymerisation were detected by FT-IR method using a JASCO FT-IR 4200 type apparatus used in reflection mode equipped with diamond ATR type PRO470-H. Two spectra were made from each fly ash and geopolymer sample. Spectra are the average of 32 measurements with 4 cm^{-1} wavenumber resolution.

1.2.5. X-ray fluorescence

X-ray fluorescence measurements (XRF) were performed by a Rigaku Supermini 200 type XRF spectrometer. After determination of L.O.I. 1.00 g burned powder sample was mixed with 6.00 g lithium tetraborate ($Li_2B_4O_7$). The melted mixture was analyzed by fusion bead oxide method built in XRF software. L.O.I. values were taken into consideration. Every sample was analyzed three times then these values were averaged.

1.2.6. X-ray diffraction

The mineral composition of the fly ash was determined by a Bruker D8 Advance XRD powder diffractometer (Cu-Kα radiation, 40kV, 40mA) in Bragg-Brenanto geometry (with detector side graphite monochromator). X-ray diffraction patterns were recorded in 2–70°(2θ) range. Sample crystalline phases are detected by Bruker DiffracPlus software package in its EVA module ICDD PDF-2 (2005) database is employed for search/matching of phases. Quantitative evaluation was made by Rietveld fitting method on TOPAS3 software using FPM based instrument convolution.

1.2.7. Leaching tests

The aim of the leaching investigation was to determine the leachability of the main structural elements (Si, Al, Na) and the toxic components (Cr, Cu, Hg, etc.) under different pH values. Additionally the effect of retention time on leachability was studied. Since the geopolymer structure is changing for a relatively long time after its production, this change does not only effect the compressive strength, but also the immobilization properties. Therefore, two measurement series were carried out. The first series were tests after 10 days of polymerization, and the second one 4 months later. It is important to compare the results of the leaching tests with the prescribed concentration limits included in the Decree No. 20 of 2006 (IV. 5.) KvVM regarding the disposal of hazardous wastes in landfills. The experimental data from the leaching tests were compared with the prescribed concentration limit values related to the inert wastes.

It is important to note that the prescribed norm values are concerned to 10 L/kg liquid/solid ratio in distilled water as medium. However, in the case of the present investigation the liquid-to-solid ratio was 50:1 L/kg. The reason of this high ratio was to reduce the concentration of the acid consumer components (which remained in geopolymers from alkali activator solution) to keep the necessary acidity of the solution in order to dissolve the toxic components from geopolymers.

In the second measurement series both investigations were carried out using 10:1 and 50:1 liquid-to-solid ratio as well.

In both series distilled water, 1M acetic acid solution or 1M hydrochloric acid was used as leaching solution. However, in the case of geopolymers the distilled water extraction corresponds to a diluted alkali solution because of their remained alkali content. In the case of acids it corresponds to an extraction made by a system in which acid content was decreased by the quantity of the remaining alkali components.

The leachable element concentration from the raw material and from the formed geopolymers were determined by ICP-OES technique (Varian 720 ES horizontal torch). Multicomponent standard and the same matrix as the leaching solution were used for the calibration. The main component elements in the leachates were determined after an appropriate dilution with distilled water.

The leaching tests were performed on geopolymer particles in the size range of 1–2 mm. The contact time was five days.

1.3. Preparation of geopolymer

Two experimental series were performed in order to optimize the production circumstances of the geopolymer. The first one was carried out in order to determine the optimal fineness of the fly ash, the geopolymer specimens were prepared from fly ash activated in a stirred media mill for 1, 3, 5, 7 and 10 minutes residence time. Other optimal geopolymer production conditions, such as concentration of activator solution, compressing pressure, concentration of NaOH solution were determined previously.

During the second experimental series a certain part of the NaOH activator solution was replaced with water glass solution. Initially, Betol 39T (Na-silicate) then Betol SB (Na-K-silicate) type water glass was added to the mixture of the alkaline activator solution which consisted of 12M NaOH, as well as 12 M NaOH and water glass mixture. The water glass concentrations (replacement ratio of NaOH solution in the alkaline activator) varied in 25, 50, 75 and 100 m/m% in it.

During the preparation of geopolymer the first step was the mixing and homogenization of the alkaline activator solution and the mechanically activated fly ash. The geopolymer paste was mould into plastic moulds, followed by compression carried out in a TXM-V1-STC-07 consolidator with various pressures: 1.86, 3.73. 5.59, 7.46 and 9.33 bar. The residence time of the consolidation was 20 sec in each test.

The molded and compressed specimens were kept under sealed conditions and ambient temperature for 24 hours. Later the specimens were removed from the moulds and were heat cured at 90°C for 6 hours. Compressive strength tests were carried out at the age of 7 days. Chemical leaching tests were also carried out as written before.

2. Results and discussion

2.1. Mechanical activation

The variation of the specific surface area (SSA) of fly ash as function of specific grinding work can be seen in Figure 2. The different stages of the fine grinding process could be distinguished clearly. In the first, the so called Rittinger stage, until 3 min residence time, the increasing rate of the specific surface area was high. The relationship between specific grinding energy and specific surface area was linear. This was followed by the aggregation stage, between 3 and 7 min grinding time, where the specific surface area and surface energy become so high that particles could attach each other by van der Waals forces. The increasing rate then moderated, but specific surface area was still increasing during this stage. The final stage of fine grinding was the so called agglomeration stage, which started after 7 min residence time in our case, when particle surface energy was even much higher, that particles agglomerated to each other by stronger bonds (for example, covalent ones). The most significant effect of this phenomenon is the coarsening of the particle size distribution. The maximum specific surface area reached after 7 min grinding time was 2.5 m^2/g.

The compressive strength of the geopolymer specimens prepared from stirred media milled fly ash increased with the increasing fineness of fly ash until 3 min grinding time. After this stage the compressive strength values decreased (Fig. 3). The highest compressive strength was 4.14 MPa.

To detect the structural changes in the fly ash due to grinding FT-IR measurements were carried out (Fig. 4). The main peaks of the raw fly ash spectra were observed at 1100, 1020 (Assymetric stretching vibration of T-O-Si T = Si/Al), 797, 675, 610 and 594 cm^{-1} related to Si-O-Si symmetric stretching vibrations. Peak at 1455 cm^{-1} related to stretching vibrations of O-C-O bonds which correspond to the presence of carbonate.

Structural changes were observed in the fly ash structure due to mechanical activation. The spectra show differences below 1500 cm^{-1} wave number. The peak at 1100 cm^{-1} detected in the case of raw fly ash disappeared after grinding even for 3 min residence time. Peak at 1020 cm^{-1} shifted to 1036 cm^{-1} (related to T-O-Si, where T = Si/Al asymmetric stretching vibration) in the ground products spectra. However, the peaks at 797, 676 and 593 cm^{-1} did remain the same. The FT-IR spectra confirmed that structural changes took place due to the mechano-chemical effect of the grinding in stirred media mill.

2.2. Effect of alkali activator on mechanical and structural properties of geopolymer

During the laboratory investigations, beside NaOH, test specimens were prepared using two types of additives in the alkaline activator: Betol 39 T and Betol SB

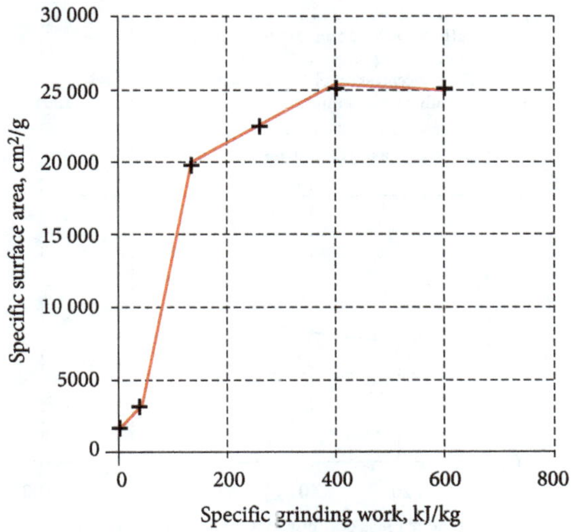

Fig. 2. Variation of specific surface area of fly ash as function of specific grinding work

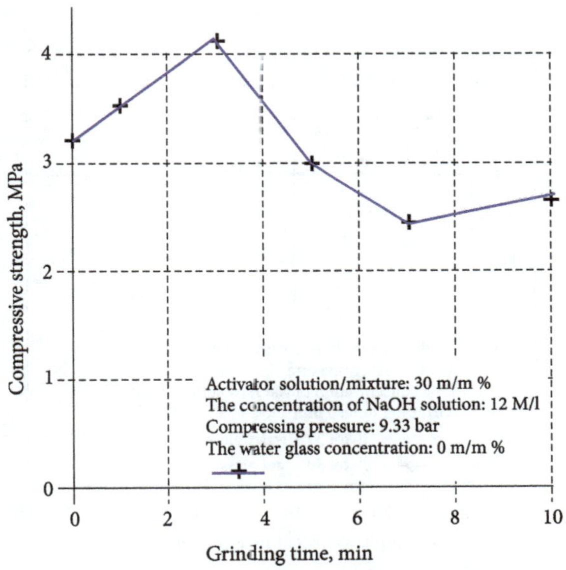

Fig. 3. Variation of compression strength as function of grinding time

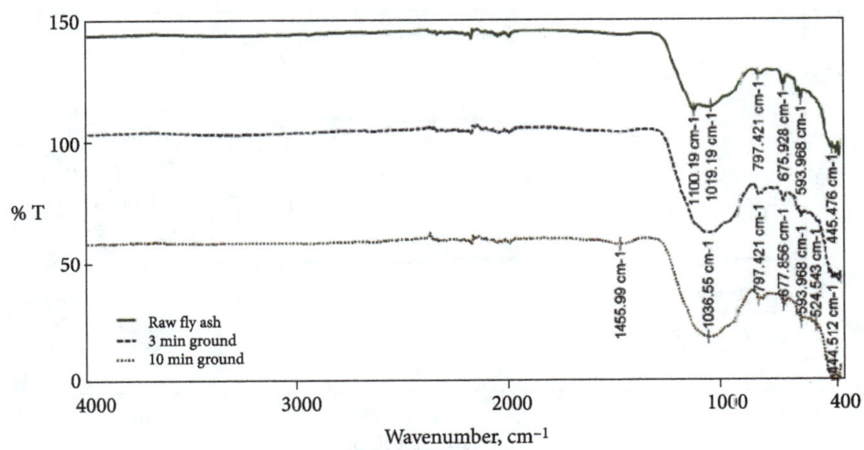

Fig. 4. FT-IR spectras of raw fly ash and mechanically activated fly ash for 3 and 10 min

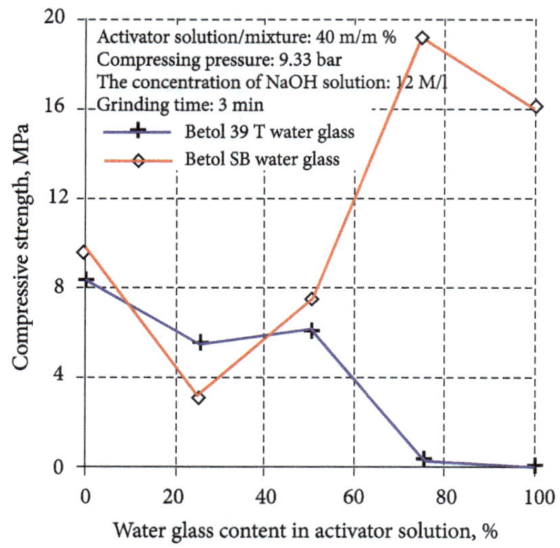

Fig. 5. Change of compression strength as function of water glass content in activator solution

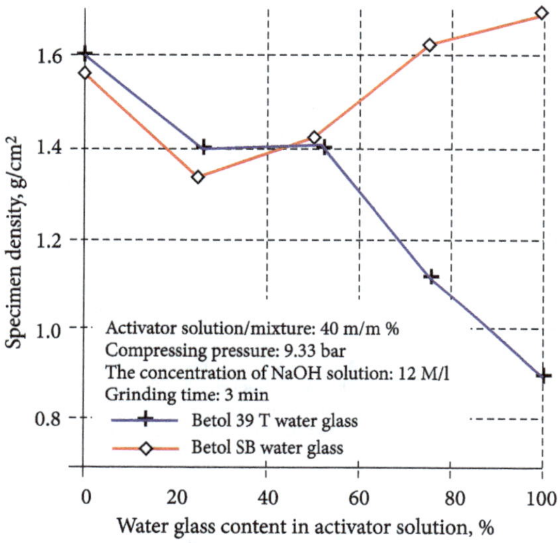

Fig. 6. Change of specimen density as function of water glass content in activator solution

(manufactured by Woellner, Austria) Na and Na-K water glass respectively. The optimal concentration of NaOH solution was determined in the previous stage of the systematic research (Research Report... 2014). It correlated well with the previous experience found in literature by other authors (i.e. Kumar *et al.* 2007a). They stated that by increasing NaOH concentration in activator solution the compressive strength of geopolymer increased. In the case of the Betol 39T water glass the fly ash particles conglomerated very fast, therefore workability of the paste reduced, a porous structure had been formed before an appropriate compaction.

The effect of different water glass types on the mechanical properties of geopolymer can be traced clearly in Figures 5 and 6. The compressive strength decreased significantly (except in case of 50 m/m % addition) when Betol 39 T was added to the alkaline activator solution (Fig. 5). The specimen density values were correlated well with the compressive strength (Fig. 6).

On the other hand, the Betol SB contains potassium in small amount beside sodium, which slows down the cross-linking of the molecules, therefore the workability of the geopolymer paste is better than that of the Betol 39 T in our case. The addition of Betol SB to the activator solution in 25 m/m% resulted in the decrease in compressive strength due to the not appropriate compaction, which fact is worth to be investigated later. However, further addition had a positive effect, the compressive strength increased dramatically up to 19 MPa, when Betol SB was added in 75m/m% as activator solution (Fig. 5). The highest specimen density value (1.69 kg/dm^3) belongs to the case, when 100m/m% Betol SB was used as activator solution (Fig. 6).

The FT-IR spectra of the geopolymer are shown in Figure 7. The peak at 1409 cm^{-1} appeared in the case of spectra belonging to the geopolymer produced without any addition of water glass (12 M NaOH only) and when

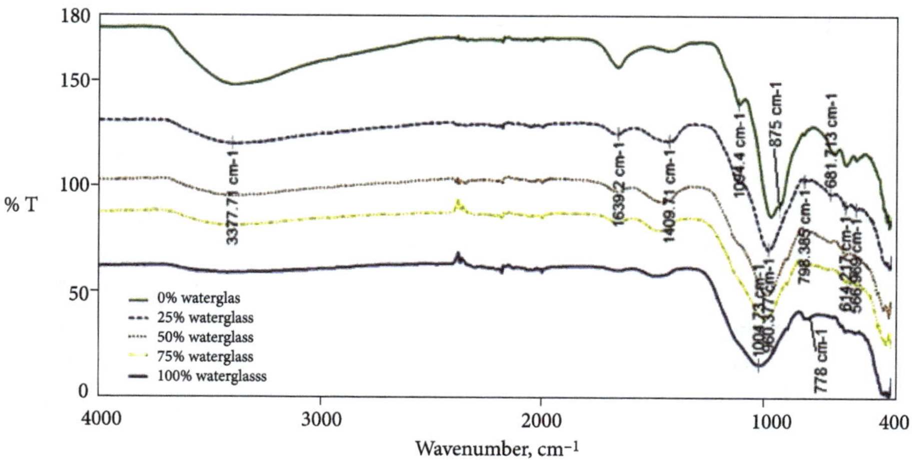

Fig. 7. FT-IR spectras of geopolymer prepared using water glass

water glass (Betol SB) was added in 25 m/m% to the activator solution it shifted to 1466 cm^{-1} (when only water glass was used). Peak (shoulder) at 875 cm^{-1} disappeared when water glass was used, while at 567 cm^{-1} (Si-O-Si and Al-O-Si symmetric stretching vibration) the addition of Betol SB resulted in a new peak. Spectrum corresponds to water glass addition in 25, 50 and 75 m/m % to activator solution shows minimal differences in the geopolymer structure. When only water glass was used peaks disappeared at 682 and 566 cm^{-1}, others shifted to other positions (the peak at 1409 cm^{-1} shifted to 1466 cm^{-1}, peak at 1093 shifted to 1004 cm^{-1}, and 798 to 778 cm^{-1}). The spectra presented in Figure 7 show structural differences between geopolymerisation performed using alkaline activator solution and the alkaline activator solution with water glass.

2.3. Leaching tests

2.3.1. Leaching of the main elements (Si, Al, Ca, Fe)

A certain part of the alkaline activator added during the formation of geopolymers remains in a leachable form after the geopolymerisation, therefore the initially planned aqueous extraction would not be an aqueous one, but approximately 0.01M NaOH extraction. This explains the reason of the mobilization of main elements of geopolymer and fly ash, such as Si, Al and Ca. They are presented in the form of alkali soluble minerals in contrast with iron. Therefore, during the leaching tests of geopolymers the mobilization of only those elements can be expected, which do not form precipitates (Si) or they are soluble in the form of hydroxo-complex(Al) due to the alkaline pH of the aqueous extraction.

The retention time of geopolymerisation before the leaching tests was 10 and 120 days after the preparation of geopolymer specimen. For the latter tests the results of the leaching of the raw material (fly ash) were also given. This data can be used for the comparison of the alteration of the elements in the initial raw material and the obtained geopolymer. Unfortunately, this comparison is restricted, since in case of raw fly ash the leaching with distilled water differs from the leaching with distilled water in geopolymer case due to the above mentioned residual leachable alkaline.

As the leaching with distilled water in the raw material (fly ash) case shown in the Figure 8, only a small dissolution of the main components could be expected. The larger amount of Si achieved indicated that a part of the Si remained unused in the cross-linking process, or located on the weakly bounded position which was then mobilized by the weak alkaline leachate (distilled water = 0.01M NaOH). Higher mobilization of the Si could be expected only by the breaking the structure which could occur when the geopolymers were treated with acids. The leached quantity of Si and other elements is demonstrated in Figures 8, 9 and 10.

The extraction results using distilled water represent only the quantities of the remaining Si which have not been built in the cross-linked network.

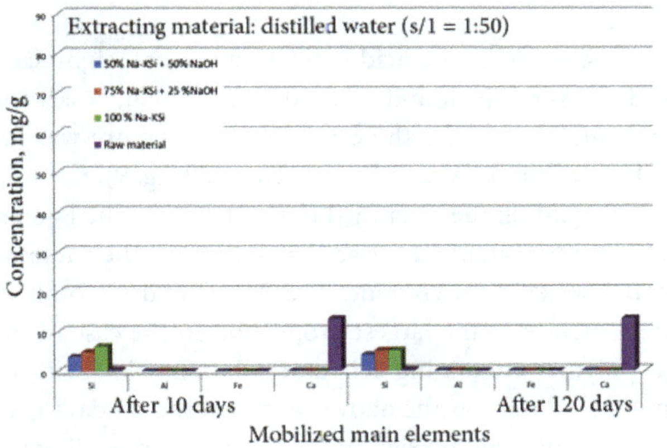

Fig. 8. Quantity of mobilized main elements in distilled water from 10 and 120 days age geopolymers formed using different alkaline activator solutions (solid/liquid ratio 1:50)

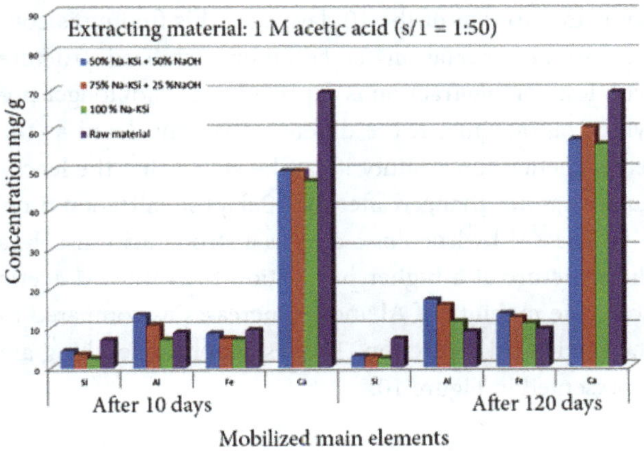

Fig. 9. Quantity of mobilized main elements from geopolymers (at 10 and 120 days age) using different alkaline activator solutions in 1 M acetic acid (solid/liquid ratio 1:50)

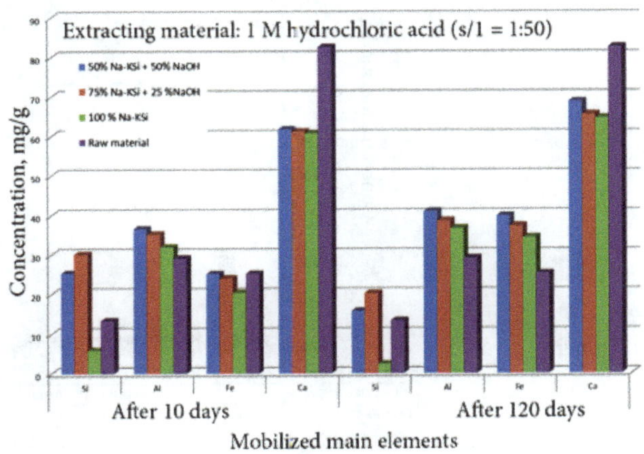

Fig. 10. Quantity of mobilized main elements from geopolymers (at 10 and 120 days age) using different alkaline activator solutions in 1 M hydrochloric acid (solid/liquid ratio 1:50)

In general, it can be observed that the increase of acidity results in a higher quantity of the soluble Si which originated from the decomposition of the structure.

The activator composition does have effect on the main component mobilization. The increasing the ratio of water glass results in a more stable and less leachable structure.

Using 1M acetic acid extracting solvent, significant acid excess can be found in the system. So that, beside the Ca-bearing minerals the cross-linked structure was attached and some Al and Fe was released (Fig. 9).

Regarding the examined elements the leached quantities shown slight decrease depending on the increase of the water glass content. The Al, Fe and Ca built in the structure in the largest proportion in the case of the geopolymer containing activator solutions Na-K silicate in 100%. Based on the above results of the 10 days geopolymer there are no significant differences regarding the different NaOH to water glass ratios based on the soluble components during the use of the 1 M acetic acid extracting solvent. However, after 4 months the soluble quantity of the examined elements was somewhat increased as compared to that of the 10 days age. This finding is contrary to the expectations, since a more stabilized structure with less soluble fraction is expected in case of longer polymerization time. If the difference is considered as the experimental uncertainty it can be stated that the leachability has not changed after the 10 days stabilization time.

The 1M hydrochloric acid as a strong acid dissolves the structure at a higher proportion than the 1M acetic acid. The mobility of Al and Fe increases as compared to the acetic acid extraction. The results of the leaching are represented in Figure 10.

The tendency of the leachability as a function of the water glass content was found being similar than that of the acetic acid leaching: the quantity of the leached components decreases.

2.3.2. Extraction of toxic elements

Regarding the environmental impacts of the geopolymer it is important to know the mobilizable toxic elements content. From this point of view, the comparison of the leached toxic element concentration to the limit values of the Decree No. 20 of 2006 (IV. 5.) KvVM will indicate the toxic or non toxic character of the geopolymers. According to the above Decree the aqueous extraction at L/S = 10 L/kg needs to be applied. It is important to note again that the soluble alkali remaining in the geopolymers turns the distilled water into a weak alkali solution, so the extraction with distilled water has to be interpreted as extraction with 0.01 M alkaline medium. Due to the presence of the alkali, minor mobility can be expected at those toxic elements which form hard-to-dissolve hydroxide precipitates. The dissolution of elements in the form of complex hydroxide may happen in this weak alkaline solution. The same situation can arise in case of oxo-anion forming elements. Regarding the measured data (see Fig. 11) the latter one can cause the environmental risk.

In the case of the examined toxic components it was found that the leached quantity of most of the elements remains under the limit values. However, arsenic, molybdenum and selenium exceeded the limit values concerning inert landfills determined in Decree No. 20 of 2006 (IV. 5.) KvVM. It can be explained as the effect mentioned above weak alkali media generated by the excess amount of alkali. Due to the unused alkali the appearance of elements

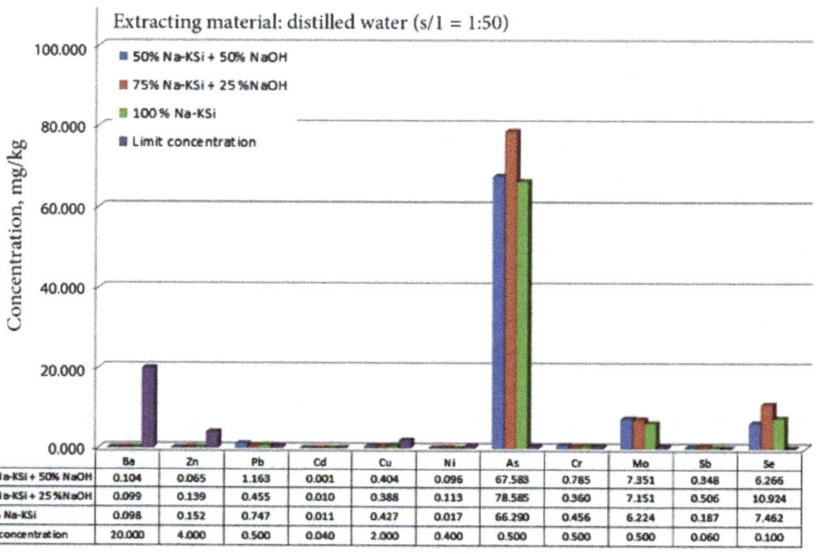

	Ba	Zn	Pb	Cd	Cu	Ni	As	Cr	Mo	Sb	Se
50% Na-KSi + 50% NaOH	0.104	0.065	1.163	0.001	0.404	0.096	67.583	0.785	7.351	0.348	6.266
75% Na-KSi + 25%NaOH	0.099	0.139	0.455	0.010	0.388	0.113	78.585	0.360	7.151	0.506	10.924
100% Na-KSi	0.098	0.152	0.747	0.011	0.427	0.017	66.290	0.456	6.224	0.187	7.462
Limit concentration	20.000	4.000	0.500	0.040	2.000	0.400	0.500	0.500	0.500	0.060	0.100

Mobilized toxic elements

Fig. 11. Quantity of mobilized toxic elements from 10 days age geopolymers using distilled water as extracting material (solid/liquid ratio 1:50)

forming oxo-anions soluble in alkali (As, Mo, Se) can be expected.

According to the Decree No. 20 of 2006 (IV. 5.) KvVM, the limit value of leachable lead is 0.5 mg/kg. In our case the results (50% Na-KSi and 50% NaOH) show 1.16 mg/kg lead leaching, however, in the case of the geopolymer containing 75% water glass and 25% NaOH the quantity of the leached Pb decreased down to 0.45 mg/kg which is under the limit. The change in Pb mobility can be explained by the phenomenon of lead building-in to the geopolymer structure.

Palomo and Palacios (2003) and Palacios and Palomo (2004) established that the main reason of the effective encapsulation of lead is the forming of insoluble Pb_3SiO_5 in fly ash based geopolymer systems activated with sodium-hydroxide.

The quantity of the leachable As significantly exceeds the limit concentrations concerning landfills (Fig. 11). It corresponds with the statement of Fernández-Jiménez *et al.* (2005). They found that the arsenic leaching from the geopolymers is higher than that of the raw fly ash.

During the second measurement, after 120 days the quantity of mobile As and Mo increases, the quantity of Se is almost the same as after 10 days ageing examination. Cd, Ni and Cr appears also, but Pb leaching could not be detected (Fig. 12).

The immobilization of Cd could correspond with the Ca-content of the geopolymer, because it substitutes the Ca. So the Ca/Cd silicate gel could form. According to Minerakova and Škvára (2006) in the low calcium-content geopolymers the immobilization of Cd does not always meet the strict environmental-protection regulations, and often the rate of its leaching depends on the applied leaching standard.

Figure 13 represents the leaching results obtained in distilled water for 120 days ageing under 1:10 L/kg liquid-to-solid ratio as proclaimed in the Decree No. 20 of 2006 (IV. 5.) KvVM.

Regarding the examined toxic elements in the case of geopolymers we identified that the As, Mo and Se exceeded the limit concentrations, but the ratio of the exceeding is much lower than that obtained in case of 50 L/kg liquid-to-solid ratio. In the case of the geopolymer produced with activator solution containing 75% Na-KSi water glass and 25% NaOH the results of the examination concerning the case of 10 L/kg liquid-to-solid ratio are as follows. For As: 84.34; for Mo: 10.35 and for Se: 0.146 mg/kg. In the case of 50 L/kg liquid-to-solid ratio these values were 100.81; 20.377 and 1.55 mg/kg respectively. It means that the dissolution depends on the solid-to-liquid ratio as well.

Conclusions

Based on the results obtained in course of the experimental research work the following conclusions can be drawn:

- Mechanical activation of the fly ash improves the geopolymer compressive strength until a certain fineness (2 m^2/g). During the further grinding the compressive strength of geopolymer decreases.
- The addition of the potassium type water glass to the NaOH activator solution significantly improves the workability of the paste and makes possible to tailor the mechanical stability.

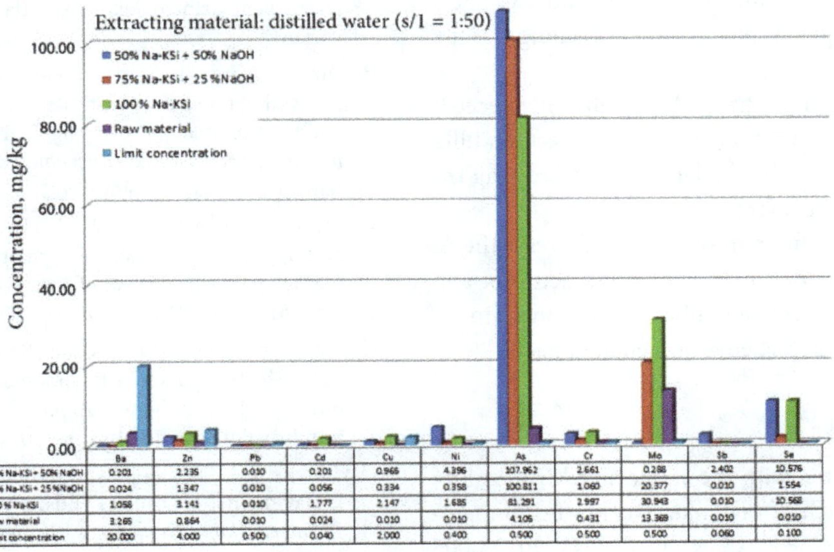

	Ba	Zn	Pb	Cd	Cu	Ni	As	Cr	Mo	Sb	Se
50% Na-KSi + 50% NaOH	0.201	2.235	0.010	0.201	0.965	4.396	107.962	2.661	0.288	2.402	10.576
75% Na-KSi + 25 %NaOH	0.024	1.347	0.010	0.056	0.334	0.358	100.811	1.060	20.377	0.010	1.554
100 % Na-KSi	1.058	3.141	0.010	1.777	2.147	1.685	81.291	2.997	30.943	0.010	10.568
Raw material	3.265	0.864	0.010	0.024	0.010	0.010	4.106	0.431	13.369	0.010	0.010
Limit concentration	20.000	4.000	0.500	0.040	2.000	0.400	0.500	0.500	0.500	0.060	0.100

Fig. 12. Quantity of mobilized toxic elements from 120 days age geopolymers using distilled water as extracting material (solid/liquid ratio 1:50)

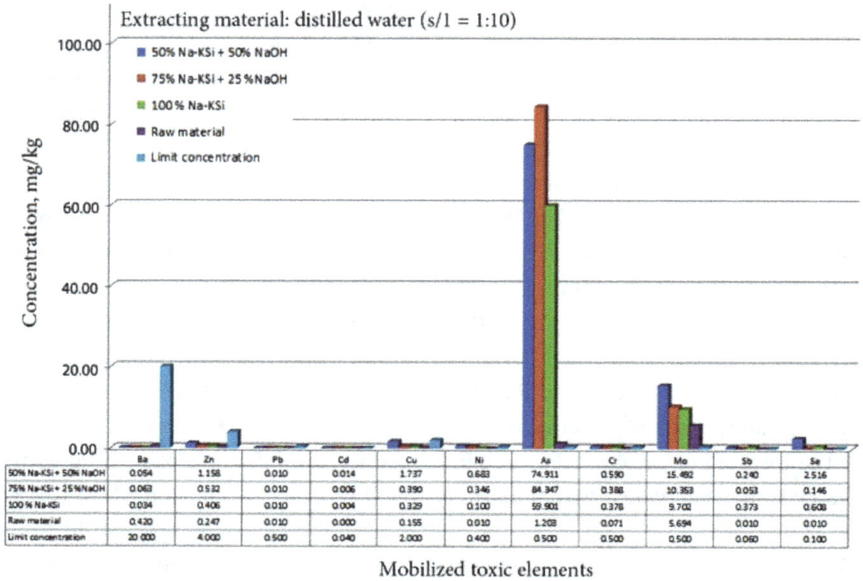

	Ba	Zn	Pb	Cd	Cu	Ni	As	Cr	Mo	Sb	Se
50% Na-KSi + 50% NaOH	0.054	1.158	0.010	0.014	1.737	0.683	74.911	0.590	15.482	0.240	2.516
75% Na-KSi + 25%NaOH	0.063	0.532	0.010	0.006	0.390	0.346	84.347	0.388	10.353	0.053	0.146
100 % Na-KSi	0.034	0.406	0.010	0.004	0.329	0.100	59.901	0.378	9.702	0.373	0.608
Raw material	0.420	0.247	0.010	0.000	0.155	0.010	1.203	0.071	5.694	0.010	0.010
Limit concentration	20.000	4.000	0.500	0.040	2.000	0.400	0.500	0.500	0.500	0.060	0.100

Fig. 13. Quantity of mobilized toxic elements from 120 days age geopolymers using distilled water as extracting material (solid/liquid ratio 1:10)

- The result of the geopolymerization reaction can be indicated by FT-IR, several new peaks appear in FT-IR spectra. It clearly proves the structural rearrangement.
- The leaching using 1 M acetic acid and 1 M hydroclorid acid has revealed that the Si, Al, Fe and Ca are built well in the structure of mechanically activated fly ash based geopolymers.
- The leaching results of toxic elements meet the criteria of the limit concentrations concerning inert landfills except As, Mo, Se.
- After 120 day ageing the quantity of mobile As and Mo increases, the quantity of Se is almost the same as after 10 day ageing. The Cd, Ni and Cr appear also in the aqueous phase, but Pb leaching was not detected.
- The quantity of leachable As significantly exceeds the limit concentration concerning inert landfills. This amount can be moderated by decreasing the residual alkali content.
- The aim of our future research is to decrease the As leachability by optimizing the alkali activator and adding components mobilizing that element in order to moderate the environmental impact.

Acknowledgements

The article was written with the support of 'Critical Elements – basic research of raw materials towards a maintainable future' TÁMOP -4.2.2.A-11/1/KONV-2012-0005 CriticEl project. A part of the work was done in the frame of the excellence of Sustainable Resource Managemnet University of Miskolc, Hungary.

Additionally, authors greatly appreciate the help in measurements of Colleagues at the Faculty of Earth Science and Engineering, University of Miskolc Ákos Debreczeni, Ferenc Kristály, Ferenc Móricz and Tamás Kántor. Furthermore, this work was partially supported by the János Bolyai Research Scholarship of the Hungarian Academy of Sciences.

References

Arioz, E.; Arioz, O.; Mete Kockar, O. 2012. Leaching of F-type fly ash based geopolymers, *Procedia Engineering* 42: 1114–1120. http://dx.doi.org/10.1016/j.proeng.2012.07.503

Barati, S. 2002. *Talajtan és talajvédelem.* Miskolc: Ökológiai Intézet a Fenntartható Fejlődésért Alapítvány. Miskolc (in Hungarian).

Bokányi, L.; Varga, T.; Mádai-Üveges, V.; Paulovics, J. 2011. Bioprocessing research in Institute of Raw Materials Preparation and Environmental Processing, University of Miskolc, in *Biotechnology & Metals 2nd International Conference*, 22–23 September 2011, Kosice, Slovakia, 5–8. ISBN 978-80-553-0723-7.

Brouwers, H.; Van Eijk, R. 2002. Fly ash reactivity: extension and application of a shrinking core model and thermodynamic approach, *Journal of Material Science* 37: 2129–2141. http://dx.doi.org/10.1023/A:1015206305942

Cheng, T. W.; Lee, M. L.; Ko, M. S.; Ueng, T. H.; Yang, S. F. 2012. The heavy metal adsorption characteristics on metakaolin-based geopolymer, *Applied Clay Science* 56: 90–96. http://dx.doi.org/10.1016/j.clay.2011.11.027

Davidovits, J. 1988. Geopolymer chemistry and properties, in J. Davidovits, J. Orlinski (Eds.). Vol. 1. *Proceedings of the 1st International Conference on Geopolymer,* 1–3 June 1988, Compiegne, France, p. 25–48.

Decree No. 20 of 2006 (IV. 5.) KvVM of the Ministry of Environmental Protection and Water Management on certain rules and conditions concerning waste tipping and landfills.

Deja, J. 2002. Immobilisation of Cr^{6+}, Cd^{2+}, Zn^{2+} and Pb^{2+} in alkali activated slag binders, *Cement and Concrete Research* 32(12): 1971–1979. http://dx.doi.org/10.1016/s0008-8846(02)00904-3

Fernández-Jiménez, A. M.; Lachowski, E. E.; Palomo, A.; Macphee, D. E. 2004. Microstructural characterisation of alkali-activated PFA matrices for waste immobilisation, *Cement and Concrete Composites* 26: 1001–1006. http://dx.doi.org/10.1016/j.cemconcomp.2004.02.034

Fernández-Jiménez, A.; Palomo, A.; Macphee, D. E.; Lachowski, E. E. 2005. Fixing arsenic in alkali-activated cementitious matrices, *Journal of the American Ceramic Society* 88: 1122–1126. http://dx.doi.org/10.1111/j.1551-2916.2005.00224.x

Füleky, Gy. 2011. *Környezetmérnöki Tudástár*. Vol. 3. Talajvédelem, talajtan. Pannon Egyetem-Környezetmérnöki Intézet (in Hungarian).

Hardjito, D.; Rangan, B. V. 2005. *Development and properties of low-calcium fly ash-based geopolymer concrete*. Curtin University of Technology.

Heidrich, C.; Feuerborn, H. J.; Weir, A. 2013. Coal combustion products: a global perspective, in *World of Coal Ash Conference*, 22–25 April 2013, Lexington, USA.

Izquierdo, M.; Querol, X.; Davidovits, J.; Antenucci, D.; Nugteren, H.; Fernández-Pereira, C. 2009. Coal fly ash-slag-based geopolymers: Microstructure and metal leaching, *Journal of Hazardous Materials* 166: 561–566. http://dx.doi.org/10.1016/j.jhazmat.2008.11.063

Kumar, S.; Kumar, R.; Alex, T. C.; Bandopadhyay, A.; Mehrotra, S. P. 2007a. Influence of reactivity of fly ash on geopolymerisation, *Advances in Applied Ceramics* 106: 120–127. http://dx.doi.org/10.1179/174367607X159293

Kumar, R.; Kumar, S.; Mehrotra, S. P. 2007b. Towards sustainable solutions for fly ash through mechanical activation, *Resources, Conservation and Recycling* 52: 157–159. http://dx.doi.org/10.1016/j.resconrec.2007.06.007

Lakatos, J. 2006. Transformation of Cr forms in solutions and high temperature processes, in *49th Hungarian Spectrometry Conference*, 10–12 July 2006, Miskolc, Hungary.

Mineriková, M.; Škvára, F. 2006. Fixation of heavy metal sin geopolymeric materials based on brown coal fly ash, *Ceramics-Silikaty* 50: 200–207.

Mucsi, G.; Molnár, Z.; Kumar, S. 2014. Geopolymerisation of mechanically activated lignite and brown coal fly ash, *Acta Physica Polonica A* 126: 994–998. http://dx.doi.org/10.12693/APhysPolA.126.994

Opoczky, L. 2001. A pernyék szilikátkémiai tulajdonságai, *A Miskolci Egyetem Közleménye A sorozat, Bányászat* 55: 97–106.

Palacios, M.; Palomo, A. 2004. Alkali-activated fly ash matrices for lead immobilisation: A comparison of different leaching tests, *Advances in Cement Research* 16: 137–144. http://dx.doi.org/10.1680/adcr.2004.16.4.137

Palomo, A.; Palacios, M. 2003. Alkali activated cementitious materials: Alternative matrices for immobilisation of hazardous wastes – Part II. Stabilisation of chromium and lead, *Cement and Concrete Research* 33: 289–295. http://dx.doi.org/10.1016/S0008-8846(02)00964-X

Pietersen, H.; Fraay, A.; Bijen, J. 1989. Reactivity of fly ash at high pH, in fly ash and coal conversion by-products: Characterization, utilization and disposal VI, in F. P. Glasser, R. L. Day (Eds.). *Materials research society symposium proceedings*. Materials Research Society, p. 139.

Research Report of the CriticEl project: Critical Elements – basic research of raw materials towards a maintainable future. 2014. University of Miskolc (in Hungarian).

Temuujin, J.; Williams, R. P.; van Riessen, A. 2009. Effect of mechanical activation of fly ash on the properties of geopolymer cured at ambient temperature, *Journal of Materials Processing Technology* 209: 5276–5280. http://dx.doi.org/10.1016/j.jmatprotec.2009.03.016

Temuujin, J.; Minjigmaa, A.; Lee, M.; Chen-Tan, N.; van Riessen, A. 2011. Characterisation of class F fly ash geopolymer pastes immersed in acid and alkaline solutions, *Cement & Concrete Composites* 33: 1086–1091. http://dx.doi.org/10.1016/j.cemconcomp.2011.08.008

Gábor MUCSI. Associate Professor at University of Miskolc, Faculty of Earth Science and Engineering. Author of more than 140 publications. Research interests: utilization of industrial by-products, mechanical activation of solids, grindability tests.

János LAKATOS. Associate Professor at University of Miskolc, Faculty of Materials Science and Engineering. Number of publications: more than 150. Research interests: immobilization of toxic elements, leaching tests.

Ágnes SZENCZI. Environmental Engineer MSc student at University of Miskolc, Faculty of Earth Science and Engineering. Research interests: development of fly ash based geopolymers, waste utilization technologies.

Zoltán MOLNÁR. PhD student at University of Miskolc, Faculty of Earth Science and Engineering. Number of publications: 33. Research interests: geopolymerization of silicate bearing waste materials.

8

MODELING OF THE SPONTANEOUS POLYPROPYLENE SORBENTS IMBIBITION WITH EMULSIONS

Jerzy SĘK, Olga S. SHTYKA, Kamila SZYMCZAK

Department of Chemical Engineering, Faculty of Process and Environmental Engineering, Lodz University of Technology, Wolczanska St. 213, 90-924 Lodz, Poland

Abstract. Oil penetration into aqueous media results in various chemical and physical processes including formation of "oil-in-water" emulsions, that have devastating effects on the environment. This research was conducted to evaluate the kinetics of permeation of "oil-in-water" emulsions into synthetic porous material. It gives the ability to develop more effective sorbents, decreasing risks of negative consequences and environmental contamination. The following publication focuses on assessment of the possibility to describe and predict a process of porous medium imbibition with emulsions by means of the Hill equation. Its adequacy is compared with the classical Lucas–Washburn equation and its modified version. The advanced approach is to substitute the commonly used parameter of the height of an imbibed emulsion ($h(t)$) by the parameter $m_h(t)$, which is the mass of an imbibed emulsion in the case of the classical and modified Washburn equations. According to the obtained results, the Hill equation provides the most appropriate and precise description of the porous material imbibition with "oil-in-water" emulsions and oil in comparison with others, and shows the highest correlation values (R_{av} = 0.995±0.001) and the lowest normalized root mean square error ($NRMSE_{av}$ 1.95±0.138).

Keywords: oil spill, emulsion, imbibition, porous material, Hill equation, classical Lucas–Washburn equation, modified Washburn equation.

Introduction

During the last several decades, the problem of environmental contamination with oil products has become one of the most serious global issues (Fingas 2011). The majority of oil and its derivatives is discharged into the aquatic ecosystems as a result of accidental spillings from tankers and underwater pipelines, emergency leakages during oil drilling process and surface runoff from land-based sources (Schnoor 2010; Fingas 2011; Valentukevičienė, Ignatavičius 2011). As a result of oil spillings in the aqueous medium, chemical and physical processes occur. Evidently, an emulsification process is considered to be constitutive due to its significant influence on the characteristics of organic contaminantion over time. The process of emulsification facilitates increase in density and corresponding viscosity (Wei *et al.* 2003). To this point, an oil emulsion is indicated as a microheterogeneous system consisting of two practically non-intersoluble liquids that differ in molecula nature (e.g. water is polar and another is non polar or low-polarity organic liquid). There are a lot of specific factors intensifying a process of "thin-layer oil" emulsification, but the dominant ones are as follows: action of the inertia force of mixing, waves and wind influence, turbulence of a stream, etc. In the open aquatic environment rapid surface spreading of the spilled oil occurs. Moreover, this process additionally stimulates and provides increase in emulsification rate (Wei *et al.* 2003; Al-Majed *et al.* 2012).

Obviously, the oil-spill-derived hydrocarbons in water ecosystems pose an extremely high risk to the environment due to general toxicity for the aquatic and terrestrial biota (Onwurah *et al.* 2007). The contaminants interrupt the food chain and without exception cause deterioration of water quality, e.g. depletion of the dissolved oxygen content due to increase in biochemical oxygen demand (BOD) for the organic components transformation into inorganic species (Shafir *et al.* 2007). Subsequently, the recovery of the spilled oil and derived hydrocarbons is considered to be the actual point at issue. Mechanical removal of spilled oil products by application of sorbents is

Corresponding author: Olga S. Shtyka
E-mail: olga.shtyka@gmail.com

regarded as the most effective method. The main principle consists of concentration of the liquid oil-spill-derived hydrocarbons in solid material commonly known as a sorbent (Adebajo *et al.* 2003; Al-Majed *et al.* 2012).

The suggested remediation approach is based on the physical phenomenon known as capillary rise. The process of material filling with a liquid can be called absorption if it occurs mostly due to the capillary suction pressure or external pressure (Masoodi *et al.* 2007). If a fluid in the porous medium is replaced by a more viscous one, this phenomenon is called imbibition (Alava *et al.* 2004). In the scientific literature, the research related to the process of liquid penetration into the porous medium and its kinetics are acute issues for discussion as well as its mathematical describing and modeling.

The equation proposed firstly by Lucas and Washburn is considered to be a basis for description of liquid rise in a capillary. However the model has some limitations, e.g. the impacts of gravitational and inertial forces are neglected (Zhmud *et al.* 2000; Fries, Dreyer 2008; Maggi, Alonso-Marroquin 2012). Recently, there has been a continuing interest in this sphere due to the fact that many phenomena are based on capillary rise. It explains that there has been a great deal of efforts to describe this process, mostly grounded on the classical and modified Lucas-Washburn equations (Zhmud *et al.* 2000; Siebold *et al.* 2000; Hamraoui, Nylander 2002; Xue *et al.* 2006; Fries, Dreyer 2008; Maggi, Alonso-Marroquin 2012).

Zhmud *et al.* in their research work pay special attention to the capillary rise of surface-active substances. Thus, the peculiarities of surfactant transport in the porous medium are reviewed as well as relaxation processes in the adsorption layer. The asymptotic solutions corresponding to both short and long time limits were also drawn in the referred publication (Zhmud *et al.* 2000). The influence of dynamic contact angle on the capillary rise of total wetting liquids was investigated in the work of Siebold *et al.* in 2000. It is worth noticing that the experiments were carried out both in a single glass capillary and in a column packed with powders (Siebold *et al.* 2000).

Hamraoui and Nylander investigated mechanisms regulating the dynamics of the capillary rise in the view of dissipation channels (Hamraoui, Nylander 2002). Factors affecting the significance of gravity within the process of fluid penetration into a porous medium are analyzed in another publication (Chan *et al.* 2004). The hydrostatic effect of the spontaneous dynamic capillary rise is a subject of discussion in the research work of Xue *et al.* The equation relevant to spontaneous capillary rise that expresses correlation between time and height of a liquid front and takes into consideration the hydrostatic effects was proposed by Xue *et al.* in 2006. The dynamics of liquid penetration into a capillary tube in the case of small inertial effects is a field of Chebbi's researches. The important issues analyzed in this publication are: effect of the contact line speed on the dynamic contact angle and gravity influence on a fluid penetration into a vertical tube (Chebbi 2007). The investigations of Fries, Dreyer connect with deriving the analytic solution for capillary rise in a cylindrical column and in a porous material. A modified form of the classical Lucas-Washburn equation is proposed which takes into consideration the gravity term as a hydrostatic pressure. It provides the possibility to define behavior of the fluid rise in porous medium during the long-term period (Fries, Dreyer 2008).

The derived analytical solutions for gas displacement by a liquid in horizontal and inclined capillary tubes are represented in Hilpert's research work. Proposed solutions permit definition of various flow scenarios for a process of a liquid withdrawal from a capillary tube depending on flow direction and acceleration of the gas-liquid interface (Hilpert 2010). In a recent paper, a model of the meniscus movement in capillaries that accounts for the effect of a gas phase is described. The gas-related effects are proved to be important in the same way as the inertia and reservoir effects (Maggi *et al.* 2012).

This work focuses on assessment of the Hill equation's ability to predict the process of spontaneous imbibition of a porous material with "oil-in-water" emulsions and to compare its adequacy with both the classical and modified Washburn equation. In the present paper, the mentioned equations are also modified due to applying the parameter of an imbibed mass as a function of time instead of describing the relation between time and a fluid height.

1. General characteristics of materials and experimental media

In practice, synthetic sorbents are made on the basis of high molecular weight polymers, mainly from polyurethane, polyethylene and polyamides fibers (Zhu *et al.* 2011). In the current experiments the polypropylene fiber "Polfilter" of regional production (obtained from "Zakład Pracy Chronionej", Lodz) was used as a synthetic sorbent of emulsions.

The oil-based liquid employed in the research was kerosene oil "Dragon" that is produced by "Zakład Chemii Budowlanej" (Kampinos, Poland). Its emulsions with different concentrations of the dispersed phase were applied as the investigated medium.

The edible gelatin "Delecta" from JSC "Rieber Foods Polska" was used as an additive to increase the viscosity of the obtained emulsions. The surfactant "Rokacet" (Chemical Plant "Rokita", Brzeg Dolny) served as an emulsifier to reduce the interfacial tension between the phases in the prepared media due to its partial solubility in both emulsion phases. The coloring agent was "Sudan III" added to the emulsions to facilitate observation of the imbibition process.

Thus, the experimental media were emulsions prepared by means of kerosene oil mixing with water, having dispersed phase concentrations (DPC) of 0.5%; 1%; 10% and 20%. The concentration of the surfactant was 2% and fractions of gelatin (GC) consequently were as follows: 0.05% and 0.1%.

In the present experimental work viscosity and density of emulsions were measured and results are represented in Table 1.

Table 1. Properties of the prepared "oil-in-water" emulsions

Property	Concentration of the dispersed phase in the investigated emulsion			
	0.5%	1%	10%	20%
Density, kg/m^3	999	998	980	960
	concentration of gelatin 0.05%			
Viscosity, mPas	0.925	0.893	0.923	0.925
	concentration of gelatin 0.1%			
	0.902	0.880	0.923	1.159

2. Experimental procedures

The imbibition capacity of the polypropylene sorbent was determined directly by dipping the material sample into the prepared emulsion. In previous scientific research concerning the capillary rise of liquids, a glass tube and/or a column packed with sorbent was used (Zhmud *et al.* 2000; Siebold *et al.* 2000; Fries, Dreyer 2008). In the current experiment a strip was used in order to investigate entirely the imbibition capacity of the offered media.

A strip of the polypropylene sorbent of width of 3 cm, length of 20 cm, and thickness of 0.5 cm was submerged with one end into the emulsion to a depth of 2 cm (see Fig. 1).

The process of imbibition was followed from the moment the polypropylene material was put in contact with the medium. The main consideration was description of the kinetics of the mass change of an imbibed emulsion, $m_h(t)$ in the determined period of time, (t). The mass difference between the initial mass of an emulsion in a beaker (see (3) in Fig. 1) and the mass after sorbent immersing was accepted as the mass of imbibed liquid.

Measurements were taken until steady state was reseved. The measurement of an imbibed mass was observed at 15 sec, 30 sec, and 1 min after the process beginning.

3. Modeling of the process of sorbent imbibition with emulsions

The process of imbibition is based generally on capillary rise, thus the Washburn equation describing this process

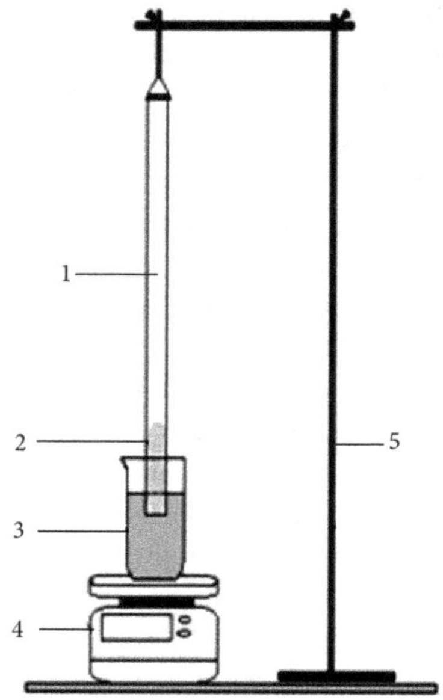

Fig. 1. Scheme of the device used to measure imbibition capacity: 1 – polypropylene stripe; 2 – imbibed emulsion; 3 – beaker with the investigated emulsion; 4 – analytical balance; 5 – stand

can also serve for analyzing issues concerning the spontaneous imbibition of porous medium with an emulsion (Zhmud *et al.* 2000; Fries, Dreyer 2008; Maggi, Alonso-Marroquin 2012). The previously mentioned basic equation is occasionally known as the Lucas–Washburn equation. In this case, it has the following form:

$$t = a_1(b_1(\ln(\frac{b_1}{b_1 - h(t)})) - h(t)), \tag{1}$$

where h is the height of an imbibed emulsion; t is the time of the process duration; a_1, b_1 are parameters of the given equation.

The height of an imbibed emulsion ($h(t)$) in Eq. 1 was replaced by the parameter $m_h(t)$, the mass of an imbibed emulsion, consequently after substitution the equation can be represented as follows:

$$t = a_1(b_1(\ln(\frac{b_1}{b_1 - m_h(t)})) - m_h(t)), \tag{2}$$

where m_h is the mass of an imbibed emulsion; t is the time of the process duration; a_1, b_1 are parameters of the equation.

Although the equation is proven as valid in a majority of cases, it can become inadequate due to various contradictions with the fundamental laws of hydrodynamics (Zhmud *et al.* 2000; Fries, Dreyer 2008). Moreover it is impossible to describe correctly the process of the capillary rise in a porous medium on account of the parameters selectivity. It takes into account viscosity, interfacial tension, and boundary pressure (Maggi, Alonso-Marroquin 2012).

Taking into consideration criticisms of the classical Washburn equation, there are many attempts to describe the capillary rise by means of modifications of this widely used equation (Zhmud *et al.* 2000; Fries, Dreyer 2008). For example, Fries, Dreyer proposed a modified version of the classical Washburn equation to describe the capillary rise of surfactant and to obtain more accurate solutions (see Eq. 3).

Therefore the modified equation may be expressed as follows:

$$h(t) = \frac{a_2}{b_2}(1 - e^{-\frac{tb_2^2}{a_2}}), \quad (3)$$

where h is the height of an imbibed emulsion; t is the time of the process duration; a_2, b_2 are parameters of the given equation. The proposed expression is considered to be an equation that evolves long time asymptotic solution (Fries, Dreyer 2008).

In Eq. 3 the height of an imbibed emulsion ($h(t)$) was also substituted by the parameter $m_h(t)$, the mass of an imbibed emulsion, and finally after rearranging one obtains:

$$m_h(t) = \frac{a_2}{b_2}(1 - e^{-\frac{tb_2^2}{a_2}}), \quad (4)$$

where m_h is the imbibed mass of an emulsion; t is the time of the process; a_2, b_2 are parameters of this equation.

Additionally, the Hill equation was applied to describe the imbibition process. It is recognized primarily for modeling of biochemical processes in physiological and pharmacological investigations and was derived from the law of mass action (Gesztelyi *et al.* 2012). The Hill equation is mainly used to analyze the binding equilibrium in ligand-receptor interactions (Goutelle *et al.* 2008), but its classical form (Gesztelyi *et al.* 2012) was found as appropriate to describe the process of porous material imbibition with various emulsions. Eventually including some alterations, the classical form of the Hill equation can be rewritten as:

$$m_h(t) = V_{\max}\frac{t^n}{k^n + t^n}, \quad (5)$$

where m_h is the imbibed mass of an emulsion; t is the time of the process duration; V_{max}, k, n are defined as parameters of the proposed equation.

The accuracy of the models was estimated via correlation coefficient, R (see Eq. 6) and normalized root mean square error, *NRMSE* (see Eq. 7) in accordance with following formulas:

$$R = \frac{\operatorname{cov}(x,y)}{\sigma_x \sigma_y} = \frac{\sum(x_i - \bar{x})(y_i - \bar{y})}{\sqrt{\sum(x_i - \bar{x})^2 \sum(y_i - \bar{y})^2}}, \quad (6)$$

$$NRMSE = \frac{\sqrt{\frac{1}{n}\sum_{i=1}^{n}(x_i - y_i)^2}}{y_{\max} - y_{\min}}, \quad (7)$$

where x_i is the calculated data; y_i is the experimental data; $\bar{x}$ and $\bar{y}$ are the mean values of data and n is the number of experiments (Gmurman 2004).

The Origin Program which applies the Levenberg-Marquardt algorithm (LMA) was used for the experimental data approximation by means of the proposed mathematical models.

4. Analysis of results and discussion

The experimental data obtained for the process of polypropylene material imbibition with emulsions during a defined time period were fitted using previously mentioned equations (see Eq. 2, Eq. 4 and Eq. 5).

According to analysis of the results, changes in the character of the process of the material imbibition with time were noticed. Thus, when correlation between a mass of an imbibed emulsion and a maximal mass (m_{im}/m_{max}) gets near 0.3 (~ 0.00682 ± 0.00026 kg), the rate of the imbibition decreases in all investigated cases. It provides the possibility to divide imbibition process into two phases. The first is characterized with a fast rate that is supposed to be conditioned by inertial forces, whereas within the second stage the process slows down and transfers into a so-called viscous phase (Siebold *et al.* 2000). The rapid reduction of the imbibition kinetics is defined as a transition point (Siebold *et al.* 2000). The time interval after the moment of transition is also accompanied with a change of the equations parameters used for fitting. Correspondingly, in this publication two phases of the process of the porous material imbibition were investigated, though the entire imbibition process was also considered and analyzed as an additional case.

The values of the equations coefficients used for fitting are given in Tables 2–4.

For the classical Washburn equation, it is difficult to see a trend with regard to coefficients *a*, and *b*. Thus, the parameter *a* lies in the scattered range from 0.241 to 14.071 and correspondingly the parameter *b* varies from 0.150 to 8.329 without determined regularities in their distribution (see Table 2).

The values of the coefficients for the modified Washburn equation are represented in Table 3. It is possible to deduce a probable tendency of the coefficients alteration within the second phase of the investigated process. Thus, values of the parameters increase uniformly with increase in the dispersed phase concentration in emulsions. This tendency is defined as typical for emulsions with the different concentrations of gelatin. On condition that concentrations of the oil phase are relatively low, e.g. 0.05% and 1%, the parameter *a* is close to 2.867±0.027 and *b* is 2.647±0.045. For emulsions with dispersed phase concentration of 10% and gelatin concentration of 0.05%, the values of *a* and *b* are higher and equal to 3.772±0.049 and

Table 2. Parameters for the classic Lucas–Washburn equation

Concentration of gelatin	Parameter	Concentration of the dispersed phase in the investigated emulsion, %			
		0.5	1	10	20
The first phase of imbibition process					
0.05%	a_1	1.183 ±0.054	1.500 ±0.135	0.380 ±0.010	0.241 ±0.005
	b_1	1.213 ±0.033	1.358 ±0.088	0.902 ±0.009	0.951 ±0.004
0.1%	a_1	0.183 ±0.005	0.168 ±0.012	0.832 ±0.058	0.380 ±0.007
	b_1	0.150 ±0.010	0.113 ±0.002	1.048 ±0.044	0.987 ±0.005
The second phase of imbibition process					
0.05%	a_1	1.638 ±0.069	3.579 ±0.523	0.719 ±0.021	0.367 ±0.017
	b_1	1.538 ±0.035	2.517 ±0.277	1.137 ±0.010	1.025 ±0.006
0.1%	a_1	2.099 ±0.427	1.756 ±0.145	0.943 ±0.038	0.505 ±0.022
	b_1	1.844 ±0.235	1.624 ±0.077	1.220 ±0.019	1.055 ±0.009
Entire imbibition process					
0.05%	a_1	1.776 ±0.047	2.615 ±0.134	0.632 ±0.012	0.328 ±0.006
	b_1	1.606 ±0.025	2.022 ±0.072	1.100 ±0.006	1.015 ±0.002
0.1%	a_1	12.258 ±5.882	14.071 ±8.057	1.054 ±0.031	0.464 ±0.007
	b_1	7.514 ±3.333	8.329 ±4.451	1.273 ±0.016	1.041 ±0.003

Table 3. Parameters for the modified Washburn equation

Concentration of gelatin	Parameter	Concentration of the dispersed phase in the investigated emulsion, %			
		0.5	1	10	20
The first phase of imbibition process					
0.05%	a_2	4.089 ±0.073	3.814 ±0.060	8.777 ±0.378	13.330 ±0.774
	b_2	5.982 ±0.144	6.427 ±0.150	12.821 ±0.635	16.155 ±1.015
0.1%	a_2	2.940 ±0.035	2.642 ±0.034	5.237 ±0.170	8.419 ±0.382
	b_2	4.429 ±0.088	3.895 ±0.086	8.982 ±0.369	10.103 ±0.516
The second phase of imbibition process					
0.05%	a_2	2.889 ±0.023	2.844 ±0.030	3.772 ±0.049	4.779 ±0.060
	b_2	2.693 ±0.038	2.601 ±0.051	3.753 ±0.064	4.768 ±0.070
0.1%	a_2	2.983 ±0.024	2.972 ±0.011	3.488 ±0.023	4.105 ±0.088
	b_2	2.671 ±0.042	2.749 ±0.018	3.475 ±0.032	4.110 ±0.105
Entire imbibition process					
0.05%	a_2	3.924 ±0.092	3.627 ±0.096	6.631 ±0.254	11.268 ±0.598
	b_2	4.299 ±0.121	4.161 ±0.141	7.485 ±0.322	12.571 ±0.719
0.1%	a_2	3.263 ±0.046	3.104 ±0.032	4.347 ±0.116	7.724 ±0.308
	b_2	3.203 ±0.062	3.012 ±0.044	4.723 ±0.150	8.543 ±0.370

3.753±0.064 correspondingly. In the case of the oil phase concentration increasing to 20%, the mentioned parameters of the modified Washburn equation also enlarge (viz. a is 4.779±0.060 and b is 4.768±0.070). An analogous tendency is observed for emulsions with the same higher value of the disperse phase concentration and gelatin content of 0.1% (see Table 3).

The parameters of the Hill equation are within the defined ranges. Thus in both cases of the separated phases, V_{max} lies in a range from 1.121 to 1.836, while k is equal to 0.214÷0.791 (see Table 4).

However the coefficient n is in a range of 0.433÷0.839 for the first phase of the process and for the second one with gelatin concentration of 0.05%. On condition of the second phase and gelatin concentration of 0.1%, a value of the parameter n increases and is in a range from 1.203 to 1.969.

The divergent situation is observed for the entire process, in this case V_{max} is larger and lies inside the boundaries of 2.012÷2.481, the coefficient k is founded within an interval of 0.952÷1.959 and n is equal to 0.412÷0.781. Nevertheless, in cases of the second phase and the entire process (gelatin concentration of 0.05% and 0.1%, the dispersed phase concentration of 20%) the determined coefficients of the Hill equation are distinguished and do not follow defined tendency (see Table 4).

The correlation coefficient is applied as one of the quality criteria for the equations fitting the experimental data. It is necessary to underline the relationship between two sets of variables. The values of the correlation coefficient are given in Table 5.

According to the results, the correlation coefficients of the modified Washburn equation are the highest for the second phase, nevertheless these values are defined as insufficient and quite low in comparison with coefficient values obtained using the Hill equation and the classical Washburn equation, besides an occurence with gelatin concentration of 0.1%.

Conforming to the results, a majority of the lowest correlation coefficient values were observed for emulsions with the dispersed phase concentration of 20%, thus in this case a perfect fit to these experimental data is not obtained.

An approximation with the Hill equation provides in general the highest values of the correlation coefficient. As a result, in this case the most appropriate degree of correlation between experimental and analytical data was assumed (see Table 5).

Table 4. Parameters for the Hill equation

Concentration of gelatin	Parameter	Concentration of the dispersed phase in the investigated emulsion, %			
		0.5	1	10	20
The first phase of imbibition process					
0.05%	V_{max}	1.153 ±0.049	1.141 ±0.067	1.438 ±0.108	1.565 ±0.102
	k	0.236 ±0.027	0.257 ±0.040	0.283 ±0.085	0.214 ±0.066
	n	0.745 ±0.023	0.728 ±0.024	0.491 ±0.023	0.433 ±0.023
0.1%	V_{max}	1.482 ±0.133	1.279 ±0.141	1.277 ±0.163	1.521 ±0.160
	k	0.544 ±0.116	0.417 ±0.102	0.308 ±0.126	0.269 ±0.114
	n	0.751 ±0.026	0.839 ±0.037	0.590 ±0.039	0.508 ±0.042
The second phase of imbibition process					
0.05%	V_{max}	1.603 ±0.268	1.836 ±0.532	1.750 ±0.370	1.017 ±0.011
	k	0.540 ±0.024	0.791 ±0.068	0.567 ±0.048	0.206 ±0.026
	n	0.792 ±0.158	0.789 ±0.147	0.501 ±0.113	2.418 ±0.444
0.1%	V_{max}	1.121 ±0.020	1.178 ±0.021	1.161 ±0.026	1.608 ±0.205
	k	0.330 ±0.004	0.319 ±0.005	0.240 ±0.006	0.262 ±0.008
	n	1.969 ±0.167	1.489 ±0.073	1.203 ±0.075	0.350 ±0.065
Entire imbibition process					
0.05%	V_{max}	2.377 ±0.181	2.200 ±0.190	2.214 ±0.125	1.358 ±0.030
	k	1.812 ±0.437	1.477 ±0.396	1.610 ±0.409	0.113 ±0.011
	n	0.562 ±0.016	0.591 ±0.019	0.412 ±0.011	0.486 ±0.014
0.1%	V_{max}	2.481 ±0.200	2.012 ±0.100	2.432 ±0.235	1.458 ±0.073
	k	1.615 ±0.333	0.952 ±0.115	1.959 ±0.687	0.227 ±0.046
	n	0.688 ±0.020	0.781 ±0.018	0.518 ±0.180	0.525 ±0.027

Accuracy of the equations fitting was estimated in addition using the normalized root mean square error (*NRMSE*). It is a non-dimensional form of the root mean square error that is used to assess the difference between actually observed data and predicted by a certain model. The values of the normalized root mean square error are represented in Table 6.

Table 5. The correlation coefficient values for applied equations

DPC, %	Concentration of gelatin					
	0.05%			0.1%		
Type of equation	Modified Washburn	Hill	Lucas–Washburn	Modified Washburn	Hill	Lucas–Washburn
The first phase of imbibition process						
0.5	0.970	0.995	0.995	0.985	0.996	0.845
1	0.979	0.997	0.994	0.983	0.994	0.906
10	0.877	0.997	0.993	0.938	0.990	0.991
20	0.811	0.994	0.993	0.859	0.979	0.996
The second phase of imbibition process						
0.5	0.983	0.994	0.994	0.982	0.990	0.943
1	0.978	0.988	0.990	0.997	0.997	0.984
10	0.963	0.993	0.989	0.991	0.995	0.988
20	0.972	0.970	0.942	0.896	0.950	0.934
Entire imbibition process						
0.5	0.947	0.996	0.998	0.979	0.996	0.989
1	0.922	0.994	0.996	0.989	0.997	0.988
10	0.840	0.997	0.994	0.925	0.995	0.995
20	0.782	0.995	0.985	0.867	0.987	0.993

Table 6. The values of the normalized root mean square error for applied equations

DPC, %	Concentration of gelatin					
	0.05%			0.1%		
Type of equation	Modified Washburn	Hill	Lucas–Washburn	Modified Washburn	Hill	Lucas–Washburn
The first phase of imbibition process						
0.5	4.3	1.7	1.9	3.1	1.6	2.0
1	3.8	1.5	2.1	3.4	2.2	-
10	7.6	1.7	2.2	5.8	2.3	2.6
20	9.4	1.6	2.3	11.4	2.5	1.9
The second phase of imbibition process						
0.5	3.8	2.4	2.3	4.1	3.2	7.2
1	5.1	3.9	3.4	1.7	1.6	3.8
10	5.7	2.5	3.3	2.8	2.1	3.4
20	5.9	5.9	7.8	9.6	6.5	7.6
Entire imbibition process						
0.5	6.2	1.6	1.3	4.2	1.8	3.1
1	6.6	1.8	1.6	3.1	1.6	3.2
10	9.3	1.2	2.2	6.9	1.8	2.0
20	9.7	1.4	3.2	10.3	2.0	2.2

Correlations between experimental data and those calculated analytically with the Washburn equation (see Eq. 1) are represented in Figures 2–4. The straight lines in the figures refer to the bounds within which solutions are found. The results are in a range of ±10% for the separated first and the second phases of the imbibition process and ±15% for the entire process.

From analysis of the graphs, it is possible to notice overlapping of the time intervals that can be explained by a difference in the transition points. This also depends on the disperse phase concentrations in the emulsions.

Figures 5–8 display the relation between experimental values and data received due to fitting by means of the modified Washburn equation (see Eq. 4), which is used to define the kinetics of the imbibed mass change per certain period of time.

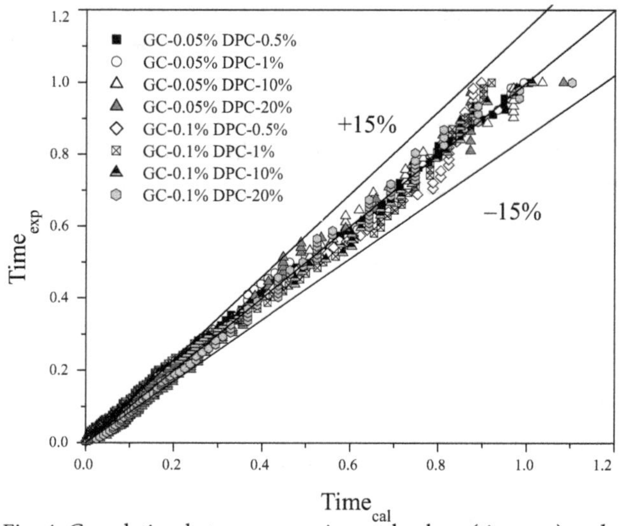

Fig. 4. Correlation between experimental values (timeexp) and calculated (timecal) by means of the classical Lucas–Washburn equation for the entire process of the imbibition process

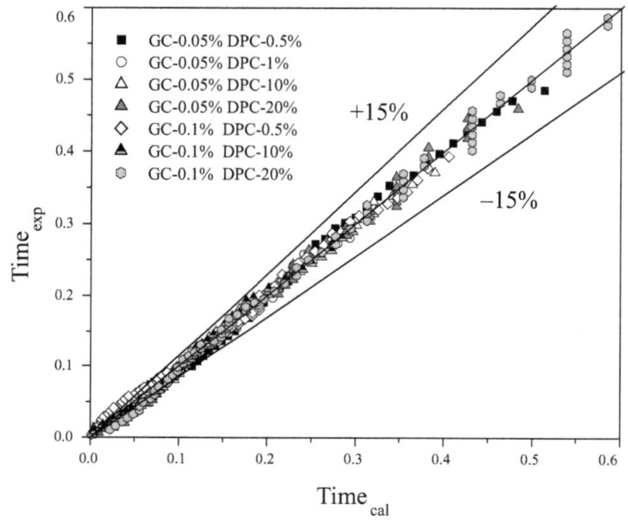

Fig. 2. Correlation between experimental values (timeexp) and calculated (timecal) by means of the classical Lucas–Washburn equation for the first phase of the imbibition process

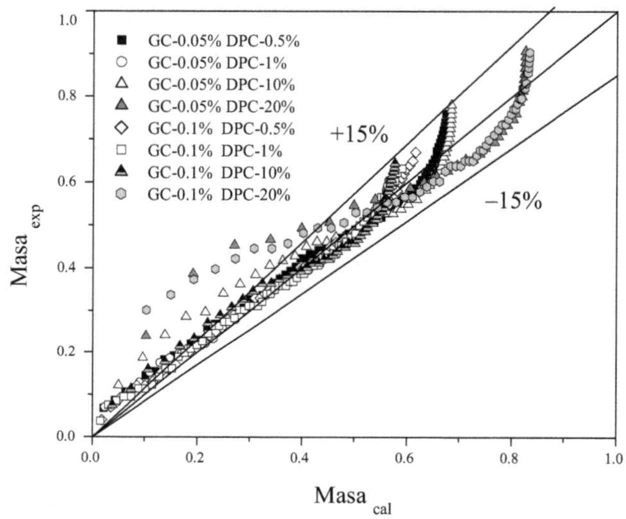

Fig. 5. Correlation between experimental values (massexp) and calculated (masscal) by means of the modified Washburn equation for the first phase of the imbibition process

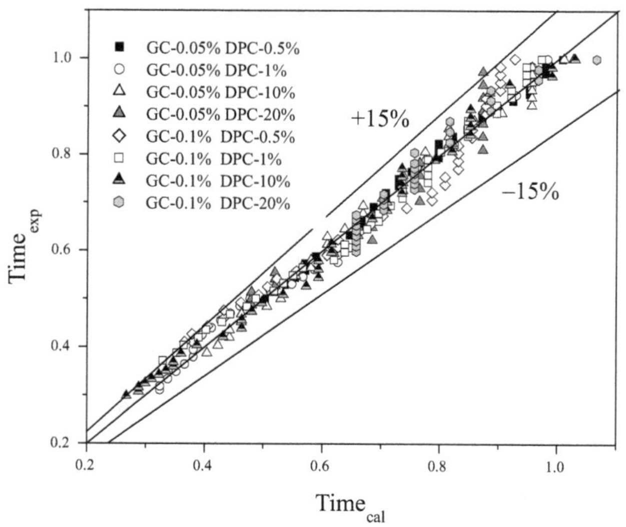

Fig. 3. Correlation between experimental values (timeexp) and calculated (timecal) by means of the classical Lucas–Washburn equation for the second phase of the imbibition process

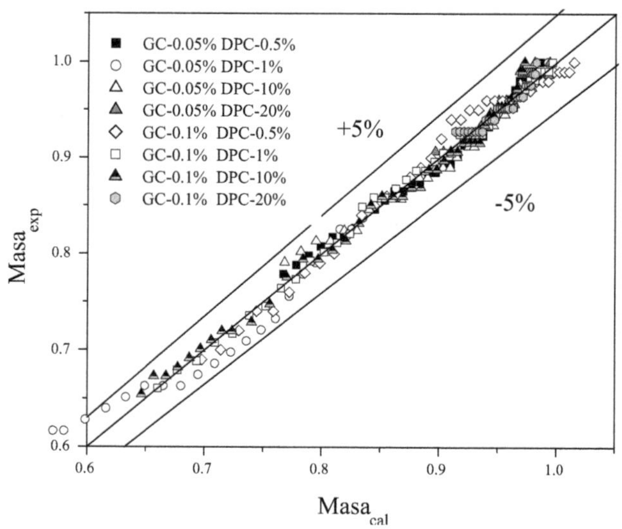

Fig. 6. Correlation between experimental values (massexp) and calculated (masscal) by means of the modified Washburn equation for the second phase of the imbibition process

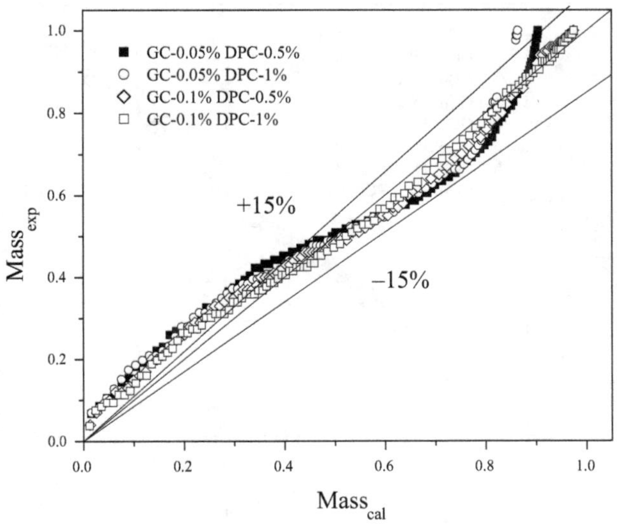

Fig. 7. Correlation between experimental values (massexp) and calculated (masscal) by means of the modified Washburn equation (entire process with gelatin content of 0.5% and 1%)

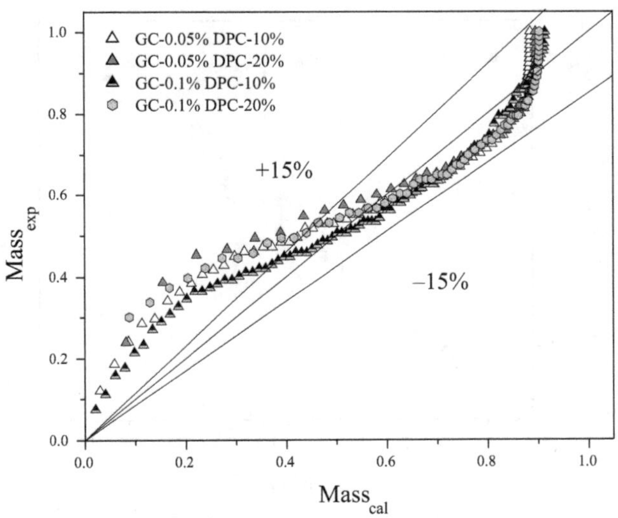

Fig. 8. Correlation between experimental values (massexp) and calculated (masscal) by means of the modified Washburn equation (entire process with gelatin content of 10% and 20%)

Thus, the values conditionally lie in a range of ±15% for the first phases of the imbibition process, while emulsions with the dispersed phase concentration of 20% (gelatin concentrations of 0.1% and 0.05%) and 10% (with emulsifier concentration of 0.05%) are not in the defined range (see Fig. 5).

The values obtained for the second phases of the imbibition process are in a range of ±5% (see Fig. 6).

In case of the entire process fitting by means of the modified Washburn equation, the obtained data were distributed into two sets: the first one for emulsions with the higher disperse phase concentrations (10% and 20%) and the second with low (0.05–1%). The main purpose of such separation was to show the specific tendencies that are representative for emulsions with the lower disperse phase concentration (see Fig. 7) and for another type of the investigated medium (see Fig. 8).

Thus, the modified Washburn equation approximates more precisely the experimental data of the entire process on condition that the disperse phase concentration of an emulsion is low and equals 0.05–1%.

In detail, Figures 9–11 describe the relation between experimental values of the mass change and those calculated by means of the Hill equation.

Considering the results, one may assume that this equation gives a better approximation of the experimental data in comparison with other applied models because of lower bounds within which solutions are found. Thus, values lie within an interval of ±10% for the first phase and the entire process, while in the case of the secondary phase the bounds are delineated in a range of ±5%.

An analysis of the porous material imbibition with oil was also conducted in order to compare with emulsions and obtained experimental data were also approximated by the referred equations. The results of fitting are represented in tabular and graphical forms.

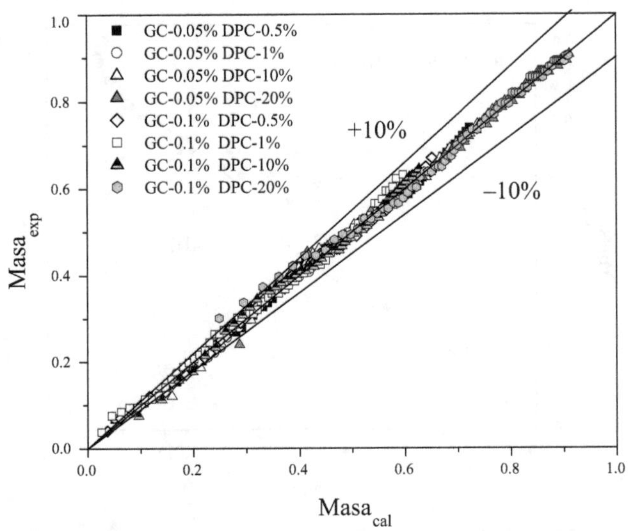

Fig. 9. Correlation between experimental values (massexp) and calculated (masscal) by means of the Hill equation for the first phase of the imbibition process

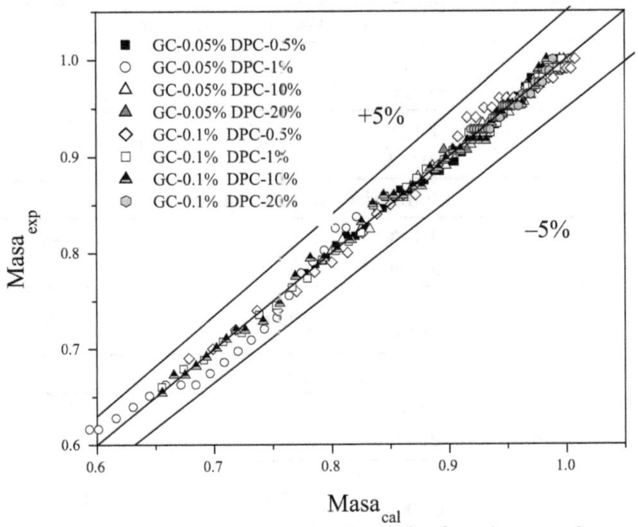

Fig. 10. Correlation between experimental values (massexp) and calculated (masscal) by means of the Hill equation for the second phase of the imbibition process

The accuracy of the models was estimated as in the previous cases via correlation coefficient and normalized root mean square error. The values are given in Table 7.

The values of parameters for the classical Washburn equation (see Eq. 2) in the case of sorbent imbibition with oil significantly differ from parameters obtained for emulsions. The analogous situation is observed for the modified Washburn equation (see Eq. 4).

In case of the Hill equation (see Eq. 5), the value of the parameter Vmax is slightly lower (viz. 1.091± 0.024), but in spite of this it lies close to the typical interval for emulsions (see Table 4 and 7). The same situation is observed for the coefficient *k*. The parameter n is equal to 1.146±0.057, but by contrast, it is mainly higher than in the case of material imbibition with emulsions.

According to the results, the correlation coefficient for the Hill equation is higher compared with other applied models and is equal to 0.992 (see Table 7). The value of the normalized root mean square error is also lower in case of the mentioned equation (viz. it is 2.4).

Thus it is possible to assume that generally approximation with the Hill equation provides the most appropriate and precise description of the process of the porous material imbibition with emulsions and oil.

Figures 12–14 display the relation between experimental data and results received due to fitting by means of the proposed equations. The values are in the range of ±10% in case of fitting via the classical Washburn equation and its modified version (see Fig. 12 and 13) and ±5% if

Table 7. Parameters of the equations applied to describe the porous material imbibition with oil

Parameter of equation	Value of parameter	Correlation coefficient, *R*	Normalized root mean square error, NRMSE
Classical Lucas–Washburn equation			
a_1	0.717 ±0.033		
b_1	1.121 ±0.015	0.989	18.4
Modified Washburn equation			
a_1	4.443 ±0.106		
b_1	4.675 ±0.132	0.983	3.4
Hill equation			
V_{max}	1.091 ±0.024		
K	0.165 ±0.008	0.992	2.4
N	1.146 ±0.057		

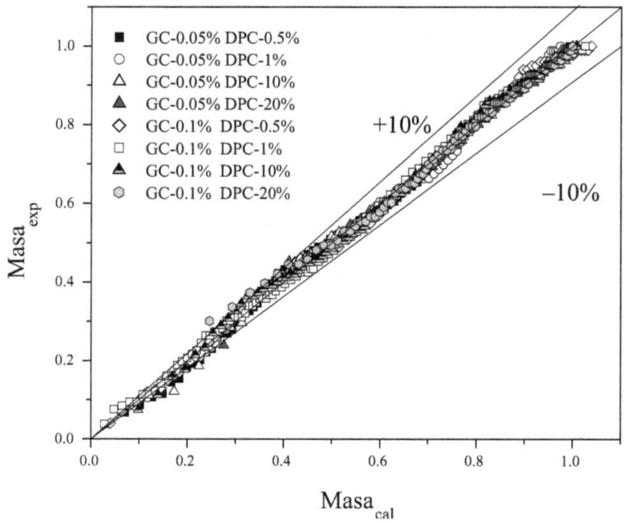

Fig. 11. Correlation between experimental values (massexp) and calculated (masscal) by means of the Hill equation for the entire process of the imbibition process

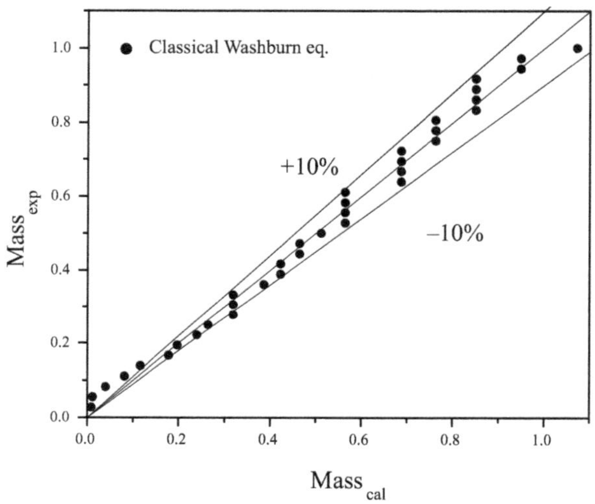

Fig. 12. Correlation between experimental values ($mass_{exp}$) and calculated ($mass_{cal}$) by means of the classical Lucas–Washburn equation for the process of porous material imbibition with oil

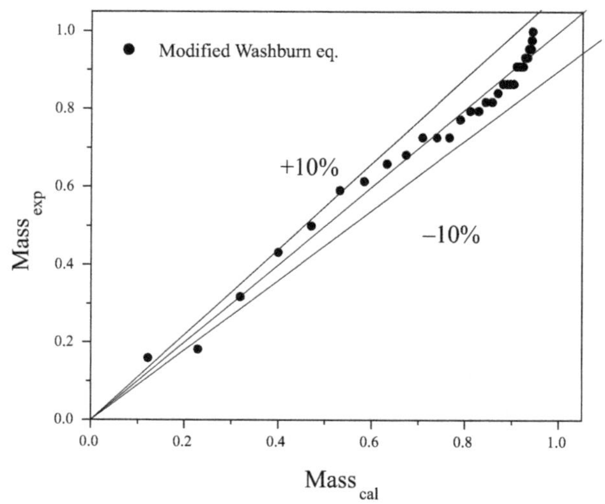

Fig. 13. Correlation between experimental values ($mass_{exp}$) and calculated ($mass_{cal}$) by means of the modified Washburn equation for the oil imbibition process

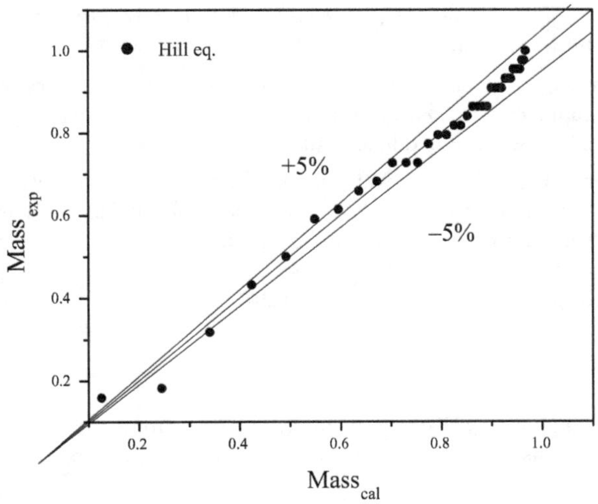

Fig. 14. Correlation between experimental values ($mass_{exp}$) and calculated ($mass_{cal}$) by means of the Hill equation for the oil imbibition process

description is fulfilled by means of the Hill equation (see Fig. 14).

Conclusions

1. A mathematical description of the porous material imbibition with oil and emulsions is discussed. The advanced approach of this publication lies in alteration of the classical and modified Washburn equations via substitution of commonly used parameter of the height, $h(t)$ by the parameter $m_h(t)$ that denotes the mass of an imbibed emulsion/oil.

2. The classical Hill equation was adapted and used to describe and predict the process of the porous material imbibition with petrochemicals for the first time.

3. The results of these investigations show that the highest correlation coefficient values for all investigated cases ($R_{av} = 0.995 \pm 0.001$) and the lowest normalized root mean square error ($NRMSE_{av}$ 1.95±0.138) are mostly defined in case of approximation by means of the Hill equation (exception is 20% emulsion). Moreover, the solutions of fitting via this equation lie within restricted intervals: ±10% for the first phase and the entire process and ±5% for the secondary phase and in case of the imbibition with oil.

4. Thus, summarizing it is possible to assume that the Hill equation provides the most appropriate and precise description of the process of the porous material imbibition with "oil-in-water" emulsions as well as with oil in comparison with other proposed equations.

Reference

Adebajo, M. O.; Frost, R. L.; Kloprogge, J. T.; Carmody, O. 2003. Porous materials for oil spill cleanup: a review of synthesis and absorbing properties, *Journal of Porous Materials* 10(3): 159–170. http://dx.doi.org/10.1023/A:1027484117065

Alava, M.; Dubéc, M. ; Rost, M. 2004. Imbibition in disordered media, *Advances in Physics* 53(2): 83–175. http://dx.doi.org/10.1080/00018730410001687363

Al-Majed, A. A.; Adebayo, A. R.; Hossain, M. E. 2012. A sustainable approach to controlling oil spills, *Journal of Environmental Management* 113: 213–227. http://dx.doi.org/10.1016/j.jenvman.2012.07.034

Chan, T. Y.; Hsu, C. S.; Lin, S. T. 2004. Factors affecting the significance of gravity on the infiltration of a liquid into a porous solid, *Journal of Porous Materials* 11: 273–277. http://dx.doi.org/10.1023/B:JOPO.0000046354.27879.9b

Chebbi, R. 2007. Dynamics of liquid penetration into capillary tubes, *Journal of Colloid and Interface Science* 315(1): 255–260. http://dx.doi.org/10.1016/j.jcis.2007.06.073

Fingas, M. 2011. *Oil spill science and technology: prevention, response, and clean up*. Elsevier, Burlington.

Fries, N.; Dreyer, M. 2008. An analytic solution of capillary rise restrained by gravity, *Journal of Colloid and Interface Science* 320: 259–263. http://dx.doi.org/10.1016/j.jcis.2008.01.009

Gesztelyi, R.; Zsuga, J.; Kemeny-Beke, A.; Varga, B.; Juhasz, B.; Tosaki, A. 2012. The Hill equation and the origin of quantitative pharmacology, *Archive for History of Exact Sciences* 66(4): 427–438. http://dx.doi.org/10.1007/s00407-012-0098-5

Gmurman, V. E. 2004. *Probability theory and mathematical statistics*. 10 ed. Higher School, Moscow (in Russian).

Goutelle, S.; Maurin, M.; Rougier, F.; Barbaut, X.; Bourguignon, L.; Ducher, M.; Pascal, M. 2008. The Hill equation: a review of its capabilities in pharmacological modeling, *Fundamental & Clinical Pharmacology* 22(6): 633–648. http://dx.doi.org/10.1111/j.1472-8206.2008.00633.x

Hamraoui, A.; Nylander, T. 2002. Analytical approach for the Lucas–Washburn equation, *Journal of Colloid and Interface Science* 250(2): 415–421. http://dx.doi.org/10.1006/jcis.2002.8288

Hilpert, M. 2010. Liquid withdrawal from capillary tubes: explicit and implicit analytical solution for constant and dynamic contact angle, *Journal of Colloid and Interface Science* 351(1): 267–276. http://dx.doi.org/10.1016/j.jcis.2010.07.016

Onwurah, I.; Ogugua, V.; Onyike, N.; Ochonogor, A.; Otitoju, O. 2007. Crude oil spills in the environment, effects and some innovative clean-up biotechnologies, *International Journal of Environmental Research* 1(4): 307–320.

Maggi, F.; Alonso-Marroquin, F. 2012. Multiphase capillary flows, *International Journal of Multiphase Flow* 42: 62–73. http://dx.doi.org/10.1016/j.ijmultiphaseflow.2012.01.011

Masoodi, R.; Pillai, K. M.; Varanasi, P. P. 2007. Darcy's law based models for liquid absorption in polymer wicks, *AIChe Journal* 53(11): 2769–2782. http://dx.doi.org/10.1002/aic.11322

Schnoor, J. L. 2010. The Gulf oil spill, *Environmental Science and Technology* 44(13): 4833. http://dx.doi.org/10.1021/es101727m

Siebold, A.; Nardin, M.; Schultz, J.; Walliser, A.; Oppliger, M. 2000. Effect of dynamic contact angle on capillary rise phenomena, *Colloids and Surfaces A: Physicochemical and Engineering Aspects Flow* 161(1): 81–87. http://dx.doi.org/10.1016/S0927-7757(99)00327-1

Shafir, S.; Van Rijn, J.; Rinkevich, B. 2007. Short and long term toxicity of crude oil and oil dispersants to two representative coral species, *Environmental Science and Technology* 41(15): 5571–5574. http://dx.doi.org/10.1021/es0704582

Valentukevičienė, M.; Ignatavičius, G. 2011. Analysis and evaluation of the effect of the solids from road surface runoff on the sediments of river bed, *Ekologija* 57(1): 39–45. http://dx.doi.org/10.6001/ekologija.v57i1.1308

Xue, H. T; Fang, Z. N.; Yang, Y.; Huang, J. P.; Zhou, L. W. 2006. Contact angle determined by spontaneous dynamic capillary rises with hydrostatic effects: experiment and theory, *Chemical Physics Letters* 432(1–3): 326–330. http://dx.doi.org/10.1016/j.cplett.2006.10.017

Wei, Q. F.; Mather, R. R.; Fotheringham, A. F.; Yang, R. D. 2003. Evaluation of polypropylene nonwoven oil sorbents in marine oil spill recovery, *Marine Pollution Bulletin* 46(6): 780–783. http://dx.doi.org/10.1016/S0025-326X(03)00042-0

Zhmud, B. V.; Tiberg, F.; Hallstensson, K. 2000. Dynamics of capillary rise, *Journal of Colloid and Interface Science* 228: 263–269. http://dx.doi.org/10.1006/jcis.2000.6951

Zhu, H.; Qiu, S.; Jiang, W.; Wu, D.; Zhang, C. 2011. A evaluation of electrospun polyvinyl chloride/polystyrene fibers as sorbent materials for oil spill cleanup, *Environmental Science and Technology* 45(10): 4527–4531. http://dx.doi.org/10.1021/es2002343

Jerzy SĘK. DSc Eng., Professor of the Department of Chemical Engineering, Faculty of Process and Environmental Engineering, Lodz University of Technology, Lodz (Poland). Publications: author/co-author of over 50 scientific papers. Research interests: emulsions, rheology, porous media, sorption.

Olga S. SHTYKA. MSc, PhD student of the Department of Chemical Engineering, Faculty of Process and Environmental Engineering, Lodz University of Technology, Lodz (Poland). Publications: author/co-author of over 20 scientific papers. Research interests: emulsions sorbtion, porous medium sorbtion, process engenering, nanomaterials.

Kamila SZYMCZAK. MSc of the Department of Chemical Engineering, Faculty of Process and Environmental Engineering, Lodz University of Technology, Lodz (Poland). Publications: author/co-author of over 5 scientific papers. Research interests: emulsion absorbtion, synthetic materials.

9

EROSIVITY INDICATORS BASED ON RAINFALL IN NORTHWESTERN MEXICO

Mariano NORZAGARAY-CAMPOS[a], Patricia MUÑOZ-SEVILLA[b], Leticia ESPINOSA-CARREÓN[a], Rosario RUÍZ-GUERRERO[c], Héctor GONZÁLEZ-OCAMPO[a], Omar LLANES-CÁRDENAS[a]

[a]*Instituto Politécnico Nacional-CIIDIR-Sinaloa, Blvd. Juan de Dios Bátiz Paredes No. 250, Colonia San Joachín, Guasave Sinaloa, México CP:81101*

[b]*Instituto Politécnico Nacional - CIIEMAD, Calle 30 de Junio de 1520 s/n, Barrio la Laguna Ticomán C.P. 07340 Del. Gustavo A. Madero, Ciudad de México, México*

[c]*Instituto Politécnico Nacional- CIITEC, Cerrada de Cecati S/N. Col. Santa Catarina Azcapotzalco Ciudad de México, México CP: 02250*

Abstract. This study is motivated by the problem of erosivity (R), exacervated in semiarid zones by intense seasonal storms. The purpose was to estimate the spatial variation of R in a coastal area covering 37500 km^2 which is one of the most important agricultural areas in northwestern Mexico. Four methods were used. Rainfall data from 11 SMN-CONAGUA weather stations (from 1966 to 2013) were used to calculate R. The annual average R_1 was 1181.08, and R_2 was 1084.51 MJ mm ha^{-1} h^{-1} with ranges of 2.35–5220.55 and 2.93–4711.38 MJ mm ha^{-1} h^{-1}. Statistical tests showed that a transformation of the data of the form $y = log(x)$, was appropriate for an ANOVA analysis of the data. The value of the test statistic was $F = 1.77$ with $p = 0.149$, showing interdependence between the indicators P ($\alpha = 0.05$). The values of the correlation coefficients for the data were P vs. $R_1 = 0.96$, P vs. $R_2 = 0.99$, P vs. $AI_m = 0.98$, P vs. $MFI = 0.99$. The classification of risk in this region showed that 2017.5 km^2 of the study area was at a very high risk of rain erosion, 2407.5 km^2 under high risk, 5662.5 km^2 under medium to high risk, and 14250 km^2 under low risk. The results are shown on 1:10,000 maps. Results are a set of useful information for soil management programs and for cultivation planning that takes the seasonal variation of R into account in this region where large volumes of extensive crops are grown.

Keywords: rainfall, risk, indicators, semiarid zones, coastal area.

Introduction

Preventing tillage, wind, and the erosive force of rain from reducing soil fertility and degrading the cultural record depends to some degree on successful management of agricultural and forest resources and of water resources to reduce the loss of sediments (FAO 2015; Morgan 2005). A major factor determining soil erosion processes by water is the erosive potential caused by raindrop impact. This issue is of special interest in the agriculture and forestry industries due to the harmful effects of rain erosion on the soil. It is also a key agent in the process of sediment production. Many models have been developed for studying raindrop erosivity (R) and its properties. Parameters such as intensity, speed, size, and kinetic energy are commonly used to develop indexes describing the behavior of R. Among indicators of R, the most common algorithms are: (1) the universal soil loss equation (USLE) (R_1), recognized as one of the best parameters for predicting erosive potential by raindrop impact (Muñoz *et al.* 2011; Oñate 2004; De Santos, De Azevedo 2001); (2) regression curves calculated from empirical pluviographic and pluviometric data for each region (R_2) (Martínez 2005; Pérez, Mesa 2002); (3) the modified Fournier index, better known as climate aggressiveness (MFI) (Yuksel *et al.* 2008) and (4) the Lal index (AI_m) or rainfall erosivity (Lal 1976). Throughout the world, determining the behavior of R is a problem related to climate, especially in agricultural areas where it implicitly has both an environmental and economic impact (Arshad, Martin 2002).

Corresponding author: Patricia Muñoz Sevilla
E-mail: npmsevilla@gmail.com

In northwestern Mexico, three of the most important agricultural valleys; Guasave Valley, El Fuerte Valley and Mocorito Valley, are intensely cultivated. Erosion produced by misuse of the soil has been reported to be an important environmental problem (Llanes *et al.* 2011).

This, together with the additional factor of intense rainfall (*P*), motivated the present study. The goal is to estimate *R* by the four indexes listed above. To date, there have been no studies of the progress of erosion, so it is not known how potential soil loss scenarios might play out in the region. The height of the rainy season (July to September) is a particularly critical period. Furthermore, the absence of any sediment management plans (soil use plans, for example), makes it difficult to counteract erosion and favors factors that promote it. It is important to note that this study was carried out in a region where large volumes of corn, tomatoes, potatoes and other crops of great domestic and international importance are grown. As in other parts of the world, the population explosion and the sharp increase in agriculture and trade in recent years have stimulated a set of environmental problems that have increased erosion levels directly related to structural soil stability (Núñez *et al.* 2007; Llanes *et al.* 2013). The results of this study will provide soil users with a fundamental understanding of the agents that govern sediment production processes. Finding the spatial distribution of *R* and its risks will enable the development of strategies to reduce erosivity in both coastal and mountain environments in this region which is influenced by the dynamics of climate, topography, vegetation and other factors that favor erosion.

1. Materials and methods

The methodologies for measuring indicators of spatial variation of *R* were applied to an area covering approximately 37 500 km^2 in northwestern Mexico. The study area included various hydric conditions and extensive coastal plains with typical rainy seasons from July to September. The main water tributaries are the Sinaloa, El Fuerte and Mocorito Rivers. They arise in the Sierra Madre Mountains in Chihuahua State where the Mohinora and Basoapa currents join (Toutcha *et al.* 2005). After approximately 380 km, they empty into the Gulf of California with annual average rates of flow (*Q*) of 700 to 2,240 m^3 s^{-1}. The high carrying capacities of these rivers readily transport their sediment loads to the final destination. In particular, the sediments of the Sinaloa River are freely carried to its delta due to the dredging carried out by the state government in 2013 that put an end to the problem of accumulated sediments, but increased output of the products of continuous erosion. There are other secondary tributaries; Arroyo De Cabrera, De Ocoroni and De San Rafael, which with their respective flow rates of 18 to 25, 12.3 to 15.2 and 125 to 150 m^3 s^{-1} also transport sediment loads (Fig. 1).

Pluviographic and pluviometric data from 1966 to 2013 were obtained from weather Data Base Northwest México of SMN-CONAGUA (Mexican National Meteorological Service-Mexican National Water Commission). The R_1 index was calculated by the method of Almoza *et al.* (2007), R_2 was calculated using the logarithmic algorithm of Pérez, Mesa (2002). The AI_m index

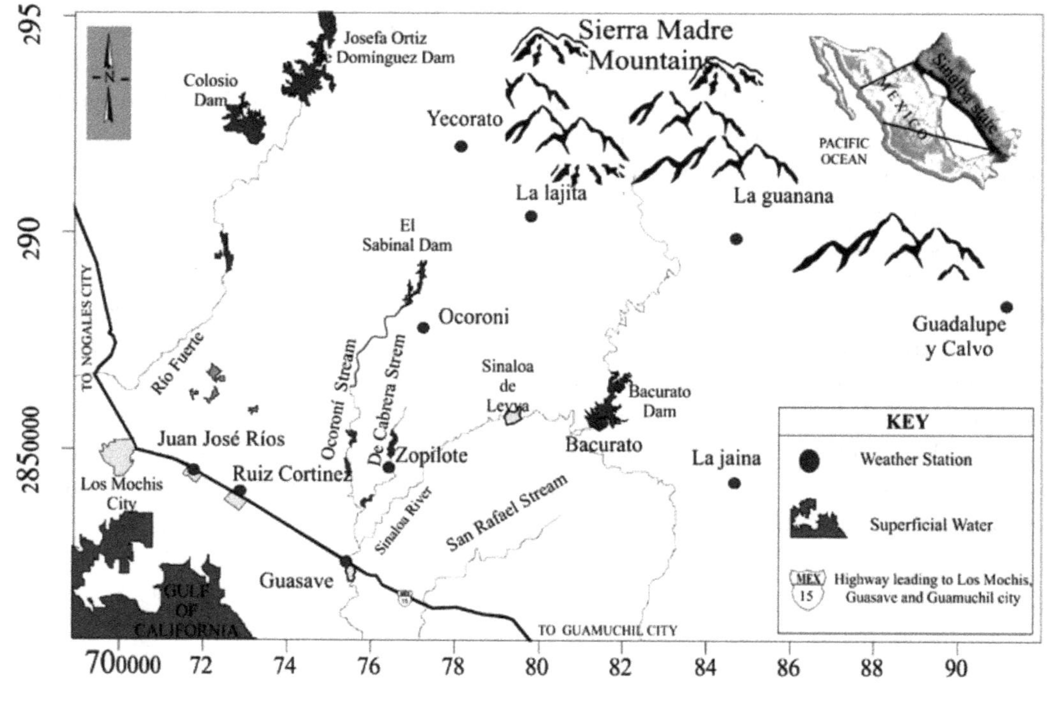

Fig. 1. Location of study area

(rainfall erosivity) was calculated using the method of Lal (1976):

$$AI_m = \sum_{J=1}^{12}\left[\sum_{1}^{n} a I_{\max}\right], \quad i = 1, ..., 12;$$

where a was daily rainfall (cm), I_{max} the maximum intensity of rainfall at 7 minutes (cm h^{-1}), n the number of storms in a month and i indicates the month. The *MFI* index or climate aggressiveness was calculated by the method of Yuksel *et al.* (2008):

$$MFI = \frac{\sum_{i=1}^{12} P^2}{P_i}, \quad i = 1, ..., 12;$$

where P_i was the mean monthly rainfall (mm) and p^2 the mean annual rainfall. The variables used for the estimates R_1 and R_2 were total kinetic energy of rain (E) in MJ ha^{-1} and the maximum intensity of each storm at 30 minutes (I_{30}) in mm h^{-1} (Khosrowpanah, Leroy 2001). R_2 was calculated using the intensity of rainfall and/or a specific increase of rain in pluviophase k (Δik), and the period of increase of the storm or duration of the pluviophase k (Δt_k), both in mm h^{-1} (for more information see Almoza *et al.* 2009). Using pluviographic data, the number of downpours per year (n), the pluviophase of the downpours (q) and unit kinetic energy ($ek = k$); or kinetic energy per mm of rain (MJ ha^{-1} mm^{-1}) were determined. For E, the logarithmic regression $E = a + b \log_{10} I_j$ proposed by Elaheh *et al.* (2012) was used, for which the constants a and b and the intensity of P (I_j) in mm hr^{-1} were estimated. The AI_m index ($cm^2 h^{-1}$) was calculated using daily P (a) in cm, maximum intensity of the rain for 7 minutes in cm h^{-1} (I_{max7}) and the number of downpours per month (n); and for the MFI index, P and its monthly average (p^2), both in mm, were used. The criteria of Da Silva (2004) were used to classify the risk of R by means of R_1 and R_2. The classes were: very low <2452, low 2452–4905, moderate 4905–7357, high 7357–9810, and very high >9810. To compare the annual average of R_1 and R_2 to values in other regions of the world, the ranges shown in Table 1 were used.

To classify risk based on *MFI* values, the categories of Lobo, Gabriels (2005) were applied: 0–60 very low, 60–90 low, 90–120 moderate, 120–160 high, >160 very high. Based on the relationship between the spatial variance of *MFI* and of AI_m (Almoza *et al.* 2007), the following classification was proposed for *MFI*: very low 0–100, low 100–200, moderate 200–450, high 450–1050, and very high >1050.

A statistical analysis was performed to compare the behavior of P with respect to the four indicators. The Shapiro-Wilk test was applied to test for normality in the distributions of the indicators, with significance level a = *0.05*. As all the distributions were found to be skewed, the transformation $y = \log(x)$ was applied to the data in order to make a linear model appropriate and enable estimates of the respective correlations between the dependent variables (R_1, R_2, *MFI* and AI_m) and P to be calculated. To determine whether there were significant differences, a one-factor ANOVA test was applied with five levels, at a significance level of $\alpha > 0.05$. Groups that showed a difference in the ANOVA ($F > 1, p < 0.05$), were tested using a Tukey HSD (Honestly Significant Difference) test, using the criterion 0.05 (95 %). For the ANOVA analysis, the PAST 2.17 b program (Hammer *et al.* 2001) was used and corroborated with the IBM SPSS Statistics 2.1 program (IBM 2012).

The maps were drawn on a scale of 1:10,000 in geographic coordinates (UTM), using kriging interpolation in the SURFER 10.0 program (Emery 2007) and a spatial variation in two regions (UTM: Zone 12 and 13). To represent these two zones, the distance between one station in Zone 12 and another in Zone 13 was measured and added to the initial coordinate, to provide continuity to the interpolations. The maps were finished using CorelDRAW X7.

2. Results and discussion

The averages of P, P^2 and a were 753.31 mm $year^{-1}$, 62.77 mm $month^{-1}$ and 2.09 mm day^{-1} respectively, and their ranges of variation were 70.9–2174.5 mm $year^{-1}$,

Table 1. Average annual variation of rainfall erosivity (R) for comparison of R_1 and R_2 with other regions of the world (MJ mm ha^{-1} h^{-1})

		Rainfall erosivity (R)		
Country	Region	Maximum Value (MJ mm $ha^{-1}h^{-1}$)	Minimum Value (MJ mm $ha^{-1}h^{-1}$)	Author
Ecuador	Ucabanba and San Cristóbal	1, 140.3	1, 231.4	Suffis 2004
Portugal	Flanders	20	3, 741.8	De Santos, De Azevedo 2001
Brazil	Center of Brazil	20, 000	24, 000	Da Silva 2004
Colombia	Caldas Department	800	860	Ramírez, Hincapié 2009
Chile	Curicó Valley	180	285	Mancilla 2008
Cuba	Cuyaguateje Valley	8, 200	18, 000	Almonza 2007
Spain	Provinces of Córdoba, Jaén and Cadiz	40	600	De la Rosa, Moreira 1987

8.90–181.20 mm month^{-1} and 0.74–6.04 mm day^{-1} respectively. Rainfall in the wettest months, July, August and September, with annual monthly averages of 6.09, 113.27 and 143.97 mm year^{-1}, accounted for 78.8 % of total *P*. February, March, April and May, with 3.6 % average annual rainfall, were the driest months (Fig. 2, part A). During the course of the year, *n* varied and was directly proportional to *P*. The highest rainfall occurred during the wettest months and was moderate to low along the coast, moderately low to high in the agricultural area and highest in the Sierra Madre Mountains.

A total of 70 downpours were recorded at six weather stations in the Sierra Madre Mountains; 25 at three coastal stations and 20 at two stations in the central agricultural region. Given that rain is the main cause of damage to structural stability of the soil during the first few minutes of a downpour, the values of *n* were used as a first approximation of the susceptibility of the soil to erosivity (*R*) (Núñez *et al.* 2007) and thus as a first indicator of the expected value of R_1, R_2, *MFI* and AI_m (Fig. 2, part B). The average annual values of I_{30} and I_{max7} were 20.72 and 4.8 mm year^{-1} and the ranges of variation in daily values were 15–30 and 3.5–7 mm. The ranges of Δik and Δt_k were 2.4–3.45 mm h^{-1} and 1.54–3.22 hours, and the average annual $e_k = k$ per mm of rain was 0.105–114 MJ ha^{-1} mm^{-1}. These environmental conditions in the region produced rainy, stormy weather with spatial variation in *E* from 211 to 4474.60 MJ ha^{-1} and an annual average of 1105.47 MJ ha^{-1}. This indicates high *P* and a high expected value for *R* (Fig. 2, part C). The variation of *E* is shown in part C of Figure 3. In the eastern part of the study region, values of *E* were small; they were larger in the central area and intermediate in the west. The Guanana and Bacurato weather stations recorded the highest values. Note in Figure 3 that the variation of *E* is similar to that of *n* and *P*, suggesting that the spatial variance of these variables is correlated (Chica 2005).

Six stations in hilly country and in the Sierra Madre Mountains had a variance of 0.99 (Yecorato, La Lajita, Guanana, Bacurato, Jaina and Guadalupe y Calvo), three

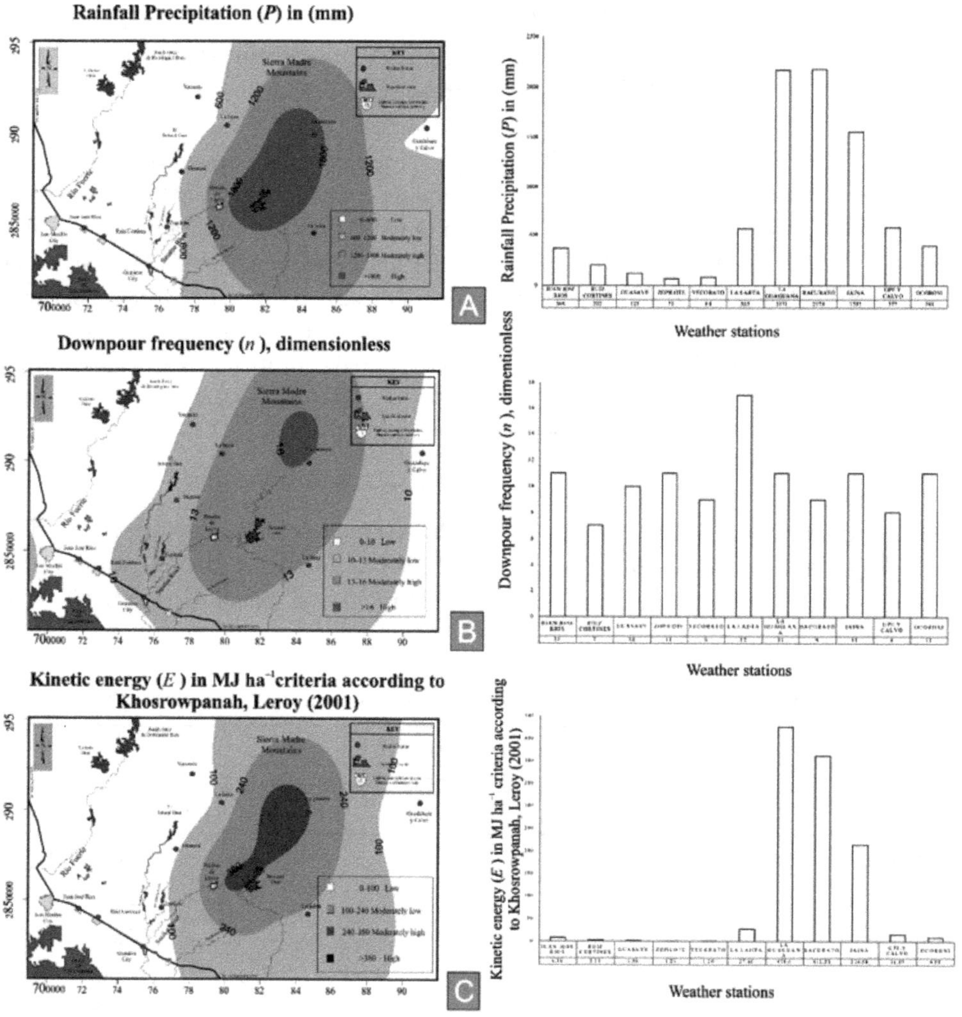

Fig. 2. (A) Spatial variation of annual average rainfall (*P*) in mm, (B) Spatial variation of annual numbers of downpours (*n*) and (C) total kinetic energy (*E*) of rainfall in MJ ha^{-1} in northwestern Mexico

weather stations on the coast had a variance of 0.007 (Juan Jose Ríos, Ruiz Cortínez and Guasave), and two stations in the agricultural region also had a variance of 0.007 (Zopilote and Ocoroni). The high values of n, E and P from localities at higher elevation are 99%, indicative of expected high values for R, particularly during the rainiest months. However, considering that R is based not only on the amount of rainfall, but also on the intensity of I_{max7}, I_{30} and E, high values of R in the agricultural and coastal areas cannot be ruled out. The average annual value of R_1 was 1161.30 and of R_2 1084.51 MJ mm ha^{-1} h^{-1}, with ranges 2.4–3701.2 and 2.93–4081.1 MJ mm ha^{-1} h^{-1} respectively. As shown in Table 1, where these results are compared to values from other regions in the world, these results are similar to those found by Suffis (2004) for Ecuador (Ucabanba and San Cristobal) with annual averages of 1140.3–1231.4 MJ mm ha^{-1} h^{-1}. Considering the influence of the climate, which is less influential than P, and except for values from Brazil where rainfall amounts are high (Da Silva 2004), the annual averages of R_1 and R_2 from northwestern Mexico can be considered high. The spatial variations of R_1 and R_2 are shown on maps A and B in Figure 4, and it can be seen that with a variance of 0.98 in the indicators, they are quite similar, as are E, P and n. The small differences between the maps in Figure 4 were attributed to use of the appropriate logarithmic regression equation that determined E and which was represented by the constants $a = 0.32$ and $b = 2.08$.

The average annual MFI was 130.30 cm^2 h^{-1} and AI_m 781.78 mm. Their ranges were 14.1–362.56 cm^2 h^{-1} and 84.50–2175.22 mm respectively. The values of AI_m and MFI are shown in Figure 5 A and B. As in the maps of R_1 and R_2, there is a spatial pattern in which values are lower in the east, higher in the central area, and intermediate in the west. The Bacurato and Guanana stations showed the highest values. The fraction of AI_m and MFI is analogous to the fraction of R_1 and R_2 in most seasons, as would be expected. The areas where each indicator reached its with the highest risk to the soil as indicated by R. These were the places with the highest values of annual precipitation and the heaviest downpours. In the total 37,500 km^2 study area, using the criteria of Da Silva (2004), the risk as measured by R_1 indicates that 5.38% of the region (2017.5 km^2) has a high to very high risk of erosion by rain, 6.42% (2407.5 km^2) a medium to high risk, 15.1% (5662.5 km^2) a low to medium risk, 38.2% (14325 km^2) low risk and 34.9% (13087.5 km^2) a very low risk. In terms of R_2, the percentage of high to very high was 10.2% (3825 km^2), 14.7% medium to high (5512.5 km^2), 13.7% low to medium (5137.5 km^2), 23.2% low (8700 km^2) and 38.2% very low (14325 km^2). The results obtained for AI_m using the territorial classification of Lobo, Gabriels (2009) were 38% very low risk (14250 km^2), 31.2% low (11 700 km^2), 13.5% moderate (5062.5 km^2), 12.3% high (4612.5 km^2) and 5% very high risk (1875 km^2). Using MFI, 28% of the area (10 500 km^2) was classified as very low risk, 24.7% low (9262.5 km^2), 20.3% moderate (7612.5 km^2), 15% high (5625 km^2) and 12% (4500 km^2) very high risk. The results of the statistical analysis are presented in Table 2, showing that P spatially resembles R_1, R_2, AI_m and MFI. The Shapiro-Wilk

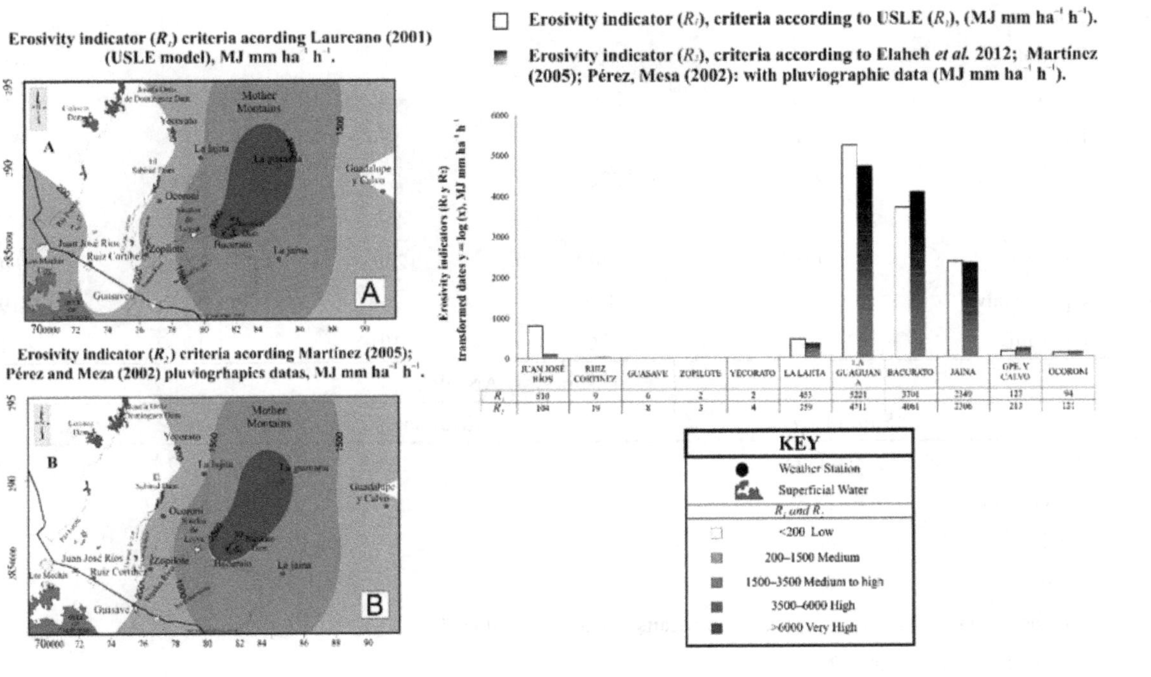

Fig. 3. (A) Spatial variation of annual average factors of rainfall erosivity according to $USLE$ (R_1), (B) annual average factor of erosivity according to the curves of logarithmic regressions (R_2) both in MJ mm ha^{-1} h^{-1} and their respective ranges of risks for soils in northwestern Mexico

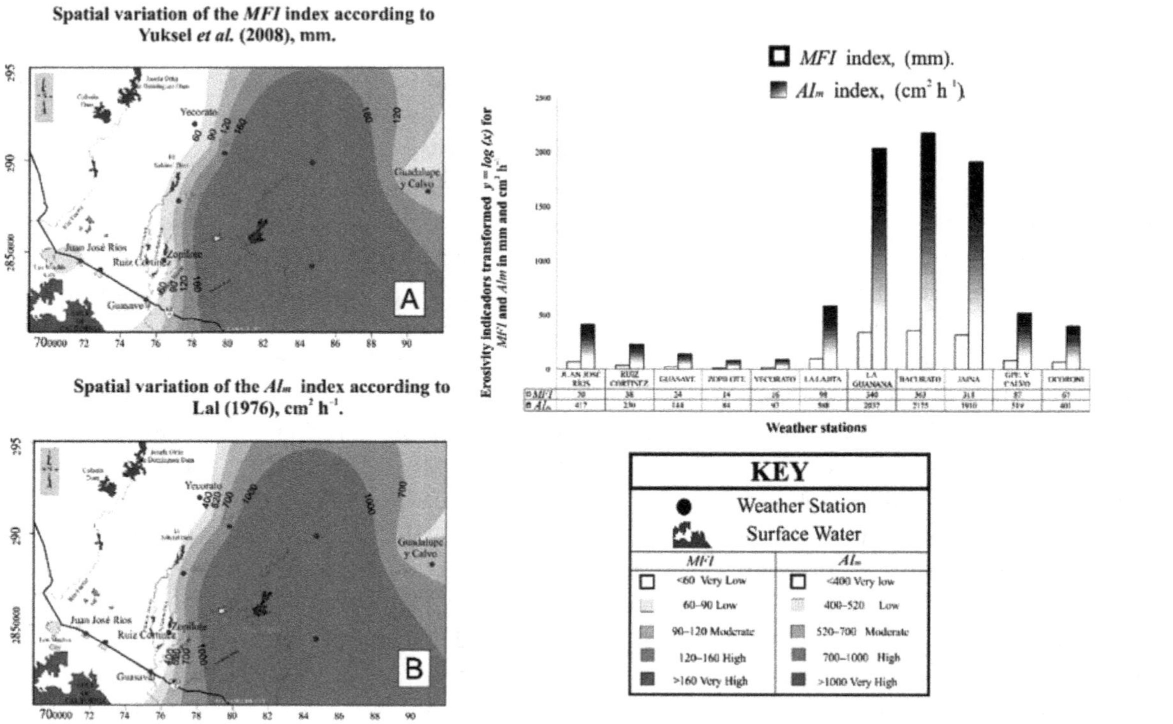

Fig. 4. Spatial variation of the *MFI* index (mm) and AI_m index (cm² h⁻¹)

Table 2. Magnitude of annual average precipitation (*P*) in mm, annual averages of indicators of erosivity: R_1 (MJ mm ha^{-1} h^{-1}), R_2 (MJ mm ha^{-1} h^{-1}), AI_m (cm^2 h^{-1}) and MFI (mm) and ANOVA results in northwestern Mexico.

Number	Weather Station	***P*** (mm year^{-1})	R_1 (MJ mm ha^{-1} h^{-1})	R_2 (MJ mm ha^{-1} h^{-1})	***MFI*** (mm)	AI_m (cm^2 h^{-1})
1	Juan José Ríos	368	810	104	70	417
2	Ruiz Cortínez	202	9	19	38	230
3	Guasave	125	6	8	24	144
4	Zopilote	71	2	3	14	84
5	Yecorato	84	2	4	16	93
6	La Lajita	565	453	359	98	588
7	La Guanana	2171	5221	4711	340	2037
8	Bacurato	2174	3701	4081	363	2175
9	La Jaina	1553	2349	2306	318	1910
10	Gpe. y Calvo	577	127	213	87	519
11	Ocoroni	394	94	121	67	401

One-way ANOVA

	Sum of *sqrs*	*df*	Mean Square	*F*	p (same)
Between groups:	5.469	4	1.367	1.772	0.149
B Within groups:	38.589	50	0.772		
Total:	44.058	54			
Omega^2:	0.0531				

Levene's test for homogeneity of variance, based on means: p (same) = 0.00062

Based on medians: p (same) = 0.00072

Welch F test in the case of unequal variances: $F = 3.961$, $df = 24.45$, $p = 0.01293$

test was applied to test for a normal distribution, $\alpha <$ 0.05. The resulting test statistics and significance levels were: R_1 ($W = 0.714$, $p = 0.0007$), R_2 ($W = 0.663$, $p =$ 0.0002), *MFI* ($W = 0.756$, $p = 0.0025$) and AI_m ($W = 0.755$, $p = 0.0024$), indicating that none of the indices were normally distributed. The data were therefore transformed by $y = log(x)$ ($\alpha > 0.05$). A one-factor ANOVA test was applied to the transformed data. The resulting statistics were R_1 ($W = 0.905$, $p = 0.21$), R_2 ($W = 0.925$, $p = 0.36$), *MFI* ($W = 0.917$, $p = 0.29$), AI_m ($W = 0.916$, $p = 0.28$), supporting the *log* transformation for the data.

The value of the ANOVA test statistic was $F = 1.772$ with $p = 0.149$. Considering the established level for significant differences, a test statistic >1 indicated that means of the groups were not significantly different. However, since the F statistic >1, the results of the Tukey HSD test indicated which of the estimated means were different. The results of the HSD test were: t (20) = 4.16, MSE (Mean square error) = 0.77 and HSD = 1.10. The pairwise comparisons of HSD between indices were: [P] vs. [R_1] = 0.553, [P] vs. [R_2] = 0.529, [P] vs. [MFI] = 0.746, [P] vs. [AI_m] = -0.03, [R_1] vs. [R_2] = -0.023, [R_1] vs. [MFI] = 0.193, [R_1] vs. [AI_m] = -0.583, [R_2] vs. [MFI] = 0.216, [R_2] vs. [AI_m] = -0.559, [MFI] vs. [AI_m] = -0.776. Using the criterion of 0.05 (95%) significance level in the pairwise comparisons, [P] vs. [MFI] and [MFI] vs. [AI_m] are the pairs with the highest differences between their averages, with HSD = 1.10. The results of the ANOVA and Tukey tests show that the 80 % of the results are dependent on P, and any differences observed can be attributed to random variation.

The correlation and interdependence between indicators of R in the region are illustrated in Figure 5, where the graphs are the result of applying a multivariate linear model with four dependent variables characterized by the following coefficients of determination: P vs. $R_1 = 0.91$, P vs. $R_2 = 0.95$, P vs. $AI_m = 0.98$, P vs. $MFI = 0.97$.

Taking into account the spatial behavior of the results in this agricultural region of northwestern Mexico, farmers should schedule tillage and irrigation to prevent damage to the soil; that is, mainly during the months when R is highest, which is manifested not only by the value of P, but also by the influence of I_{30}, I_{max7} and E.

The rainfall in the region has a high erosive potential that is highest in the Guanana and Bacurato region, at elevations between 250 and 350 meters above sea level. This range is the source of both surface water and groundwater in the valley. The variations illustrated in the maps are of great utility for farmers since they indicate which places have greater and lesser R risk and enable planning for sustainable agricultural systems, which is of great importance for the regional and national economy.

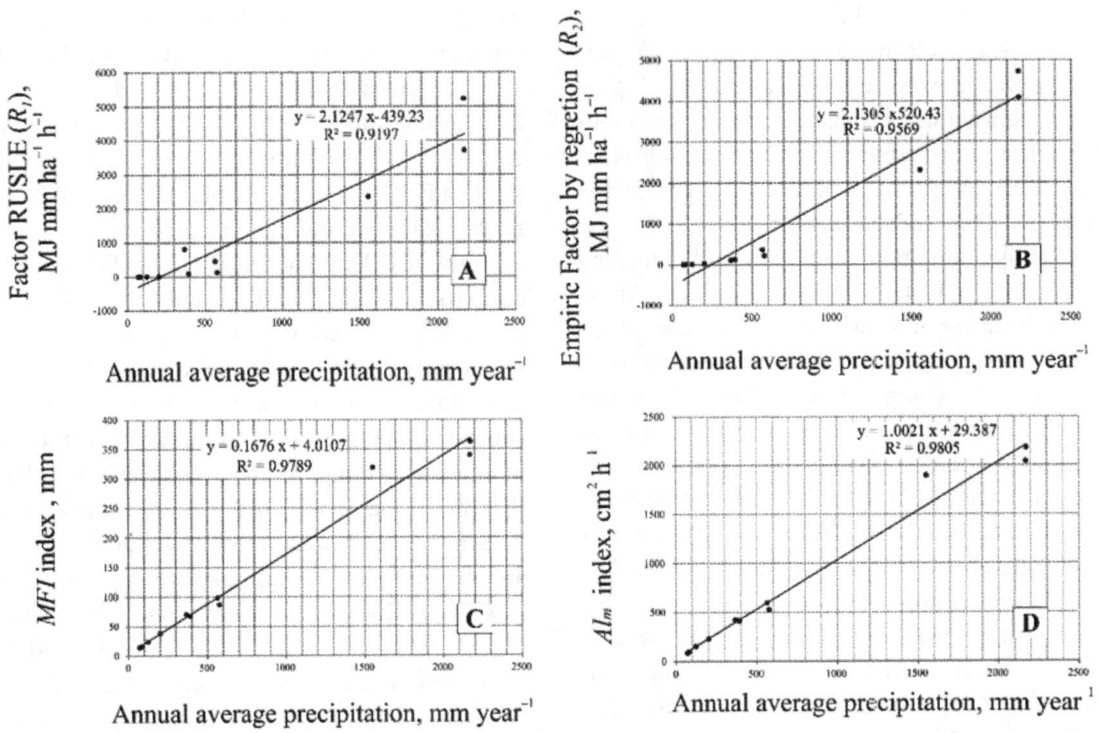

Fig. 5. Linear regression of annual average precipitation (P) in mm with respect to indicators of erosivity: (A) R_1 in MJ mm ha^{-1} h^{-1}, (B) R_2 in MJ mm ha^{-1} h^{-1}, (C) *MFI* index in mm and (D) AI_m index in cm^2 h^{-1}; applied in northwestern Mexico.

Conclusions

The most important cause of degradation that puts soil fertility at risk is erosion caused by water (Zhu *et al.* 2011). It reduces sustainable sources of income in every country and threatens the principles of Conservation Agriculture (CA) (Morgan 2005). Any methodology that can contribute to conserving soil stability is important; and studies such as the present one that can define the erosivity risk of soils in a given agricultural region (here northwestern Mexico) enables management plans that can focus on the first of the three CA principles: (1) Direct planting of crop seeds with minimal mechanical disturbance of the soil; (2) permanent soil cover, especially by crop residues and cover crops and (3) crop diversity.

In spite of increasing interest in CA, there are still sustainable soil management approaches that focus mainly on an input-product approach. Studies such as the present one can help change this by directing attention to a comprehensive vision based on sustainable management. To achieve this, appropriate models and baseline indexes of erosion rates are required, which are rarely considered in developing countries (Seyed, Shahla 2015).

In the context of CA, the results of spatial variation of the four indicators in these agricultural valley regions enabled the areas with the largest values of *R* to be identified. These coincided with the areas showing the greatest erosion problems, and vice versa. This reciprocal behavior between erosion and *R* is attributed to the effect of other factors including the slope of the land, and soil texture and porosity. On the behavior of the slope, note in Figure 3 that in the west where *R* is lower (agricultural area), gentle, rolling hills predominate. In the west where *R* is higher (less erosion), there are steep, rugged slopes (mountainous areas). Integrating information from the four indicators that define soil risk in terms of *R* enabled inference of the spatial nature of a phenomenon that is released in discrete "jumps" and is therefore highly energetic and damaging the longer it interacts with the soil system. Given these properties, the behavior of *R* should be investigated in agricultural valleys around the world and at the same time validated by the methodology described here. The results obtained from applying this methodology can be used as a first indicator of soil loss. However, the reciprocal erosion-erosivity behavior points out the need to take other USLE factors into account in studies predicting total water erosion, as suggested by Wischmeier, Smith (1960). It is urgently recommended that this methodology be applied in areas with gentle slopes where large quantities of rain fall in a short time, in order to develop emerging plans for sites where *R* is high, to recover the dynamic equilibrium of the soil and to conserve dynamic equilibrium where it is low, avoiding losses caused by high *E* when raindrops fall on the soil. This is due to the fact that when *R* is of long duration, the risk of breaking the dynamic equilibrium of the soil is greater than when *R* is frequent and of short duration. When the soil system is broken, it is difficult to restore, and the losses that occur are irreversible. Considering that each system has a different threshold of resistance, that *R* is different in each environment and if *R* surpasses the soil resistance threshold it can result in loss, and that systems tend to evolve, it is important to know and monitor these thresholds in the agricultural valleys of the world. It is therefore important to quantify the risk and the spatial variation of *R* – as in this study – in different scenarios around the world in order to carry out erosion management to control soil loss caused by rain, especially during rainy seasons. In light of this problem, appropriate measures should be taken to minimize the erosive effect of the rainy season, such as maintaining a permanent vegetative cover and refraining from working the land to remove or expose soil during the seasons of heaviest rain. In addition, further studies are needed to estimate *R* at a finer scale. A monitoring network should be established involving more weather stations to provide indexes calculated from rainfall data and/or more complete pluviographic data to create a comprehensive specialized database that will enable numerical modeling of *R* using geographic information systems and specialized software.

In this study, data from eleven weather stations were available, limiting the available data in time and space. Many parts of the world lack a suitable meteorological infrastructure to generate sufficient pluviographic and pluviometric data for such studies, making the necessary data difficult or impossible to obtain. This is commonly the situation in developing countries (Yu *et al.* 2001).

In some countries, the lack of daily and long term data on the intensity of precipitation means that any attempt to model soil erosion and sediment yield based on the USLE-R, USLE and RUSLE factors will be more difficult to calibrate and often inapplicable.

The calculation of actual values of *R* depends on precise estimation of *E* and I_{30}, and the original USLE and USLE-RUSLE method requires pluviographic and pluviometric records from numerous weather stations over a considerable amount of time for calculating EI_{30}. In this context, the present study had data that met these criteria in time but not in geographical distribution, which means that the estimates of *R* are not as precise in space as might be desired. More weather stations should be established and monitored if future studies are to produce better estimates of *R*. The present study was designed following the criteria of Petrovsek, Mikos (2004), who stated that broad approximations can be made with the available data, so it was deemed feasible to proceed with the study and estimate the magnitude and behavior of *R* in space with the existing data. However, it must be acknowledged that the results of this study reflect the variation of the available data in time and space.

González *et al.* (2010) and Taguas *et al.* (2013) report for precipitation regimes in the Mediterranean zone, that high variability in long term records of rain intensity makes the treatment of data less appropriate for calculating the correct values of *R*. Analysis using data ranging over shorter periods tends to result in more accurate estimates of the behavior of *R*. Given this, future work should focus on the use of annual, monthly and daily pluviographic and pluviometric data, which is the form in which these data are normally reported (Shamshad *et al.* 2008). This is the format in which SMN-CONAGUA has always handled the data in Mexico, and if other private and public agencies did so too, it would contribute to soil conservation and CA. Moreover, it would help avoid slow, laborious treatment of long term rain intensity data year after year (Diodato 2005; Da Silva 2004). The application of this methodology to shorter time periods (days, months, one year) would be even better under the recommendations of Ziadat, Taimeh (2013), who state that rain intensity is the most important factor affecting soil erosion, which can occur in relatively small-scale areas of moist soil caused by previous rainfall events. In consequence, locating and identifying areas of moist soil and measuring annual, monthly and daily P can help protect against erosion, and potentially reduce future soil fertility loss and the effects of seasonal dry periods, with resulting benefits to food security and world food subsistence, contributing to meeting the challenges of CA.

Acknowledgments

The authors thank the two anonymous reviewers whose suggestions and recommendations contributed to enhancing the quality of this research. This manuscript is the result of research undertaken by the first author since the 1980s in northwestern Mexico. He is very grateful to all the people who in any way, directly or indirectly, contributed to achievement of the goals of this work. The authors are fellows of EDI and COFAA; and of SNI-CONACYT. They acknowledge the support of these institutions.

References

Almoza, Y.; Ruiz, E.; Alonso, G. 2007. Determinación de la erosividad de las precipitaciones en el Oeste de Cuba, *Zonas Áridas* 11(1): 139–149.

Almoza, Y.; Ruíz, M.; Medina, H. 2009. Cambios climáticos y sus efectos en las pérdidas de suelo en la cuenca del río Cuyaguateje, *Revista Ciencias Técnicas Agropecuarias* 18(2): 32–39.

Arshad, M.; Martin, S. 2002. Identifying critical limits for soil quality indicators in agro-ecosystems, *Agriculture Ecosystems and Environment* 88(1): 153–160. http://dx.doi.org/10.1016/S0167-8809(01)00252-3

Chica, O. M. 2005. La Geoestadística como herramienta de análisis espacial de datos de inventario forestal: análisis espacial de datos de inventario. En: Actas de la I Reunión de Inventario y Teledetección Forestal, *Cuadernos de la Sociedad Española de Ciencias Forestales* 19(1): 47–55.

SMN-CONAGUA. 2015. Weather data base Northwest México. *Mexican National Meteorological Service- Mexican National Water Commission* [online], [cited 15 January 2015]. Available from Internet: http://peac-bc.cicese.mx/datosclim/dcbc.php

Da Silva, A. M. 2004. Rainfall erosivity map for Brazil, *Catena* 57: 251–259. http://dx.doi.org/10.1016/j.catena.2003.11.006

De la Rosa, D.; Moreira, J. M. 1987. *Evaluación ecológica de Recursos Naturales de Andalucía.* Ed. Agencia de Medio Ambiente. Junta de Andalucía. 192 p.

De Santos, L. N.; De Azevedo, C. M. 2001. A new procedure to estimate the RUSLE EI_{30} Index based on monthly rainfall data and applied to the Algarve region, Portugal, *Journal of Hydrology* 250: 12–18. http://dx.doi.org/10.1016/S0022-1694(01)00387-0

Diodato, N. 2005. Predicting RUSLE (revised universal soil loss equation) monthly erosivity index from readily available rainfall data in Mediterranean area, *Environmentalist* 25: 63–70.

Elaheh, M.; Davood, N.; Mohammad, H.; Ebrahim, P. 2012. Investigating rainfall erosivity indices in arid and semiarid climates of Iran, *Turkish Journal of Agriculture and Forestry* 36: 365–378.

Emery, X. 2007. Conditioning simulations of Gaussian random fields by ordinary kriging, *Mathematical Geology* 39(6): 607–623. ISSN 0186-2979.

FAO. 2015. *Principios básicos de la Agricultura de Conservación.* Organización de las Naciones Unidas para la Agricultura y la Alimentación. Departamento de Agricultura y Protección del Consumidor-Agricultura de conservación [online], [cited 28 April 2015]. Available from Internet: http://www.fao.org/ag/ca/es/1b.html.

González, H. J. C.; Batalla, R. J.; Cerda, A.; De Luis, M. 2010. Contribution of the largest events to suspended sediment transport across the USA, *Land Degradation and Development Journal* 21(1): 83–91. http://dx.doi.org/10.1002/ldr.897

Hammer, O.; Harper, D.; Ryan, J. 2001. Paleontological statistics software package for education and data analysis, *Paleontología Electronica* 4: 9–9.

IBM. 2012. *IBM*SPSS* Statistics.* International business machines and statistical product and service solutions. United States. 25 p.

Khosrowpanah, S.; Leroy, F. 2001. *Rainfall erosivity factors (R-factors) for selected islands in the federated states of Micronesia.* Water and Environmental Research Institute of the Western Pacific University of Guam. Technical Report Number 92. 28 p.

Lal, R. 1976. *Analysis of factors affecting rainfall erosivity, in International Conference on Soil Conservation and Management in the Humid Tropics*, 10 September 1976, International Institute of Tropical Agriculture and Agricultural Research Council of Niger, Ibadan, Nigeria. 54 p.

Llanes, C. O.; Norzagaray, C. M.; Muñoz, S. P. 2013. *Análisis espectrales en acuíferos del Noroeste de México: El proceso de desertificación en zonas semiáridas.* Editorial Publicia. 120 p.

Llanes, C. O.; Norzagaray, C. M; Maya, D. Y.; Muñoz, S. P.; Beltrán, M. F. A.; Murillo, A. B.; Troyo, D. E. 2011. Efectos hidroambientales de la extracción de agua del acuífero del Río

Sinaloa, *Universidad y Ciencia* 27(3): 239–249.

Lobo, D.; Gabriels, D. 2005. Assessing the rain erosivity and rain distribution in different agro-climatological zones in Venezuela, *Sociedad y Naturaleza Special Issue* 1: 16–29.

Mancilla, G. A. 2008. *Uso de la ecuación universal de pérdidas de suelo en el campo forestal: apuntes docentes*. Facultad de Ciencias Forestales, Departamento de Silvicultura, Universidad de Chile, Chile.

Martínez, M. 2005. *Estimación de la erosión del suelo.* Secretaría de Agricultura, Ganadería, Desarrollo Rural, Pesca y Alimentación (SAGARPA). México. 156 p.

Morgan, R. P. C. 2005. *Soil erosion and conservation*. 3rd ed. UK: Blackwell Publishing Ltd. 269 p.

Muñoz, S. P.; Menanteu, L.; Escobedo, U. 2011. *Changes in land use and their impact on the coastal zone in northern Sinaloa (Gulf of California, Mexico)*. Southampton, Boston: Wessex Intitute of Technology. 206 p.

Núñez, C.; Pérez, M.; Cuesta, S.; Cabrera, J. 2007. *Procesos de degradación del suelo*. Grupo de conservación y recuperación de suelos contaminados. Centro De Investigaciones Energéticas Medioambientales y Tecnológicas (CIEMAT). 67 p.

Oñate, V. 2004. Metodología para la evaluación del riesgo de erosión hídrica en zonas áridas y su aplicación en el manejo y protección de los proyectos hidráulicos, *Revista electrónica de la REDLACH* 1(1): 6–12.

Pérez, A.; Mesa, S. 2002. *Estimación del factor de erosividad de la lluvia en Colombia, in Conferencia Internacional: Memorias I Simposio Latinoamericano de Control de Erosión. Corporación Autónoma Regional para la Defensa de la Meseta de Bucaramanga-Sociedad Colombiana de Geotecnia*, 12 March 2002, Bucaramanga, Colombia, 18–22.

Petrovsek, G.; Mikos, M. 2004. Estimating the R factor from daily rainfall data in the sub-Mediterranean climate of southwest Slovenia, *Journal of Hydrology* 49: 869–877.

Ramírez, O. F. A.; Hincapié, G. E. 2009. Riesgo a la erosión en la zona cafetera central del departamento de caldas, *Cenicafé* 60(2): 173–189.

Seyed, H. S.; Shahla, T. 2015. Development of stational models for estimation of rainfall environment, *Land Degradation and Development Journal* 24: 582–590.

Shamshad, A.; Azhari, M. N.; Isa, M. H.; Wan, H. W. M.; Parida, B. P. 2008. Development of an appropriate procedure for estimation of RUSLE EI30 index and preparation of erosivity maps for Pulau Penang in Peninsular Malaysia, *Catena* 72: 423–432. http://dx.doi.org/10.1016/j.catena.2007.08.002

Suffis, J. 2004. *Alternative land use systems for controlling water erosion on steep slopes in small watersheds of the Paute River Basin (Ecuador)*. Universiteit Gent, Gent.

Taguas, E. V.; Carpintero, E.; Ayuso, J. L. 2013. Assessing land degradation risk through the long-term analysis of erosivity: a case study in southern Spain, *Land Degradation and Development Journal* 24(1)9: 179–187.

Toutcha, L.; Sosa, M.; Soto, R. 2005. La flora de las barrancas del Cobre, Chihuahua, México, *Ecología Aplicada* 4(12): 17–23.

Wischmeier, W. H.; Smith, D. D. 1960. A universal soil-loss equation to guide conservation farm planning, in *International Congress of Soil Science, Trans,* November 1960, Madison, Wisconsin, United States, 418–425.

Yu, B.; Hashim, G. M.; Eusof, Z. 2001. Estimating the R-factor with limited rainfall data: a case study from peninsular Malaysia, *Journal of Soil and Water Conservation* 56: 101–105.

Yuksel, A.; Gundogan, R.; Akay, A. 2008. Using the remote sensing and GIS technology for erosion risk mapping of Kartalkaya dam watershed in Kahramanmaras, Turkey, *Sensors* 8: 4851–4865. http://dx.doi.org/10.3390/s8084851

Zhu, Q.; Chen, X. W.; Fan, Q. X. 2011. A new procedure to estimate the rainfall erosivity factor based on Tropical Rainfall Measuring Mission (TRMM) data, *Science China Technological Sciences* 54: 2437–2445.
http://dx.doi.org/10.1007/s11431-011-4468-z

Ziadat, F. M.; Taimeh, A. Y. 2013. Effect of rainfall intensity, slope, land use and antecedent soil moisture on soil erosion in an arid environment, *Land Degradation and Development* 24: 582–590. http://dx.doi.org/10.1002/ldr.2239

Mariano NORZAGARAY-CAMPOS. PhD. Specializing in groundwater, climate change and water resources. Research Professor at the National Polytechnic Institute at the Interdisciplinary Center of Research for Regional Integral Development (CIIDIR-IPN-Sinaloa). Professor in the Master of Natural Resources and Environment and Conservation, PhD in natural heritage.

Patricia MUÑOZ-SEVILLA. PhD. Specializing in water resources and climate change. Research Professor at the National Polytechnic Institute and Ministry of Research and Graduate Studies. Professor in the Master of Natural Resources and Environment and Conservation, PhD in natural heritage.

Teresa Leticia ESPINOSA-CARREÓN. PhD. Specializing in physical-biological connections. Research at National Polytechnic Institute. Professor in the Master of Natural Resources and Environment, and Environment and Conservation, PhD in natural heritage.

Rosario RUIZ-GUERRERO. PhD. Specializing in water resources and analytical chemistry. Research Professor at the National Polytechnic Institute. Professor in the Master in Advanced Technology and PhD in Advanced Technology.

Héctor Abeardo GONZÁLEZ-OCAMPO. PhD. Specializing in environmental impact of human activities and sustainable aquaculture. Research Professor at the National Polytechnic Institute at the Interdisciplinary Center of Research for Regional Integral Development (CIIDIR-IPN-Sinaloa). Professor in the Master of Natural Resources and Environment and Conservation, PhD in natural heritage.

Omar LLANES-CÁRDENAS. PhD. Specializing in groundwater, climate change and water resources. Research Professor at the National Polytechnic Institute at the Interdisciplinary Center of Research for Regional Integral Development (CIIDIR-IPN-Sinaloa). Professor in the Master of Natural Resources and Environment Conservation.

10

A SIMPLIFIED MULTI-CRITERIA EVALUATION MODEL FOR LANDFILL SITE RANKING AND SELECTION BASED ON AHP AND GIS

Debishree KHAN, Sukha Ranjan SAMADDER

Department of Environmental Science & Engineering, Indian School of Mines, Dhanbad-826004, India

Abstract. This study used GIS based Multi-criteria Decision Analysis (MCDA) approach for evaluating the most environmentally suitable landfill sites in the study area. The weights of relative importance of the factors guiding landfill siting were estimated using pair-wise comparisons in AHP. The maps showing suitable landfill sites were generated applying a weighted linear combination (WLC) in GIS using a comparison matrix to aggregate different significant scenarios associated with environmental and economic objectives. To determine the appropriate areas where landfill sites can be located, thematic maps for all the criteria were generated using GIS. A final map was produced showing suitability for the location of the landfill sites. The suitable sites having an area equal to or above 4 ha at one place and 90% of which is barren land were considered suitable for landfill. The selected candidate sites were ranked to get the most desirable sites for landfill.

Keywords: analytic hierarchy process, Dhanbad, geographical information system, landfill siting, multi-criteria decision making, municipal solid waste, weighted linear combination.

Introduction

Municipal Solid Waste Management (MSWM) in developing nations is one of the biggest environmental challenges due to rapid population growth that demands more and more suitable sites for municipal solid waste disposal. The generation of MSW is an unavoidable result of human activity which involves the generation, onsite storage, collection, transfer and transportation, recycling, treatment and disposal (Chiueh, Yu 2006). Waste generation is the result of inefficient and unsustainable uses of natural resources and energy in production processes, which leads to economic and environmental problems, such as large additional expenses for the collection, processing and waste landfilling and elimination (Schiopu *et al.* 2012; Simion *et al.* 2013). Unfortunately, materials resulting from construction and demolition are often considered as waste for disposal, rather than resources for processing and reuse. They have negative impacts on environment, economy, public health, and social life (Ionescu *et al.* 2011; Robu *et al.* 2007). The management of Municipal Solid Waste (MSW) is going through a critical phase due to the unavailability of suitable facilities for treatment and disposal of large amounts of MSW generated in metropolitan cities. Unscientific disposal causes adverse impacts on all aspects of the environment and human health (Rathi 2006; Sharholy *et al.* 2008). In India, it is very often observed that due to limited availability of land and lack of regulatory measures, the MSW is disposed-off on low-lying areas without taking any precautions or operational controls. The present practice of MSWM demands an environment friendly and suitable landfill sites for waste management in fast growing regions of developing nations as the landfill is the most common method for disposal of MSW. The "not in my backyard" (NIMBY) and "not in anyone's backyard" (NIABY) concepts (Chang *et al.* 2008; Kontos *et al.* 2003) are popular in most cities of developing nations that create problems for the decision makers involved in MSWM.

An inappropriate landfill site may have negative impacts on the environment. Therefore, it should be selected carefully by considering both regulatory requirements and other constraints (Şener 2004). The siting of MSW landfill is a spatial multi-criteria decision analysis (SMCDA) for which both GIS and MCDA methods should be used (Kontos *et al.* 2005). Some relevant studies on landfill siting and solid waste management have integrated GIS and MCDA in various ways which also include analytic hierarchy process(AHP) using weighted linear combination (WLC) (Carver 1991; Siddiqui *et al.* 1996; Charnpratheep *et al.* 1997; Kao *et al.* 1997; Lukasheh *et al.* 2001; Kontos *et al.* 2003, 2005; Higgs 2006; Şener *et al.* 2006;

Corresponding author: Sukha Ranjan Samadder
E-mail: sukh_samadder@yahoo.co.in

Chang *et al.* 2008; Delgado *et al.* 2008; Wang *et al.* 2009; Şener *et al.* 2010; Kara, Doratli 2012; Yildirim 2012; Achillas *et al.* 2013; Allesch, Brunner 2014; Demosouka *et al.* 2014).

The present paper has considered the ranking of selected candidate landfill sites using linear combination of weights derived from AHP which has not been done in earlier studies on landfill site selection. This study will help the municipal engineers to get the most precise environmentally sound landfill sites with suitability ranking. Siting decisions are governed by the pre-existing land use dynamics of the urban area as well as the nature of potential interactions of the landfill with the pre-existing environmental, geologic, hydrological, and socio-economic parameters of the area (Sumathi *et al.* 2008). Siddiqui *et al.* (1996) were among the first to integrate GIS and AHP for landfill siting. This integration plays a significant role in siting landfills for solid waste management. Complex geographical data can be represented by thematic layers and correlation among large volumes of discrete geographical information can be attributed together to provide the demand of the investigators. Hence GIS and remote sensing techniques can be applied in the field of MSWM using a multi-criteria decision making technique to provide the organized approach for assessing and integrating the impact of various factors as indicators of suitability of landfill sites.

On the other hand AHP is a powerful tool for solving complicated problems that may have interactions and correlations among multiple objectives. This technique provides a means of decomposing the problem into a hierarchy of sub-problems that can be more easily comprehended and subjectively evaluated. The subjective evaluations are converted into numerical values that are ranked on a numerical scale (Bhushan, Rai 2004). This present study examined the applicability of the GIS techniques in combination with a multi-criteria approach for analyzing the land suitability for landfill. The most prevalent procedure for integrating multi-criteria evaluation and multi-objective evaluation (MOE) in GIS for land suitability analysis is WLC approach (Hopkins 1977). The WLC procedure allows full trade off among all factors and offers much more flexibility than the Boolean approaches. Most of these methods were developed for case studies by considering the regionalised social and geo-environmental criteria. The generalised global method might not be an appropriate method for landfill suitability at regional level as there are lots of evaluation criteria mentioned in the literature but their existence vary from region to region. Previous studies have considered many more criteria to make the selection process more precise but the main challenges in this study were the unavailability and proper applicability of the data. The purpose of this study was to locate the suitable landfill sites in Dhanbad city, Jharkhand, India for MSW disposal using AHP and WLC approach by considering various localized factors that may influence locating landfill sites in the study area. The criteria which were having more environmental impact and appropriate for the present study area were only considered in the study to keep the methodology simple so that decision makers can easily adopt the proposed methodology. Different standardized scales were used in the process of assigning the weightage to each one of the criteria of landfill siting and a ranking process was adopted to get the candidate site ranking order.

1. Materials and methodology

1.1. Study area

Dhanbad is one of the mineral rich (especially coal reserves) districts of India located in the state of Jharkhand. The region lies on the eastern part of Chhotanagpur plateau and has an undulating topography. The study area extends from 86° 06′ 11″ E to 86° 50′ 26″ E longitude and from 23°38′ 58″ N to 24° 03′ 30″ N latitude (Fig. 1). Present study area covers the main agglomerated urban region of the city within Dhanbad Municipal Corporation (DMC), consisting of 24 administrative wards. The total area of the present study region is about 101 km^2 that supports about 0.49 million population as per census of India, 2011. The density of population was estimated to be 2947 persons/km^2 and it is rising at an annual rate of 1.5%. The study area mainly consists of residential area, agricultural area, scrubs and quarry areas. It has considerable scope of expansion and may alter the land-use pattern in the near future for which overall planning of municipal facilities including municipal solid waste management in the city as a whole is required. The economic growth in the study area due to the extraction of mineral reserves has led to a rapid growth in population and change in land-use pattern that resulted in increase of solid waste generation. At present the study area has neither any proper demarcated landfill sites, nor any organized method of solid waste disposal.

1.2. Pre-processing of data

The present study requires satellite images, toposheets containing ground information, local administrative maps, and field observation data. Prior to the beginning of suitability analysis, the data were corrected and processed to facilitate GIS functionality of the data. To perform a land-use/landcover mapping, atmospheric correction and precise geometric registration of images with location accuracy is a pre-requisite. Atmospheric scattering contributes some additional spectral values to the ground reflectance (Jensen 1986). In this study, the LISS IV image was corrected for atmospheric path radiance using dark object subtraction method (Chavez 1988). These images were geometrically corrected by taking well distributed Ground Control Points (GCPs) that were extracted from topographical maps of the study area. The PAN images were registered with the LISS IV images up to the sub-pixel accuracy. The pre-processed images were pan sharpened before the land-use classification method was used as a pan sharpened image gives best of both the images with high spectral resolution (multi-spectral image) and high spatial resolution (panchromatic image). In this study interactive visual interpretation method was applied to classify the pan-sharpened LISS IV image into existing land-use. Some land-use units were recorded using the reference from high resolution Google Earth image to make the land-use map more precise. The land-use map of the present study area is shown in Figure 2.

Slope, aspect and flow accumulation were derived from Carto-DEM of the study area, road map, groundwater depth map and soil map were extracted from PAN image, CGWB (Central Ground Water Board, India) map and NBSS (National Bureau of Soil Survey, India) map respectively. Land-use related criteria layers were extracted from the land-use map prepared from the pan sharpened LISS IV image.

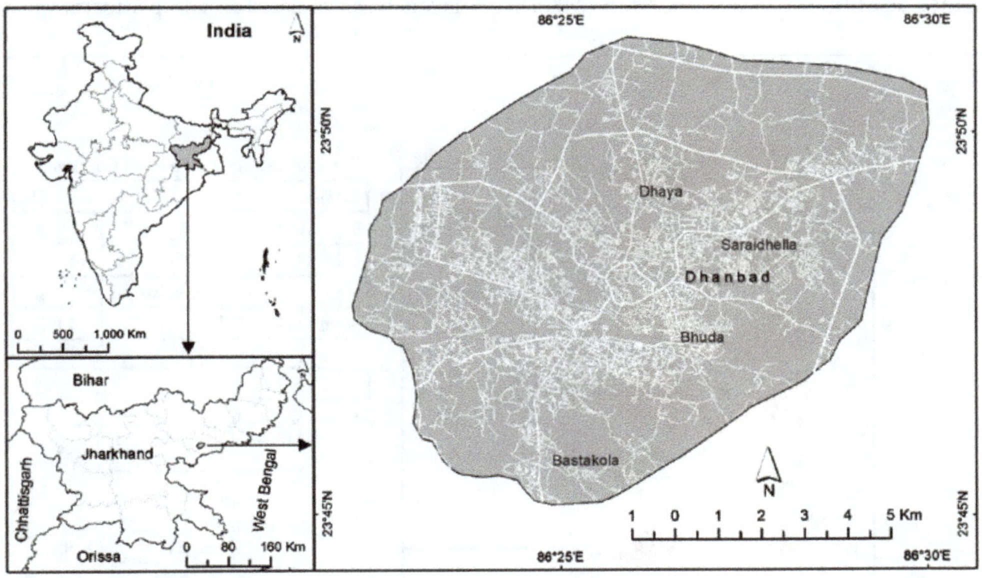

Fig. 1. Location of the study area

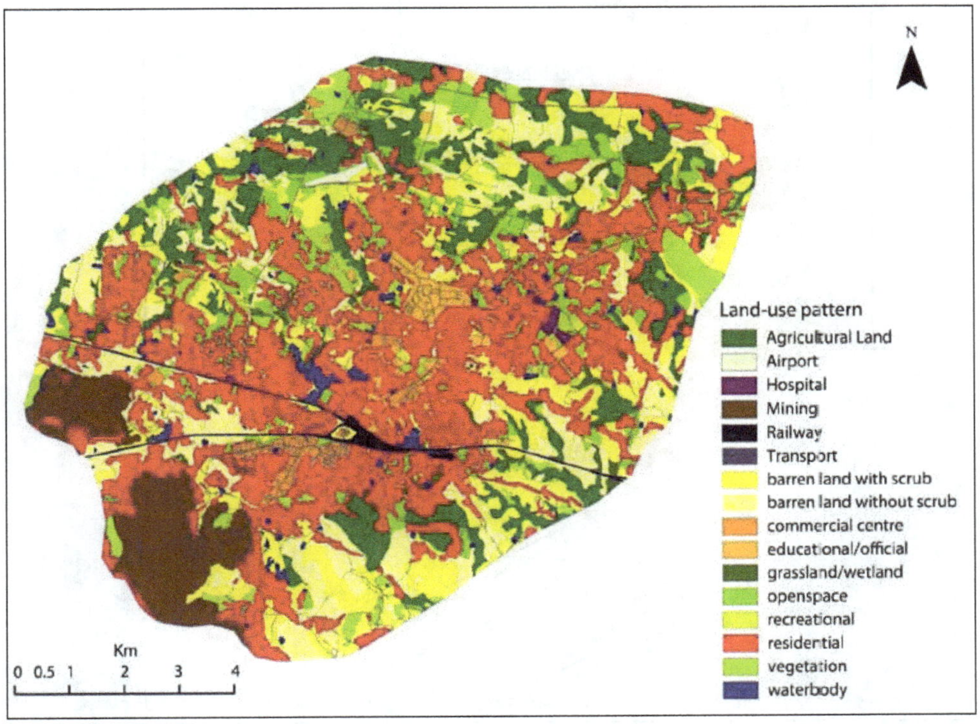

Fig. 2. Land-use of the study area

1.3. Methodology

The stepwise operational methodology used in this study is portrayed in Figure 3 and explained in detail in the following sections:

1.3.1. Multi-criteria evaluation

Spatial multi-criteria decision making (MCDM) is a process where geographical data are combined and transformed into a decision. It involves input data, the decision makers' performances and manipulation of both information using specified decision rules. The actual process of applying the decision rule is evaluation. In order to meet the objective, several criteria need to be evaluated and is termed as Multi-Criteria Evaluation (MCE). A variety of scales on which criteria and sub-criteria can be measured.

MCDA requires the values contained in the various criterion map layers to be transformed into comparable units (Moeinaddini *et al.* 2010). Criteria maps can be classified on the basis of the types of available information for map construction. In order to compare criteria with each other, all units were transformed into the same unit of measurement scale. Standardization was done by assigning a numeric value between 1 and 5, where 1 signified the maximum suitability rate and 5 as the not suitable for each criterion (Table 1). Each of the criteria was standardized based on expert opinions considering the potentiality of the land for landfill site and then each of the prepared layers was evaluated in GIS platform.

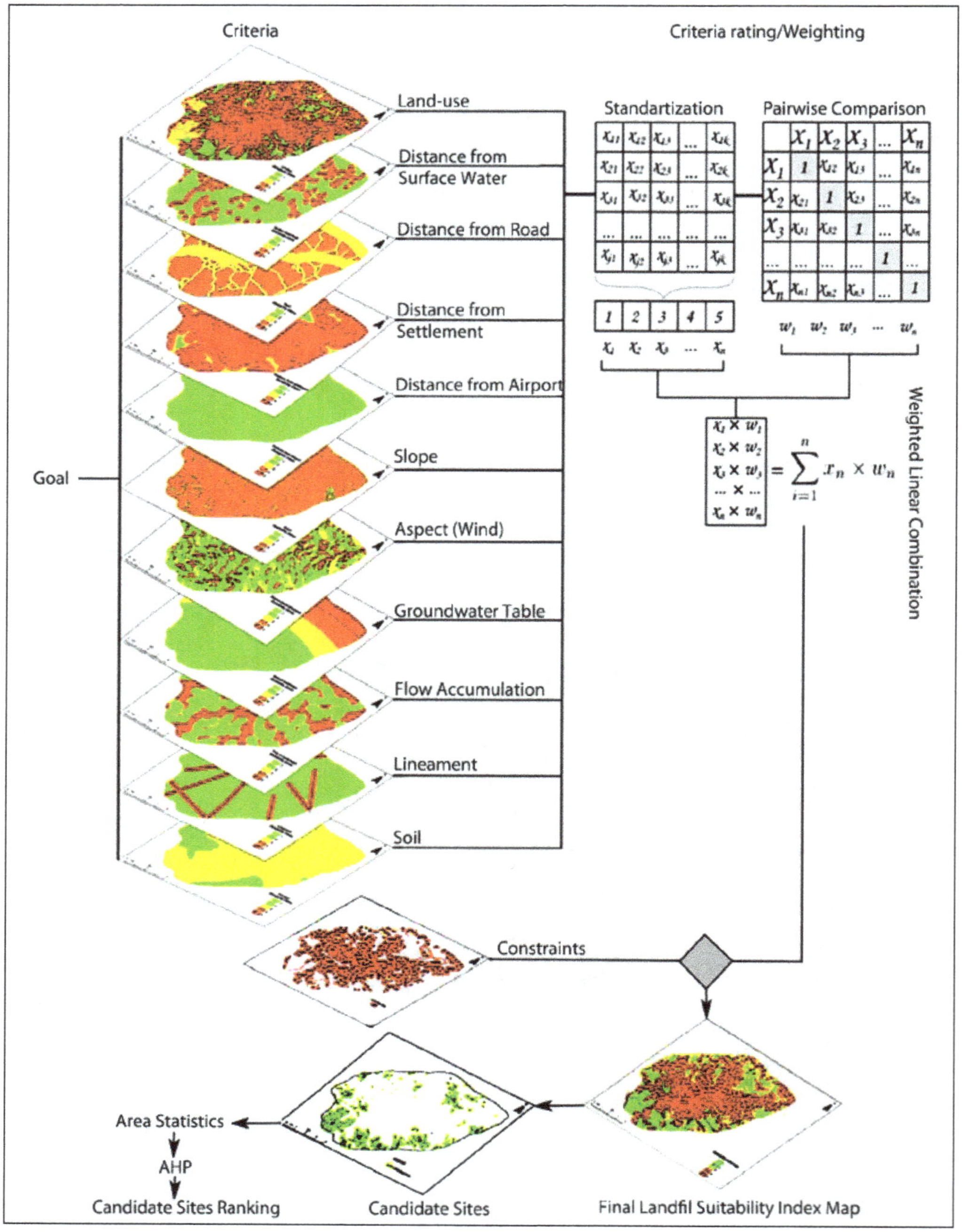

Fig. 3. Methodology of the study

1.3.2. Assigning criteria weights

AHP is one of the multiple criteria decision-making methods that were originally developed by Saaty (1977). It has the capability to simplify preference ratings among decision criteria using pair-wise comparisons. The AHP technique is used to evaluate each criterion as multiple important objectives are difficult to prioritize. The weights of specific criteria are assigned using Saaty's technique (Saaty 1977), in which a ratio (reciprocal) matrix is constructed where each criterion is compared with the other criteria, relative to its importance on a scale with values from a set {1/9,1/8, 1/7, 1/6, 1/5, 1/4, 1/3, 1/2, 1, 2, 3, 4, 5, 6, 7, 8, 9}, ranging from the minimum value of 1/9 representing the least important, to 1 for equal importance and to 9 for the most important.

The weights are calculated by normalizing the eigen-vector associated with the maximum eigen-value of the matrix.

The calculation involves the following steps: (a) computation of sum of values in each column of pair-wise matrix, (b) normalization of the matrix by dividing each element by its column total, (c) computation of mean of the elements in each row of the normalized matrix.

Then the consistency ratio (CR) is computed to check the consistency of comparisons using the Eq. (1) and Eq. (2).

$$CI = \frac{\lambda_{\max} - n}{n-1}; \quad (1)$$

$$CR = \frac{CI}{RI}, \quad (2)$$

where $\lambda_{\max}$ = Principal eigen value = sum of the products between each element of the priority vector and column total, n = number of criteria. CI = consistency index and RI = random inconsistency index.

If the consistency ratio (CR) is greater than 0.10, then some pair-wise values need to be reconsidered till the desired value of $CR < 0.10$ is reached.

After standardization of criteria and determination of relative weights of each of the criteria in landfill site selection process, the next step is to aggregate them to obtain the suitability of the land for the purpose. WLC method is the widely used method for aggregation of standardized criteria and their weights in suitability assessment process. Weighted maps and constraints maps were aggregated in the present study to obtain the final suitability due to the obvious reasons that these are the land-use patterns that cannot be disturbed for such planning. Empirically see Eq. (3):

$$S = \sum_{i=1}^{n} w_i \cdot x_i \prod_{j=1}^{m} c_j, \quad (3)$$

where S = Composite suitability score, x_i = Criterion score, w_i = Weight assigned to each criterion, c_j = Constrains, $\sum$= Sum of Weights and $\prod$= Products of the constraints.

Final candidate sites were chosen from final suitability map which had area greater than a threshold value of 4 ha, the minimum area required for a service period of 20 years (Manual on Municipal... 2000). So, the areas below the threshold value were considered as not suitable for landfill. Then all candidate sites were ranked based on comparative statistical analysis of area occupied by each attributes of three main criteria i.e. land-use pattern, distance from settlement, and distance from water bodies. Further an AHP was constructed to pool three rank vectors in one rank vector to prioritize the selected candidate sites.

2. Results and discussion

2.1. Site selection criteria

In this study, eleven possible criteria were selected for evaluating landfill suitability which includes the major environmental and socio-economic parameters of the study area (land-use pattern, slope, distance from the settlement, distance from water bodies, depth of groundwater table, lineament, wind orientation, flow accumulation, distance from roads, distance from airport and soil type). Figure 4 shows suitability maps for all the criteria that were considered for the present study and Table 1 shows the classes of each criterion and their standardized score (rating).

The weights were assessed taking the feasibility and possibility of modifying the natural conditions of the sites by appropriate engineering intervention, so as to increase their suitability (as was done by Delgado *et al.* 2008). These criteria are discussed below:

2.1.1. Slope

Lin, Kao (1998) have suggested that the appropriate slope for constructing a landfill is about 8–12%, because too steep slope would make it difficult to construct and maintain and too flat of a slope would affect the drainage. A slope above 12% creates high runoff rates for precipitation. With higher runoff rate and decreased infiltration, contaminants may travel greater distances from the containment area. Topographically the present study area is a lower plateau having relatively little undulations. The general slope is from North West to South East direction. The criteria were graded as per the suitability class and ranked accordingly. A slope range of 8–16% was considered to be the most preferable for the landfill site and hence assigned grade1 in the present study.

2.1.2. Distance from roads

The landfill must not be located within 200 m of any major highways and city streets and 500 m of a railway line (Şener *et al.* 2006). On the other hand, the landfill site should not be placed too far from the existing road networks for reducing the cost of new road construction, transportation and collection costs of solid wastes. In the present study, buffer zones were created taking the radial distance of 100, 200, 300, 400, 500 meters respectively and suitable grades were assigned.

2.1.3. Land-use

The land-use characterizes a degree of economic activities and population density associated with the area under study. The different land-use types were assigned different weights considering their respective economic activities and the population density. The residential or other settlement areas, recreational areas, vegetation and the airport were considered inappropriate for MSW landfill siting and therefore, were assigned relatively high grade to discard the area for landfilling and the area which is exclusively barren land and barren land with some scrubs were considered as suitable for such activity and hence was assigned grades 1 and 2 respectively. The other land-use types were assigned grades as per their relative importance.

2.1.4. Distance from settlement

The landfill site should not be in the vicinity of the residential or an urban area to avoid adverse impact on land value and future development and to protect the people from possible environmental hazards of landfill sites. The landfill must be located within 10 km of an urban area (Baban, Flannagan 1998) but should not be within 1 km of

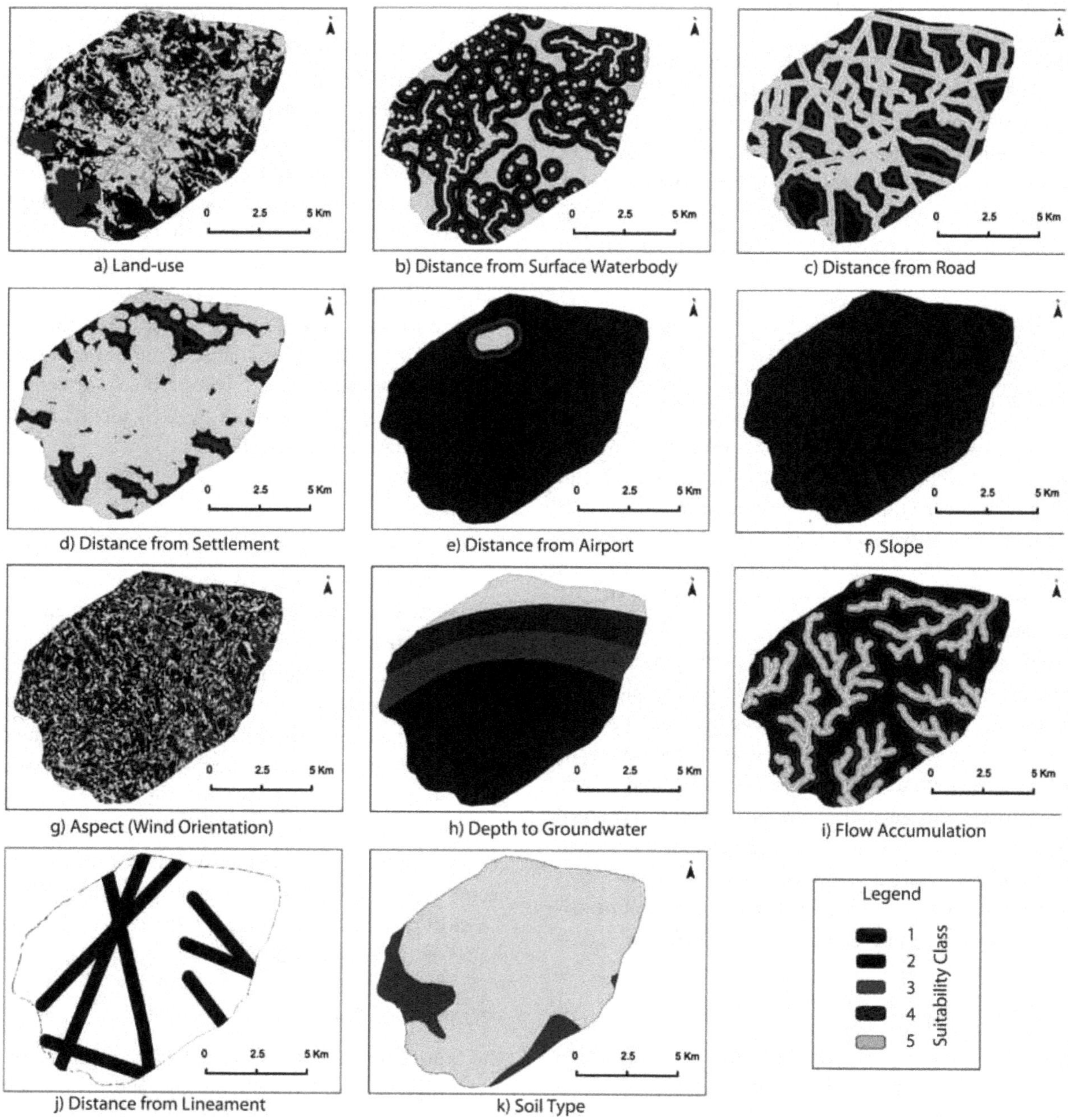

Fig. 4. Suitability map of all the selected criteria

an urban area as per Turkish Solid Waste Control Regulations (TSWCR 2002). A distance of more than or equal to 500 m from settlement was taken as suitable for landfill siting in the present study (consistent with Demesouka *et al.* 2014).

2.1.5. Depth to groundwater

A landfill site should be located in an area where groundwater is sufficiently deep, so that the groundwater quality is not affected due leachate movement from landfill sites. Keeping this in mind, zones with varying water-table depth were demarcated and grades were assigned in such a way that the zone with maximum depth was considered most suitable and vice-versa to avoid the area with groundwater table near the surface. The depth of groundwater table from ground surface was reclassified in a depth range varying from 0–2 m, 2–2.5 m, 2.5–3 m, 3–3.5 m, 3.5–4 m and each range was standardise based on their suitability. Groundwater depth more than 3.5 metres from ground surface was considered safe for constructing landfill (Manual on Municipal... 2000) in the present study.

2.1.6. Lineament

Faults are geological structures that restrict the siting of a safe landfill site (Gemitzi *et al.* 2007) as these allow the leachate to flow quickly to reach groundwater. The landfill site should be located at least 100–200 m away from any faults, fissures, joints, stream/lakes and other shear zones avoid the groundwater pollution of any area (Demesouka *et al.* 2014). All the lineaments were buffered by a distance interval of 0–100 m. According to the calculations in the present study, the areas located at more than 400 m away from the lineaments had the lowest weight values (most suitable), but those <100 m from the lineaments had the highest gradevalues (least suitable).

2.1.7. Wind orientation

Settlements near the landfill get affected by odour generated from the landfill; hence it is always preferable to site the location of a landfill at a place where wind direction is not towards the settlement and away from the settlement zone. To evaluate the wind direction of the study area, an aspect map was prepared using DEM (as suggested by Şener *et al.* 2010). The hillsides located in the windward direction relative to the settlements were assigned a lower grade. The northeast (NE) and southwest (SW) winds were the dominant wind direction; hence the hillsides located in NE–SW and flat areas were assigned a grade 5 in the present study area, because they were more exposed to the wind and thus unsuitable. The other hillsides were assigned lower grades, depending on their wind frequency and suitability in the study area.

Table 1. Reclassified criteria and their rating

Criteria	Class	Rating
Land-use	Barren land without scrub	1
	Barren land with scrub	2
	Mining, Open space	3
	Vegetation, Grassland, wetland	4
	Agriculture, Surface water body, Settlements	5
Slope	8–12%	1
	12–16%	2
	4–8%	3
	16–20%	4
	>20%	5
Distance from settlement	0–200m	5
	200–400m	4
	400–600m	3
	600–800m	2
	>800m	1
Distance from surface water body	0–100m	5
	100–200m	4
	200–300m	3
	300–400m	2
	400–500m	1
Depth of groundwater	0–2m	5
	2–2.5m	4
	2.5–3m	3
	3.3–5m	2
	3.5-4m	1
Lineament density	0–100m	5
	100–200m	4
	200–300m	3
	300–400m	2
	>400m	1
Aspect	South West	1
	North East	2
	South East	3
	North West	4
	North, South, East, West	5
Flow accumulation	0–600 cells	1
	600–2500 cells	2
	2500–6000 cells	3
	6000–15000 cells	4
	15000–32473 cells	5
Distance from road	500m	1
	400m	2
	300m	3
	200m	4
	100m	5
Distance from airport	0–200m	5
	200–400m	4
	400–600m	3
	600–800m	2
	800–1000m	1
Soil type	Loamy to fine	2
	Loamy to Loamy	3

2.1.8. Flow accumulation

A flow accumulation raster calculates the number of cells that will contribute flow into each cell of the study area (O'Callaghan, Mark 1984). It determines the size of the region over which water from precipitation (rainfall, snowfall, etc.) can be accumulated. It also depicts the contributing area of a watershed and is highly correlated with both its surface runoff and peak discharge. The high flow accumulated areas were demarcated as not suitable to avoid the direct impact of surface runoff in the present study.

2.1.9. Distance from water bodies

Landfills release noxious gases and leachate. So they should not be in proximity to water wells and surface water bodies (Dorhofer, Siebert 1998). Weights were assigned using AHP in similar way as was done for roads. The distance from the water bodies was buffered at a distance varying from 0–100 m, 100–200 m, 200–300 m, 300–400 m, 400–500 m. The distance from the water bodies more than 500 metres was considered safe for constructing landfill in the present study area (Manual on Municipal... 2000).

2.1.10. Distance from airport

The distance between an airport and landfill site should be a minimum of 3km for safety measures (Wang *et al.* 2009). Another reason for locating a landfill away from the airport is to avoid birds that easily get attracted to landfill sites due to availability of different types of food wastes (Moeinaddini *et al.* 2010). Recently in a study, Gbanie *et al.* (2013) formed a 0.5 km buffer zone from helipads in the examined area. The distance of landfill from airport was taken 1km as the most suitable one in this area as the airport is not in use on the regular basis and the area proposed for the airport is not sufficient, hence further development of the airport may not be possible.

2.1.11. Soil type

Soil should be of sufficiently low permeability to ensure very slow movement of leachate from the site towards groundwater. Thus, sites on clayey soil having low permeability should be preferred. A landfill constructed over a permeable formation such as gravel, sand or fractured bedrock can pose a significant threat to groundwater quality. The present study area has two dominant types of soils and they are fine loamy to fine and fine loam to loamy. Fine loamy to fine was considered to be more suitable for landfill site and grade accordingly due to its low permeability over the other type soil present in the study area.

2.2. Final landfill site suitability analysis

For selection of potential landfill sites in the present study, AHP was used after detail evaluation of the various criteria of landfill siting. Pair-wise comparison method was used only to assign weights and to set up the significance of environmental criteria (Table 2a) using AHP where highest weightage was assigned to the land-use pattern. Then the column total is derived and the cell values of each column are divided by its respective column total to get a normalized value matrix as given in Table 2b. Finally the weights were calculated by taking the average values of the row for each criteria. The detail calculation steps were same as discussed in the methodology section.

The CR value of all comparisons was lower than 0.1 that indicated the use of weights was suitable (consistent with Eastman 2003).

WLC was applied to aggregate all criteria and display potential landfill sites considering all the criteria and their relative weights for each of the criteria and accordingly an intermediate suitability maps were

Table 2a. Comparison matrix for the site selection criteria

Parameters	x_1	x_2	x_3	x_4	x_5	x_6	x_7	x_8	x_9	x_{10}	x_{11}
Land-use (x_1)	1	2	3	3	4	4	5	4	5	8	9
Slope (x_2)	1/2	1	1/3	2	4	5	6	3	6	8	8
Distance from settlement (x_3)	1/3	3	1	2	4	3	4	3	6	7	9
Distance from waterbody (x_4)	1/4	1/2	1/2	1	3	2	6	4	7	6	8
Groundwater (x_5)	1/4	1/4	1/4	1/3	1	3	2	3	3	3	8
Lineament (x_6)	1/5	1/5	1/3	1/2	1/3	1	3	3	4	3	7
Aspect (wind) (x_7)	1/4	1/6	1/4	1/6	1/2	1/3	1	3	3	2	5
Flow accumulation (x_8)	1/4	1/3	1/3	1/4	1/3	1/3	1/3	1	3	2	4
Transport (x_9)	1/5	1/6	1/6	1/5	1/3	1/4	1/3	1/3	1	2	3
Distance from Airport (x_{10})	1/8	1/8	1/7	1/6	1/3	1/3	1/3	1/2	1/2	1	3
Soil type (x_{11})	1/9	1/8	1/9	1/8	1/8	1/7	1/5	1/4	1/3	1/3	1
Column total	3.55	7.87	6.42	9.68	17.96	19.39	28.37	25.08	38.83	42.33	65.00

Table 2b. Normalized matrix and significance weight of each criteria

Parameters	x_1	x_2	x_3	x_4	x_5	x_6	x_7	x_8	x_9	x_{10}	x_{11}	Weight (%)
Land-use (x_1)	0.28	0.25	0.47	0.31	0.22	0.21	0.18	0.16	0.13	0.19	0.14	23.03
Slope (x_2)	0.14	0.13	0.05	0.21	0.22	0.26	0.21	0.12	0.15	0.19	0.12	16.40
Distance from settlement (x_3)	0.09	0.38	0.16	0.21	0.22	0.15	0.14	0.12	0.15	0.17	0.14	17.58
Distance from waterbody (x_4)	0.09	0.06	0.08	0.10	0.17	0.10	0.21	0.16	0.18	0.14	0.12	12.95
Groundwater (x_5)	0.07	0.03	0.04	0.03	0.06	0.15	0.07	0.12	0.08	0.07	0.12	7.70
Lineament (x_6)	0.07	0.03	0.05	0.05	0.02	0.05	0.11	0.12	0.10	0.07	0.11	7.06
Aspect (wind) (x_7)	0.06	0.02	0.04	0.02	0.03	0.02	0.04	0.12	0.08	0.05	0.08	4.86
Flow accumulation (x_8)	0.07	0.04	0.05	0.02	0.02	0.02	0.01	0.04	0.08	0.05	0.06	4.22
Transport (x_9)	0.06	0.02	0.03	0.01	0.02	0.01	0.01	0.01	0.03	0.05	0.05	2.67
Distance from Airport (x_{10})	0.04	0.02	0.02	0.02	0.02	0.02	0.02	0.02	0.01	0.02	0.05	2.24
Soil type (x_{11})	0.03	0.02	0.02	0.01	0.01	0.01	0.01	0.01	0.01	0.01	0.02	1.28

generated using GIS. The lower values of suitability cl-ass1and 2 on the maps indicated most suitable areas for landfill siting (Fig. 5). The area in the study region with value 1 was negligible hence the next class of area with suitability class value 2 (more suitable than 3, 4 and 5) was considered for extraction of the suitable areas for landfill site.

The final analysis showed that about 12 km^2 of the total area (101 km^2) is suitable for locating landfill sites. Eleven candidate sites were finalized on the basis of the size of a site greater than 4 ha and are shown in Figure 6.

To make the selection more precise, all the eleven candidate sites were further compared with the three most significant criteria i.e. land-use pattern, distance from the settlement and the distance from surface water bodies. This study proved to be efficient and provide more accurate results. Figure 7 shows the area covered by each type of the land-use for comparative study of the individual suitability of all eleven candidate sites. Each bar in a graph (Fig. 7) represents the area coverage of each candidate site in percentage to make comparison with each other. Candidate sites 7, 8 and 10 had maximum quantity of suitable land and candidate sites 1, 2, and 9 had maximum area located at suitable distance from settlement whereas candidate sites 6, 4 and 1 had enough land situated at a suitable distance from surface water bodies.

These sites were ranked based on the area statistics of each attribute of these three criteria individually. Finally these ranks were pooled in a single rank using linear combination of weight derived from AHP (Table 3) of the three criteria and their ranks. The calculation step is same as followed for Table 2.

The candidate sites 6, 7 and 8 (Fig. 8) were the most suitable landfill sites in the study area as these sites had high ranking order 1, 2, and 3 respectively (Table 4).

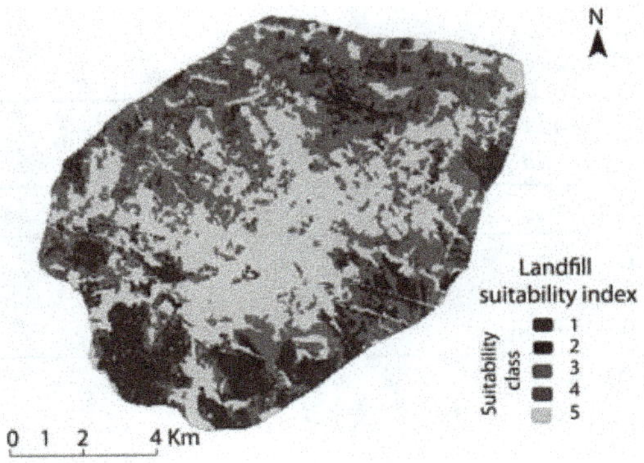

Fig. 5. Landfill suitability index

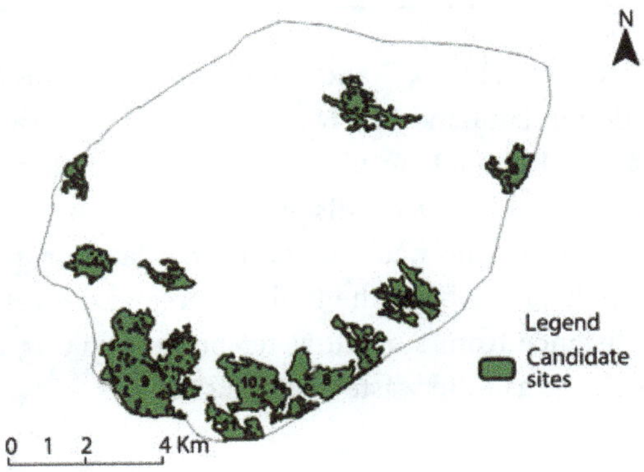

Fig. 6. Selected candidate sites

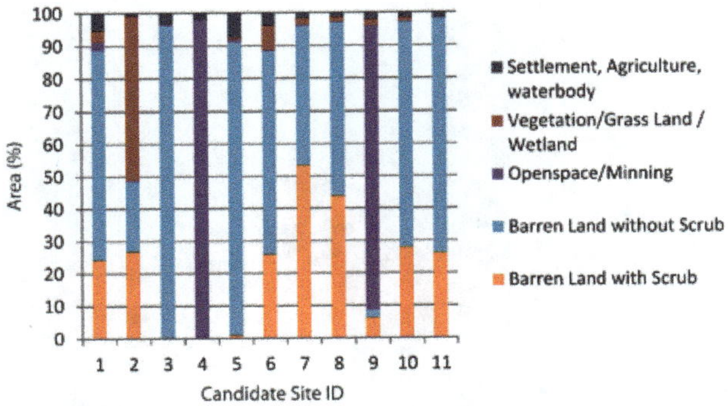

Fig. 7. Comparison of all the candidate sites with respect to land-use

Table 3(a). AHP for candidate site ranking

Criteria	(s_1)	(s_2)	(s_3)
Land-use (s_1)	1	4	5
Distance from Settlement (s_2)	1/4	1	3
Distance from Waterbody (s_3)	1/5	1/3	1
Column Sum	1.45	5.33	9

Table 3(b). Normalized matrix and significant weight of each criteria for candidate site ranking criteria

Criteria	(s_1)	(s_2)	(s_3)	Weight
Land-use (s_1)	0.69	0.75	0.56	0.67
Distance from Settlement (s_2)	0.17	0.19	0.33	0.23
Distance from Waterbody (s_3)	0.14	0.06	0.11	0.10

Table 4. Rank of Candidate Sites

	Candidate Site ID	1	2	3	4	5	6	7	8	9	10	11
Rank based on	Land-use	6	7	8	11	9	5	1	2	10	3	4
	Distance from Settlement	1	2	7	6	11	4	5	9	3	8	10
	Distance from Waterbody	3	5	8	2	11	1	7	4	6	9	10
	Overall Rank	4	6	8	10	11	3	1	2	9	5	7

Candidate sites 6, 7 and 8 had more than 90% area under barren land and nearly 30% area was in between a distance of 400 to 600 m from settlement and more than 40% area was in between a distance of greater than 400 m. So, the selected sites had enough barren land or quarry, substantially away from the built up area and located at a safe distance from surface water bodies making them more suitable as solid waste disposal sites.

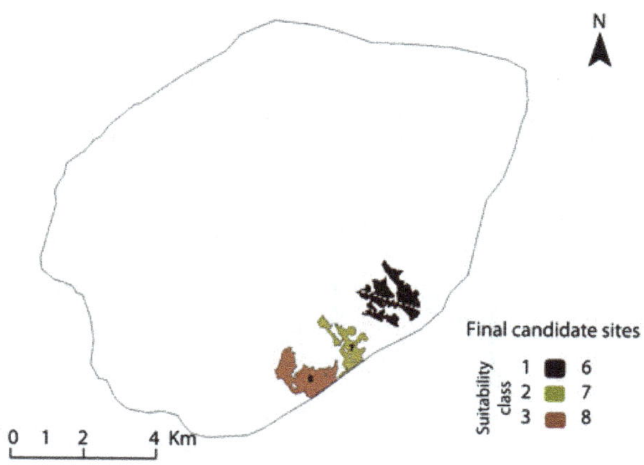

Fig. 8. Final candidate sites

Conclusions

Landfill site selection is one of the most important components of any MSWM system and needs to be planned scientifically using recent advancements in the spatial sciences. GIS is a very powerful tool and provides a rapid assessment for suitable landfill siting for MSW disposal in a rapidly growing city like Dhanbad. An unscientific siting of a landfill may result in adversely affected surrounding environment. This situation is observed in most developing nations including India. Most cities of developing countries are faced with the improper management of data as one of the limitations. For the selection of landfill sites, eleven criteria were considered relevant to Dhanbad city. Generally more criteria give more precise results, but selecting less important criteria which have negligible impact, make the system very complex. So, the evaluation criteria must be selected based on their impact on the suitability analysis. Environmental criteria are more sensitive due to its control over the leachate transport rate than socio-economic criteria which are important for limiting financial expenditure. In this study, eleven possible landfill sites were identified using GIS that satisfied most of the desirable criteria of locating an environmentally suitable landfill site. The study area is rich in coal reserve and there is a limitation of using geological data as it is not accessible for the present study area and therefore the depth to the bedrock was not considered. Suitable landfill sites were identified using overlay analysis of different layers selected on the basis of AHP applicable for the characteristics of the region. Final landfill suitability maps were generated using WLC and the prioritizing candidate sites have been done by repeating the whole process considering the most desirable factors. The factors which were assigned more weights were compared to select the most appropriate landfill site. The factors considered for comparison study were land-use pattern, distance from the settlement, and the distance from the road network. Out of the 11 candidate sites identified in this study, three "potential landfill" sites were selected as the most acceptable landfill sites with suitability ranking using repetitive application of AHP and WLC considering three most influential parameters affecting landfill siting. The most suitable identified sites for landfill were confirmed by field visits. They were found to be feasible with respect to the environmental criteria. Furthermore, the study could be validated by considering site owernership and the public acceptance study. Insitu field survey such as litholog analysis, soil permeability test, hydrogeological survey should be conducted before finalising the landfill location. This study used secondary coarse resolution data for soil and lineaments data, fine resolution data for the same could give more accurate results.

Acknowledgements

The authors acknowledge the support provided by the Department of Environmental Science and Engineering, Indian School of Mines, Dhanbad for carrying out this research work.

Reference

Achillas, C.; Moussiopoulos, N.; Karagiannidis, A.; Banias, G.; Perkoulidis, G. 2013. The use of multi-criteria decision analysis to tackle waste management problems: A literature review, *Waste Management & Research* 31(2): 115–129. http://dx.doi.org/10.1177/0734242X12470203

Allesch, A.; Brunner, P. H. 2014. Assessment methods for solid waste management: A literature review, *Waste Management & Research* 32(6): 461–473. http://dx.doi.org/10.1177/0734242X14535653

Baban, S. J.; Flannagan, J. 1998. Developing and implementing GIS-assisted constraints criteria for planning landfill sites in the UK, *Planning Practice and Research* 13(2): 139–151. http://dx.doi.org/10.1080/02697459816157

Bhushan, N.; Rai, K. 2004. Strategic decision making: applying the analytic hierarchy process. New-York: Springer-Verlag.

Carver, S. J. 1991. Integrating multi-criteria evaluation with geographical information system, *Journal of Geographical Information System* 5(3): 321–339. http://dx.doi.org/10.1080/02693799108927858

Chang, N.; Parvathinathan, G.; Breeden, J. B. 2008. Combining GIS with fuzzy multi criteria decision making for landfill siting in a fast-growing urban region, *Journal of Environmental Management* 87(1): 139–153. http://dx.doi.org/10.1016/j.jenvman.2007.01.011

Charnpratheep, K.; Zhou, Q.; Garner, B. 1997. Preliminary landfill site screening using fuzzy geographic information systems, *Waste Management and Research* 15(2): 197–215. http://dx.doi.org/10.1177/0734242X9701500207

Chavez, P. S. Jr. 1988. An improved dark object subtraction technique for atmospheric correction of multispectral data, *Remote Sensing of Environment* 24(3): 459–479. http://dx.doi.org/10.1016/0034-4257(88)90019-3

Chiueh, P. T.; Yu, Y. H. 2006. Assessment on the solid waste management information system in Taiwan, *Journal of Environmental Engineering and Management* 16(6): 427.

Delgado, O. B.; Mendoza, M.; Granados, E. L.; Geneletti, D. 2008. Analysis of land suitability for the siting of inter-municipal landfills in the Cuitzeo Lake Basin, Mexico, *Waste Management* 28(7): 1137–1146. http://dx.doi.org/10.1016/j.wasman.2007.07.002

Demesouka, O. E.; Vavatsikos, A. P.; Anagnostopoulos, K. P. 2014. GIS-based multicriteria municipal solid waste landfill suitability analysis: A review of the methodologies performed and criteria implemented, *Waste Management & Research* 32(4): 270–296. http://dx.doi.org/10.1177/0734242X14526632

Dorhofer, G.; Siebert, H. 1998. The search for landfill sites—Requirements and implementation in Lower Saxony, Germany, *Environmental Geology* 35(1): 55–65. http://dx.doi.org/10.1007/s002540050292

Eastman, J. R. 2003. IDRISI Kilimanjaro: *Guide to GIS and image processing*. Worcester: Clark Laboratories, Clark University. 328 p.

Gbanie, S. P.; Tengbe, P. B.; Momoh, J. S.; Medo, J.; Kabba, V. T. S. 2013. Modelling landfill location using geographic information systems (GIS) and multi-criteria decision analysis (MCDA): Case study Bo, Southern Sierra Leone, *Applied Geography* 36: 3–12. http://dx.doi.org/10.1016/j.apgeog.2012.06.013

Gemitzi, A.; Tsihrintzis, V. A.; Voudrias, E.; Petalas, C.; Stravodimos, G. 2007. Combining geographic information system, multicriteria evaluation techniques and fuzzy logic in sitting MSW landfills, *Environmental Geology* 51(5): 797–811. http://dx.doi.org/10.1007/s00254-006-0359-1

Hopkins, L. D. 1977. Methods for generating land suitability maps: a comparative evaluation, *Journal of the American Institute of Planners* 43(4): 386–400. http://dx.doi.org/10.1080/01944367708977903

Higgs, G. 2006. Integrating multi-criteria techniques with geographical information systems in waste facility location to enhance public participation, *Waste Management & Research* 24(2): 105–117. http://dx.doi.org/10.1177/0734242X06063817

Ionescu, S.; Deac, V.; Corpodean, C. 2011. Risk of exposure to asbestos in case of the mechanical degradation of some asbocement composites, *Environmental Engineering and Management Journal* 9: 1773–1777.

Jensen, J. R. 1986. *Introductory digital image processing: a remote sensing perspective*. New Jersey: Prentice-Hall.

Kao, J. J.; Lin, H. Y.; Chen, W. Y. 1997. Network geographic information system for landfill siting, *Waste Management & Research* 15(3): 239–253. http://dx.doi.org/10.1177/0734242X9701500303

Kara, C.; Doratli, N. 2012. Application of GIS/AHP in siting sanitary landfill: a case study in Northern Cyprus, *Waste Management & Research* 30(9): 966–980. http://dx.doi.org/10.1177/0734242X12453975

Kontos, T. D.; Komilis, D. P.; Halvadakis, C. P. 2003. Siting MSW landfills on Lesvos island with a GIS based methodology, *Waste Management & Research* 21(3): 262–277. http://dx.doi.org/10.1177/0734242X0302100310

Kontos, T. D.; Komilis, D. P.; Halvadakis, C. P. 2005. Siting MSW landfills with a spatial multiple criteria analysis methodology, *Waste Management* 25(8): 818–832. http://dx.doi.org/10.1016/j.wasman.2005.04.002

Lin, H. Y.; Kao, J. J. 1998. A vector-based spatial model for landfill siting, *Journal of Hazardous Materials* 58(1): 3–14. http://dx.doi.org/10.1016/S0304-3894(97)00116-7

Lukasheh, A. F.; Droste, R. L.; Warith, M. A. 2001. Review of expert system (ES), geographic information system (GIS), decision support system (DSS), and their applications in landfill design and management, *Waste Management & Research* 19(2): 177–185. http://dx.doi.org/10.1177/0734242X0101900209

Manual on Municipal Solid Waste Management. 2000. 1st ed. Central Public Health and Environmental Engineering Organisation (CPHEEO), Ministry of Urban Developement, Govenrment of India, New Delhi.

Moeinaddini, M.; Khorasani, N.; Danehkar, A.; Darvishsefat, A. A. 2010. Siting MSW landfill using weighted linear combination and analytical hierarchy process (AHP) methodology in GIS environment (case study: Karaj), *Waste Management* 30(5): 912–920. http://dx.doi.org/10.1016/j.wasman.2010.01.015

O'Callaghan, J. F.; Mark, D. M. 1984. The extraction of drainage network from digital elevation data, *Computer Vision, Graphics and Image Processing* 28(3): 328–344. http://dx.doi.org/10.1016/S0734-189X(84)80011-0

Rathi, S. 2006. Alternative approaches for better municipal solid waste management in Mumbai, *Waste Management* 26(10): 1192–1200. http://dx.doi.org/10.1016/j.wasman.2005.09.006

Robu, B. M.; Caliman, F. A.; Betianu, C.; Gavrilescu, M. 2007. Methods and procedures for environmental risk assessment, *Environmental Engineering and Management Journal* 6(6): 573–592.

Saaty, T. L. 1977. A scaling method for priorities in hierarchical structures, *Journal of Mathematical Psychology* 15(3): 234–281. http://dx.doi.org/10.1016/0022-2496(77)90033-5

Schiopu, A. M.; Piuleac, G. C.; Cojocaru, C.; Apostol, I.; Mamaliga, I.; Gavrilescu, M. 2012. Reducing environmental risk of landfills: leachate treatment by reverse osmosis, *Environmental Engineering and Management Journal* 11(12): 2319–2331.

Şener, B. 2004. Landfill site selection by using geographic information systems. M.Sc Thesis. Middle East Technical University METU, Ankara, Turkey, 114 p. [online], [cited 12 March 2013]. Available from Internet: http://www.rsgis.metu.edu.tr

Şener, B., Süzen, M. L., & Doyuran, V. 2006. Landfill site selection by using geographic information systems, *Environmental Geology* 49(3): 376–388. http://dx.doi.org/10.1007/s00254-005-0075-2

Şener, Ş.; Şener, E.; Nas, B.; Karagüzel, R. 2010. Combining AHP with GIS for landfill site selection: a case study in the Lake Beysehir catchment area (Konya, Turkey), *Waste Management* 30(11): 2037–2046. http://dx.doi.org/10.1016/j.wasman.2010.05.024

Sharholy, M.; Ahmad, K.; Mahmood, G.; Trivedi, R. C. 2008. Municipal solid waste management in Indian cities – a review, *Waste Management* 28(2): 459–467. http://dx.doi.org/10.1016/j.wasman.2007.02.008

Siddiqui, M. Z.; Everett, J. W.; Vieux, B. E. 1996. Landfill siting using geographic information systems: a demonstration, *Journal of Environmental Engineering* 122(6): 515–523. http://dx.doi.org/10.1061/(ASCE)0733-9372(1996)122:6(515)

Simion, I. M.; Ghinea, C.; Maxineasa, S. G.; Taranu, N.; Bonoli, A.; Gavrilescu, M. 2013. Ecological footprint applied in the assessment of construction and demolition waste integrated management, *Environmental Engineering and Management Journal* 12(4): 779–788.

Sumathi, V. R.; Natesan, U.; Sarkar, C. 2008. GIS-based approach for optimized siting of municipal solid waste landfill, *Waste Management* 28(11): 2146–2160. http://dx.doi.org/10.1016/j.wasman.2007.09.032

Turkish Solid Waste Control Regulations (TSWCR) 2002. *Solid waste control regulation*. Ministry of Environment and Forestry of Turkey, Official Gazette No. 24736, Ankara, Turkey.

Wang, G.; Qin, L.; Li, G.; Chen, L. 2009. Landfill site selection using spatial information technologies and AHP: a case study in Beijing, China, *Journal of Environmental Management* 90(8): 2414–2421. http://dx.doi.org/10.1016/j.jenvman.2008.12.008

Yildirim, V. 2012. Application of raster-based GIS techniques in the siting of landfills in Trabzon Province, Turkey: a case study, *Waste Management & Research* 30(9): 949–960. http://dx.doi.org/10.1177/0734242X12445656

Debishree KHAN, M. Sc., Senior Research Fellow, Department of Environmental Science and Engineering, Indian School of Mines, Dhanbad, India. Research interests: application of remote sensing and GIS, solid waste management.

Sukha Ranjan SAMADDER, PhD, Assistant Professor, Department of Environmental Science and Engineering, Indian School of Mines, Dhanbad-826004, India. Research interests: application of remote sensing and GIS in broad based environmental management, solid waste management, groundwater pollution.

11

SOCIO-ECONOMIC ASSESSMENT IN ENVIRONMENTAL IMPACT ASSESSMENT: EXPERIENCE AND CHALLENGES IN LITHUANIA

Renata DAGILIŪTĖ, Gintarė JUOZAPAITIENĖ

Department of Environmental Science, Vytautas Magnus University, Vileikos g. 8, LT-44404 Kaunas, Lithuania

Abstract. In 2011, the Environmental Impact Assessment (EIA) Law amendment was approved and a socio-economic assessment formally became an integrated part of EIA in Lithuania. Therefore, this study examines a socio-economic assessment in the environmental impact statements (EIS) and the perceptions of the EIA experts towards a socio economic assessment. Although formally a socio-economic assessment has been validated only recently, 30% of specialists claimed having conducted a socio-economic assessment in detail prior to the amendments to the law. Thus, the EIS analysis has shown a rather poor consideration of these issues as mainly creation of work places was addressed. The survey has shown that preparation of the EIA documents after the formalisation of a socio-economic assessment would change mainly nominally. The EIA experts working in the private sector were more optimistic than those from state institutions. Reluctance of changes and personal attitudes have to be addressed, especially those particular to the experts of state institutions. In addition, methodological guidance, integrative approach and public involvement into the decision-making process could change the current situation and increase the effectiveness of the EIA process and the social impact assessment in general.

Keywords: socio-economic assessment, social impact assessment, environmental impact assessment, Lithuania.

Introduction

The Environmental Impact Assessment (EIA) has been in existence for more than 40 years already (Morgan 2012) and has been playing a particularly important role in the concept of sustainable development provisions. The EIA procedures help achieving the main objective of sustainable development, that is, harmonization of the environmental, economic and social interests of different social groups. EIA is the main tool for the integration of environmental aspects of projects and helping to implement them in the environmentally friendly manner. Since first provisions in the National Environmental Policy Act (1970), an environmental impact assessment is recognized a main tool for managing environment in a variety of international and national policy documents (Morgan 2012).

Although the socio-economic aspects are identified as one of the EIA objectives, the EIA reports are usually dominated by the environmental issues (Kruopiene *et al.* 2009; Değirmenci, Evcimen 2013). The EIA directive (2011/92/EU) is aiming to protect the environment and quality of life and at the same time to ensure that national legislation is aligned with respect to the environmental impacts of public and private projects. Despite little attention given in practice, a socio-economic assessment is recognized as important as other components considered in the process of environmental impact assessment. A social impact assessment together with the range of different impact assessments has evolved in line with EIA (Morrison-Saunders *et al.* 2014). A social impact assessment (SIA) or a socio-economic impact assessment[1] as such was developed in the 1970s and 1980s mainly in relation to the assessment of the impacts of major resource development projects, such as nuclear power stations in the US, hydro-electric schemes in Canada and the UK's North Sea oil and gas-related developments (Glasson 2009). At the

[1] Though a social impact assessment is considered to be interconnected with a socio-economic impact assessment, in this paper these two assessments are treated as similar ones although they have different methodologies and rely on different expertise (Lockie 2001). Attention in this paper is driven not to the terms but more to the content they cover regarding EIA.

Corresponding author: Renata Dagiliūtė
E-mail: r.dagiliute@gmf.vdu.lt

beginning, SIA was (and in some countries still is) an integral part of EIA (Esteves *et al.* 2012). Currently SIA is also used as an independent tool contributing to the development (separately or in conjunction with EIA (Franks, Vanclay 2013)). As indicated by C. J. Barrow (2002), EIA and SIA are completely different in many aspects but at the same time they are overlapping. Social and economic impacts are closely intertwined with the biophysical effects; therefore, it is almost impossible to completely separate the processes of impact assessment (Stanley *et al.* 2004). Environmental impacts may determine social impacts and those can trigger environmental ones (Burdge, Vanclay 1996).

Socio-economic impacts occur in most activities and their inclusion in a document facilitates a more balanced view of the range of impacts and provides a greater transparency of the process. Socio-economic impacts are relevant because the economic success, lifestyles, and values of people are important (Glasson 2009). The social impact assessments emphasize the human aspects of environment and seek to determine the impacts on people – who benefits and who loses. SIA helps to ensure that needs and opinions of different groups of people and communities, especially those of vulnerable and disadvantaged (Esteves *et al.* 2012), are taken into account and included into the environmental impact statement (EIS). Additionally, a socio-economic impact assessment is the process of providing social and economic impacts, allowing the decision makers an early understanding of the possible effects of planned activities on people. It could also be considered an efficient tool for the environmental conflict management (Barrow 2010).

In general, a sustainable development is the core of SIA, aiming for the "ecologically, socio-culturally and economically sustainable and equitable environment" (Vanclay 2003). However, this aim often lacks real considerations in practice (Suopajärvi 2013; Pope *et al.* 2013). Practice shows that only reports of major investment focus on the socio-economic issues in detail, and the EIA practitioners try to interpret this assignment as an opportunity to describe the favourable socio-economic impacts of planned investment, such as job level rise (COWI 2009). However, usually in such cases a social context is not taken into account (Rowan 2009).

Furthermore, a social impact assessment is a broader approach than the limited issues assessed in the EIA process (such as demographic changes, labour problems, financial protection, and the impact on a family life). Therefore, a limited approach to SIA limits the detection of social impacts compared to what is considered in related areas, such as the impact assessment on health, culture, heritage, and aesthetics (Vanclay 2003). A poor analysis of social and economic impacts within the EIA process may be stipulated by several reasons (Glasson 2009): (1) socio-economic impacts occur seldom; (2) their inclusion may understate the biophysical effects; (3) they are always negative; (4) they cannot be easily identified. There are also myths restricting the application of social assessment in the planning/decision making process (Burdge 2003): (1) social impacts cannot be measured, so they can be ignored; (2) social impacts are common, and everyone knows what they are; (3) social impacts occur rarely and therefore they should not be assessed; (4) social impacts are associated with the cost rather than the benefit, therefore, a social impact assessment muffles or suspends the project; (5) a social impact assessment is not important. Also, one of the main failures of SIA is a limited participation of local people in the process and lack of effort analysing possible advantages and disadvantages for them (Suopajärvi 2013).

Nevertheless, some authors (Morrison-Saunders, Fischer 2006) are sceptic about integrating socio and economic considerations into EIA, as, first, socio-economic issues could overcome environmental ones and, second, too aggregated assessment might oversee some important aspects and in this case might miss the main aims of EIA. In general, proliferation of various assessment types and practices might hamper integration of the results into one decision-making process and challenge promotion of sustainable development in general (Morrison-Saunders *et al.* 2014).

This paper aims to figure out the experiences of SIA in the EIA in Lithuania focusing on some reports and perceptions of the EIA experts. The paper is structured as follows. At first, we will discuss some methodology issues; and further, the results of our research based on SIA in relation to the EIA legislation in Lithuania, EIS analysis and survey data will be presented (section 3). Finally, conclusions will be drawn and some recommendations for the socio-economic impact assessment development in Lithuania will be suggested.

1. Methodology and data issues

Only several studies (Kruopiene *et al.* 2008, 2009) on EIA are present in Lithuania. Therefore, there is a need for research and discussion on improvement of the whole EIA process. This study aims to contribute to the debate on EIA effectiveness in Lithuania.

Research is both qualitative and quantitative, covering the EIA legislation analysis and focusing on SIA, prepared EIA reports and survey of the EIA experts.

In order to analyse the EIA reports and examine which environmental components are included in the environmental assessment study, 10 EIA reports (covering a period from 2006 till 2011) were randomly selected from Kaunas Regional Environmental Protection Department (REPD). During this period, Kaunas REPD has approved

36 EIA reports and issued 36 positive decisions for the planned economic activity. Though this part of the research has some limitations (e.g., small sample, not all activities are covered), it still might give some insights on prevailing trends. The analysis of environmental components covered by the EIA reports was conducted in two steps. At first, we examined whether the EIA report assessed the impacts of planned economic activity (PEA) on the components of the environment as foreseen in the EIA Law: public health, wildlife, soil, land surface and its entrails, air, water, climate, landscape and biodiversity, socio-economic environment, material values, and immovable cultural values. Later we looked in more detail at the socio-economic aspects and the elements that were considered in the EIA report: well-being of people, business and job opportunities, income and quality of life, land use, appropriate infrastructure, protection of cultural and heritage resources, suitable services.

In order to evaluate the opinion of EIA experts on EIA and socio-economic assessment in the EIA process, the survey was conducted. A prepared questionnaire consisted of closed and open questions and was placed on the website. A first part of questions was related to EIA in general and a second part focused on a socio-economic assessment, including knowledge about the EIA Law amendment, a possible influence of this change to the attitudes towards EIA and a socio-economic assessment, most important aspects of a socio-economic assessment, etc. A third block of questions was addressed to the socio-demographic variables, that is, age, gender, education, work experience, and type of institution. Some questions related to the EIA document preparation were addressed only to the EIA experts from consulting companies. In October 2011, links to the questionnaire were sent to 97 EIA experts, i.e. specialists from state institutions and randomly selected EIA documents preparers (consultants). Thirty six (36) of EIA experts answered the questionnaire, i.e. 37% of those who received it. Forty seven percent (47%) of respondents were the EIA practitioners from consulting firms and 53% of them were the EIA experts from competent state authorities. The interviewed specialists equally distributed by age: 50% of the specialists were under 40 years of age and 50% of them were over 40 years of age. Most of the respondents were female (69%). All the EIA experts had a university degree: as much as 39% of them had an environmental science background. Most of the experts had been working in the field of EIA from 5 to 10 years or 10 years or more, 40% and 37%, respectively.

Analysis is based mainly on descriptive statistics. The influence of various factors on certain attitudes and positions is assessed using a chi-square (χ^2) test. Significance level (α) of 0.05 is applied.

2. Results

2.1. A socio-economic assessment in Lithuanian EIA: legal basis

Compared to a rather long EIA and SIA practice in other countries, Lithuania, like other Central and Eastern European countries (for example, Poland (Woloszyn 2004)), has experienced an evolution of impact assessment in a rather short time.

The Lithuanian Law on Environmental Impact Assessment was adopted in 1996 and until today two revisions of the document have been made: one revision in 2000 and a second one in 2005. J. Kruopiene *et al.* (2009) presented a general EIA process in Lithuania, and since then there have not been any major changes in the process itself, except some changes in a public information provision, time for comments and revision of some EIA stages and competences and responsibility of governmental institutions depending on the level of projects.

The Law on Environmental Impact Assessment (2005) does not directly refer to the investigation of the impacts of a planned activity on a socio-economic environment. Article 4 says that one of the objectives of the EIA is to "identify, characterize and assess the potential direct and indirect impacts of planned economic activities on the public health, wildlife, soil, land surface and its entrails, air, water, climate, landscape and biodiversity, material assets and immovable cultural heritage and the interaction between these environmental components". However, Guide for the Environmental Impact Assessment (Ministry of Environment of the Republic of Lithuania 2009) says that the EIA process focuses not only on the impacts on the natural environment but also on the man-made environment, including human health and social, cultural and economic well-being. The Guide also lists the components of a socio-economic environment: (1) economic factors, such as characteristics of the labour market, labour supply and demand, etc.; (2) demography, such as population structure and population trends; (3) housing, such as housing demand and supply; (4) local services, such as services supply and demand, health care, education, etc.; (5) social and cultural factors, such as quality of life, social problems, potential conflicts with the public.

Provisions for Preparation of Programmes and Reports on the Environmental Impact Assessment (2005) also indicate that the following aspects should be examined in the EIA report regarding socio-economic environment:

- Information about the area (activity impact zone): population, permanent and visiting residents, their changes, migration trends, population characteristics, demographic indicators in the area (birth rate, mortality rate, etc.), investments, labour mar-

ket and unemployment (in the area), area development, the cost of land, houses and premises (residential areas, commercial and industrial areas), cost variation.
- Potential impacts for economic conditions, labour market, investment, prices of real estate and land, demographics, industries, transport, mining, construction of residential houses (low-rise, multi-storey), trade (wholesale and retail), living conditions, potential conflicts within society.

Despite that, J. Kruopiene *et al.* (2008) confirm that a socio-economic assessment is a problematic area in Lithuania, showing that a socio-economic assessment is the weakest one compared to the biophysical components of EIA. One of the reasons might be a lack of consistent legal basis and proper guidelines. Aiming to overcome this, on the 9th of June, 2011, amendment to the Law on EIA was adopted amending Paragraph 1 of Article 4 and including the assessment of impacts of a planned economic activity on the social and economic environment. As the result, one of the objectives of the EIA became identification, description and evaluation of potential direct and indirect effects of planned economic activities on the public health, wildlife, soil, land surface and its entrails, air, water, climate, landscape and biodiversity, socio-economic environment and material values, immovable cultural heritage and the interaction between these environmental components. Thus, after the amendment to the Law came into force, a socio-economic environment assessment became a component requiring its evaluation in EIA. Therefore, it is important to disclose how the impacts on a socio-economic environment were assessed in selected EIA reports prior to the amendment and possible changes following the amendment to the Law on Environmental Impact Assessment. Of course, only formal changes to the Law, without methodological and institutional support, are insufficient. A third section of this paper presents possible areas for improvement.

2.2. The EIA report analysis

Evaluation of such environmental components as public health, wildlife, soil, surface and underground, air, water, climate, landscape and biodiversity, socio-economic environment and material values, immovable cultural heritage is one of the objectives of the EIA. Analysis of the EIA reports of Kaunas REPD (2006–2011) revealed that none of the randomly selected 10 reports had evaluated effects on climate, material assets or flora and fauna. All EIA reports comprehensively analysed impacts of a planned activity on air and water.

Impact of a planned economic activity on a socio-economic environment has been examined in more detail only in two out of ten analysed EIA reports. Those two reports focus on the activities of logistics centre and grain processing. In the case of logistics centre, it was assessed that there would be no impact on demography and that the centre would have a positive impact on the economy, service sector and employment; activity would use services of the local surrounding businesses, that is, catering, vehicle maintenance, etc. However, the report lacks a more particular analysis on human well-being, quality of life and infrastructure. In the case of grain processing enterprise, effects on economic conditions, labour and real estate markets and influence on other economic sectors were analysed. The assessment also included analysis of the potential public dissatisfaction and potential conflict. Nevertheless, EIS lacks assessment on human well-being, quality of life, land use and services.

A socio-economic environment is not considered at all in three reports. These are reports on the heat insulation material factory, bypass construction and expansion and modernization of fertilizers' factory. The remaining reports describe only job creation as a socio-economic impact of a planned activity. Hence, analysis of socio-economic aspects in the EIA reports shows a rather narrow understanding of SIA. Despite the discussed limitations, the competent authority had no comments on the climate, material properties, flora and fauna or socio-economic assessment and all discussed reports were approved.

2.3. Attitudes of the EIA experts towards the environmental impact assessment

Thirty six (36) EIA specialists participated in the survey: 47% of them were the EIA practitioners (consultants) and 53% of them were experts from state institutions. In order to evaluate the EIA specialists' opinion on the preparation of EIA reports, only practitioners (consultants) were included; in other words, the questions were answered only by those who represented private companies and prepared but did not assess the quality of the EIA documents (screening, scoping documents and final report).

Most companies represented by the respondents specialize in preparation of the EIA documents for engineering constructions (53%), waste management, as well as the extraction and processing industry (both 29%). Such activities as agriculture and aquaculture, forestry, metal production and processing industry, rubber industry are much less dominating (under 18%). Forty one percent (41%) of companies indicate that they do not specialize in document preparation for a certain planned economic activity (PEA) and that they perform assessment of the impacts of various activities. Usually documents are of a higher quality when the specialist of a consulting company preparing the EIA has the knowledge and accumulated experience in some particular area or when a company has several specialists who are experts in certain areas – this contributes to a qualitative EIA.

Compared to other countries, specialization regarding country peculiarities is also prevailing. In Bulgaria, the EIA cases are mostly related to wind farms and holiday villages on the Black Sea coast and hydropower companies. In the Czech Republic, most cases are related to the industry business, infrastructure projects, fossil fuel and wind energy equipment, whereas in France most EIA are prepared for agriculture and infrastructure projects (COWI 2009). Therefore, the document preparation of EIA according to the economic activity is highly dependent on the country where the activities are developed, the specifics of activity, thresholds set and economic development.

More than half (65%) of respondents personally assess environmental impact for individual (particular) components of the environment, while less than half (35%) assess impacts for all environmental components. The statistical analysis results have shown that as much as 88% of respondents assessing individual environmental components are under 40 years of age ($\chi^2 = 2.784$, $p < 0.1$) (Table 1). This could be related to the fact that younger professionals specialize in the individual environmental components and older ones, being more experienced, are able to evaluate the impact on all components of the environment.

Table 1. Factors influencing the variety of environmental components covered in EIA. Survey results: the EIA practitioners

Factors	χ^2	p
Age	2.784	0.095*
Gender	2.269	0.132
Work experience	**0.625**	**0.732**
Education:		
Environmental science	**0.714**	**0.398**
Chemistry	1.111	0.292
Engineering	0.227	0.634
Other	0.096	0.756

Note: * – statistically significant when $p < 0.1$.

The EIA practitioners indicate that most often considered components of the environment are water (78%), immovable cultural heritage (67%), landscape (56%), biodiversity, socio-economic environment, weather conditions (44%), climatic conditions (44%), and air (44%). Other elements of the environment, such as climate, air pollution or noise (around 11%), are given less consideration. The fact that 44% of EIA practitioners claim to be evaluating socio-economic impacts contradicts the EIA reports' analysis which shows that a socio-economic aspects are estimated poorly and incompletely. As already mentioned, job creation is the most frequently considered aspect of a socio-economic environment.

Question about three most important elements of the environment in EIA was addressed to all EIA specialists. Air, water (both 22%) and public health (19%) are distinguished as elements of the greatest importance. The least attention is given to wildlife, soil, surface and depth, climate, landscape, biodiversity, socio-economic environment, material and immovable cultural heritage (22% all together). Nevertheless, even 53% of the respondents say that the significance of the environmental components depends on the planned economic activity and 50% say that all elements of the environment are equally important. In 2008, J. Kruopiene *et al.* conducted a study on the quality of EIA documents and found that issues related to air, water and waste management were overviewed well enough, but socio-economic environment, biodiversity and natural resources were the least analysed areas. These results are also consistent with the results of EIA reports from Kaunas REPD covering the period from 2006 to 2011, where it was found that a socio-economic environment was assessed poorly; however, air-related and water-related aspects were assessed most comprehensively.

2.4. The EIA specialists' approach to a socio-economic assessment

Questions related to a socio-economic assessment were answered both by the EIA practitioners and by the state institutions' experts. Ninety four percent (94%) of the respondents knew about the amendment to the Law on EIA, adopted on June 9, 2011. Those who had completed environmental studies and had been working in the field of EIA up to 5 years were more often aware of the amendment to the Law. This might be related to the fact that young specialists are more actively interested in news of their pursued area, and specialists who work longer are working automatically, as usual, and are not interested in or not paying attention to the alterations that do not change the procedures in general.

A rather worrying fact is that more than half (59%) of respondents think that the attitude of EIA practitioners will not change after the amendment to the Law. Those with the environmental science background are more often sceptical ($\chi^2 = 4.97$, $p < 0.05$) (Table 2). Those who work in state institutions more often think that the attitude of document preparers is not going to change, and those who work in the private sector and prepare the EIA documents believe that attitude will change ($\chi^2 = 3.927$, $p < 0.05$) (Table 2). In our case, we can assume that specialists from state institutions indicate that the amendment to the Law will not bring any practical benefits. Specialists from state institutions are older compared to those in private companies ($\chi^2 = 7.255$, $p < 0.05$) and probably have more experience in this field and are pessimistic about possibilities for quick changes in practice. On the other hand, this could also be showing their reluctance for changes.

Despite the fact that part of the respondents (41%) think that the attitude of EIA practitioners is going to change, more than half of them believe that this will only be a formal change (57%); and only about one third (29%) think that the amendment to the Law will improve the quality of the reports. Some specialists (14%) think that the process of EIA will be more complicated. That is one more indication showing that a socio-economic environment will not gain appropriate attention during preparation of the EIA reports.

Table 2. Dependence of the EIA practitioners' attitude towards changes on different factors (bold values are significant at $p < 0.05$). Survey results: all respondents

Factors	χ^2	p
Age	1.943	0.163
Gender	0.008	0.928
Work experience	4.297	0.117
Type of company	**3.927**	**0.048**
Education:		
Environmental science	**4.970**	**0.026**
Chemistry	0.455	0.500
Engineering	3.331	0.068*
Other	0.010	0.919

Note: * – statistically significant when $p < 0.1$.

Table 3. Dependence of the EIA specialists' attitude on different factors (bold values are significant at $p < 0.05$). Survey results: all respondents

Factors	χ^2	p
Age	0.965	0.617
Gender	0.640	0.726
Work experience	**10.341**	**0.035**
Type of company	3.329	0.189
Education:		
Environmental science	**6.993**	**0.030**
Chemistry	0.640	0.726
Engineering	0.972	0.615
Other	2.288	0.318

Those specialists who have completed the environmental studies more often think that the process of EIA will become more complicated following the amendment to the Law ($\chi^2 = 6.993$, $p < 0.05$) (Table 3). Of those who say that the attitude will change only formally, as much as 88% have been working in the EIA area for more than 10 years ($\chi^2 = 10.341$, $p < 0.05$). None of the experts working in a state institution thinks that the Law amendment will improve the quality of EIA reports, while 45% of specialists in private companies believe that the quality of the reports will improve following the Law amendment. This fact to some extent builds grounds for positive changes in EIS; however, competent institutions have to change their attitude to encourage these changes regarding SIA.

For 64% of respondents, legitimation of a socio-economic assessment in the EIA only partially changed their personal attitude to a socio-economic assessment. The fact that the changes are only formal and there are no clear approved guidelines for a socio-economic assessment could influence this result. Only 6% of respondents admit that their own attitude has changed, so we could expect that from now on at least some EIA specialists will pay more attention to the socio-economic issues.

More than half (56%) of respondents believe that the significance of socio-economic components depends on a specific planned economic activity, and 36% think that all components of a socio-economic environment are equally important (Fig. 1). Among the individual components of a socio-economic environment, the well-being of people (25%), business and job opportunities (25%) were mentioned most often. The least important components of a socio-economic environment were income and quality of life (8%), land use (11%), and appropriate infrastructure (14%). Such components of a socio-economic environment as protection of cultural and heritage resources, suitable services were not mentioned at all.

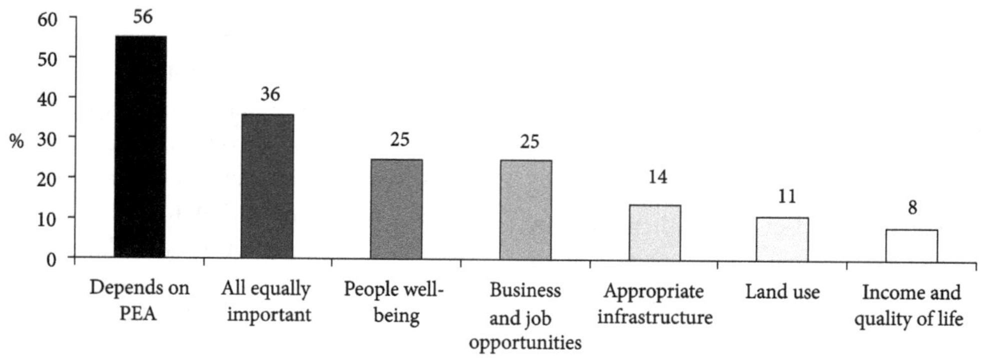

Fig. 1. The most important components of a socio-economic environment. Survey results

In general, analysis revealed that there was no consensus on the importance of the valued environmental components. It is worrying that more than half (59%) of the specialists do not see significant changes in practise even after the amendment to the Law on EIA. It is clear that new measures are required to improve the quality of EIA and to change specialists' attitudes towards the EIA process as well as expand a rather narrow understanding of SIA.

3. Discussion and recommendations

Aside from a rather long practice in other countries (e.g., USA (Burdge, Vanclay 1996)), Lithuania has no significant experience in SIA, taking into account even incorporation of SIA into EIA. Though SIA is recognized as a tool for predicting social consequences of various projects, minimizing negative impacts and increasing positive ones (Esteves *et al.* 2012), it is still only gaining attention in Lithuania. Nevertheless, socio-economic aspects were always included (though not so firmly) in the EIA legislation, and in some cases attempts to include the socio-economic issues exist.

Though some might argue that "socio-economic" is a rather narrow understanding of a social impact assessment, the Law amendment could be treated as a positive step for further improvement in planning and decision-making. However, threats of losing the value and effectiveness should be taken into account applying interdisciplinary approach and avoiding preclusion of sustainability goals (Morrison-Saunders *et al.* 2014).

The survey results show that specialists who have been working with EIA for more than 10 years often believe that the attitudes of EIA practitioners would change only formally after the amendment to the Law on EIA. The EIA experts who have completed environmental studies and are working at state institutions more often believe that the attitudes towards the socio-economic environment assessment in the process of EIA will not change at all after the adoption of changes in the EIA Law. While the formal socio-economic assessment has been validated relatively recently, 30% of experts claim to have been conducting a socio-economic assessment in detail, and more than half of the experts believe that socio-economic components assessed depend on the significance of the PEA. Human well-being, business and job opportunities are considered the most important socio-economic environment components. The least important socio-economic components are income and quality of life, while cultural and heritage resources as well as appropriate services have not been mentioned at all by the specialists. Though SIA is considered a useful tool for environmental management and planning facilitation (Barrow 2002), these results highlight problems to be resolved regarding EIA and inclusion of SIA aspects into EIA in Lithuania.

In general, effectiveness of impact assessment covers procedural, substantive, transactive, and normative effectiveness as presented by Ch. Chanchitpricha and A. Bond (2013). Therefore, whilst reviewing and improving the EIA process and incorporating SIA, all the factors behind that have to be addressed. Despite the importance of substantive and normative effectiveness, procedural issues are dominating in our research. Having a number of factors behind that might influence procedural effectiveness of EIA in general (e.g., Zhang *et al.* 2013), recommendations for improvement of the EIA process and socio-economic assessment quality would focus mostly on some procedural factors: policy, guidelines, experiences, political context, finance, and public participation.

Firstly, the EIA legislation and process itself have to be reviewed and improved. The EIA experts should be involved in the development of laws and subordinate legislation, based on their experience. In addition, personal activity and interest in law amendments, proper guidance of already adopted provisions of the Law is an issue. Here a better coordination within different institutions would be beneficial. Of course, political context and will play a crucial role in ensuring policy's consistency and succession. There is still a need to consider EIA as a tool for the achievement of economic, social and environmental benefits, and it should be seen not as a bureaucratic impediment, especially during the economic downturn.

Irrespective of the project size, it is also important to ensure the quality of the EIA documents and include into assessment not only direct but also indirect impacts on the environment components of planned activities. The competent institutions must carefully evaluate and pay attention to the effects on all components of the environment in EIA, including socio-economic issues. Not of least importance is a necessity to prepare methodological guidance for a socio-economic assessment or expand the existing one on EIA with topics related to social capital, people needs and perceptions. Clear distinction in terms and definitions could help the EIA consultants to properly assess the impacts and for the competent institution this would help to ensure the quality of the process and make a sound decision. Otherwise, the threat that no changes will take place remains. In addition to methodological guidance, problems like poor databases (especially of the socio-economic variables on the local level) and insufficient qualification of consultants may hamper a proper SIA (Barrow 2002) or impact an assessment in general. In our case, the EIA experts mostly have nature and engineering science background, thus they lack social science skills significant in conducting social impact assessment. In addition, currently existing attempts to assess the socio-economic impacts are limited to quantitative assessment mainly focusing on a labour force. Development of competencies through constantly organized compulsory training courses for the EIA specialists could also contribute to the improvement of all EIA stages. Inclusion of social science specialists in the EIA process might be another option; this could contribute to the recognition of SIA values and expertise (Rowan 2009). It seems to be necessary to increase significance and transparency of the process, to share experiences and enhance the capacity building, and to provide an access to the on-going EIA through some national level database on conducted EIAs.

Follow-up monitoring is an issue, too. Whilst incorporating SIA into EIA, relevant legislation and practical implementation of monitoring have to be carefully addressed. So far, follow-up monitoring programs do not gain much attention in EIS and application is only a formality in most cases. No responsibilities in case of monitoring absence are foreseen and only provisions for monitoring economic activities are applied. If regional environmental protection departments are doing some monitoring of biophysical environment, there is no institution monitoring social changes arising as the result of certain economic activities.

Despite some recent changes in public involvement into the EIA process, fulfilment of one of the most important precondition for the effective impact assessment, that is, ensuring equal and effective public participation and involvement into discussions (Lockie 2001; Vanclay 2003; Esteves *et al.* 2012), remains rather challenging. The need to move from mechanistic participation approach (formal information provision) to a more collaborative one (influencing decision-making) is obvious (Morgan 2012). Though it is not only the case of Lithuania (e.g., Turkey (Değirmenci, Evcimen 2013)), passiveness of public is particularly persisting with some rare exceptions in Lithuania. Therefore, not only formal legislative changes are required but also transformational changes in impact assessment practise (methods, approaches) (Esteves *et al.* 2012) and communities' involvement are urgent in order to increase effectiveness. The necessity to involve local communities in a decision-making is already apparent (e.g., the case of shale oil extraction project).

Inclusion of other stakeholders (e.g., business, education, and health care institutions) also remains an issue. Creation of possibilities for the discussion at the earlier stages of EIA, that is, not only during EIA preparation but also already during the screening or scoping stage, holding public meetings (Woloszyn 2004; Değirmenci, Evcimen 2013) and aiming that the public concerns are taken into account in the programme content, would be an option.

Regarding future, development of SIA as a separate tool for specific projects could be discussed and experiences of other countries could be employed (e.g., Finland (Kauppinen, Nelimarkka 2004), Australia (Holm *et al.* 2013)). Nevertheless, all proposals should be adapted according to the country's peculiarities, its experiences and possibilities, as different countries, different jurisdictions and institutions might differently interpret even the same goals (Morgan 2012; Savan, Gore 2015). This might affect the assessment significance to the decision-making, influence the effectiveness of assessment in terms of time and other resources allocation as well as an overall change in attitudes towards impact assessment. Meanwhile, a better integration of SIA in EIA seems to be a most credible option in case of Lithuania; however, a question of overlapping issues (public health assessment and socio-economic assessment) should be resolved.

Conclusions

Study reveals situation in a social impact assessment in EIA in Lithuania by examining a current situation and possible integration directions and challenges:

Some references to the social aspects were existent in the EIA policy documents but not gaining proper attention in practise. Only recently, SIA was formalized more firmly to some extent in the national legislation and now it has become an integral part of EIA.

Despite the aim to identify, describe and assess the potential direct and indirect impacts on a socio-economic environment as listed in the Law on EIA, a wider definition or particular guidelines how this has to be done are not provided in more detail. Therefore, proper databases and guidelines are needed.

Engineers or environmental science specialists perform most of the assessments having a limited experience in social science methods, techniques and peculiarities. Development of competencies through routinely organized compulsory training courses for the EIA specialists or inclusion of social science specialist in the EIA process could contribute to the recognition of SIA.

Orientation to a more integrative assessment from the science and public as well as other stakeholders' participation perspectives should be taken into account to increase the effectiveness of EIA and SIA as an integral part of it.

Competent state authorities and other state institutions (municipalities, cultural heritage protection, health protection institutions, etc.) should take more responsibilities (not only formally) in the process of EIA. Monitoring and inspection responsibilities, especially related to the SIA issues, should be assigned to the separate state institution that is not making a decision on PEA.

If these aspects are overlooked, the integration of SIA might be seen as one more bureaucratic impediment or just additional costs and the significance of the impact assessments on the planning and decision-making might remain low. In addition, a progress of EIA process, normative changes and related issues is the object of further research and discussion.

References

Barrow, C. J. 2002. Evaluating the social impacts of environmental change and the environmental impacts of social change: an introductory review of social impact assessment, *International Journal of Environmental Studies* 59(2): 185–195. http://dx.doi.org/10.1080/00207230210922

Barrow, C. J. 2010. How is environmental conflict addressed by SIA?, *Environmental Impact Assessment Review* 30: 293–301. http://dx.doi.org/10.1016/j.eiar.2010.04.001

Burdge, R. J. 2003. The practice of social impact assessment – background, *Impact Assessment and Project Appraisal* 21(2): 84–88. http://dx.doi.org/10.3152/147154603781766356

Burdge, R. J.; Vanclay, F. 1996. Social impact assessment: a contribution to the state of the art series, *Impact Assessment* 14(1): 59–86. http://dx.doi.org/10.1080/07349165.1996.9725886

Chanchitpricha, Ch.; Bond, A. 2013. Conceptualising the effectiveness of impact assessment processes, *Environmental Impact Assessment Review* 43: 65–72. http://dx.doi.org/10.1016/j.eiar.2013.05.006

COWI. 2009. A/S, European Commission, Department of the Environment Study concerning the application and the effectiveness of the EIA Directive, *Kongens Lyngby: COWI.* 222 p.

Değirmenci, K.; Evcimen, O. 2013. A critical analysis of the SIA process in Turkey: current problems and recommendations for improving the practice of SIA, *Impact Assessment and Project Appraisal* 31(4): 305–317. http://dx.doi.org/10.1080/14615517.2013.827542

Directive 2011/92/EU of the European Parliament and of the Council of 13 December 2011 on the assessment of the effects of certain public and private projects on the environment. 21 p.

Esteves, A. M.; Franks, D.; Vanclay, F. 2012. Social impact assessment: the state of the art, *Impact Assessment and Project Appraisal* 30(1): 34–42. http://dx.doi.org/10.1080/14615517.2012.660356

Franks, D. M.; Vanclay, F. 2013. Social impact management plans: innovation in corporate and public policy, *Environmental Impact Assessment Review* 43: 40–48. http://dx.doi.org/10.1016/j.eiar.2013.05.004

Glasson, J. 2009. Socio-economic impacts 1: overview and economic impacts, in P. Morris, R. Therivel. *Methods of environmental impact assessment.* 3rd ed. Routledge. 577 p.

Holm, D.; Ritchie, L.; Snyman, K.; Sunderland, Ch. 2013. Social impact management: a review of current practice in Queensland, Australia, *Impact Assessment and Project Appraisal* 31(3): 214–219. http://dx.doi.org/10.1080/14615517.2013.782704

Kauppinen, T.; Nelimarkka, K. 2004. A review of Finnish social and health impact assessments, *Journal of Environmental Assessment Policy and Management* 1(6): 1–17. http://dx.doi.org/10.1142/S1464333204001584

Kruopiene, J.; Zidoniene, S.; Dvarioniene, J. 2009. Current practice and shortcomings of EIA in Lithuania, *Environmental Impact Assessment Review* 29: 305–309. http://dx.doi.org/10.1016/j.eiar.2009.02.003

Kruopiene, J.; Zidoniene, S.; Dvarioniene, J.; Mikalauskas, A. 2008. Evaluation of environmental impact assessment effectiveness in Lithuania, *Environmental Research, Engineering and Management* 2(44): 28–33.

Law on Environmental Impact Assessment of the Proposed Economic Activity. 15 August 1996, No. I-1495 (as last amended on 30 June 2008 – No X-1654). Republic of Lithuania.

Lockie, S. 2001. SIA in review: setting the agenda for impact assessment in the 21st century, *Impact Assessment and Project Appraisal* 19(4): 277–287. http://dx.doi.org/10.3152/147154601781766952

Ministry of Environment of the Republic of Lithuania. 2009. *Guide of Environmental Impact Assessment.* Vilnius. 106 p.

Morgan, R. K. 2012. Environmental impact assessment: the state of the art, *Impact Assessment and Project Appraisal* 30(1): 5–14. http://dx.doi.org/10.1080/14615517.2012.661557

Morrison-Saunders, A.; Pope, J.; Gunn, J. A. E.; Bond, A.; Retief, F. 2014. Strengthening impact assessment: a call for integration and focus, *Impact Assessment and Project Appraisal* 32(1): 2–8. http://dx.doi.org/10.1080/14615517.2013.872841

Morrison-Saunders, A.; Fischer, T. B. 2006. What is wrong with EIA and SEA anyway? A sceptic's perspective on sustainability assessment, *Journal of Environmental Assessment Policy and Management* 8(1): 19–39. http://dx.doi.org/10.1142/S1464333206002372

Pope, J.; Bond, A.; Morrison-Saunders, A.; Retief, F. 2013. Advancing the theory and practice of impact assessment: setting the research agenda, *Environmental Impact Assessment Review* 41: 1–9. http://dx.doi.org/10.1016/j.eiar.2013.01.008

Provisions for preparation of Programmes and Reports of Environmental Impact Assessment. Approved by the Minister of Environment, 23 December, 2005, No. D1-636 (as last amended on 6th May 2010 – No. D1-370).

Rowan, M. 2009. Refining the attribution of significance in social impact assessment, *Impact Assessment and Project Appraisal* 27(3): 185–191. http://dx.doi.org/10.3152/146155109X467588

Savan, B.; Gore, C. 2015. Translating strong principles into effective practice: environmental assessment in Ontario, Canada, *Journal of Environmental Planning and Management* 58(3): 404–422. http://dx.doi.org/10.1080/09640568.2013.859572

Stanley, J.; Clouston, B.; Binney, J. 2004. *Conducting social and economic impact assessment: a practical guide for regional NRM bodies.* Department of Natural Resources, Mines and Water. 19 p.

Suopajärvi, L. 2013. Social impact assessment in mining projects in Northern Finland: Comparing practice to theory, *Environmental Impact Assessment Review* 42: 25–30. http://dx.doi.org/10.1016/j.eiar.2013.04.003

Vanclay, F. 2003. International principles for social impact assessment, *Impact Assessment and Project Appraisal* 21(1): 5–12. http://dx.doi.org/10.3152/147154603781766491

Woloszyn, W. 2004. Evolution of environmental impact assessment in Poland: problems and prospects, *Impact Assessment and Project Appraisal* 22(2) 109–119. http://dx.doi.org/10.3152/147154604781765950

Zhang, J.; Kørnøv, L.; Christensen, P. 2013. Critical factors for EIA implementation: Literature review and research options, *Journal of Environmental Management* 114: 148–157. http://dx.doi.org/10.1016/j.jenvman.2012.10.030

Renata DAGILIŪTĖ, Dr Lecturer, Department of Environmental Sciences, Faculty of Natural Sciences, Vytautas Magnus University (VMU), Lithuania. Author of several scientific publications and participator of number of international conferences. Research interests: eco-efficiency, sustainable development, sustainable production and consumption, environmental policy and law.

Gintarė JUOZAPAITIENĖ, MSc, in Environmental Management, Department of Environmental Sciences, Faculty of Natural Sciences, Vytautas Magnus University (VMU), Lithuania. Research interests: sustainable development, environmental policy and law, environmental impact assessment.

12

WATERSHED ROAD NETWORK ANALYSIS WITH AN EMPHASIS ON FIRE FIGHTING MANAGEMENT

Mohsen MOSTAFA[a], Shaban SHATAEE JOUIBARY[b], Majid LOTFALIAN[c], Amir SADODDIN[d]

[a, b]*Department of Forestry, Faculty of Forest Sciences, Gorgan University of Agricultural Sciences and Natural Resources, P.O. Box 386, Gorgan, Iran*

[c]*Department of Forest Engineering, Faculty of Natural Resources, Sari University of Agricultural Sciences and Natural Resources, P.O. Box 737, Sari, Iran*

[d]*Department of Watershed Management, Faculty of Rangeland and Watershed, Gorgan University of Agricultural Sciences and Natural Resources, P.O. Box 386, Gorgan, Iran*

Abstract. The aim of this study is fire hazard zoning the Chehel-Chay watershed and analysis of road network in order to fire-fighting management. Using effective factors on fire occurrence, the fire hazard map of the study area produced by support vector machine algorithm and then was divided into four hazard classes. The road length and type were investigated in the each fire hazard classes. The results showed that most of occurred fires are located in the close distances of roads and forest areas. The results showed that road types and land cover are important in fire occurrences and suppression. In high dangerous zone, the roads pass through forestlands, but in low dangerous zone, the roads are passing from farmlands. The roads do not cover the half of area and do not pass at two third of high hazard class zones. Therefore, appreciate road network planning is necessary according to fire-fighting management.

Keywords: Forest road, fire hazard map, fire-fighting vehicles, Support Vector Machine (SVM).

Introduction

Fire is one of the important components of natural ecosystems such as forests and rangelands (Sakr *et al.* 2010; Dlamini 2010). Forest fires can lead to destruction of environment, damages to human health and property, and endanger life, cause environment deterioration and soil losses (Bowyer *et al.* 2009; Fox *et al.* 2015) and create economic problems (Ozbygol, Bozer 2012). In Iran, fire is one of the most destructive occurring phenomena in forests and rangelands (Adab *et al.* 2013; Pourghasemi *et al.* 2016; Pourtaghi *et al.* 2015). The discussion about forest fire ignition can be divided into wildfire and anthropogenic fire (Verma *et al.* 2013). Most of fires occurring in Hyrcanian forests of Iran are anthropogenic fire (Pourghasemi *et al.* 2016).

Fire hazard mapping is important to proper planning to fire hazard reduction, management, and control of fire (Jaiswal *et al.* 2002). Several studies have been conducted to analyses forest fire using parametric and nonparametric algorithms (Table 1). Many studies have been used parametric algorithms such as linear regression (Dong *et al.* 2006; Oliveira *et al.* 2012), logistic regression (Hernandez-Leal *et al.* 2006; Martinez *et al.* 2009; Zhang *et al.* 2010; Arndt *et al.* 2013; Guo *et al.* 2016), spatial clustering (Wu *et al.* 2015), generalized linear mixed models (Kwak *et al.* 2012; Boubeta *et al.* 2015), Monte Carlo simulations (Carmelm *et al.* 2009; Reineking *et al.* 2010), kernel methods (Amatulli *et al.* 2007; Kuter *et al.* 2011), and nonparametric algorithms as Random Forests (RF) (Massada *et al.* 2013; Pierce *et al.* 2012; Oliveira *et al.* 2012; Arpaci *et al.* 2014), Artificial Neural Networks (ANN))Bisquert *et al.* 2012; Safi, Bouroumi 2013; Sahu 2015), and SVM for fire hazard mapping. Cortez and Morais (2007) used data mining (DM) approaches (SVM and RF) to predict the forest fires in the northeast region of Portugal using meteorological data. These studies showed that SVM algorithm had satisfactory performance in forest fire hazard mapping compared to other algorithms in general. Sakr *et al.* (2011) compared performance of ANN and SVM algorithms to reduce number of parameters and cost required system especially for developing countries and reported that SVM outperformed ANN based on cumulative precipitation and relative humidity.

Corresponding author: Shaban Shataee Jouibary
E-mail: shataee@yahoo.com

Most human-caused fires occur in proximity to human settlements and road corridors (Soto 2012; Fox *et al.* 2015; Arpaci *et al.* 2014). Relationship among road network, fire occurring and fire management has been recognized for some times. Density of road and distance to roads influence the fire ignition (Nunes 2012; Chas-Amil *et al.* 2013; Guo *et al.* 2016). Roads provide access option inside forests in order to perform management practices, timber harvesting, and other services. Roads separate the ecosystem, create high contrast edges, (Forman, Alexander 1998), and reduce forest areas (Schonewald-Cox, Buechner 1992). Forest roads play significant role on wildfire prevention and fire suppression (Narayanaraj, Wimberly 2013), and creating linear gaps in forest (Spellerberg 2002) which changes microclimates closed to road (Chen *et al.* 1993). Relations between fire occurrences, fire management, and road have been studied in several researches (Table 2).

Çoban and Eker (2010) investigated the forest road before and after fire in burnt area, Turkey. Their analysis has explained that roads should be planned in fire sensitive regions and be facilitated management activities after fire event. Narayanaraj and Wimberly (2013) described effects of edge roads on the spatial pattern of burn severity in Okanogan–Wenatchee National Forest in Washington State. Nasiri (2012) examined the plan of skidding routes on fire fighting in DarabKola forest, north of Iran. Tian *et al.* (2013) investigated effective factors of forest fires in China; they explained 58% of occurred fires were located less than 5 km to roads. Roman and Martinez (2006) and Martinez *et al.* (2009) reported significant relationships between road density and forest fire in Mexico and Spain.

Table 1. Previous studies on fire hazard mapping

Author	Country	Algorithms		Variables
		parametric algorithm	nonparametric algorithms	
Dong *et al.* (2006)	China	Linear regression	–	Fuel, topography, anthropogenic
Hernandez-Leal *et al.* (2006)	Spain	Logistic regression	–	Slope, altitude, insulation, proximity to road, vegetation cover
Amatulli *et al.* (2007)	Spain	Kernel methods		Occurred fire, cover type, fire caused, weather
Zhang *et al.* (2010)	China	Logistic regression	–	Topography, weather, human factor
Kuter *et al.* (2011)	Turkey	Kernel methods	–	Information of occurred fire
Sakr *et al.* (2011)	Lebanon	–	ANN, SVM	Precipitation, relative humidity, solar radiation, average wind speed
Soto (2012)	Chile	Fuzzy logic		Spatial localization of the fires, road network, areas with an urban- wild land interface
Oliveira *et al.* (2012)	Portugal, Spain, France, Italy and Greece	Linear regression	Random forest	Topography, land cover, climate, infrastructures, demographic, socio-economic
Kwak *et al.* (2012)	Korea	Generalized linear mixed models	–	Topography, human factor, distance from road, forest cover
Bisquert *et al.* (2012)	Spain	Logistic regression	ANN	Monitor vegetation Status, land surface temperature
Arndt *et al.* (2013)	Austria	Logistic regression	–	Socio-economic
Pourghasemi *et al.* (2016)	Iran	Mamdani fuzzy logic, analytical hierarchy process		Topography, climate, land use, vegetation index
Arpaci *et al.* (2014)	Austria	–	Max Ent and Random Forests	Topography, vegetation, climate , socio-economic datasets
Wu *et al.* (2015)	China	Spatial clustering	–	Topography, climate, vegetation, occurred fire
Boubeta (2015)	Spain	Generalized linear mixed models		Days without rain, Number of fires, Population size, Cadastral parcel, Scrub area, Percentage of wet lands, Percentage of wood lands
Sahu (2015)	Portugal	–	ANN	Fine fuel moisture code, duff moisture code, drought code, initial spread index, fire behavior index
Guo *et al.* (2016)	China	Logistic regression		Topographic, vegetation, climatic, infrastructure, socioeconomic

Maingi and Henry (2007) studied important factors that have effects on wildfire in eastern Kentucky, USA. Their investigation showed that proximity to roads is an affective factor on forest fire ignition. Hoyo *et al.* (2011) reported most of fires start close to the roads in Madrid, Spain. Verma *et al.* (2013) found that high road connection led to more forest fire in Missouri forest division in India during 2000 to 2011.

The Chehel-Chay watershed is one of the areas that have different natural disasters such as fire occurring due to special ecological, economical, and social conditions and containing various land covers and human activities. In the study area, the road network with different road types are spatially distributed in the whole of watershed. In addition to preparing accessibility to human settlements for different needs, the roads and its network are playing important role both in fire occurring due to preparing accessing people to whole area and play key role in suitable fire fighting due to preparing facilities for fire fighters and vehicles. The investigation on fire risk assessment of a watershed and its zoning based on fire probability as well as performance analysis of current road network in order to achieve fire-fighting management is very necessary for watershed management. Therefore, the aim of the current study is (1) zoning the Chehel-Chay watershed based on fire hazard probability using support vector machine (SVM((2), analysis of road network in order to Fire Fighting Management (FFM) in different land covers.

Table 2. Different studies, which are showed relationships between fire occurrences, fire management, and roads

Author	Year	Country	Subject	
			fire occurrences	fire management
Martinez *et al.*	2009	Spain	•	–
Arienti *et al.*	2009	Canada	•	–
Çoban and Eker	2010	Turkey	•	•
Avila-Flores *et al.*	2010	Mexico	•	–
Hoyo *et al.*	2011	Spain	•	
Narayanaraj and Wimberly	2011	USA	•	–
Zumbrunnen *et al.*	2012	Switzerland	•	–
Massada *et al.*	2013	USA	•	
Nasiri	2012	Iran	–	•
Tian *et al.*	2013	China	•	–
Narayanaraj and Wimberly	2013	USA	•	–
Liu *et al.*	2015	China	•	–
Barni *et al.*	2015	Brazil	•	–

1. Materials and methods

The Chehel-Chay Watershed is located in Minoodasht County in the Golestan Province, Northern Iran. The area of watershed is about 25680 hectares and is located between latitudes 36°, 59′ to 36°, 17′ N, and longitudes 55°, 22′ to 55°, 37′ E (Fig. 1). The elevation above sea level is ranging 190 to 2750 m. The dominant land-cover vegetation of watershed is forest in the northern part, and mixed rangeland and forest in southern part. The climate type varies from temperate to semi-humid (Pourghasemi *et al.* 2016). Average annual precipitation of the study area is approximately 750 mm. The main land uses in the watershed is include forest (67%), croplands (28.8%), grassland (4.09%), and human residential areas (0.53%).

The datasets and methodology are shown in (Fig. 2). The flowchart shows two steps of the study: (1) Fire hazard

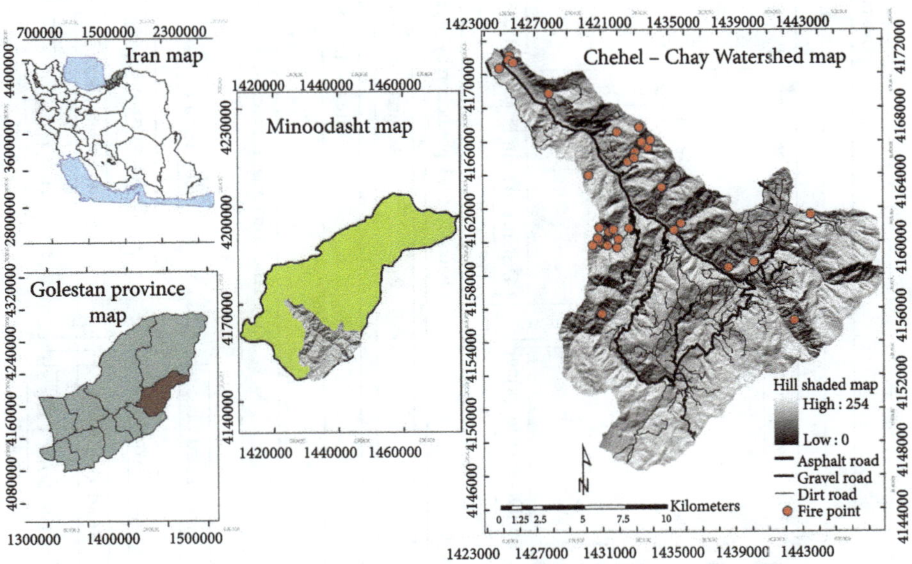

Fig. 1. Location the study area in the Golestan Province, Northern Iran, and location of occurred fires in the past years

mapping by SVM (2) assessing the road network distribution in the study area according to FFM.

In order to produce a fire hazard probability map, the effective factors on fire occurrences including climate, topography, vegetation, human factors, distance to water recourse were identified and prepared from different sources and using various techniques in addition OFFP in the recent past years (Table 3).

Climate influences the fire regime through several processes (Deng *et al.* 2013). Data related to Climatology is complex in forest areas and many microclimates can be observed in these areas (Cottle 2007). In the current study, wind direction, and speed, maximum absolute temperature, monthly average evaporation, minimum relative humidity, and monthly average precipitation of autumn were used as climate related data because of accruing about 70% of fires in October, November, and December. The climate maps were created using continues data (Fig. 3).

Topography is a significant physiographic factor that influences the fire spreading and this occurrence of fire indirectly (Jaiswal *et al.* 2002) and affects the climate on the landscape and fuel availability (Vasilakos *et al.* 2009). Elevation, slope, and aspect as topographical factors influence the ignition of forest fires (Lee *et al.* 2004; Kwak *et al.* 2012). In this research, topographic variables including elevation, slope, and aspect were used to create fire hazard map. The contour lines from, National Cartographic Center (NCC) were used to obtain Digital Elevation Model (DEM), slope, and aspect maps. Topographical characteristics of the study area are shown in (Fig. 4).

Vegetation is one of the main environmental variables that affect fire-regime (Amatulli *et al.* 2007; Tuček, Majlingová 2009). Vegetation (dead or alive covers) prepares the fuel for fire (Wu *et al.* 2015) and causes some

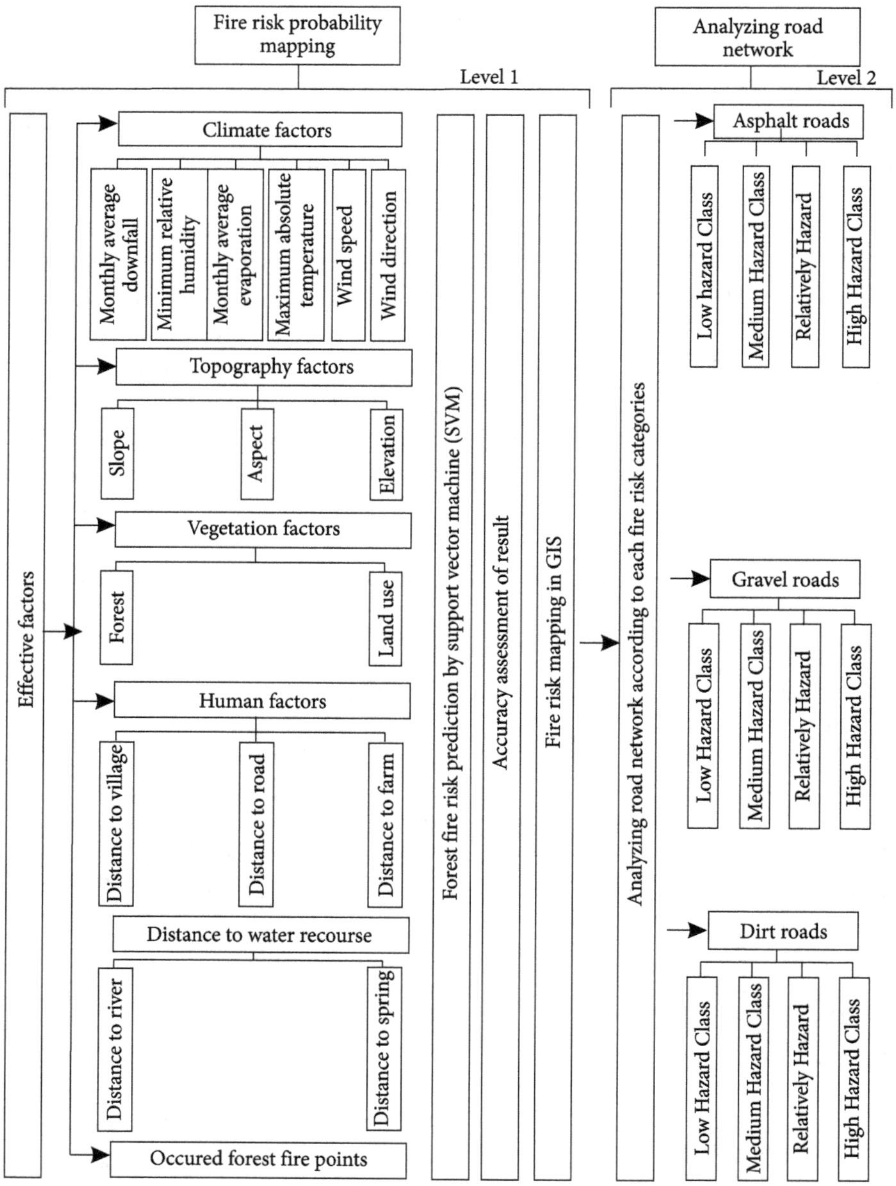

Fig. 2. Flowchart of the implemented methodology

Table 3. Variables used for fire hazard probability mapping (FRPM)

Variable type	Variable name	Data format/ Resolution/ Scale	Source of data
Climatic	Monthly dominant wind direction Monthly maximum wind speed Monthly maximum absolute temperature Monthly average evaporation Monthly minimum relative humidity Monthly average precipitation	Excel data	Iranian Meteorological Organization(IRIMO) Golestan Regional Water Authority (GRWA)
Topographic map	Elevation Slope Aspect	Raster/10 m	National Cartography Center (NCC)
Vegetation	Forest type Forest and non-forest map	Vector/ 1:25000	1 – Forest management plan. 2 – Integrated management of the Chehel-Chay Watershed. 3 – Extensive fieldwork. 4 – Satellite image, National Cartographic Center (NCC)
Infrastructure	Distance to road Distance to village Distance to recreational sites Distance to agricultural areas	Vector/ 1:25000	1 – Forest management plan. 2 – Integrated management of the Chehel-Chay Watershed. 3 – Extensive fieldwork. 4 – Satellite image, National Cartographic Center (NCC)
Water recourse	Distance to river Distance to spring	Vector/ 1:25000	1 – Forest management plan. 2 – Integrated management of the Chehel-Chay Watershed. 3 – Extensive fieldwork. 4 – Satellite image, National Cartographic Center (NCC)
Occurred forest fire	–	Point, Polygon, Excel data	1 – Extensive fieldwork. 2. Archive of Department of Natural Resources and Watershed (DNRW), Golestan province

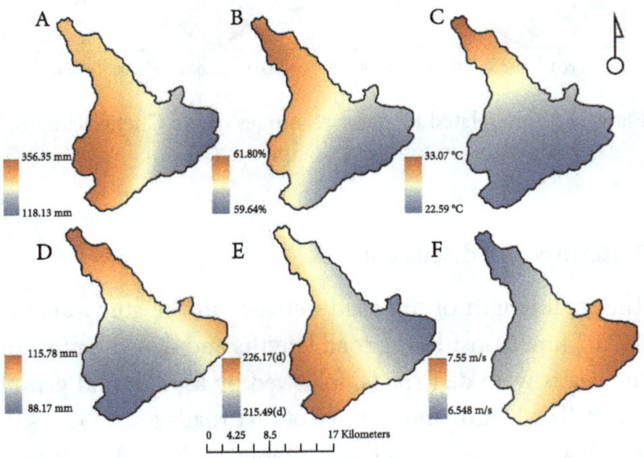

Fig. 3. Climatology characteristics of the Chehel-chay Watershed for period of October-December: Precipitation (mm) (A), Minimum relative humidity (%) (B), Maximum absolute temperature (Celsius) (C), Evaporation (mm) (D), Wind direction (Degree) (E), Wind speed (km/h) (F)

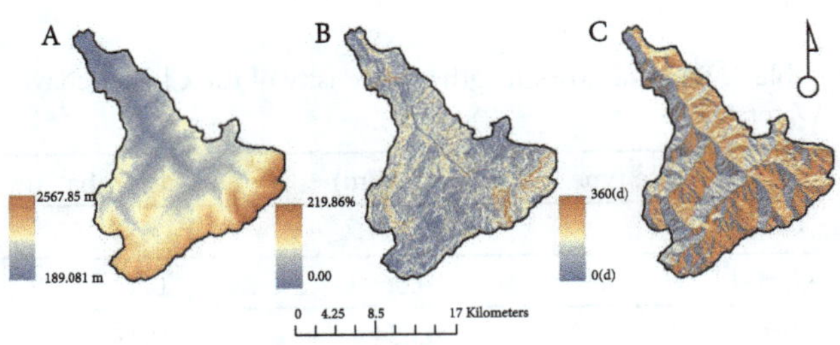

Fig. 4. Topographical characteristics of the Chehel-chay Watershed: Elevation (m) (A), Slope (%) (B), Aspect (Degree) (C)

problems for forest fire fighting (Chen, Di 2015). The vegetation factor in the study area was used in two maps: land use map and forest type map. The land use map was classified into two classes, forest and non-forest, and land cover was classified into nine classes, *Carpineo-Parrotium*, *Carpinetum-Betulus*, *Quereteo-Castaneifolia*, *Quereteo-Carpinetum*, *Carpinetum-schuschaensis*, *Petrocaryo- Alnetum*, *Pinus-Cupressus* (reforestation), Grassland and non-forest (Fig. 5).

Human activities such as agricultural, human settlements and recreational activities are very important in fire occurrence (Amatulli *et al.* 2007). The main causes of forest fires in Japan, China, USA, Brazil, and Europe are related to human activities (Martinez *et al.* 2009; Barni *et al.* 2015). Fire regimes are affected by human activities during last century (Chen *et al.* 2015) and more than 80% of fires are caused by human activity (FAO 2007). The human

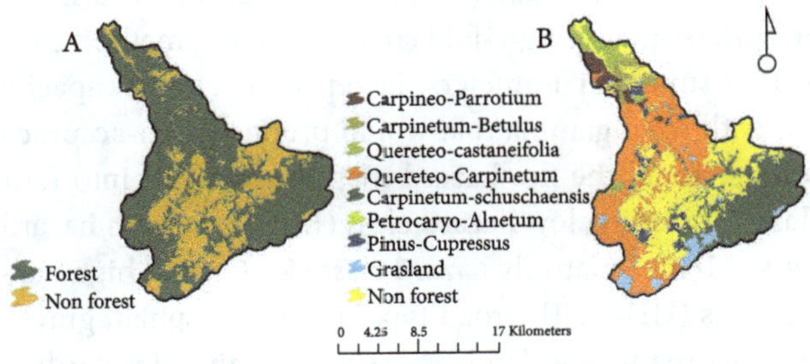

Fig. 5. Vegetation characteristics of the Chehel-chay Watershed; Landaus map (A), Landover maps (B)

factors considered in this research are distance to roads, villages, recreational sites and agricultural areas (horticultures, irrigated and dry farms, rice paddies, livestock) (Fig. 6).

Water sources and access to its are one of the main factors that affect the forest fire (Jaiswal *et al.* 2002) as well as an important factor that prevent the spread of forest fire (Češljar, Stevović 2015). In this research, distance to rivers and springs were used as water factor (Fig. 7).

The existing road network map was created on 1:25,000 scale (NCC), using satellite image, and mapping by GPS. The road network was classified into three classes including asphalt, gravel, and dirt roads by through filed visit road (Fig. 1).

Road network density is one of factors used to find correlation between lightning fire and road network (Arienti *et al.* 2009). The road network density was calculated for the study area, fire hazard classes, and land use classes in whit in fire hazard probability map classes (Eq. (1)):

$$ND = \frac{L}{A}, \quad (1)$$

where ND – road density; L – length of road network (meters); A – area (ha).

Forest roads provide access for ground and aerial firefighting equipment (Kapusniak, Majlingová 2014). In the study area, all road types can be used for fire extinguishing, the maximum distance was 500 meters from the roads. Deployment of ground mobile fire-fighting equipment to fight the fire was possible on 100 meters to asphalt and gravel roads. In order to find the area that covered with the road network considering firefighting buffer map, 500 meters from both sides of the roads generated using ArcGIS 10.1. Then the cover percentage of road network was calculated (Eq. (2)):

$$E(\%) = \frac{a}{A} \cdot 100, \quad (2)$$

where $E(\%)$ – cover percentage of road network for firefighting; A – total area of study area; a – area covered with road network.

Using effective factors on fire occurrence including climate, topography, vegetation, human factor, factors related to water sources and the occurred forest fire points, the fire hazard probability map of the study area produced by support vector machine (SVM) algorithm. The specified grid search using v-fold cross-validation method used to find suitable parameters, i.e. epsilon (ε) and capacity (C) with fixed gamma that would produce high-accuracy results. Then, the fire hazard map was divided into four classes including low hazard class (LHC), medium hazard class (MHC), relatively hazard class (RHC), and high hazard class (HHC). The road type network (asphalt, gravel, and dirt roads) and land covers were analyzed according to each fire hazard categories.

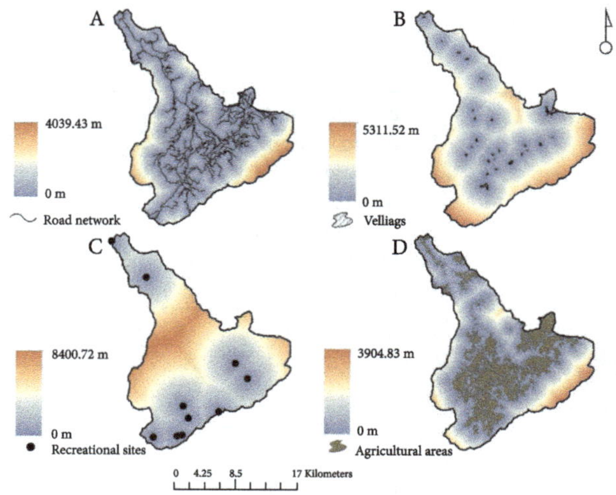

Fig. 6. Human factors of the Chehel-chay Watershed: Distance to road (A), Village (B), Recreational sites (C), Agricultural areas (D)

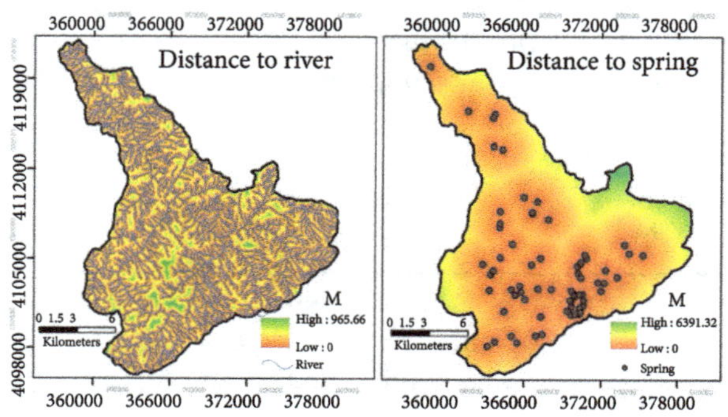

Fig. 7. Factors related to water recourses of the Chehel-chay Watershed: Distance to river (A), Springs (B)

2. Results and discussion

The total length of the road network in the study area is 337.07 km. Considering road lengths and density; most of the roads were dirt roads, followed by asphalt and gravel roads. The length and density of dirt roads are 243.54 km and 9.47 mha^{-1}, respectively (Table 3). The asphalt and gravel roads as public roads were constructed in order to access to villages whereas all dirt roads were used to access farmlands, livestock, and recreational sites. There are not roads that not constructed specifiable for forest operation in the study area.

Table 3. Road network length and density of the Chehel-chay Watershed

Road surface type	Length (km)	Density (mha^{-1})
Asphalt	82.87	3.23
Gravel	10.66	0.42
Dirt	243.54	9.47
Total	337.07	13.12

More than half of the study area (58%) is covered by road network for the purpose FFM. On other hand about 14900 ha of the study area is accessible in term of purpose. Figure 8a indicates the areas were accessible and inaccessible by road network. In other hand, 42% of watershed dose not covered by roads. Therefore, road network planning is necessary for this area according to FFM. As discussed by Kapusniak and Majlingová (2014) new road planning suitable to fighting fire in natural conditions is necessary. The results indicated that the LHC class is the largest class in fire hazard probability map (10640.5 ha, 42%), then followed by HHC (7995.8 ha, 31.5%), MRHC (3959.8 ha, 15%) and RHC (3083.8 ha, 12%) (Table 4 and Fig. 8b). The results show that 83% of past fires are located in HHC class and in the distances of less than 1000 m to roads. The finding of this research in line with what has been stated by (Spellerberg 2002; Narayanaraj, Wimberly 2013) that the roads have important effects on ignition in forest.

Table 4. Fire hazard classes area of the Chehel-chay Watershed

Fire hazard	LHC	MRHC	RHC	HHC
Area (ha)	10640.49	3959.853	3083.813	7995.843

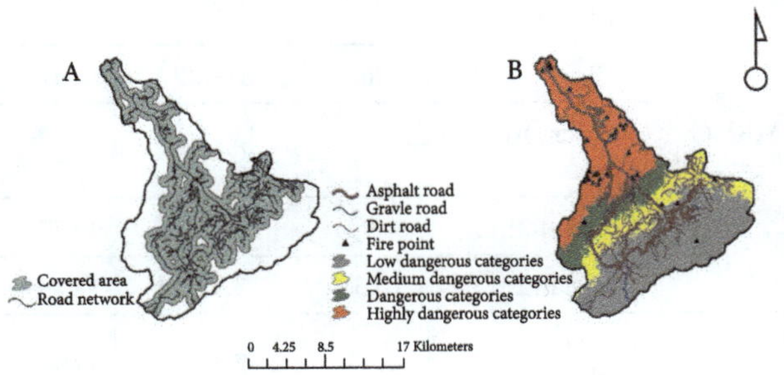

Fig. 8. Cover area of road network (A), Road network on the fire hazard probability map (B)

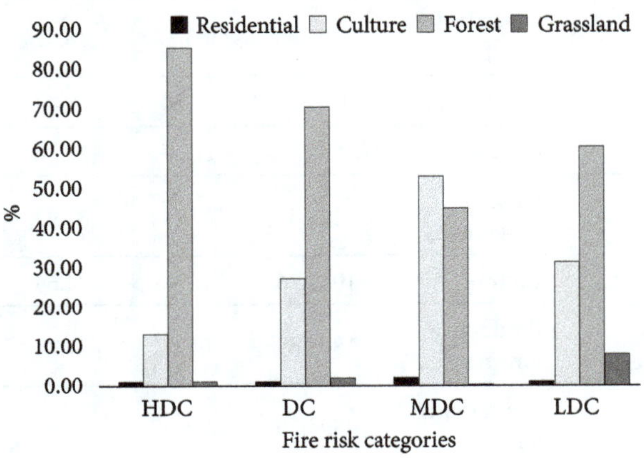

Fig. 9. Percentage of land use on the fire hazard classes of the Chehel-chay Watershed

The results indicate that 85.51 percent of HHC and 70.18 percent of RHC were covered by forest (productive forest, conservative forest, and reforestation). The conservative forests, agricultural lands were the largest land use in the LCD (69.09%) and MRHC (53.2%), respectively (Table 5, Fig. 9). The analysis of land use classes according to the fire hazard probability map fire were in line with previous studies (Carmo *et al.* 2011; Pourghasemi *et al.* 2016). Forests were the more fire prone land use; human settlements areas and agriculture land use (dry farming, Rice field, and garden) were the less susceptible land uses, respectively.

Table 5. Area of land uses on the fire hazard classes of the Chehel-chay Watershed

Land use/Fire hazard classes	LHC (ha)	MRHC (ha)	RHC (ha)	HHC (ha)
Human settlements	62.91	53.51	7.45	22.68
Culture	3330.36	2106.46	842.94	1036.26
Productive forest	0	0	56.82	1803.69
Conservative forest	6394.29	1787.42	1965.38	893.58
Reforestation	0	1.95	142.05	139.62
Grassland	852.93	10.51	69.17	100.01

The largest and lowest road densities belong to MRHC and HHC, respectively. The largest densities of asphalt roads density belong to HHC (3.74 mha^{-1}) and RHC (4.41 mha^{-1}). Table 6 indicates the detail of each road on the fire hazard probability map.

Table 6. Detailed length and density of road network on the fire hazard probability map of the Chehel-chay Watershed

Road/Fire hazard classes		LHC	MRHC	RHC	HHC
Length of road network on the fire hazard classes	Asphalt road (km)	29.6	9.83	13.62	29.93
	Gravel road (km)	7.85	0.32	0.25	2.22
	Dirt road (km)	113.21	69.82	22.56	37.94
Density of road network on the fire hazard classes	Asphalt road (M/Ha)	2.79	2.49	4.41	3.74
	Gravel road (m/Ha)	0.74	0.08	0.08	0.28
	Dirt road (m/Ha)	10.64	17.64	7.32	4.75
	Total	14.17	20.13	11.81	8.77

About 111.39 km of roads were located in agriculture areas (dry farming, Rice field, and garden) and 29.81 km of roads were located in forest areas (conservative forest) in LCD classes and density of these roads were 33.54 km/ha and 4.66 m/ha, respectively. In MRHC classes, the road density in agriculture areas and forest areas (conservation forest) was 31 m/ha and 7.89 m/ha, respectively. In RHC classes, the road density in agriculture areas are 28 m/ha and in forest areas (conservative forest, productive forest, and reforestation) are 64.3 m/ha. In HHC classes, the road density in agriculture areas are 18.62 m/ha and in forest areas (conservative forest, productive forest, and reforestation) are 39.19 m/ha (Table 7, Fig. 10). Most of the roads in MRHC and LHC were dirt roads, which are the basic communication networks among agriculture areas. However, the most of fire occurred and asphalt roads are located in the HHC and RHC that are forest areas land use. In fact, the asphalt roads are the principle route to connect the study area to center of county (Minoodasht city). It can be concluded that type of road and land cover are important in fire ignition in wild land. The main roads increases human pressure in natural resources lands and lead to fire ignition by accident and negligence (Soto 2012; Renard *et al.* 2012).

The MRHC class has the largest cover percentage of road network (64.24%) that road network covered 2543.65 ha of the area. The road network covered the 34.5% of the HHC class area (Table 7).

Table 7. Cover percentage of road network in fire hazard classes of the Chehel-chay Watershed

Fire hazard classes	HHC	RHC	MRHC	LHC
Road network cover area (ha)	2758.30	1156.74	2543.65	4674.69
Road network cover percentage	34.50	37.52	64.24	43.93

The lowest cover percentage of road network in forest areas are belonged to LHC class. In MRHC, class 37.31% conservation forest and 99.75% of reforestation areas are covered by road network. In RHC class, 91.69% of productive forest, 18.51% of conservation forest, and 67.44% of reforestation area covered by road network. In HHC class, 33.74% of productive forest, 24.43% of conservation forest, and 55.75% of reforestation areas covered by road. Table 8 and Figure 11 show the detail cover percentage of road network in land use class in fire hazard classes. Roads provide access to forest and help to fire ignitions but they reduce burn severity and prevent the fire spread (Renard *et al.* 2012; Narayanaraj, Wimberly 2013). The lowest cover percentage of rod network in our study area belong to forestland use in HHC, this results are in line with (Narayanaraj, Wimberly 2011).

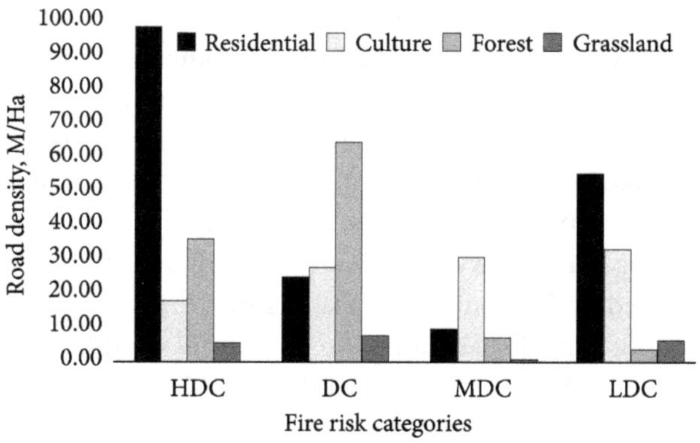

Fig. 10. Road density in fire risk categories

Table 8. Detailed cover percentage of road network in land use class in fire hazard classes of the Chehel-chay Watershed

Fire hazard classes	Land use	Area (ha)	Area coverage (ha)	Cover (%)
LHC	Human settlements	62.91	62.48	99.32
	Culture	3330.36	2874.49	86.31
	Conservation forest	6394.29	1552.15	24.27
	Grassland	852.93	185.57	21.76
MRHC	Human settlements	53.51	51.62	96.47
	Culture	2106.46	1819.29	86.37
	Conservative forest	371787.42	666.87	37.31
	Reforestation	1.95	1.95	99.75
	Grassland	10.51	3.92	37.29
RHC	Human settlements	7.45	7.45	99.96
	Culture	842.94	625.75	74.23
	Productive forest	56.82	52.10	91.69
	Conservative forest	1965.38	363.77	18.51
	Reforestation	142.05	95.80	67.44
	Grassland	69.17	11.87	17.16
HHC	Human settlements	22.68	19.94	87.93
	Culture	1036.26	773.46	74.64
	Productive forest	1803.69	608.62	33.74
	Conservative forest	4893.58	1244.42	25.43
	Reforestation	139.62	77.83	55.75
	Grassland	100.01	34.03	34.02

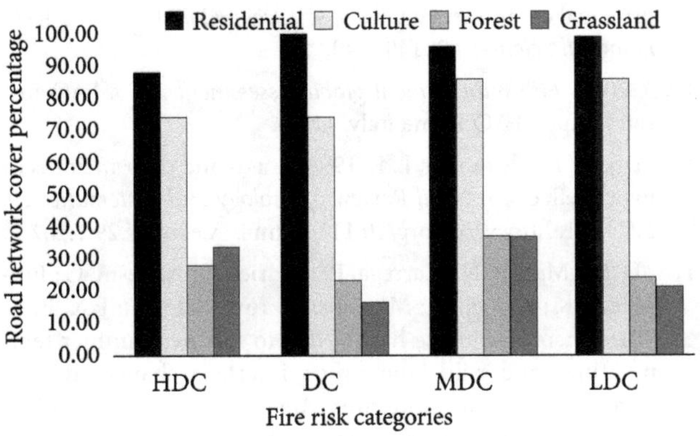

Fig. 11. Cover percentage of road network in land use class in fire risk categories

Conclusions

This study is about fire risk, management, and road network in a watershed (i.e. Chehl-Chay watershed) with special ecological, economical, and social conditions. In the study area, the numerous wild fires were occurred in the past years and fire suppression had very difficulties. By the way, in forested watershed areas, for road designing and constructing are generally considered only settlement accessing and traveling need without covering other needs such as disaster (like wildfires) management. The aim and novelty of the study is assessment the current road network situation and its spatial position for fire management.

According to results, the about two third of occurred fires are located in the distances of close to roads and in HHC class that they happened in fall season. Although, based on road network analysis results, the road density in forests is less than the other land uses, but the most of forest land cover is located in the high and relative hazard zones. This result indicates that forests are dangerous areas for fire and be the more fire prone land use due to preparing fire fuel.

As it is stated in the introduction, the most of forest fires occurred in the bottom to middle of watershed, and the forest land cover where people and eco-tourists are interesting to settle in these places. In addition, the studied area is a mountainous watershed and elevation above sea level is ranging 190 to 2750 m. Therefore, the temperature and humidity is very hot and low in the bottom to middle of watershed in the summer and fall. With these causes, in the forest land covers, the risk of fire is always high due to be prepare fire fuel and flammability.

The road type analysis showed that most of roads in the forest areas are asphalt roads. Based on these results, it can be concluding that types of road and land cover are important both in fire occurring, and in fire prevention and fire suppression in wild land. The one reason for fires occurrence in the areas close to the asphalt roads is fast and comfortable accessing and travelling the people and passing with all of vehicles and high ratio of traffic and high density of population (local community and tourists) inhabits the edge of roads. However, there are some difficulties and limitations for traveling in the gravel and dirt roads. In other hand, based on road passing analysis results, the roads do not cover close to half of watershed area as well as don't pass at the two third of high hazard class zone. Therefore, appreciate road network planning and designing for studied area is necessary according to fire-fighting management or using of suitable fighting fire vehicles for forest mountainous region.

This study concluded that in the forest and watershed road network designing and planning in watershed and landscape levels, the integrated watershed management goals such as wildfire fighting accessibility and other roads services should be considered particularly in the spatial domains. It is a work that can be done everywhere with the same kind of problems and can help to solve problems of fire risk and extinction to the fire fighters or/and authorities.

References

Adab, H.; Kasturi, Devi, K; Solaiman, K. 2013 Modeling forest fire hazard in the northeast of Iran using remote sensing and GIS techniques, *Nat Hazards* 65: 1723–1743. https://doi.org/10.1007/s11069-012-0450-8

Amatulli, G.; Per´ez-Cabelloa, F.; Rivaa, J. 2007. Mapping lightning/human-caused wildfires occurrence under ignition point location uncertainty, *Ecological Modeling* 200: 321–33.

Arienti, M. C.; Cumming, S. G.; Krawchuk, M. A.; Boutin, S. 2009. Road network density correlated with increased lightning fire incidence in the Canadian western boreal forest, *International Journal Wild Land Fire* 18(8): 970–982. https://doi.org/10.1071/WF08011

Arndt, N.; Vacik, H.; Koch, V.; Arpaci, V.; Gossow, H. 2013. Modeling human-caused forest fire ignition for assessing forest fire danger in Austria, *iForest – Biogeosciences and Forestry* 6: 315–325.

Arpaci, A.; Malowerschni, B.; Sass, O.; Vacik, H. 2014. Using multi variate data mining techniques for estimating fire susceptibility of Tyrolean forests, *Applied Geography* 53: 258–270. https://doi.org/10.1016/j.apgeog.2014.05.015

Avila-Flores, D.; Pompa-Garcia, M.; Antonio-Nemiga, X.; Rodriguez-Trejo, D. A.; Vargas-Perez, E.; Perez, J. 2010. Driving factors for forest fire occurrence in Durango State of Mexico: a geospatial perspective, *China Geogra Science* 20(6): 491–497. https://doi.org/10.1007/s11769-010-0437-x

Barni, P. E.; Pereira, V. B.; Manzi, A. O.; Barbosa, R. I. 2015. Deforestation and Forest fires in Roraima and their relationship with phytoclimatic regions in the northern Brazilian Amazon, *Environmental Management* 55(5): 1124–1138. https://doi.org/10.1007/s00267-015-0447-7

Bisquert, M.; Caselles, E.; Sánchez, J. M.; Caselles, V. 2012. Application of artificial neural networks and logistic regression to the prediction of forest fire danger in Galicia using MODIS data, *International Journal Wild land Fire* 21(8): 1025–1029. https://doi.org/10.1071/WF11105

Boubeta, M.; Jose Lombardía, M.; Marey-Perez, F. M.; Morales, D. 2015. Prediction of forest fires occurrences with area-level Poisson mixed models, *Journal Environmental Management* 154: 151–158. https://doi.org/10.1016/j.jenvman.2015.02.009

Bowyer, C.; Withana, S.; Fenn, I.; Bassi, S.; Lewis, M.; Cooper, T. 2009. Land degradation and desertification, *Policy Department Economic and Scientific Policy*, 34–50.

Carmelm, Y.; Paz, S.; Jahashan, F.; Shoshany, M. 2009. Assessing fire hazard using Monte Carlo simulations of fire spread, *Forest Ecology and Management* 257(1): 370–377. https://doi.org/10.1016/j.foreco.2008.09.039

Carmo, M.; Moreira, M.; Casimiro, P.; Vaz, P. 2011. Land use and topography influences on wildfire occurrence in northern Portugal, *Landscape and Urban Planning* 100(1–2): 169–176. https://doi.org/10.1016/j.landurbplan.2010.11.017

Češljar, G.; Stević, S. 2015. Small reservoirs and their sustainable role in fires protection of forest resources, *Renewable and Sustainable Energy Reviews* 47: 496–503. https://doi.org/10.1016/j.rser.2015.03.071

Chas-Amil, M. L.; Touza, J.; García-Martínez, E. 2013. Forest fires in the wild land-urban interface: a spatial analysis of forest fragmentation and human impacts, *Applied Geography* 43: 127–137. https://doi.org/10.1016/j.apgeog.2013.06.010

Chen, H.; Yuanman, H.; Yu, C.; Rencang, B.; Yuehui, L.; Miao, L. 2015. Changes of forest fire regime and landscape pattern under different harvesting modes in a boreal forest of Northeast China, *Journal Arid Land* 7(6): 841–851. https://doi.org/10.1007/s40333-015-0015-2

Chen, J. Q.; Franklin, J. F.; Spies, T. A. 1993. Contrasting microclimates among clear-cut, edge, and interior of old-growth Douglas-fir forest, *Agricultural and Forest Meteorology* 63: 219–237. https://doi.org/10.1016/0168-1923(93)90061-L

Chen, J.; Di, X. Y. 2015. Forest fire prevention management legal regime between China and the United States, *Journal of Forest Research* 26(2): 447–455. https://doi.org/10.1007/s11676-015-0067-3

Çoban, H. O.; Eker, M. 2010. Analysis of forest road network conditions before and after forest fire, in *FORMEC 2010, Forest Engineering: Meeting the Needs of the Society and the Environment,* 11–14 *July* 2010, Padova, Italy. 11 p.

Cortez, P.; Morais, A. 2007. A data mining approach to predict forest fires using meteorological data, in J. Neves, M. F. Santos, J. Machado (Eds.). *New trends in artificial intelligence. Proceedings of the 13th EPIA 2007 – Portuguese Conference on Artificial Intelligence*, December 2007, Guimarães, Portugal, 512–523.

Cottle, P. 2007. Insuring Southeast Asian commercial forests: fire hazard analysis and the potential for use of data in hazard pricing and reduction of forest fire hazard, *Mitig Adapt Strat Glob Change* 12: 181–201. https://doi.org/10.1007/s11027-006-9042-9

Deng, O.; Su, G. F.; Huang, Q, Y.; Li, Q. Y. 2013. *Forest fire hazard mapping based on spatial logistic model of northeastern China forest zone.* Springer-Verlag Berlin Heidelberg, 181–192.

Dlamini, W. M. 2010. Application of Bayesian networks for fire hazard mapping using GIS and remote sensing data, *Geo Journal* 76: 283–296. https://doi.org/10.1007/s10708-010-9362-x

Dong, X.; Shao, G.; Limin, D.; Zhanqing, H.; Lei, T.; Hui, H. 2006. Mapping forest fire risk zones with spatial data and principal component analysis, *Science in China: Series E Technological Sciences* 49: 140–149.

FAO. 2007. *Fire management global assessment 2006.* FAO forestry paper. FAO Rome Italy. 119 p.

Forman, R. T.; Alexander, L. E. 1998. Roads and their major ecological effects, *Annual Review of Ecology and Systematics* 29: 207–231. https://doi.org/10.1146/annurev.ecolsys.29.1.207

Fox, D. M.; Martin, N.; Carrega, P.; Andrieu, J.; Adnesm, C.; Emsellem, K.; Ganga, O.; Moebius, F.; Tortorollo, N.; Fox, E. A. 2015. Increases in fire hazard due to warmer summer temperatures and wild land urban interface changes do not necessarily lead to more fires, *Applied Geography* 56: 1–12. https://doi.org/10.1016/j.apgeog.2014.10.001

Guo, F.; Su, Z.; Wang, G.; Sun, L.; Lin, F.; Liu, A. 2016. Wildfire ignition in the forests of southeast China: identifying drivers and spatial distribution to predict wildfire likelihood, *Applied Geography* 66: 12–21. https://doi.org/10.1016/j.apgeog.2015.11.014

Hernandez-Leal, P. A.; Arbelo, M.; Gonzalez-Calvo, A. 2006. Fire hazard assessment using satellite data, *Advances in Space Research*, 37: 741–746. https://doi.org/10.1016/j.asr.2004.12.053

Hoyo, L, V.; Martin Isabel, M. P.; Martınez Vega, F. J. 2011. Logistic regression models for human–caused wildfire hazard estimation: analyzing the effect of the spatial accuracy in fire occurrence data, *European Journal Forest Research* 130: 983–996. https://doi.org/10.1007/s10342-011-0488-2

Jaiswal, R. K.; Mukherjee, S.; Raju, D. K.; Saxena, R. 2002. Forest fire hazard zone mapping from satellite imagery and GIS, *International Journal of Applied Earth Observation and Geoinformation* 4: 1–10. https://doi.org/10.1016/S0303-2434(02)00006-5

Kapusniak, J.; Majlingová, A. 2014. Results of forest opening–up analysis for ground mobile fire-fighting equipment deployment in LS Maluzina territory – Case study, *European Journal of Environmental and Safety Sciences* 2(2): 37–47.

Kuter, N.; Yenilmez, F.; Kuter, S. 2011. Forest fire hazard mapping by kernel density estimation, *Croatian Journal Forest Engineering* 32(2): 599–610.

Kwak, K.; Leea, W. K.; Saborowskibm, J.; Leecm, S. Y.; Soo, M. 2012. Estimating the spatial pattern of human-caused forest fires using a generalized linear mixed model with spatial autocorrelation in South Korea, *International Journal Geographical Information Science* 26(9): 1589–1602. https://doi.org/10.1080/13658816.2011.642799

Lee, S. Y.; Won, M. S.; An, S. H.; Lee, M. B .2004. Developing of forest fire occurrence probability model by using the meteorological characteristics in Korea, *Korean Journal of Agricultural and Forest Meteorology* 6(4): 242–249.

Liu, X.; Zhang, J.; Tong, T. 2015. Modeling the early warning of grassland fire hazard based on fuzzy logic in Xilingol Inner Mongolia, *Nat Hazards* 75: 2331–2342. https://doi.org/10.1007/s11069-014-1428-5

Maingi, K. J.; Henry, M. C. 2007. Factors influencing wildfire occurrence and distribution in eastern Kentucky, USA, *International Journal Wildland Fire* 16(1): 23–33. https://doi.org/10.1071/WF06007

Martinez, J.; Vega-Garciam, C.; Chuvieco, E. 2009. Human-caused wildfire hazard rating for prevention planning in Spain, *Journal of Environmental Management* 90(2): 1241–1252. https://doi.org/10.1016/j.jenvman.2008.07.005

Massada, A. B.; Syphard, A. D.; Stewart, S. I.; Radeloff, V. C. 2013. Wildfire ignition-distribution modeling: a comparative study in the Huron-Manistee National Forest, Michigan, USA, *Journal of Environmental Management* 22(2): 174–183.

Narayanaraj, G.; Wimberly, M. C. 2011. Influences of forest roads on the spatial pattern of wildfire boundaries, *International Journal of Wildland Fire* 20(6): 792–803.

Narayanaraj, G.; Wimberly, M. C. 2013. Influences of forest roads and their edge effects on the spatial pattern of burn severity, *International Journal of Applied Earth Observation and Geoinformation* 23: 62–70. https://doi.org/10.1016/j.jag.2012.12.006

Nasiri, M. 2012. Skidding routes simulation for opening access to high-hazard fire areas, *World Applied Sciences Journal* 16(6): 791–798.

Nunes, A. N. 2012. Regional variability and driving forces behind forest fires in Portugal an overview of the last three decades, *Applied Geography* 34: 576–586. https://doi.org/10.1016/j.apgeog.2012.03.002

Oliveira, S.; Oehler, F.; San-Miguel-Ayanz, J.; Camia, A.; Pereira, J. M. C. 2012. Modeling spatial patterns of fire occurrence in Mediterranean Europe using Multiple Regression and Random Forest, *Forest Ecology and Management* 275: 117–129. https://doi.org/10.1016/j.foreco.2012.03.003

Ozbygol, M. A., Bozer, R. 2012. Estimation of the burned area in forest fires using computational intelligence techniques, *Procedia Computer Science* 12: 282–287. https://doi.org/10.1016/j.procs.2012.09.070

Pierce, A. D.; Farris, C. A.; Taylor, A. H. 2012. Use of random forests for modeling and mapping forest canopy fuels for fire behavior analysis in Lassen Volcanic National Park, California, USA, *Forest Ecology and Management* 279: 77–89. https://doi.org/10.1016/j.foreco.2012.05.010

Pourghasemi, H. R.; Beheshtirad, M.; Pradhan, B. 2016. A comparative assessment of prediction capabilities of modified analytical hierarchy process (M-AHP) and Mamdani fuzzy logic models using Netcad-GIS for forest fire susceptibility mapping, *Geomatics, Natural Hazards and Hazard* 7(2): 861–885. https://doi.org/10.1080/19475705.2014.984247

Reineking, B.; Weibel, P.; Conedera, M.; Bugmann, H. 2010. Environmental determinants of lightning- v. human-induced forest fire ignitions differ in a temperate mountain region of Switzerland, *International Journal of Wildland Fire* 19(5): 541–557. https://doi.org/10.1071/WF08206

Renard, Q.; Pelissier, R.; Ramesh, B. R.; Kodandapani, N. 2012. Environmental susceptibility model for predicting forest fire occurrence in the Western Ghats of India, *International Journal of Wildland Fire* 21(4): 368–379. https://doi.org/10.1071/WF10109

Roman, C. R.; Martinez, V. M. J. 2006. Effectiveness of protected areas in mitigating fire within their boundaries: case study of Chiapas, Mexico, *Conservation Biology* 20(4): 1074–1086. https://doi.org/10.1111/j.1523-1739.2006.00478.x

Pourtaghi, Z. S.; Pourghasemi, H. R.; Rossi, M. 2015. Forest fire susceptibility mapping in the Minudasht forests, Golestan province, Iran, *Environ Earth Science* 73(4): 1515–1535. https://doi.org/10.1007/s12665-014-3502-4

Safi, Y.; Bouroumi, A. 2013. Prediction of Forest Fires Using Artificial Neural Networks, *Applied Mathematical Sciences* 7(6): 271– 286. https://doi.org/10.12988/ams.2013.13025

Sahu, M. 2015. Integrating artificial neural network with machine learning in forecasting problem domain, *International Journal of Recent Research Aspects* 2(2): 8–11.

Sakr, G. E.; Elhajj, I. H.; Mitri, G. 2010. Artificial intelligence for forest fire prediction: a comparative study, in D. X. Viegas (Ed.). 6th *International Conference on Forest Fire Research*, 15–18 November 2010, Coimbra, Portugal, 1–11. https://doi.org/10.1109/aim.2010.5695809

Sakr, G.; Elhajj, E. I. H.; Mitri, G. 2011. Efficient forest fire occurrence prediction for developing countries using two weather parameters, *Engineering Applications of Artificial Intelligence* 24: 888–894. https://doi.org/10.1016/j.engappai.2011.02.017

Schonewald-Cox, C. M.; Buechner, M. 1992. Park protection and public roads, in P. L. Fiedler, S. Jain (Eds.). *Conservation biology*. The theory and practice of nature conservation, preservation and management. Chapman and Hall, New York, 373–396. https://doi.org/10.1007/978-1-4684-6426-9_15

Soto, M. E. C. 2012. The identification and assessment of areas at hazard of forest fire using fuzzy methodology, *Applied Geography* 35: 199–207. https://doi.org/10.1016/j.apgeog.2012.07.001

Spellerberg, I. F. 2002. *Ecological Effects of Roads*. Science Publisher, Inc., Enfield, USA.

Tian, X.; Zhao, F.; Shu, L.; Wang, M. 2013. Distribution characteristics and the influence factors of forest fires in China, *Journal of Forest Ecology and Management* 310: 460–467. https://doi.org/10.1016/j.foreco.2013.08.025

Tuček, J.; Majlingová, A. 2009. Forest fire vulnerability analysis, in Střelcová K. *et al.* (Eds.). *Bioclimatology and natural hazards*. Springer: Dordrecht, 219–230. https://doi.org/10.1007/978-1-4020-8876-6_19

Vasilakos, C.; Kalabokidis, K.; Hatzopoulos, J.; Matsinos, I. 2009. Identifying wild land fire ignition factors through sensitivity analysis of a neural network, *Nat Hazards* 50: 125–143. https://doi.org/10.1007/s11069-008-9326-3

Verma, A. K.; Singh, D.; Sharma, S. D.; Kamlesh, K. 2013. Forest Fire Hazard Zonation in Raipur Range, Missouri Forest Division using GIS and Remote Sensing Technology, *International Journal of Advanced Scientific and Technical Research* 3(6): 141–150.

Wu, Zh.; Hem, H. S.; Yang, J.; Liang, Y. 2015. Defining fire environment zones in the boreal forests of Northeastern China, *Science of the Total Environment* 518–519: 106–116. https://doi.org/10.1016/j.scitotenv.2015.02.063

Zhang, Z. X.; Zhang, H. Y.; Zhou, D. W. 2010. Using GIS spatial analysis and logistic regression to predict the probabilities of human-caused grassland fires, *Journal of Arid Environments* 74: 386–393. https://doi.org/10.1016/j.jaridenv.2009.09.024

Zumbrunnen, T.; Menendez, P.; Bugmann, H.; Conedera, M.; Gimmi, U.; Burgi, M. 2012. Human impacts on fire occurrence: a case study of hundred years of forest fires in a dry alpine valley in Switzerland, *Regional Environmental Change* 12: 935–949. https://doi.org/10.1007/s10113-012-0307-4

Mohsen MOSTAFA. PhD student of forest sciences. He is studying on forest road network designing based on watershed management needs and goals particularly in relationship between road and fire. Gorgan University of Agricultural Sciences and Natural Resources (GUASNR), Iran.

Shaban SHATAEE JOUIBARY. PhD. Specializing in forest remote sensing. Associate Professor, GUASNR. He received his PhD from Tehran University in 2002. His research interesting is including using remote sensing in natural resources and natural hazard. He published more than 80 papers in national and international scientific journals in remote sensing, GIS and forestry categories. Now he is dean of faculty of forest sciences, Gorgan University of Agricultural Sciences and Natural Resources (GUASNR), and chief editor of Journal of Conservation and Utilization of Natural Resources in GUASNR.

Majid LOTFALIAN. PhD. Specializing in forest engineering. Associate Professor at Sari University of Agricultural Sciences and Natural Resources, Sari, Iran. He received his PhD from Tehran University in 2000. His research interesting is including forest road and forest harvesting. Now he is head of forestry department, natural resources faculty, University of Agricultural Sciences and Natural Resources, Sari, Iran.

Amir SADODDIN. PhD. Specializing in watershed management. Associate Professor, GUASNR. He received his PhD from National University, Australia, in 2006. His research interesting is including watershed management. He was dean of Faculty of Rangeland and Watershed Management of GUASNR. Currently he is manger of the National Mega Project on the integrated watershed management in Iran.

13

VEGETATION IMPACT ON THE DUNE STABILITY AND FORMATION ON THE LITHUANIAN COAST OF THE BALTIC SEA

Ramunė URBONIENĖ[a], Loreta KELPŠAITĖ[b], Ingrida BORISENKO[c]

[a]*Landscape Architecture and Environmental Planning, Klaipėda University, Herkaus Manto g. 84, LT-92294 Klaipėda, Lithuania*

[b]*Geophysical Department, Klaipėda University, Herkaus Manto g. 84, LT-92294 Klaipėda, Lithuania*

[c]*Computer Science Department, Klaipėda University, Herkaus Manto g. 84, LT-92294 Klaipėda, Lithuania*

Abstract. The aim of this study was to estimate vegetation coverage impact on the sandy beach and protective beach dune ridge (PBD) formation. Studies of the Lithuanian Baltic Sea coastal morphometric parameters and vegetation coverage projection were carried out from 1999 to 2011. In order to evaluate the dependence of the beach and the PBD morphometric parameters on the projection of vegetation coverage changes, correlation analyses were performed. A very strong link between vegetation projection coverage of the PBD top and beach width and incline of the eastern slope of PBD, and western slope vegetation projection coverage with the incline of the western slope of PBD were observed. A strong correlation relationship was observed between vegetation projection coverage of the PBD top and width of the PBD top. A medium relationship was noted between vegetation projection coverage of the PBD top with the width of the eastern slope. A negative relationship was noticed between the PBD western slope vegetation projection coverage and beach width.

It may be admitted that the projection coverage of vegetation may be used as PBD's parameter of stability as well as in the implementation of integrated coastal zone management.

Keywords: beach, vegetation, sand dune, environmental sustainability.

Introduction

Beaches are formed by waves, currents and tidal action, with waves generally being the predominant force. Within the surf zone, deep water waves begin to interact with the seabed. These result in changes in the direction and height of the incoming waves, which tend to align themselves in a direction parallel to the shoreline. Depending on the actual direction from which these waves approach, sand or other material may be transferred along the coast or in an onshore/offshore direction, or in a combination of both (Komar 1998). Dunes form when sand is carried by the wind from a beach towards the land. Although most sand particles are moved by saltation, surface creep may account for 20–25% of the moved sand. Transplanting vegetation will not prevent erosion, but it will accelerate recovery after storm damage by creating a trap for sand within the fore-dunes. This makes the dunes better able to survive the following period of erosion. Dune vegetation promotes the large-scale trapping of sand. The stems of dune grasses reduce the wind velocity near the surface, causing deposition of sand. Plant roots also serve to bind and consolidate the sand. Dune grasses thrive on incoming sand and accelerate their growth to keep up with the increasing height of the dune. The vegetation coverage represents the difference between a mobile pile of sand and a stabilized dune (Boak, Turner 2005; Stallins 2005).

Unfortunately, sand dunes, like wetlands, have been perceived as "useless" areas and their many functions, such as sand reserves for the beach and protection for land areas from harmful salt laden winds have been little understood (Komar 1998; Heslenfeld *et al.* 2004; Maun 2004).

Vegetation is one of the most important environmental factors which forms the beach and protects the beach

Corresponding author: Ramunė Urbonienė
E-mail: ramune.uo@gmail.com

dune ridge on sandy coastlines. Dense vegetation coverage protects the coast against erosion caused by the eolian processes (Vilà *et al.* 2007; Isermann 2008), and human activities (Provoost *et al.* 2002). Dunes overgrow with various herbs, which are typical for the sandy coastline. This includes the *Honckenya peploides* (L.) Ehrh., *Cakile baltica* Jord. ex. Pobed., *Ammophila arenaria* (L.) Link., *Leymus arenarius* (L.) *Hochst.*, which are stable, however, a very important parameter is the projection coverage of vegetation and species composition of plant communities (Fontana 2005; Łabuz 2007; Hilton, Konlechner 2011).

A vegetation communities analysed at the Baltic seaside of Lithuania are attributed to the complex of vegetation of the sands *Ammophiletum arenariae* Br.-Bl. et R. Tx. 1943 of the white dunes. It consists of plants growing in unfavorable meteorological conditions on the free-flowing sand constantly being blown and which also has a low amount of nutritious materials. The order *Ammophiletalia arenariae* Br.-Bl. 1933 is represented by one row, which combines primary communities of the white dunes, while the Ammophila arenaria of different development levels of the white dunes belong to the union. Communities prevalent on the coastline of the Baltic Sea according to the growing conditions and species composition are attributed to the association *Leymo-Ammophiletum arenariae* Br.-Bl. et De Leeuw 1936 (Stankevičiūtė 2006).

Seaside flora not only depends on natural and anthropogenic environmental factors, but they also affect the microclimate and soil of a specific region (Remke, Blindow 2011). On the sand dunes, the natural change of the plant communities is that they are growing; however, their progress is disturbed at places by the artificial reinforcement of dunes, beautification processes and recreational activity. Sufficient plant cover protects fragile dunes against: wind erosion by decreasing wind speed at ground level and by providing a protective cover over the dune; reduce but not prevent damage from wave erosion; build up sand dunes and thereby increase the sand reserve for storm waves. Dune vegetation has the ability to regenerate naturally after storm damage and facilitate natural dune recovery (NSW 2001). The aim of this study is to identify the impact of sandy beach and PBD vegetation coverage on their morphometric parameters in the south-eastern Baltic Sea region.

1. Materials and methods

1.1. Study area

The south-eastern Baltic Sea coast (Lithuanian coastal zone) covers 90.6 km of beach and PBD ridge along the whole western side of the sea coast of the Curonian Spit and mainland coast (Fig. 1). The length of the dune ridge at the Curonian Spit is about 51.01 km. The mainland coast's morphologically is more diverse: the beach and its protective dune ridge covers 38.42 km, including moraine cliff, cascading sand cliffs (5.62 km) and natural coastal dunes (3.73 km) (Jarmalavičius *et al.* 2012a). The intensive change of meteorological conditions and effects, which impacts the change of beach and PBD indicators, is constantly taking place on the Baltic Sea coast of Lithuania. Moreover, intensive anthropogenic activity, which unevenly affects the analysed territory, is taking place here as well (Žilinskas, Jarmalavičius 2007; Povilanskas *et al.* 2009; Jarmalavičius *et al.* 2012b).

The first article related to PBD formation and the protection of the Curonian Spit (SE Baltic Sea coast) was published by S. Biörn, 1807. Further PBD regulation works were described in several German scientists' works at the end of 19th century (Stankevičiūtė 2006).

The need to create PBD at the Curonian Spit became evident between the fifteenth and nineteenth century. During that time, a number of settlements were established along the Curonian Lagoon, seeking protection from the Baltic winds behind the high dunes. Only a few small areas remained forested by the end of the eighteenth century. A natural low PBD in the Southern part of the Curonian Spit (width of the spit ~700 m) wasn't sufficient to protect the beach during storms, and the Baltic Sea water would pass along the PBD to the Curonian Lagoon. Further north, free masses of sand evoke dune migrations and a number of Curonian villages were buried (Buynevich *et al.* 2007). In 1807, under the S. Biörn initiative the work of the forming and planting of the protective foredune was commenced at a settlement in Southern section of the Curonian Spit seashore to curb the sand drift and the seas reaching the plains beyond the dunes. The introduction of woody and grassy plants to the drifting sand of Nida and its approaches were started in 1825 under the initiative of G. D. Kuvert. That work of forming and planting the protective dune in the Curonian Spit was finished in 1904.

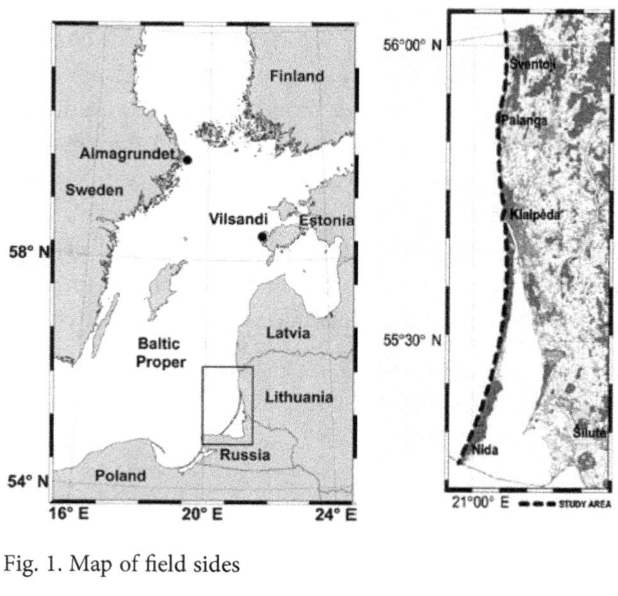

Fig. 1. Map of field sides

The protective dune supervision is being continued to this day (Olšauskas 2009).

1.2. Characteristics of the research area

The studies of coastline's morphometric parameters and projection coverage of vegetation were carried out during the 1999–2011 period. While evaluating long-term data of the water level on the Lithuanian coast, it can be suggested that the lowest rise in the total annual water level (only 2%) is in August and the water level is equal or close to "zero" of the perennial water level (Dailidienė *et al.* 2006).

Measurements of the cross profiles of the beach and PBD (in total 139 profiles, Fig. 1) were made in each 500–1000 m from the state border with the Republic of Latvia to the border with the Russian Federation (Kaliningrad region), following recommendations for measurement of heights and distances provided in the work of Skeivalas (2001) (Fig. 2).

The following parameters of the beach and PBD were measured: the beach width and height (m), the width and height of the PBD western slope (m), the top width (m), and the width and height of the eastern slope (m). Beach: the beach incline (°), the incline of the western slope (°), and the incline of the eastern slope (°) were calculated from the width and height data. The theodolite (T-30 (N 12511)) was used for the beach profiling. If there was a need to measure additional distance to the supplementary points, the calibrated tape-measure (50 m) was used (see Fig. 2).

1.3. Vegetation research description

During the last 200 years there have been various reforestation programs for the outer coastal regions, with the aim of establishing a proper vegetation coverage and protecting settlement combined with a reduction in soil erosion and sand dune movement (Savukynienė *et al.* 2003). While undertaking research on morphometric parameters of the beach and PBD, research on the projection coverage of vegetation was also carried out. The vegetation coverage

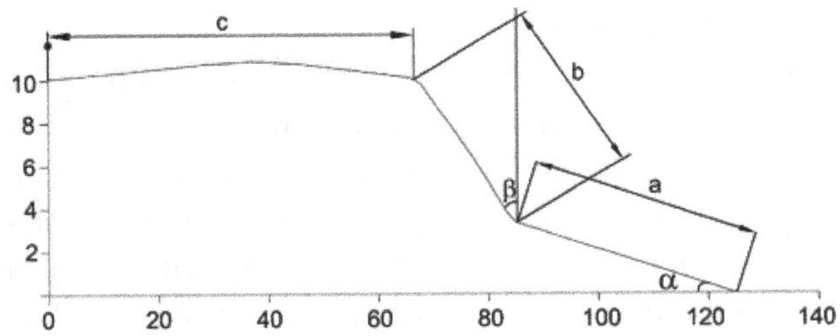

Fig. 2. Layout of the profile measurement

Note: Distances: a – beach width; b – width of the western slope of the protective beach dune ridge; c – width of the top of the protective beach dune ridge. Distances are measured using a calibrated roulette and expressed in meters. Inclines measured in degrees: α – beach incline, β – incline of the western slope of the protective beach dune ridge.

research was carried out at the site in a 4×4 m area. This area was chosen due to the fact that the seacoast sand plant communities are located in a mosaic way and take up a small area. In order to carry out a more detailed analysis of the projection coverage of vegetation on the PBD, research on vegetation in all profiles was carried out across the PBD – that is, on the western slope and on the top and on the eastern slope. Descriptions of the plant communities of the PBD were made by applying French-Swiss principles (that of the Zurich-Montpellier school) principles for the vegetation research. Latin names of plants are provided based on inventories of botanical names. To evaluate the abundance and cover of species of vegetation communities, the Braun-Blanquet scale was used (Walker *et al.* 2004). The projection coverage of grassy plants has been estimated according to 10 gradation of covering (0, 10, 20, 30…100% of the investigating area).

The coverage and abundance of vegetation at the sites is evaluated as a percentage using the frame divided into 25 small squares, each of which conforms to 4% of the total coverage. Firstly, coverage of each species of plant is fixed in each square; later on, after counting them, the total coverage of each species of plants at the site is received. For systematic analysis of flora, the system of vascular plants made by Gudžinskis (1999) was used. Vascular plants were described according to Snarskis (1968) and Lekavičius (1989). More detailed description of the procedure is described at Olšauskas, Urbonienė (2009).

1.4. Statistical data analysis

To evaluate the obtained morphometric parameters of the PBD and the results of the vegetation research, a statistical analysis was carried out using R software package. To analyse the data, a correlation analysis was made.

Correlation analysis. The correlation analysis was made to indicate an impact of the analysed factors on the parameters and interrelationship intensity between the parameters. To evaluate the relationship cohesion between variables, Pearson's correlation rate is most often used:

$$r = \frac{\sum_{i=1}^{n}(x_i - \overline{x})(y_i - \overline{y})}{(n-1)s_x s_y}; \quad (1)$$

where: x_i and y_i – values of the sample of the variables; y and x – the average values of the sample of these variables; s_x and s_y – standard tolerances of the variables; n – the number of pairs of the variable values (Mukaka 2012). In the correlation regressive analysis, the independent (cause) variable is marked as x and the dependent (output) variable is marked as y.

Pearson's correlation coefficient may be applied only in the case of linear and close-to-linear dependence when the variables are distributed under the

Gaussian distribution. If these conditions are not met, Spearman's correlation coefficient is used for assessment of cohesion of relationship between the variables. Pearson's correlation coefficient, as well as Spearman's, may have values in the range [−1, 1]. Dependence (relationship between the variables) is as strong as it is close to the value of the correlation coefficient of 1 or −1. Positive value shows that there is positive (direct) dependence between the variables, and vice versa a negative (inverse) dependence. In statistical literature, different scales for evaluation of the correlation coefficient are provided. In this work, the following interpretation of values of the correlation coefficient was used:

- 0.00–0.19 – very poor relationship;
- 0.20–0.39 – weak relationship;
- 0.40–0.69 – medium relationship;
- 0.70–0.89 – strong relationship;
- 0.90–1.00 – very strong interrelationship.

Even if there is no dependence between the variables, the correlation coefficient calculated based on the sample data will hardly be equal to zero. Thus, in the case of low values of the correlation coefficient (let's say $r < 0.2$), the question arises whether non-equality of the correlation coefficient to zero is accidental (Mukaka 2012). When evaluating the reliability of the correlation coefficient r (statistical significance) and the existence of the relationship between the analysed variables, hypotheses H_0: $R = 0$, H_1: $R \neq 0$ is being checked and Student's distribution is used. When using the statistical package in the calculations, it is suggested that the correlation coefficient is reliable when the calculated p (Sig. (2-tailed)) < 0.05 (hereafter 0.05 is the default significance level used in the work).

2. Results and discussion

2.1. Analysis of vegetation and species composition's projection coverage

The western slope of the protective beach dune ridge is affected by visitors, prevailing winds and during storms by waves. *Ammophila arenaria* and *Leymus arenarius* can be found on the dune, which is closest to the shoreline. It has been noticed that *Gypsophila paniculata* L., *Lathyrus maritimus* (L.) Bigelow also appears on the seaside dune part further to the dune top. Growing this species stabilizes the structure which traps sand being blown by the wind from the sea to the coast. They also dominate at the dune foot and in that way constitute a convenient environment for other species which occur further on in the dune. *Ammophila arenaria* and *Leymus arenarius* prefer dry and low-nutritious sandy soils, and tolerate saline sandy soils. Together with the *Gypsophila paniculata* and *Lathyrus maritimus* they form coastal sand dunes and stabilize sand drifting (Fontana 2005; Hilton, Konlechner 2011; Lucas, Carter 2013). Plants with their powerful root system make matts of various densities, well reinforcing the PBD and in this way reducing the sand transfer (Stankevičiūtė 2006; Olšauskas, Urbonienė 2009). On the western edge of the PBD beside *Leymus arenarius*, *Ammophila arenaria*, *Gypsophila paniculata* and *Lathyrus maritimus* there appears to be *Calamagrostis epigejos* (L.) Roth. These plants resist drought, strong wind and the flow of visitors.

On the top and eastern slope of the PBD there are mainly plants sensitive to intense sand movement which grow there, such as *Festuca sabulosa* H. Lindb., *Tragopogon heterospermus* Schweigg, *Hieracium umbellatum* L., *Pilosella officinarum* F. W. Schultz et Sch. Bip., *Viola littoralis Spreng.*, *Linaria loeselli* Schweigg., *Carex arenaria* L., *Lathyrus maritimus* and *Corynephorus canescens* (L.) P. Beauv. It should be noticed that *Eryngium maritimum* L. appears only in the Nida-Pervalka (Curonian Spit) area. The sandy soil of the eastern slope of the PBD contains larger amount of humus than the western side. It is also the lee side for the stormy Western and Northwestern winds, and it is characterized by a steeper slope in the dune ridge. All these components create a favorable environment to the *Jasione montana* L., *Artemisia campestris* L., *Carex arenaria*, *Lathyrus maritimus* and *Gypsophila paniculata*, and also in the Curonian Spit *Eryngium maritimum* and some other species.

Plants growing on the beach and PBD may be divided into four groups:

- Plants "building" a PBD – *Ammophila arenaria*, *Leymus arenarius*, *Cakile maritima* L, *Honckenya peploides* and *Agropyron dasyanthum* Lebed. These plants form the fore-dune and grow on the foot of the western slope of the PBD.
- Plants which tolerate carrying up of the sand and stabilize the PBD – *Lathyrus maritimus*, *Carex arenaria*, *Calamagrostis epigeios*, *Gypsophila paniculata*, *Eryngium maritimum* and *Petasines spurius* (Retz.) Reichb. These plants grow on and stabilize the western slope of the PBD and the western part of the dune top.
- Plants which do not tolerate carrying up of the sand and stabilize the top of the protective beach dune ridge, including *Festuca areniaria* L., *Corynephorus canescens*, *Tragopogon heterospermus*, *Hieracium umbellatum*, *Hieracium pilosella* L., *Viola littoralis*, *Linaria loeselii*, *Artemisia campestris* L., *Jasione montana* L., *Thymus serpillius* L., *Trifolium arvense* L., *Carex arenaria* and *Lathyrus maritimus*. These plants stabilize the eastern part of the top and the eastern slope of the PBD.
- Plants that tolerate the beach visitors. These are plant species which form a favorable environment for leisure (plants with a soft over-ground part):

Lathyrus maritimus, Gypsophila paniculata, Petasites spurius, Tragopogon heterospermus, Viola littoralis, Jasione montana and *Thymus serpillius.*

2.2. Impact of the PBD vegetation projection coverage on the PBD morphometric parameters

In order to evaluate the dependence of morphometric parameters of the beach and its PBD on the projection coverage of vegetation, in the 1999–2011 time period correlation analysis was done. It was assumed that the analysed variables distributed according to a Gaussian distribution. After carrying out the analysis between projections coverage of vegetation and morphometric parameters of the research object, the Pearson correlation coefficient and the significance value were calculated (Table 1).

Table 1. The linear correlation relationships between the vegetation projection coverage and the beach morphometric parameters (made by the authors)

Dependent variable	Independent variable	R	Equation of linear dependence	R^2
Beach width	Vegetation of the PBD top	0.69	$y = 21.70 + 0.73x$	0.480
Width of the PBD top	Vegetation of the PBD top	0.66	$y = 6.68 + 0.26x$	0.436
Width of the eastern slope of PBD	Vegetation of the PBD top	–0.64	$y = 20.85 - 0.02x$	0.407
Incline of the eastern slope of PBD	Vegetation of the PBD top	0.64	$y = 14.55 + 0.13x$	0.412
Width of the western slope of PBD	Vegetation of the western slope of PBD	0.90	$y = 7.94 + 0.23x$	0.806
Incline of the western slope of PBD	Vegetation of the western slope of PBD	–0.93	$y = 57.55 - 1.41x$	0.872
Beach width	Vegetation of the western slope of PBD	–0.51	$y = 52.97 - 0.37x$	0.256

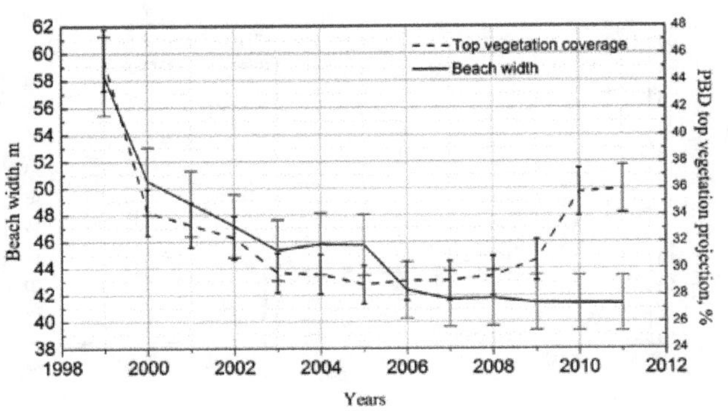

Fig. 3. Changes of beach width and PBD top vegetation coverage

It is noticeable that a very strong relationship is associated with an abundance of vegetation at the western slope and the width of the western slope of PBD. While an abundance of vegetation on the western slope is associated with the incline of the western slope with strong negative relationship. A medium correlation relationship is associated with a richness of vegetation on the top and the (i) beach width, (ii) width of the PBD top and (iii) incline of the eastern slope of PBD. It has been noticed that there is inverse dependence with medium correlation between the richness of vegetation on the PBD top and width of the eastern slope; vegetation of the western slope of PBD and beach width.

Analysis of the calculated values of determination coefficients (Table 1) shows that the change of projection coverage of vegetation on the PBD top impacts on the change of the other PBD sections: by a 48.% on average value of the beach width, by 43.61% on width of the PBD top; by the 40% on width of the eastern slope; by the 41% on incline of the eastern slope. Changes of projection coverage of vegetation of the western slope of PBD impacts: by 80% on variation of the width of the western slope; by 87% on variation of the incline of the western slope of PBD and by 25% on variation of the width of the beach.

On the PBD's western slope, the plant species *Ammophila arenaria, Leymus arenarius, Lathyrus maritimus* and *Festuca sabulosa* mainly occur, which together with other plant species form a stable combination of herbaceous vegetation. This stable combination of vegetation and projection coverage reduce or highly limit the transfer of free sand mass on the western slope of the PBD and other elements of its relief. Therefore, the fore-dune is formed, the width of the beach and the western slope remains stable and the incline of the western slope of the PBD decreases.

Festuca sabulosa, Gypsophila paniculata and *Lathyrus maritimus* (together with *Corynephorus canescens* and *Ammophila arenaria* grow on the top of the western PBD and make a very stable composition of vegetation, which traps the sand on the western part of the PBD and does not allow it to transfer onto the eastern slope. Projection coverage of vegetation and the composition of its species also increases the width of the top as free sand mass is withheld on the top, the sand is not transferred onto the eastern slope, plants growing on it are not covered up and due to that fact the eastern slope is quite stable as well and its width even decreases, while the incline remains unchanged or even increases.

During the analysed period, the beach width decreased on average from 58.3 m to 41.4 m. The most significant average decrease of the width of the overall beach of Lithuania was after Hurricane Anatoly in 1999 when its width decreased by up to 8 m. At the same time PBD top vegetation projection decreased by 8 percentage units in one year (Fig. 3). The distribution of data values, here and further, is represented by showing a single data point, representing the mean value of the data, and error bars to represent the overall distribution of the data. The error bars shown in the line graph above represent a description of how confident you are that the mean represents the true beach width (top vegetation coverage and other parameters) value. In our cases the error bars shows 5 percent positive and negative potential error amounts in the results of a scientific experiment.

From 2000 until 2003, vegetation coverage on the dune top was still decreasing, and later it stabilized. From 2009 to 2010 PBD top vegetation coverage increased by 6%. During the last five years, the beach width has stabilized; this was the result of the increased abundance of vegetation on the PBD top. Correlation regression analysis showed that projection coverage of vegetation of the PBD top and the change of parameters of the beach width depending on it, (R = 0.69), therefore, it may be presumed that in the future the beach width even after strong storms will recover quite quickly as abounding vegetation on the top will trap the transfer of free sand masses.

The average width of the PBD top was not constant and has decreased from 19.5 meters to 13.5 meters (Fig. 4).

The width of the PBD top as well as the beach was destroyed by Hurricane Anatoly in 1999 (Jarmalavičius *et al.* 2011). *Festuca sabulosa*, *Gypsophila paniculata*, *Lathyrus maritimus*, *Corynephorus canescens* and other plants growing on the PBD top make a very stable composition, which holds sand on the top. Though the projection coverage of vegetation on the top has decreased from 45% to 29% in 9 years, from 2008 the projection coverage of vegetation has been increasing every year. As the free sand masses are more available on the beach, it recovers faster in comparison with the PBD top, where vegetation coverage should first recover. That is noticeable in Figure 4 where PBD top widths become stable just 3 years after a stabilized quantity of vegetation coverage appears. While analysing the projection coverage of vegetation of the PBD top and the change of morphometric parameters of the width of the PBD top depending on the former, a strong correlation (R = 0.66) can be noticed between these two parameters, therefore, it may be presumed that in the future the width of the PBD top will be stable as well.

During the 1999–2011 period, the width of the eastern slope was constant. Its width variations were from 19.9 meters in 1999 to 20.5 meters in 2005. While analysing the projection coverage of vegetation of the PBD top and the change of morphometric parameters of the width of the eastern slope of PBD, a strong inverse correlation (R = −0.64) was observed; thus, it may be presumed that the width of the eastern slope of the PBD will also remain stable in future and will slightly decrease as increasing vegetation on the top will trap the transfer of free sand masses (Fig. 5).

It was noticed during the research that projection coverage of vegetation of the eastern slope of PBD does not affect the change of the eastern slope of the PBD (R = 0.23).

Regardless the fact that the width of the PBD's eastern slope was changing unevenly, its incline (Fig. 6) during the analysed period was changing evenly and decreased in ~4°, from 21.1° to 17.8°. The decrease of the incline of the eastern slope of the PBD, as well as the

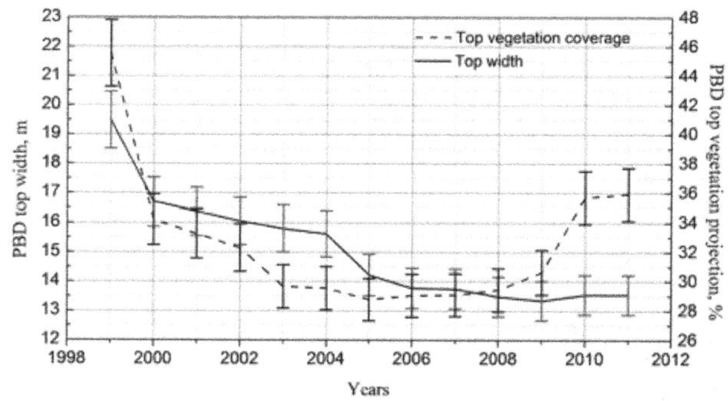

Fig. 4. Changes of PBD top width and vegetation coverage

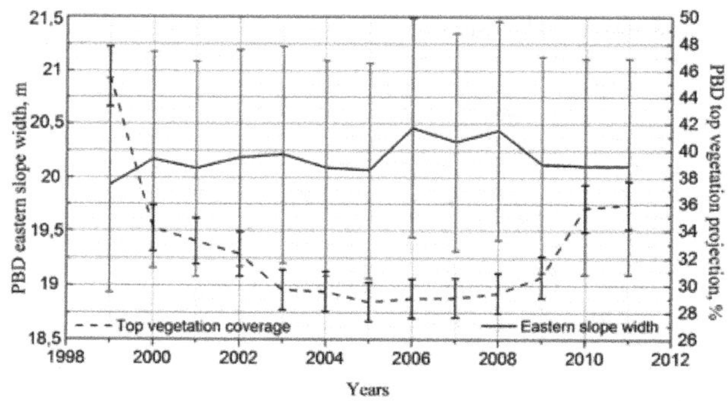

Fig. 5. Changes of PBD eastern slope width and PBD top vegetation coverage

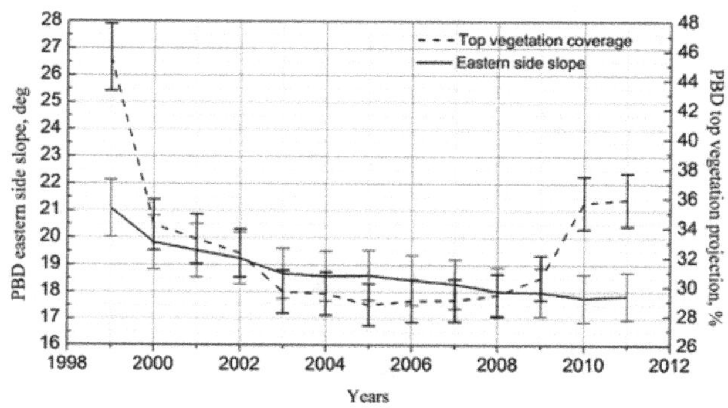

Fig. 6. Changes of the PBD eastern side slope and top vegetation coverage

change of other PBD morphometric parameters, was impacted by Hurricane Anatoly in 1999 (Jarmalavičius *et al.* 2011). From 2003 to 2005, the incline of the eastern slope of the PBD stabilized, however, after Hurricane Ervin in 2005, the incline started decreasing again, and the dune was becoming still more inclined and wider (Figs 5, 6). In this way, the sand reserve on the eastern slope of the PBD was increasing.

The incline change on the PBD's eastern slope, as well as its width, was impacted by projection coverage of vegetation on the top, which has been constantly increasing since 2006. While analysing dependency of the projection coverage of vegetation of the PBD top and the change of morphometric parameters of the incline of the eastern slope of the PBD, a strong positive correlation was noticed (R = 0.64) (Fig. 6).

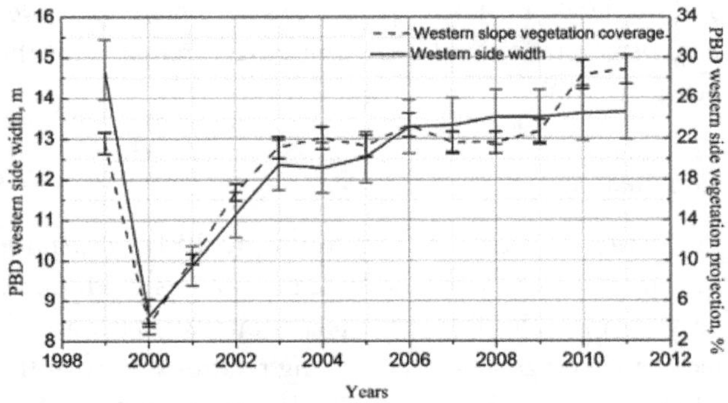

Fig. 7. Changes of the PBD western side width and vegetation coverage

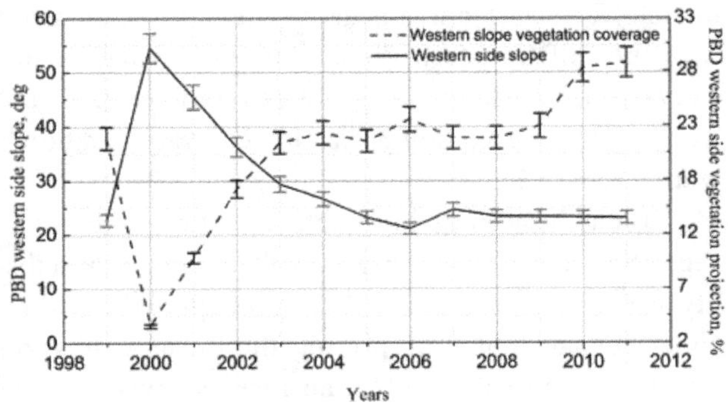

Fig. 8. Changes of the PBD western side slope and vegetation coverage

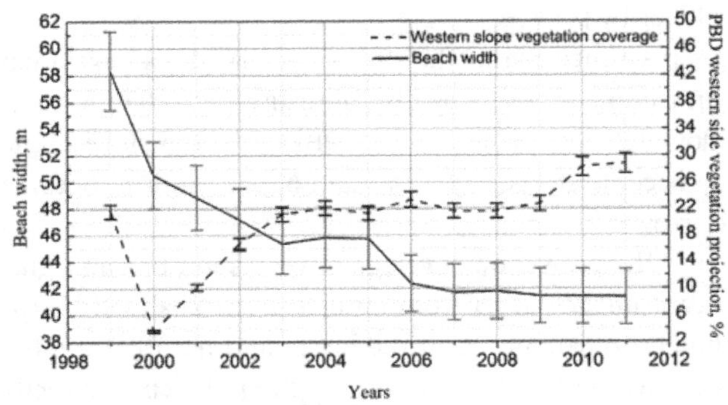

Fig. 9. Changes of beach width and the PBD western side vegetation coverage

On the PBD's western slope the Ammophila arenaria, Leymus arenarius, Lathyrus maritimus, Festuca sabulosa and other plants grow, forming a stable composition of herbaceous vegetation and strongly limiting the transfer of free sand masses onto the western slope of the PBD. The projection coverage of the vegetation on the western slope of the PBD has slightly increased from 21% to 28% during the research period. However, after Hurricane Anatoly, due to washed away part of the western slope of the PBD together with vegetation growing on it, the projection coverage of vegetation also decreased down to 3%. Starting from 2003, the situation has been stabilizing and the projection coverage has been increasing every year.

From 1999 to 2011, the width of the western slope of the PBD slightly decreased from 14.7 m to 13.7 m, however the most significant decrease of the width was after Hurricane Anatoly in 1999 (the width decreased up to 6 m). Changes in the incline of the western slope were impacted by the projection coverage of vegetation of the western slope. A very strong correlation (R = 0.90) has been noticed between the projection coverage of vegetation of the western slope of the PBD and morphometric parameters of the width of the western slope of the PBD; thus, it may be presumed that the width of the western slope of the PBD will also remain stable in the future if extreme Anatoly-type hurricanes do not occur again (Fig. 7).

A very strong negative correlation relationship (R = −0.93) exists between the incline of the western slope of the PBD and the projection coverage of vegetation of the western slope of the PBD. Due to the narrowing western slope of the PBD (Fig. 7), its incline increased (Fig. 8).

The most significant increase of the incline was recorded after Hurricane Anatoly in 1999 when, after washing away a part of the western slope of the PBD, the side of the dune ridge became steeper (it was 22° in August 1999 and 54.5° after the hurricane). However, due to eolian processes, the incline of the western slope of the PBD returned quickly to the quasi-stable profile and its incline was up to 25° (Fig. 8).

During the study period, the beach width significantly decreased from 58.3 m at the beginning of the research to 41.4 m at the end of the research. These changes were impacted by the highly decreased amounts of sand masses which were being washed away by the sea. Also, changes of the beach width are impacted by the projection coverage of vegetation of the western slope, as it was slightly increasing during the research period. While analysing the projection coverage of vegetation of the western slope of the PBD and the change of parameters of the beach width dependent on it, a moderate invert correlation was noticed between these two parameters (R = −0.51) (Fig. 9).

Therefore, it may be presumed that vegetation is one of the most important environmental factors forming the beach and the PBD in the sandy coasts and that it protects back-dune areas, which are more distant from the coastline, against the movement of free sand masses.

Sediment on the beach, under the effects of hydrodynamic and eolian processes, in the course of time may travel both back to the sea and among the PBD, where it stays for a longer time caught by vegetation, in this way forming new elements of the beach and the dune (Komar 1998). From 1999 to 2011, the beaches of Lithuania withstood the impact of two hurricanes – that of Anatoly in 1999 and Ervin in 2005. The former caused greater damage both to the beach and to the PBD. The impact of Hurricane Anatoly on vegetation is both significant and negative. Due to the washing away of part of the western slope and the top of the PBD, vegetation on them was also destroyed. Consequently, after 1999, the projection coverage of vegetation on the western slope of the PBD has decreased from 21% to 3%. Furthermore, sediment masses were liberated from the western slope of the PBD, which later on migrated both to the beach, PBD top and the eastern slope of the PBD. Hurricane Ervin in 2005 was different in its power and the prevailing wind direction from that of Anatoly, its damage to the coastline of both the mainland and the Curonian Spit was thus less significant. Also, its negative impact on the projection coverage of vegetation both on the western slope of the PBD and its top was insignificant – the projection coverage of vegetation on these elements of the PBD decreased on average only by 1%. The projection coverage of the eastern slope of the PBD has constantly been slightly increasing; from 1999 to 2009 it ranged from 65% to 67% and from 2010 the projection coverage of vegetation has increased up to 73%. The eastern slope of the PBD stabilized and became less affected by eolian processes.

While interpreting the equations of the simple linear regression, it may be suggested that

In the case of the average increase of vegetation of the PBD top by 1%:

- the beach width would increase on average by 0.73 meters;
- the width of the PBD top would increase on average by 0.26 meters;
- the width of the western slope of the PBD would decrease on average by 0.02 meters;
- The incline of the eastern slope of the PBD would increase on average by 0.13°.

It may be also suggested that in the case of the average increase of vegetation of the western slope of the PBD by 1%:

- the width of the western slope of the PBD would increase on average by 0.23 meters.
- the incline of the western slope of the PBD would increase on average by 1.41°;
- the beach width would decrease on average by 0.37 meters.

Thus, the presumption that the projection coverage of vegetation impacts the change of the beach width, the width and the incline of the western slope of the PBD, the width of the PBD top and the width of the eastern slope of the PBD has been proved as there is a very strong, strong and moderate relationship between the projection coverage of vegetation and morphometric parameters of the beach and PBD.

Conclusions

Projection coverage of vegetation on PBD's morphometric parameters, except beach incline, is medium or very strong. The PBD's western slope vegetation projection coverage has a very strong relationship with the PBD's western slope width (R = 0.90) and incline (R = −0.93). The PBD top vegetation projection coverage has a medium relationship with beach width (R = 0.69), the PBD top width (R = 0.66), and the PBD's eastern slope incline (R = 0.64), and a negative medium relationship with the PBD's eastern slope width (R = 0.64). A negative relationship was also noticed between the PBD western slope vegetation projection coverage and beach width (R = −0.51).

It may be admitted that the projection coverage of vegetation may be used as the PBD's parameter of stability as well as in implementation of integrated coastline zone management. Interpreting the linear regression results, we can suppose that an increase by 1% of the PBD top vegetation projection coverage, beach width on average will expand by 0.7 m, PBD top width will enlarge by 0.3 m, PBD eastern slope width will decrease just by 0.02 m, PBD eastern slope incline will become planer by 0.13°.

An increase in the PBD's western slope vegetation projection coverage by 1% can cause the PBD's western slope to grow by 0.23 m, the PBD's western slope inline to become steeper by 1.14°, and beach width to become narrower by 0.37 m.

The projection coverage of vegetation has no significant impact, apart from on one of the analysed morphometric parameters of the beach and the PBD – more exactly, the beach incline (R = 0.23), which is directly affected by hydrological parameters.

Acknowledgements

The authors are grateful to the Institute of Botany of the Nature Research Centre, Vilnius, Lithuania for their help with plant identification.

References

Boak, E. H.; Turner, I. L. 2005. Shoreline definition and detection: a review, *Journal of Coastal Research* 21(4): 688–703. http://dx.doi.org/10.2112/03-0071.1

Buynevich, A.; Bitinas, A.; Pupienis, D. 2007. Reactivation of coastal dunes documented by subsurface imaging of the Great Dune Ridge, Lithuania, *Journal of Coastal Research* 50: 226–230.

Dailidienė, I.; Davulienė, L.; Tilickis, B.; Stankevičius, A.; Myrberg, K. 2006. Sea level variability at the Lithuanian coast of the Baltic Sea, *Boreal Environment Research* 11(2): 109–121.

Fontana, S. L. 2005. Coastal dune vegetation and pollen representation in south Buenos Aires Province, Argentina, *Journal of Biogeography* 32: 719–735. http://dx.doi.org/10.1111/j.1365-2699.2004.01221.x

Gudžinskis, Z. 1999. *Lietuvos induočiai augalai*. Vilnius: Botanikos institutas. 209 p.

Heslenfeld, P.; Jungerius, P. D.; Klijn, J. A. 2004. European coastal dunes: ecological values, threats, opportunities and policy development, in M. L. Martinez, N. P. Psuty (Eds.), *Coastal dunes: ecology and conservation* 171: 335–350. Springer.

Hilton, M. J.; Konlechner, T. M. 2011. Incipient fore dunes developed from marine-dispersed rhizome of Ammophila arenaria, *Journal Coast* 64: 288–292.

Isermann, M. 2008. Classification and habitat characteristics of plant communities invaded by the non-native Rosa rugosa Thunb. in NW Europe, *Phytocoenologia* 38(1–2): 133–150. http://dx.doi.org/10.1127/0340-269X/2008/0038-0133

Jarmalavičius, D.; Satkūnas, J.; Žilinskas, G.; Pupienis, D. 2012a. The influence of coastal morphology on wind dynamics, *Estonian Journal of Earth Sciences* 61(2): 120–130. http://dx.doi.org/10.3176/earth.2012.2.04

Jarmalavičius, D.; Žilinskas, G.; Pupienis, D. 2012b. Impact of Klaipeda port jetties reconstruction on adjacent sea coast dynamics, *Journal of Environmental Engineering and Landscape Management* 20(3): 240–247. http://dx.doi.org/10.3846/16486897.2012.660884

Jarmalavičius, D.; Satkūnas, J.; Žilinskas, G.; Pupienis, D. 2011. Dynamics of beaches of Lithuanian coast (the Baltic Sea) for the period 1993–2008 based on morphometric indicators, *Environmental Earth Sciences* 65(6): 1727–1736. http://dx.doi.org/10.1007/s12665-011-1152-3

Komar, P. D. 1998. *Beach processes and sedimentation*. Prentice-Hall. 544 p.

Łabuz, T. A. 2007. Evaluation of past and present sea holly (Eryngium maritimum) habitats on Polish coastal dunes, *Acta Universitatis Latvensis* 273: 99–114.

Lekavičius, A. 1989. *Vadovas augalams pažinti* [Guide to know plants]. Vilnius: Mokslas. 437 p. (in Lithuanian).

Lucas, K. L.; Carter, G. A. 2013. Change in distribution and composition of vegetated habitats on Horn Island, Mississippi, northern Gulf of Mexico, in the initial five years following Hurricane Katrina, *Geomorphology* 199: 129–137. http://dx.doi.org/10.1016/j.geomorph.2012.11.010

Maun, M. A. 2004. Burial of plants as a selective force in sand dunes, in M. L. Martinez, N. P. Psuty (Eds.), *Coastal dunes: ecology and conservation* 171: 19–135. Springer.

Mukaka, M. M. 2012. Statistics corner: a guide to appropriate use of correlation coefficient in medical research, *Malawi Medical Journal* 24(3): 69–71.

NSW Department of Land and Water Conservation 2001. *Coastal dune management: a manual of coastal dune management and rehabilitation techniques.* Newcastle: Coastal Unit, DLWC.

Olšauskas, A. M. 2009. Woody and grassy vegetation development on different landscape elements of the Curonian Spit, *Environment Research, Engineering and Management* 50(4): 33–36.

Olšauskas, A. M.; Urbonienė, R. 2009. Vegetation development in different landscape element of the Lithuanian mainland seacoast, in *Proceedings of the International Scientific Conference: Rural Development*, 15–17 October 2009, Akademija, Kaunas region, Lithuania, *Rural Development* 4(2): 171–175. ISSN 1822-3230.

Povilanskas, R.; Baghdasarian, H.; Arakelyan, S.; Satkūnas, J.; Taminskas, J. 2009. Secular morphodynamic trends of the Holocene dune ridge on the Curonian Spit (Lithuania/Russia), *Journal of Coastal Research* 25(1): 209–215.

Provoost, S.; Ampe, C.; Bonte, D.; Cosyns, E.; Hoffmann, E. 2002. Ecology, management and monitoring of dune grasslands in Flanders, Belgium, in *Littoral 2002, The Changing Coast,* 22–26 September 2002, Porto, Portugal. Eurocoast/EUCC, 11–22.

Remke, E. S.; Blindow, I. 2011. Site specific factors have an overriding impact on Baltic dune vegetation change under low tomoderate N-depositiona case study from Hiddensee Island, *Journal of Coastal Conservation* 15: 87–97. http://dx.doi.org/10.1007/s11852-010-0123-2

Savukynienė, N.; Moe, D.; Ūsaitytė, D. 2003. The occurrence of former heathland vegetation in the coastal areas of the South-East Baltic Sea, in particular Lithuania: a review, *Vegetation History and Archaeobotany* 12(3): 165–175. http://dx.doi.org/10.1007/s00334-003-0008-5

Skeivalas, J. 2001. *Metrologinių ir geodezinių matavimų apdorojimas* [Processing of metrological and geodetic measurements]. Vilnius: Technika. (in Lithuanian).

Snarskis, P. 1968. *Vadovas Lietuvos augalams pažinti* [Guide to know Lithuanian plants]. Vilnius: Mintis. 502 p. (in Lithuanian).

Stallins, J. A. 2005. Stability domains in barrier island dune systems, *Ecological Complexity* 2(4): 410–430. http://dx.doi.org/10.1016/j.ecocom.2005.04.011

Stankevičiūtė, J. 2006. The succession of sand vegetation at the Lithuanian sea coast, *Botanica Lithuanica* 12(3): 139–156.

Vilà, M.; Pino, J.; Font, X. 2007. Regional assessment of plant invasions across different habitat types, *Journal of Vegetation Science* 18(1): 35–42. http://dx.doi.org/10.1111/j.1654-1103.2007.tb02513.x

Walker, D. A.; Epstein, H. E.; Gould, W. A.; Kelley, A. M.; Kade, A. N.; Knudson, J. A.; Krantz, W. B. *et al.* 2004. Frost-boil ecosystems: complex interactions between landforms, soils, vegetation and climate, *Permafrost and Periglacial Processes* 15: 171–188. http://dx.doi.org/10.1002/ppp.487

Žilinskas, G.; Jarmalavičius, D. 2007. Interrelation of morphometric parameters of the submarine shore slope of the Curonian Spit, Lithuania, *Baltica* 20(1–2): 46–52.

Ramunė URBONIENĖ, Lecturer at Landscape Architecture and Environmental Planning Department, Klaipeda University PhD student in environmental engineering, 20 publications (ISI proceedings – 2); participated in 3 international conferences with oral and poster presentations. Research interests: dynamics of the vegetation scatter of the protective dune, dynamics of the seaside, recreational influence of the protective dune.

Loreta KELPŠAITĖ, Assoc. Prof. Dr. at Geophysical Department, Klaipeda University, PhD in Civil and Environmental Engineering, 11 publications (ISI WOS –7); participated in 20 international conferences with oral and poster presentations, Fellow of the Lithuanian Academy of Science 2010–2011, awarded for the best poster in the section "Conference of Baltic Oceanographers", 7th Baltic Sea Science Congress, 17–21 August 2009, Tallinn, Estonia, Marie Curie Fellow, 2007–2009. Research interests: coastal engineering, wave dynamics, nonlinear interaction, climate change, sediment transport.

Ingrida BORISENKO, Assoc. Prof. Dr. at Computer Science Department, Klaipeda University, PhD in Mathematics, 12 publications (ISI WOS – 1); participated in 11 international conferences with oral and poster presentations. Research interests: statistics, geostatistics, geographic information systems, distance learning.

14

HETEROGENEOUS SOLAR PHOTO-FENTON DEGRADATION OF REACTIVE BLACK 5 USING FOUNDRY SAND AND FLY ASH: VALUE ADDITION TO WASTE

Himadri RAJPUT[a], Anoop VERMA[a], Manpreet KAUR[a], Taranjeet KAUR[b], Amrit Pal TOOR[b]

[a]*School of Energy and Environment, Thapar University, Thapar Technology Campus, Patiala 147004, India*
[b]*Dr. S. S. Bhatnagar University Institute of Chemical Engineering and Technology, Panjab University, Sector 14, Chandigarh 160014, India*

Abstract. This work evaluates the scope of alternative sources of iron i.e. Foundry Sand (FS), Fly Ash (FA) and mixture of FS and FA in heterogeneous photo-Fenton process for the degradation and decolourization of the dye Reactive Black5 (RB5). SEM-EDS characterization of FS and FA confirmed the presence of iron and aluminium which are required for photo-Fenton. In case of FS as an iron source, results confirmed the degradation and decolourization efficiency of 90% in 70 min and 45 min respectively when the reaction parameters were H_2O_2 = 2.2 mM, FS dose = 0.5 g, pH 3. With FA 90% degradation was achieved within 30 min and 93% decolourization rate within 15 min. When a mixture of FS (0.1 gm) and FA (0.05 gm) is used as source of iron 99% degradation and decolourization of the dye was achieved under the reaction conditions where H_2O_2 = 2.2 mM, FA:FS = 1:2, pH 3.

Keywords: foundry sand; fly ash; dye Reactive Black 5; photo-Fenton; degradation; decolourization.

Introduction

Dye effluents from textile industries are significant sources of environmental pollution as it is among one of the most water demanding sectors (Chaco *et al.* 2006; Wang *et al.* 2008). This water consumption, allied to high dosage of dyes, originates extremely colorized effluents. Release of these wastewaters to natural environment is described as very problematic to aquatic life as transmission of light, photosynthesis and aquatic life diversity will be affected and is also mutagenic to humans (Chung *et al.* 1992; Chung, Stevens 1993). The dyes do not bind to the fabric completely and its loss in wastewaters depending on the class of the dye can vary from 2% for the basic dyes to 50% for the reactive dyes, causing serious environmental problems (Park, Choi 2003; Mass, Chaudhari 2005; Pandey *et al.* 2007). They have high solubility and stability, stability, due to which are widely used in many applications (Mahmoodi, Arami 2009). Reactive Black 5 (RB5) is one of the oldest reactive dyes, which is consumed very heavily in textile industries worldwide for the dying of woollen, cotton and nylon fabrics (Arslan *et al.* 1999). It is reported to be toxic and can cause allergic reactions of the respiratory tract (Lars, Mallika 1997).

A variety of physical and chemical processes, such as elimination by adsorption on activated carbon, ozone oxidation, coagulation by a chemical agent, electrochemical method, hypochlorite oxidation etc. are available for the treatment of dye waste effluents (Yoshida *et al.* 1999; Tanaka *et al.* 2000). Nevertheless, these methods are usually inefficient, costly, non-destructive and resulted in the production of secondary waste products (Sun *et al.* 2008). On the other hand, biological treatment usually takes long time besides producing a large quantity of sludge.

Advanced oxidation technologies (AOTs) are innovative methods and are extremely useful where substances are resistant to conventional treatment technologies. AOTs are oxidation processes, where hydroxyl radicals (•OH) generates that are very effective in degrading organic pollutants due to their strong oxidant power and non selective species (Toor *et al.* 2006). Among these AOTs, Fenton's reagent is particularly gained attention because of its low cost, the lack of toxicity of the reagents, and absence of a mass transfer limitation due to its simplicity of the technology and homogeneous catalytic nature (Liao *et al.* 2009). However, it should be pointed out that the homogeneous Fenton process has significant disadvantage of

Corresponding author: Anoop Verma
E-mail: anoop.kumar@thapar.edu

sludge-containing Fe ion at the end of the wastewater and its removal/treatment is very expensive and needs large amount of chemicals and manpower.

Various waste materials of different kinds have been tested as an iron source in heterogeneous photo-Fenton degradation (Lucking *et al.* 1998; Chaudhuri, Sur 2000; Mecozzi *et al.* 2006). Photo-Fenton activity in each of the above case can be related to the activation of hydrogen peroxide (H_2O_2) by the ions of iron leached from the waste materials into the solution. Iron ions release is slow in the case of heterogeneous photo-Fenton reactions, which limits its undesired reaction in comparison with the conventional homogenous catalysis.

To overcome the disadvantages of the homogeneous Fenton process, and also considering the possibility of recovering catalyst, attempt has been made to use foundry sand (FS) and fly ash (FA) as an alternative iron source. The key motivating factor for using FS and FA is to identify its iron leaching potential for photo-Fenton studies along with reuse opportunities. Thus, the study proposes alternatives to landfill disposal and ultimately cost saving for materials that can be reused beneficially.

In this context, the present study investigates the degradation of RB5 using FS, FA and their mixture as a new low-cost iron source for the heterogeneous photo-Fenton process. To the best of our knowledge, this is the first reported study towards the use of foundry sand and fly ash as a heterogeneous catalyst in photo-Fenton degradation of dyes. This would be of economic interest keeping in mind the high costs required for disposal of FS and FA which we are using as a catalyst that is commonly considered as a waste.

1. Experimental

1.1. Reagents and chemicals

Reactive Black 5 (RB5), technical grade (55%), was purchased from Sigma- Aldrich and used as such without any further purification. Waste FS was received from Munjal Industries, Gurgoan (India) and FA was received from thermal plant, Ropar, Punjab (India) as a gift sample and used as received without any further modifications. H_2O_2 (30% w/v) was obtained from SD Fine Chemicals Limited, India. The pH of the solution was adjusted using 0.1 N H_2SO_4. Throughout the study, all the chemicals were of analytical grade and used as such without any further purification. For experimental study, 100 mgL^{-1} RB5 solution was prepared and double distilled water was used for the preparation of all solutions.

1.2. Photoreactor and experimental procedure

The shallow pond batch slurry reactor used in this study (Fig. 1) was made up of borosil glass, 160 mm in diameter and 52 mm in height having a capacity of 1200 mL. For heterogeneous solar photo-Fenton studies, 200 ml of RB5 solution (100 mgL^{-1}) was taken in this batch reactor at pH 3 and known amount of Fenton reagent (FS+H_2O_2), (FA+H_2O_2) and (FS+FA+H_2O_2) were added. The suspensions were kept in dark under stirring for 30 min for attaining any adsorption-desorption equilibrium between FS, FA and dye. For photo-Fenton studies, the above samples were kept in sunlight under same conditions of stirring and aeration. Samples were taken after regular intervals of time, filtered and analyzed for decolourization and degradation studies using an UV-visible spectrophotometer.

All the reported solar experiments were carried during timings 11.00 a.m. - 3.00 p.m. in months of March-May, 2014 at Patiala, Punjab. To check the reproducibility of results, experiments were repeated thrice and average values are reported with standard deviation varying from 1–3%.

1.3. Analytical analysis

The decolourization and degradation studies were performed with UV- visible (Hitachi V- 500 UV/VIS (Japan)) double- beam spectrophotometer with RB5 at 310 nm and 595 nm.

For the characterization of FS and FA, scanning electron microscope (SEM) and energy-dispersive X-ray spectroscopy (EDS) was performed as shown in Figure 2(a) and (b), Figure 3(a) and 3(b) respectively. Chemical composition of foundry sand and fly ash is reported in Table 1 and Table 2 (with accuracy of ± 0.5–1.0%). The analysis indicated the absence of any heavy metal in FS and FA, thus eliminated any risk of their leaching. A considerable amount of iron (23.16%) in FS and aluminium (17.66%) in FA was reported which could be utilized in photo-Fenton process. Although amount of iron detected in FA was less as compared to FS, but the presence of considerable amount of aluminium confirmed its participation in photo-Fenton reactions which could be used in the place of iron.

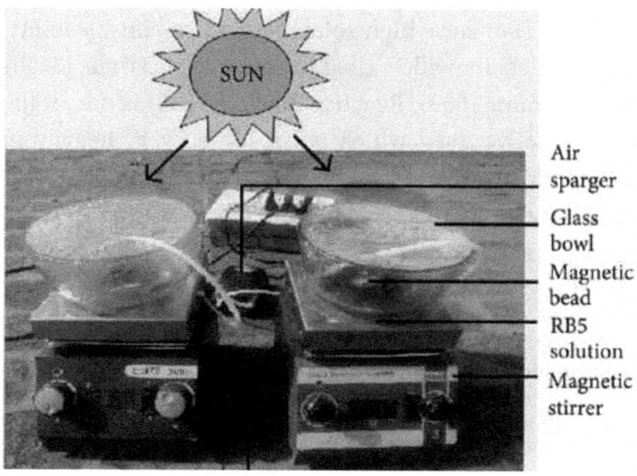

Fig. 1. Shallow pond slurry batch reactor

2. Results and discussion

2.1 Photo-Fenton studies using foundry sand as an iron source

2.1.1 Preliminary experiments

Preliminary experiments were performed to assess the catalytic activity of FS as a catalyst and the effect of individual parameter in the degradation of RB5 by heterogeneous photo-Fenton process. Low and negligible degradation and decolourization (less than 20%) in 120 min was achieved in the presence of H_2O_2 only, FS only and sunlight only as shown in Figure 4(a) and 4(b). This could be ascribed to limited production of OH radicals due to the insufficient availability of the resources in all three above cases required for the photo-Fenton process. In Fenton process (H_2O_2+ FS), 35% degradation in 90 min and 50% decolourization rates in 60 min were achieved. It is worth noting that 75% degradation in 90 min and decolourization efficiency of 85% in 60 min was achieved in the presence of FS as a heterogeneous catalyst, H_2O_2 and sunlight. The photo-Fenton results indicated that catalyst reacts with H_2O_2 in the presence of photons to generate OH radicals for the degradation of RB5. In the present work, generation of OH radicals for degradation of RB5 was proposed via photo-Fenton mechanism using FS and FA as an iron source.

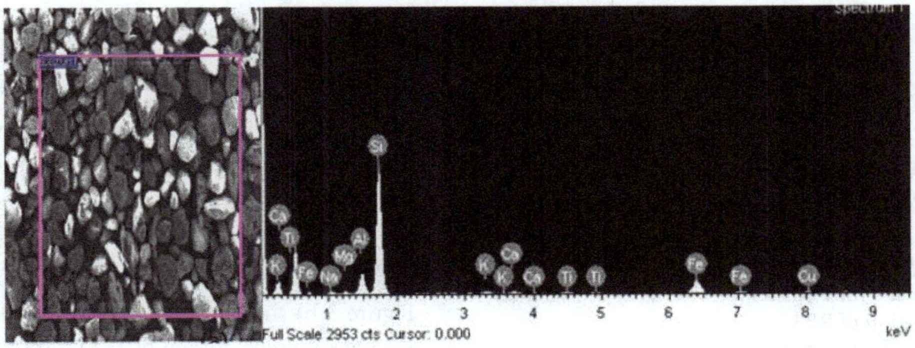

Fig. 2. a) SEM micrograph, b) EDS pattern of FS

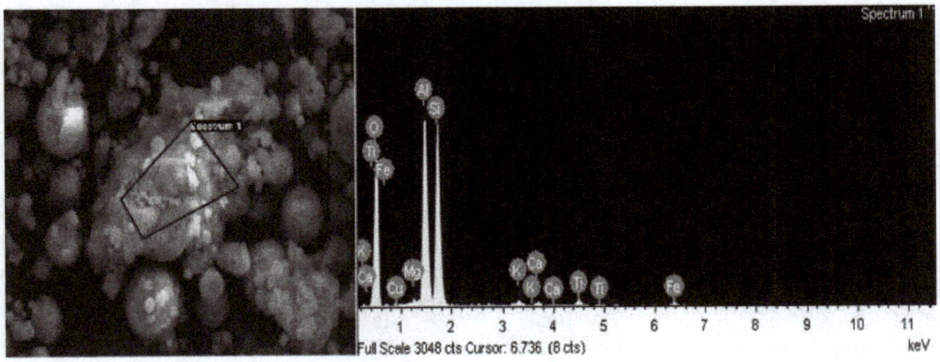

Fig. 3. a) SEM micrograph, b) EDS pattern of FA

Table 1. Composition of FS

Element	Na K	Mg K	Al K	Si K	K K	Ca K	Ti K	Fe K	Cu K	Total
Weight %	1.93	1.21	7.85	58.10	1.77	1.49	1.36	23.16	3.13	100
Atomic %	2.73	1.62	9.48	67.43	1.48	1.21	0.93	13.52	1.60	

Table 2. Composition of fly ash

Element	O K	Mg K	Al K	Si K	K K	Ca K	Ti K	Fe K	Cu K	Total
Weight %	55.87	0.42	17.66	21.85	0.67	0.50	1.33	1.00	0.70	100
Atomic %	69.45	0.35	13.02	15.47	0.34	0.25	0.55	0.35	0.22	

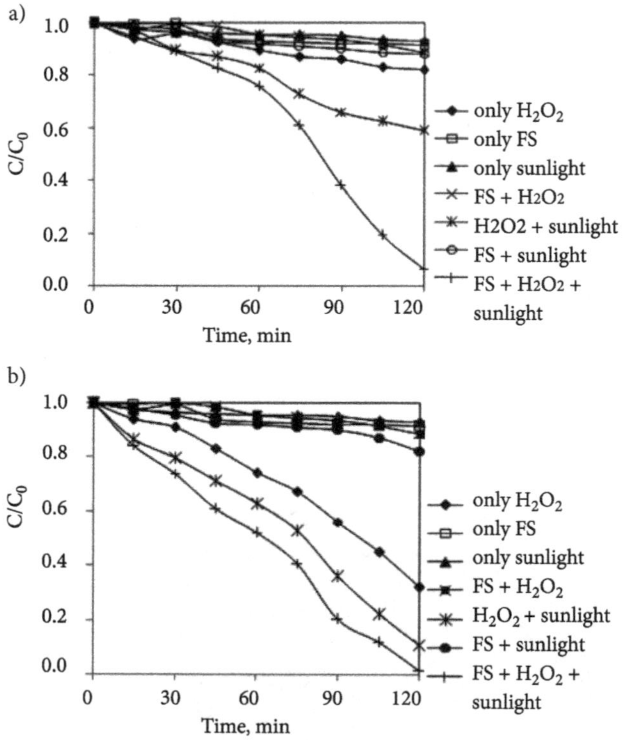

Fig. 4. Preliminary experiments performed under different conditions to evaluate their efficiency towards a) degradation and b) decolourization of RB5 using FS (where, C = concentration of dye at time *t* and C_0 = initial concentration of dye)

2.1.2. Effect of variation of FS

FS was used as an iron source in varying concentrations ranging from 0.1–1.0 gm during the photo-Fenton treatment process while maintaining the other parameters constant. The results indicated that initial increase of FS till 0.5 gm increases degradation and decolourization of the dye as shown in Figure 5(a). The lower degradation capacity of FS at small concentrations was probably due to the low OH radical production and their availability for oxidation. Further increase of FS dosage from 0.5 gm–1.0 gm, degradation and decolourization of dye decreased. This reduction can be explained by OH radicals scavenging either by the reaction with hydrogen peroxide or by the reaction with Fe^{+2} as expressed in the equations (Djeffal *et al.* 2013):

$$H_2O_2 + OH^{\bullet} \rightarrow H_2O + OH_2^{\bullet}$$

$$Fe^{+2} + OH^{\bullet} \rightarrow OH^{\bullet} + Fe^{+3}$$

The effective degradation (90%) within 90 min and decolourization (90%) within 60 min was achieved at 0.5 gm FS dosage.

2.1.3. Effect of H_2O_2 concentration

Figure 5(b) shows the degradation and decolourization of RB5 at different dosage of H_2O_2. To optimize the dosage, H_2O_2 was varied from 0.0–8.82 mM of the dye solution at fixed FS and pH. Dye removal increases with the increasing dosage of H_2O_2 till 2.2 mM, after that percentage of removal

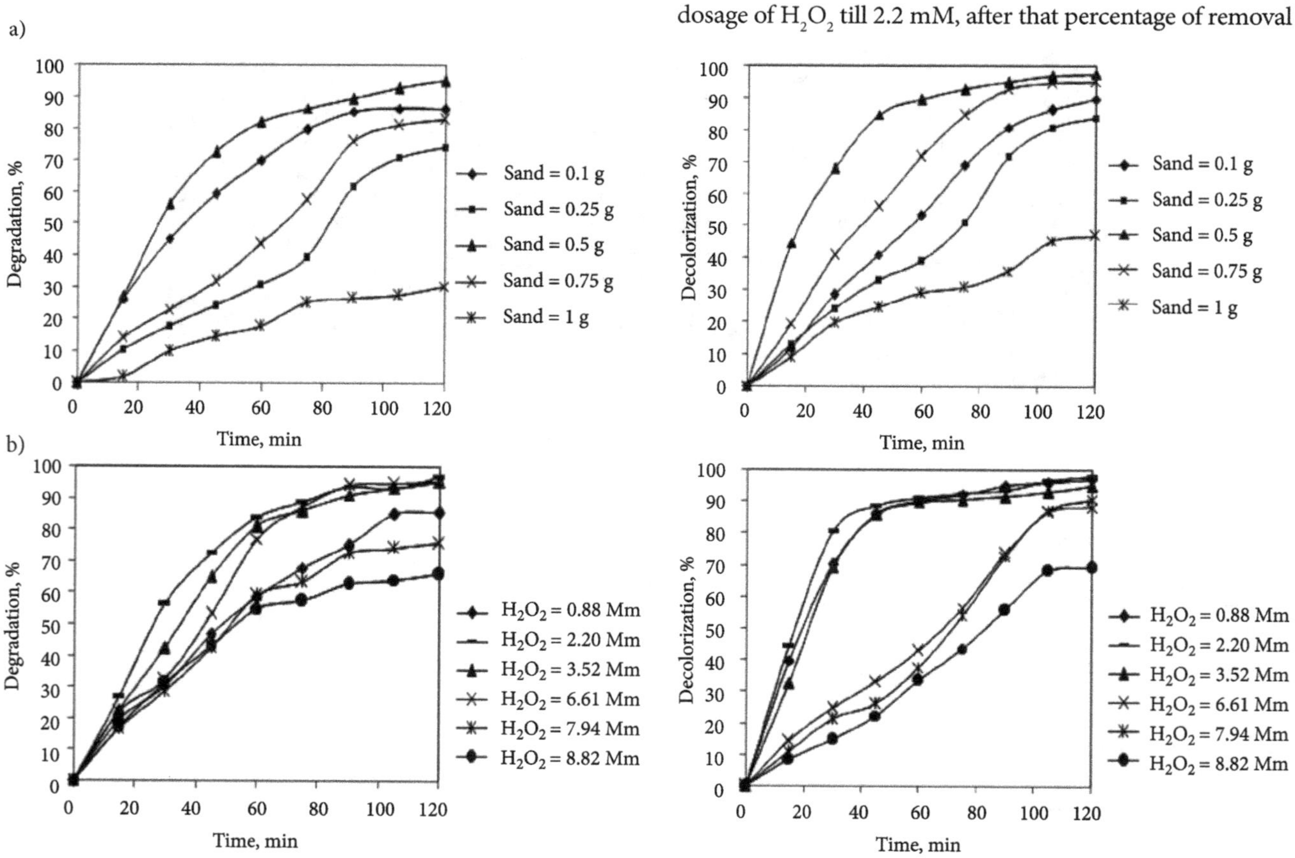

Fig. 5. Percentage degradation and decolourization of RB5 at a) varying concentration of FS b) varying concentration of H_2O_2

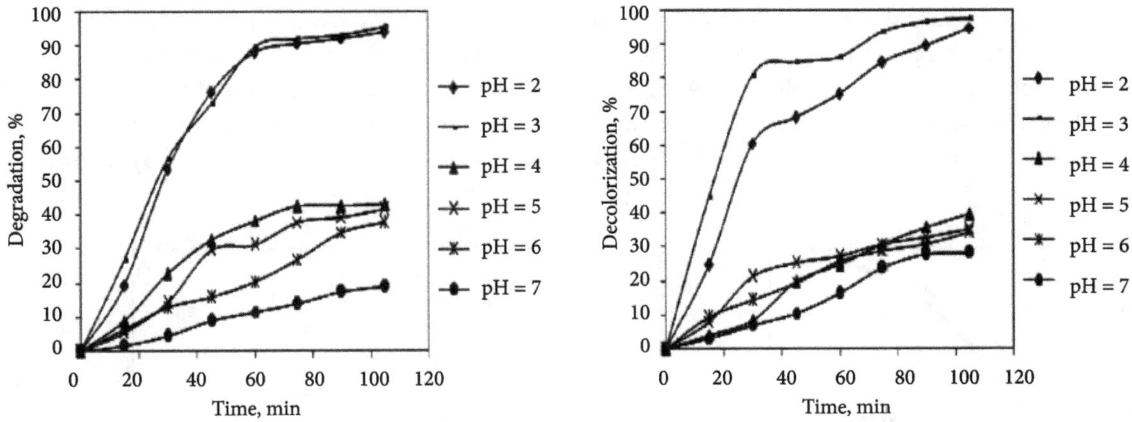

a) b)

Fig. 6. Percentage a) degradation and b) decolourization of RB5 at different initial pH of solution

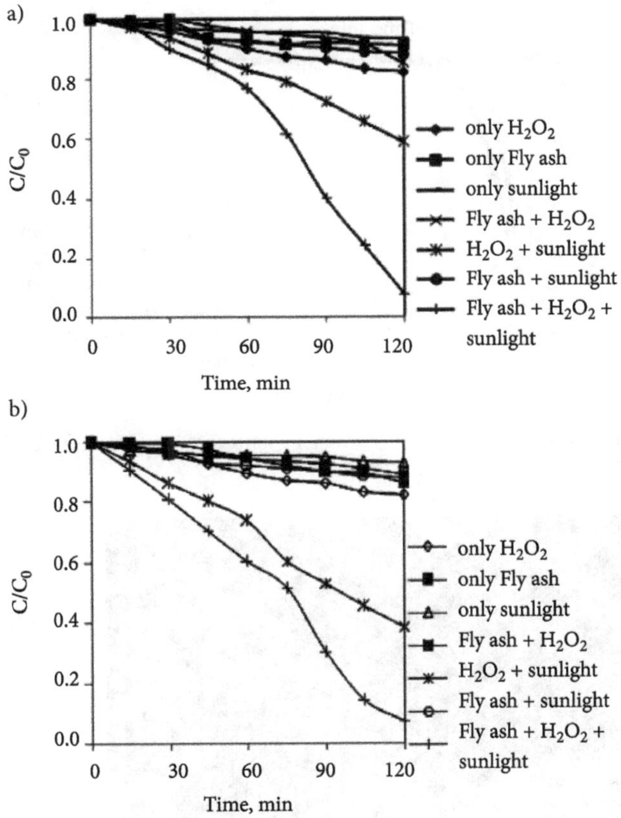

Fig. 7. Preliminary experiments performed under different conditions to evaluate their efficiency towards a) degradation and b) decolourization of RB5 using FA

becomes constant till 6.61 mM and then decreases with the increasing dosage of H_2O_2. This decrease is due to the fact of scavenging of $OH^{\cdot}$ by H_2O_2 (Walling 1975; Madhavan *et al.* 2010), can be expressed by the equation:

$$OH^{\cdot} + H_2O_2 \rightarrow H_2O + HO_2^{\cdot}$$

$$HO_2^{\cdot} + OH^{\cdot} \rightarrow H_2O + O_2$$

Effective degradation (90%) was achieved within 70 min while decolourization (90%) of dye solution was achieved within 45 min of reaction at optimum dose 2.2 mM of H_2O_2.

2.1.4. *Effect of pH*

The effect of pH on RB5 degradation and decolourization is shown in Figure 6(a) and 6(b). The pH of the reaction was varied from 2–7. 90% degradation and decolourization efficiency was achieved at pH 3, for the reaction time of 70 min and 45 min respectively. At pH 2, nearly similar results were observed. Further increment in the pH from 3 leads to reduction in the degradation (18%) and decolourization (28%) efficiencies. This behaviour could be explained by the formation of ferric hydroxo complexes during the reaction, which blocks the decomposition of hydrogen peroxide catalyzed by the ferrous irons (Faust, Hoigne 1990; Kitis, Kalpana 2007).

2.2. Photo-Fenton studies using FA as an iron source

2.2.1. *Preliminary experiment*

The similar preliminary studies were performed with FA using different conditions as in the case of FS. As shown in Figure 7(a) and (b), the degradation of dye was less than 20% in the presence of H_2O_2, around 15% in the presence of FA, negligible when exposed to sunlight, 22% in the presence of FA+ H_2O_2, 50% with sunlight + H_2O_2, very less in the presence of FA and sunlight, 87% in the presence of fly ash+ H_2O_2 and sunlight within 120 min.

2.2.2. *Effect of variation of FA*

The effect of FA dose on the degradation of the dye was observed in the range of 0.05–1 gm and shown in Figure 8(a). With the increase of FA dose, degradation and decolourization efficiency increased till certain limit (0.1 gm) after that it was decreased. The optimum amount of FA was 0.1 gm resulting in 92% degradation in 75 min and 97% decolourization in 15 min. At higher

FA concentration, a relative decrease in degradation and decolourization rates were observed may be due to the adsorption of both H_2O_2 and dye molecule on separate particles rather than at the adjacent sides of the same particle and thereby not reacting to each other (Feng *et al.* 2003; Catalkaya, Sengul 2005; Tony *et al.* 2008). This reduction can also be explained due to the fact that, at higher concentration of iron than the optimum, the hydroxyl radical initial formation rate was so high that most of them were consumed by the side reactions

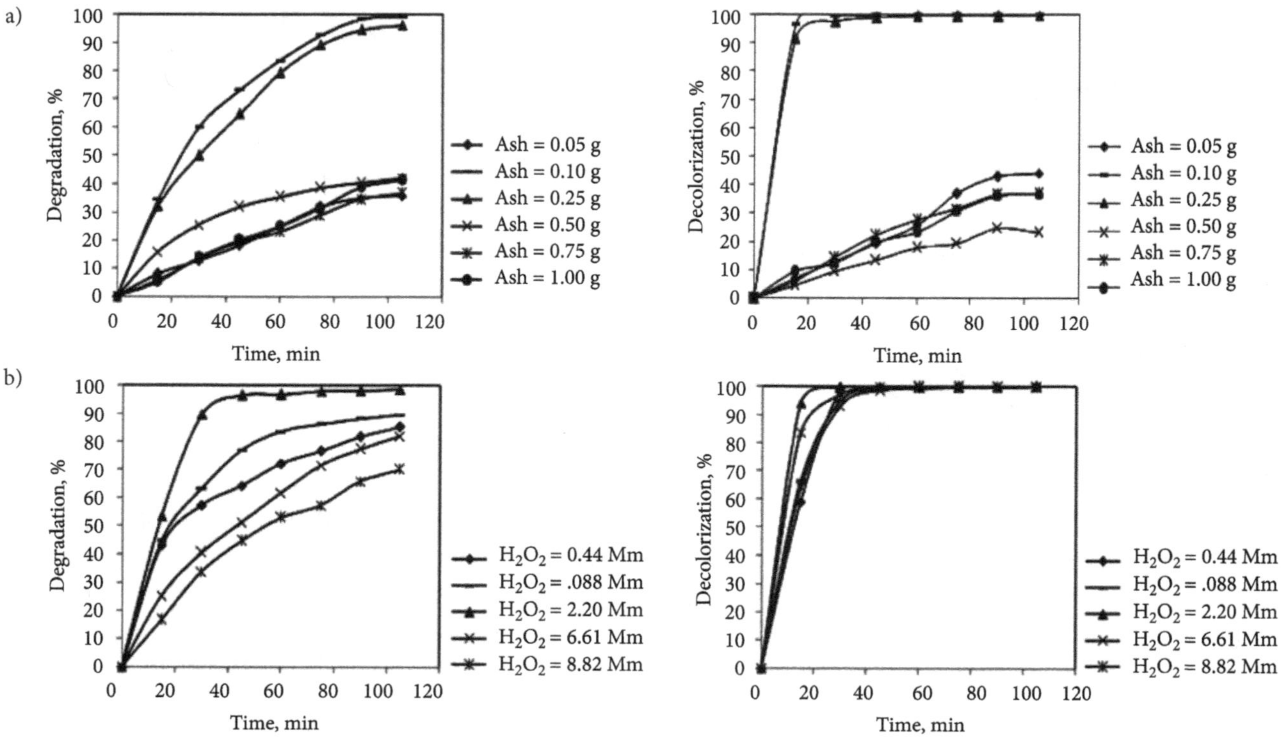

Fig. 8. Percentage degradation and decolourization of RB5 at (a) varying concentration of FA (b) varying concentration of H_2O_2

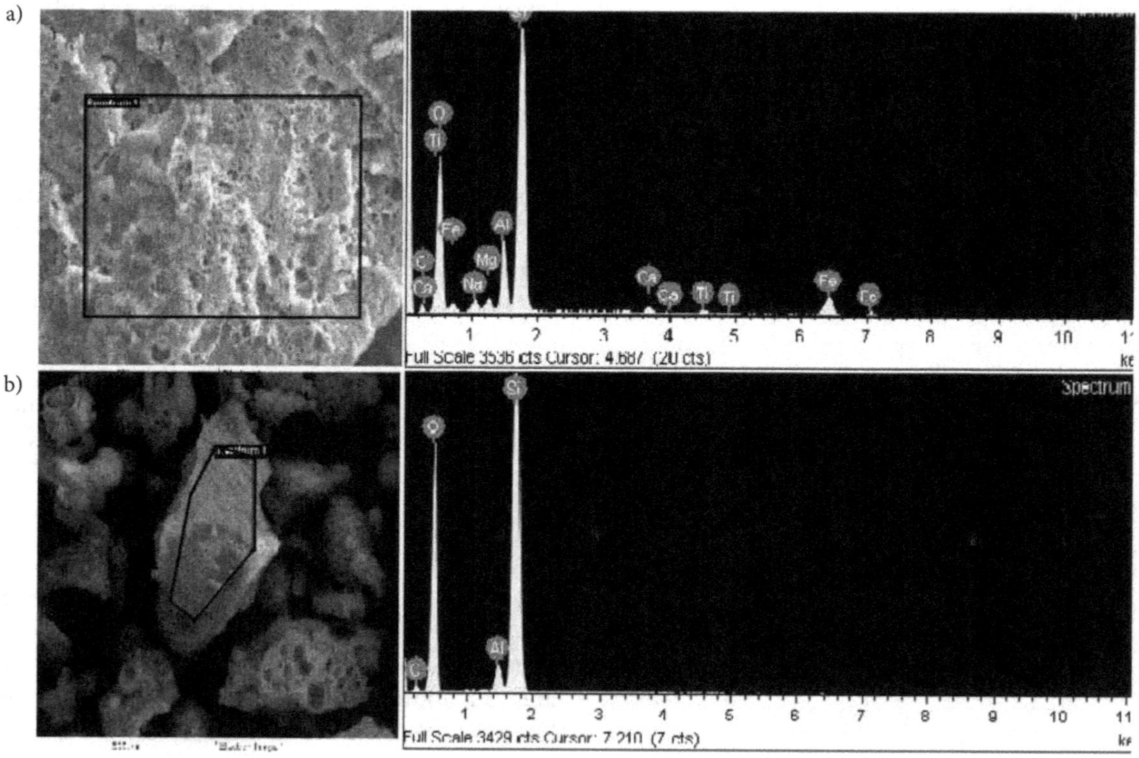

Fig. 9. SEM micrograph and EDS pattern of a) used FA b) used FS, during photo-Fenton treatment of RB5

before they could be utilized for the mineralization of the compound (Hsueh *et al.* 2005).

2.2.3. *Effect of H_2O_2 concentration*

In the presence of FA, the influence of H_2O_2 concentration on the heterogeneous photo- Fenton degradation of the dye was also investigated with different levels of H_2O_2 concentration (0.44, 0.88, 2.2, 6.61 and 8.82 mM) while keeping other operating parameters constant (FA = 0.1 gm, pH 3, C_0 = 100 mgL^{-1}). From the Figure 8 (b) it can be illustrated that the effective degradation (90%) was achieved within 30 min and decolourization (95%) within 10 min with optimum dose of 2.2 mM. Further increment in the dosage of H_2O_2 leads to decrease in the degradation rates. This reduction may be explained by the so-called hydroxyl radical scavenging effect.

The above study indicated the potential use of FS and FA as an alternative source of iron in photo-Fenton process for the treatment of RB5 dye. Effect of various reaction parameters on the degradation and decolourization efficiency was observed. The SEM- EDS for used FS and FA is shown in Figure 9(a) and (b) and data is shown in Table 3 and 4. The reduction in the composition of iron and aluminium confirmed their participation in the photo-Fenton reactions.

Table 3. Composition of FS after used in photo-Fenton process

Element	C K	O K	Na K	Mg K	Al K	Si K	Ca K	Ti K	Fe K	Total
Weight %	10.87	50.50	1.42	1.06	5.49	23.37	0.64	0.51	6.15	100
Atomic %	16.95	59.12	1.16	0.82	3.81	15.58	0.30	0.20	2.06	

Table 4. Composition of fly ash after used in photo-Fenton process

Element	C K	O K	Al K	Si K	Total
Weight %	8.33	61.12	1.54	29.02	100
Atomic %	12.38	68.17	1.02	18.44	

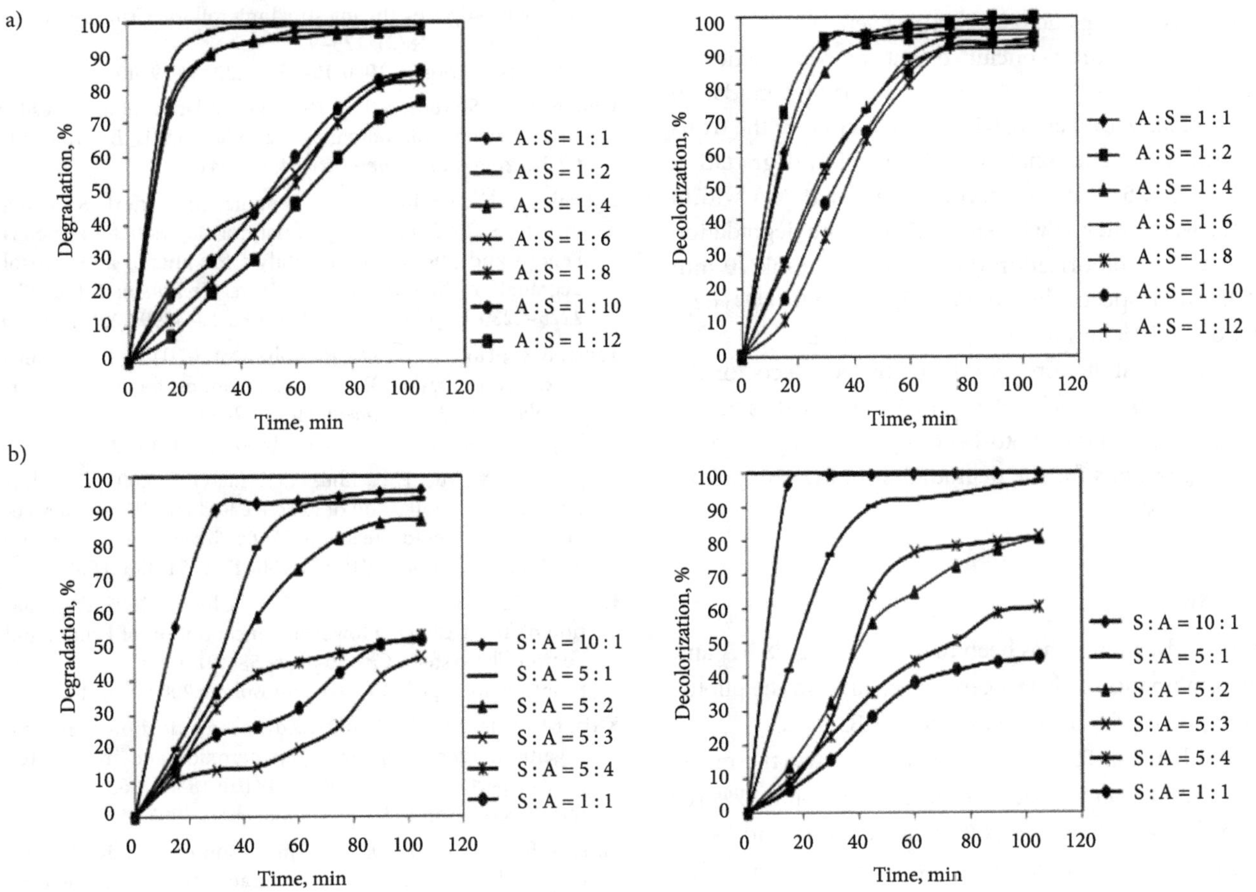

Fig. 10. Percentage degradation and decolourization of RB5 at a) different FA to FS ratio b) different FS to FA ratio

2.3. Photo-Fenton studies using mixture of FS and FA as iron source

2.3.1. Effect of FS concentration keeping FA constant

To ascertain the effect of FS in oxidizing RB5, a series of experiments were conducted by varying FS concentration in the ratio of Ash : Sand = 1:1, 1:2, 1:4, 1:6, 1:8, 1:10 and 1:12. The influence of FS dosage on degradation and decolourization efficiency is illustrated in Figure 10(a). The degradation and decolourization rate of RB5 decreased with increasing amounts of FS. More than 95% degradation and decolourization rates were achieved with the ratio Ash: Sand = 1:2, where amount of Ash and FS were 0.05 gm and 0.1 gm respectively.

2.3.2. Effect of FA concentration keeping FS constant

On similar trend, the effect of varying FA keeping FS and other parameters constant on the degradation and decolourization of RB5 was investigated in the ratio of Sand : Ash = 1:1, 5:1, 5:2, 5:3, 5:4 and 10:1 and the results were presented in Figure 10(b). The results indicated that the extent of degradation and decolourization was significantly influenced by the dosage of FA. Decrease in the degradation and decolourization efficiency was observed with increasing amount of fly ash. The optimal ratio value for the degradation (92%) in 60 min and decolourization (90%) in 30 min for the dye was achieved at Sand : Ash = 10:1 where FS and FA were taken 0.5 g and 0.05 g respectively.

The above study concluded that the combination of FS and FA gave better results in the terms of degradation and decolourization along with the reduction in the treatment time. For combination of FS and FA, degradation (90%) after 25 min and decolourization (97%) within 10 min was achieved whereas, for the same degradation (90%) and decolourization (90%) FS alone took 70 min and 45 min respectively and FA took 30 min for 90% degradation and 15 min in 93% decolourization.

Thus, combination gave the better prospects for the use of waste materials like FS and FA to be used as potential iron source in the photo-Fenton process as pre or post treatment options for the degradation of priority pollutants like dyes.

Conclusions

In this study, attempt has been made for using FS, FA and their combination as heterogeneous catalyst in the photo-Fenton process for the degradation of RB5.

While using FS as heterogeneous catalyst the maximum degradation and decolourization 95% and 97% respectively was achieved and 99% degradation and decolourization was achieved using FA, thus proposing value addition to waste.

When combination of FS and FA was tried, 99% degradation and decolourization of RB5 was achieved with different ratios of FA and FS. Thus, presents viable options for using waste materials as source for environmental remediation.

The advantages are high degradation rates (99%), simple handling, environmental friendly, and low cost (waste as an iron source), which otherwise consumes expenses towards its disposal/land filling.

References

Arslan, A.; Balcioglu, T.; Tuhkanen. 1999. Oxidative treatment of simulated dye house effluent UV and near-UV light-assisted Fenton's reagent, *Chemosphere* 39(15): 2767–2783. http://dx.doi.org/10.1016/S0045-6535(99)00211-8

Catalkaya, E. C.; Sengul, F. 2005. Application of Box-Wilson experimental design method for the photo degradation of bakery's yeast industry with UV/H_2O_2 and UV/H_2O_2/Fe(II) process, *Journal of Hazardous Materials* 128(2–3): 201–207. http://dx.doi.org/10.1016/j.jhazmat.2005.07.052

Chaco, J. M.; Leal, M. T.; Sanchez, M.; Bandala, E. R. 2006. Solar photocatalytic degradation of azo-dyes by photo-Fenton process, *Dyes and Pigments* 69: 44–150.

Chaudhuri, S. K.; Sur, B. 2000. Oxidative decolorization of reactive dye solution using Fly ash as catalyst, *Journal of Environmental Engineering* 126(7): 583–594. http://dx.doi.org/10.1061/(ASCE)0733-9372(2000)126:7(583)

Chung, K. T.; Stevens, S. E. J.; Cerniglia, C. E. 1992. The reduction of azo dyes by the intestinal microflora, *Critical Reviews in Microbiology* 18(3): 175–97. http://dx.doi.org/10.1080/19443994.2013.799440

Chung, K. T.; Stevens, S. E. J. 1993. Degradation of azo dyes by environmental microorganisms and helminths, *Environmental Toxicology and Chemistry* 54: 435–41.

Djeffal, L.; Abderrahmane, S.; Benzina, M.; Siffert, S.; Fourmentin, S. 2013. Efficiency of natural clay as heterogeneous Fenton and photo-Fenton catalyst for phenol and tyrosol degradation, *Desalination and Water Treatment* 52(10–12): 2225–2230. http://dx.doi.org/10.1080/19443994.2013.799440

Faust, B. C.; Hoigne, J. 1990. Photolysis of Fe(II)-hydroxyl complexes as sources of OH radicals in clouds, fog and rain, *Atmosphere and Environment* 24(1): 79–89. http://dx.doi.org/10.1016/0960-1686(90)90443-Q

Feng, J.; Hu, X.; Yue, P. L.; Zhu, H. Y.; Lu, G. Q. 2003. Discoloration and mineralization of Reactive Red HE-3B by heterogeneous photo-Fenton reaction, *Water Research* 37(15): 3776–3784. http://dx.doi.org/10.1016/S0043-1354(03)00268-9

Hsueh, C. L.; Huang, Y. H.; Wang, C. C.; Chen, S. 2005. Degradation of azo dyes using low iron concentration of Fenton and Fenton-like system, *Chemosphere* 58(10): 1409. http://dx.doi.org/10.1016/j.chemosphere.2004.09.091

Kitis, M.; Kalpana, M. M. 2007. Advanced oxidation of natural organic matter using hydrogen peroxide and iron-coated pumice particles, *Chemosphere* 68(10): 1846–1853. http://dx.doi.org/10.1016/j.chemosphere.2007.03.027

Lars, B. R.; Mallika, I. 1997. Optimization of Reactive Black 5 dye and Reactive Red 120 dye degradation, *Chemosphere* 35: 585–597.

Liao, Q.; Sun, J.; Gao, L. 2009. Degradation of phenol by heterogeneous Fenton reaction using multi-walled carbon nanotube supported Fe_2O_3 catalysts, *Colloids and Surfaces A: Physicochemical and Engineering Aspects* 345(1–3): 95–100. http://dx.doi.org/10.1016/j.colsurfa.2009.04.037

Lucking, F.; Koser, H.; Jank, M.; Ritter, A. 1998. Iron powder, graphite and activated carbon as catalysts for the oxidation of 4-chlorophenol with hydrogen peroxide, *Water Research* 32(9): 2607–2614. http://dx.doi.org/10.1016/S0043-1354(98)00016-5

Madhavan, J.; Grieser, F.; Kumar, M. A. 2010. Combined advanced oxidation processes for the synergistic degradation of ibuprofen in aqueous environments, *Journal of Hazardous Materials* 178(1–3): 202–208. http://dx.doi.org/10.1016/j.jhazmat.2010.01.064

Mahmoodi, N. M.; Arami, M. 2009. Degradation and toxicity reduction of textile waste water using immobilized titania nano photocatalysis, *Journal of Photochemistry & Photobiology B: Biology* 94(1): 20–24. http://dx.doi.org/10.1016/j.jphotobiol.2008.09.004

Mass, R.; Chaudhari, S. 2005. Adsorption and biological decolorization of azo dye Reactive Red-2 in semi continuous anaerobic reactors, *Process Biochemistry* 40(2): 699–705. http://dx.doi.org/10.1016/j.procbio.2004.01.038

Mecozzi, R.; Palma, L. D.; Pilone, D.; Cerboni, L. 2006. Use of EAF dust as heterogeneous catalyst in Fenton oxidation of PCP contaminated wastewaters, *Journal of Hazardous Materials* 137(2): 886–892. http://dx.doi.org/10.1016/j.jhazmat.2006.03.002

Pandey, A.; Singh, P.; Iyengar, L. 2007. Bacterial decolorization and degradation of azo dyes, *International Biodeterioration & Biodegradation* 59(2): 73–84. http://dx.doi.org/10.1016/j.ibiod.2006.08.006

Park, H.; Choi, W. 2003. Visible light and Fe(III)-mediated degradation of Acid Orange in the absence of H_2O_2, *Journal of Photochemistry and Photobiology A: Chemistry* 159(3): 241–7. http://dx.doi.org/10.1016/S1010-6030(03)00141-2

Sun, J. H.; Sun, S. P.; Wang, G. L.; Qiao, L. P. 2008. Degradation of azo dye Amido black 10B in aqueous solution by Fenton oxidation process, *Dyes and Pigments* 74(3): 647–652. http://dx.doi.org/10.1016/j.dyepig.2006.04.006

Tanaka, K.; Padermpole, K.; Hisanaga, T. 2000. Photocatalytic degradation of commercial azo dyes, *Water Research* 34(1): 327–333. http://dx.doi.org/10.1016/S0043-1354(99)00093-7

Tony, M. A.; Zhao, Y. Q.; Fu, J. F.; Tayeb, A. M. 2008. Conditioning of aluminium-based water treatment sludge with Fenton's reagent: effectiveness and optimizing study to improve dewaterability, *Chemosphere* 72(4): 673–677. http://dx.doi.org/10.1016/j.chemosphere.2008.03.032

Toor, A. P.; Verma, A.; Singh, V.; Jotshi, C. K.; Bajpai, P. K. 2006. Photocatalytic degradation of Direct Yellow 12 dye using UV/TiO_2 in a shallow pond slurry reactor, *Dyes and Pigments* 68(1): 53–60. http://dx.doi.org/10.1016/j.dyepig.2004.12.009

Walling, C. H. 1975. Fenton's reagent revisited, *Accounts of Chemical Research* 8(4): 125–131. http://dx.doi.org/10.1021/ar50088a003

Wang, K. S.; Chen, H. Y.; Huang, L. C.; Su, Y. C.; Chang, S. H. 2008. Degradation of Reactive Black 5 using combined electrochemical degradation- solar- light/immobilized TiO_2 film process and toxicity evaluation, *Chemosphere* 72(2): 299–305. http://dx.doi.org/10.1016/j.chemosphere.2008.02.012

Yoshida, Y.; Ogata, S.; Nakamatsu, S.; Shimamune, T.; Kikawa, K.; Inoue, H. 1999. Decolorization of azo dye using atomic hydrogen permeating through a Pt-modified palladized Pd sheet electrode, *Electrochimica Acta* 45(3): 409–414. http://dx.doi.org/10.1016/S0013-4686(99)00277-7

Anoop VERMA, Dr, is Assistant Professor in School of Energy and Environment, Thapar University, Patiala, Punjab, India. PhD in Environmental Engineering. Basic research interests are in the field of advanced oxidation processes, photo-Fenton treatment using alternate iron sources, fixed-bed photocatalysis etc. Verma and his team has published more than 18 papers in various international journals and presented more than 25 papers in international and national conferences.

Amrit Pal TOOR, Dr Professor in Dr. SSB University Institute of Chemical Engineering and Technology, Panjab University, Chandigarh. She is having more than 25 years of experience in the field of chemical engineering, waste treatment and published more than 30 papers in SCI journals. Presently, she is working of production of bio-diesel from non-edible plants with process kinetics. Research interests basically include advanced oxidation processes, bio-diesel production, etc.

Himadari RAJPUT, M. Tech. student, research area: Advanced Oxidation Processes for the removal of persistent pollutants from wastewater. A participant of four conferences.

Manpreet KAUR, M. Tech. student, research area: Advanced Oxidation Processes for the removal of persistent pollutants from wastewater. A participant of four conferences.

Taranjeet KAUR, PhD student perusing her degree in the field of pesticides degradation using Advanced Oxidation Processes. She has published few papers on pesticide degradation using photocatalysis. Research interests: field of photocatalysis with doped catalyst.

15

STUDIES ON COLONISATION OF FLY ASH DISPOSAL SITES USING INVASIVE SPECIES AND AROMATIC GRASSES

Deblina MAITI, Bably PRASAD

Natural Resources and Environment Management, CSIR – Central Institute of Mining and Fuel Research, Dhanbad, India
Academy of Scientific and Innovative Research (AcSIR), CSIR – Central Institute of Mining and Fuel Research, Dhanbad, India

Abstract. Fly ash disposal activities by coal based thermal power plants will continue to be a serious issue across the globe due to its hiked generation every year. To obviate the hazardous effects of fly ash disposal sites on the surrounding ecosystems, rapid stabilization of the dumps is essential. This paper conglomerates the past activities, challenges; present scenario of vegetation establishment on these sites as well as future research requirements based on various experimental case studies. An insight has been presented on the usefulness of native, tuft, aromatic grasses which can reduce the length of successive phases in reclamation programmes and also enhance the fertility of the substrate as found from the significantly increased nitrogen content in the present field sites. Metal bioaccumulation studies depicted that by virtue of high biomass production potential of *Saccharum spontaneum* it can also be used as a phytoextractor of toxic metals, thus helping in phytoremediation of the metals in fly ash. Field studies allude the fact that knowledge of phytodiversity of old fly ash deposits is essential for a right choice of species before every reclamation programme. Secondly, application of amendments is conjointly a prerequisite for establishment of plants on fly ash. In a pot scale study it was found that lower rate of amendment application (2–5% farmyard manure and 5–10% topsoil on weight basis) in fly ash improves the growth and biomass of *Cymbopogon citratus*. Extensive root system of the grass was substantiated by high root: shoot biomass which stabilized the surface of the ash. To investigate the possibilities of ground water contamination due to amendments leaching studies were carried out. An initial high concentration of some ions marginally near permissible limit as per Indian drinking water standards was observed but their concentrations were below acceptable limit during harvestable stages. Above studies can contribute significantly in field studies through a properly planned restoration programme.

Keywords: nitrogen, biomass, farmyard manure, *cymbopogon citratus, saccharum spontaneum,* bioaccumulation.

Introduction

Fly ash is the main industrial waste product generated by coal based thermal power plants. India ranks third across the globe in fly ash production (Ram *et al.* 2008). Fly ash generation is likely to surpass 300 MT by the year 2017 due to very high ash content of the coal, which is about 35–45% (Mathur *et al.* 2003). Year 2012–2013 experienced a standing amount of 63 million tonnes of unutilized ash (37% of the total ash produced) in India which was dumped, encroaching a land area of more than 40 000 Ha (Jain, Gaggar 2013). Fly ash is generally being disposed off in low lying areas, landfills, basins around thermal power plants and in abandoned mine pits (CEA 2012). These dumps have deleterious effects and contaminates nearby aquatic and terrestrial ecosystems. Furthermore in the long run, trace metals in fly ash contaminate ground water and get into living organisms through food chain (Belyaeva, Haynes 2012). Thus there is an urgent need for stabilization of the dumps.

In this context, phytomanagement is the most effective and eco-friendly approach as it stabilizes ash dumps, reduces erosion and incurs gradual restoration of the site. It also reduces gradual leaching of water and solutes into ground water and initiates carbon sequestration. They also

Corresponding author: Deblina Maiti
E-mail: dmhellodeblina@gmail.com

pave the pathway for future generation of bio-resource (Pandey 2012). This paper conglomerates the past and present status of revegetation of fly ash disposal sites. It also highlights the challenges and future research needs which should be focused on in this aspect. Furthermore an insight has been presented in the usefulness of grasses in stabilization of fly ash dump surfaces. Native hardy and tuft grasses are early colonizers as they can thrive on adverse conditions. With a view of long term management the information collected here will contribute a substantial part on an effective strategy for fly ash dump stabilization.

1. Earlier studies on reclamation of fly ash disposal sites

Vegetation cover development on fly ash disposal sites is difficult at early stages. Fly ash consists of glass-like spherical particles, with a silty loam texture and a natural compaction tendency which inhibits water infiltration as well as restricts root growth to some extent (Haynes 2009). Though, it has high moisture retention capacity, and considerable amounts of plant nutrients (Ca, K, Mg, Cu, Fe, Mn, Mo, Zn, B) in the form of oxides, hydroxides, carbonates and bicarbonates; it lacks microbial activity, and a consistent supply of nitrogen crucial for plant growth (Page *et al.* 1979). Besides, it also contains various toxic metals such as V, Cr, Ni, As, Se, Rb, Sr and Pb which can get its path into higher trophic level through food chain. Various authors have carried out their studies on a number of herb, shrub, grass as well as tree species as enumerated in Table 1. Generally, efficient, stress tolerant, fast growing species which can help in development of a quick cover are recommended. Zolnierz *et al.* (2016) has recently reported the phytodiversity of an abandoned mine filled with fly ash, 11 years after reclamation. Inclusion of native plants in presence of microbial amendments has been reported by many workers (Krzaklewski *et al.* 2012; Babu, Reddy 2011; Hrynkiewiez *et al.* 2009). Few workers have studied the effect of sandy topsoil/farmyard manure/mill mud/compost/biosolids/sewage sludge/vermiculite on ash to support revegetation (Cheung *et al.* 2000). Studies on field practices of revegetation usually involve high rate of fertilizer addition in absence of topsoil. According to agronomic techniques NPK (nitrogen:phosphorus:potassium) fertilizers are generally used in the ratio of 15:15:15 or 19:19:19 to a root tilled depth of 15 cm, (Punshon *et al.* 2002). But high rate fertilizer addition incorporates a great deal of ions and solutes which can erode into the nearby water streams or seep through the ground to contaminate water below in the long run. A cost effective amendment approach is still lacking at the current state to be implemented in the field. Thus, an effort has been made by us in a recent study to grow aromatic grasses as initial colonizers in presence of low rate of amendment application.

Table 1. Plant species used for vegetation establishment on fly ash dumps

Species name	Habitat	Reference
Parthenium hysterophorus	Herb	Dwivedi *et al.* (2008)
Solanum nigrum	Herb	Dwivedi *et al.* (2008)
Limnanthes spp	Herb	Dwivedi *et al.* (2008)
Eclipta alba	Herb	Dwivedi *et al.* (2008)
Typha latifolia	Herb	Maiti and Jaiswal (2008)
Fimbristylis dichotoma	Herb	Maiti and Jaiswal (2008)
Amaranthus defluxes	Herb	Maiti and Jaiswal (2008)
Cassia tora	Herb	Gupta and Sinha (2008)
Chenopodium album	Herb	Gupta and Sinha (2008)
Blumea lacera	Herb	Gupta and Sinha (2008)
Equisetum ramosysma	Grass	Dwivedi *et al.* (2008)
Saccharum munja	Grass	Dwivedi *et al.* (2008)
Saccharum spontaenum	Grass	Maiti and Jaiswal (2008)
Cynodon dactylon	Grass	Maiti and Jaiswal (2008)
Calamagrostis epigejos	Grass	Mitrovic *et al.* (2008)
Festuca rubra	Grass	Mitrovic *et al.* (2008)
Calotropis procera	Shrub	Gupta and Sinha (2008)
Ipomoea Carnea	Shrub	Pandey (2012)
Ricinus Communis	Shrub	Pandey (2013)
Sida cardifolia	Shrub	Gupta and Sinha (2008)
Acacia auriculiformis	Tree	Cheung *et al.* (2000)
Alnus glutinosa	Tree	Jamil *et al.* (2009)
Alnus incana	Tree	Jamil *et al.* (2009)
Cassia seamea	Tree	Juwarkar and Jambulkar (2009)
Jatropha curcas	Tree	Krzaklewski *et al.* (2012)
Leucaena leucocephala	Tree	Cheung *et al.* (2000)
Prosopis juliflora	Tree	Rai *et al.* (2004)

2. Material and method

2.1. Field investigations and sampling

Field surveys were conducted on different aged fly ash disposal sites in and around Jharkhand and West Bengal state of India during dry periods of the year 2015. Approximately, six to seven visibly dominant plant species were selected for sampling. Three replicates for each of the dominant plant species and their rhizospheric fly ash samples (depth = 30 cm) were collected from each site. The replicates for each plant species were chosen on the basis of same age group. All the samples were collected in clean labeled plastic bags and transported to the laboratory for analysis. Encountered plants were identified with the help of related flora. Descriptions of the sites chosen for study are given below.

Study site 1 is located in low lying lands around Tenughat Thermal Power station. It is situated near

Tenughat Dam in the district of Bokaro, Jharkhand India. The total capacity of the plant is 1550 MW and it encroaches an area of 2200 acres which also includes low lying areas for fly ash disposal. Fly ash disposal is generally followed by compaction through heavy machinery and covering with live topsoil acquired from nearby areas. Presence of rich biodiversity in the nearby areas helps in succession of disposal sites with native invasive species in a period of 1 year which generally accelerates after monsoon. Site 2 is located in the abandoned, opencast fly ash filled mines of Damoda located in Dhanbad district of Jharkhand state in India. Fly ash used for filling the mines was taken from nearby Chandrapura thermal power plant of 890 MW capacities located 16 km away from the mines. The process of filling was carried out for 3 years to create a dump of fly ash which is now covered with native and invasive grasses. Study site 3 is another fly ash dump located inside Crescent thermal power plant premises near Asansol, West Bengal India. The thermal power station is of 40 MW and the primary fuel is washery rejects obtained from the same premises. The dump is present in an area of 4–5 hectare of land. The reclamation programme of the dump was commenced with technical steps such as topsoiling to a thickness of about 60 cm, construction of permanent drainage channels along the slopes, garland drain at dump foot ultimately leading to a reservoir. In pit plantation of trees were done along with seeding with native grass legume species. The average annual rainfall of the entire zone is 1200–1300 mm while the temperature varies from 10–40 °C annually. Site 1 and 2 was chosen to account for the natural succession of fly ash disposal sites and identify the dominant plant species. Site 3 was taken as a reference site in contrast to site 1 and 2 wherein a technical restoration programme had been followed. The results of site 1 and 2 were compared to site 3 in this study.

2.2. Experimental design for pot scale study

Fly ash (FA) for the pot scale study was collected from the ash ponds near Chandrapura thermal power plant of Damodar Valley Corporation located at Chandrapura town in Jharkhand state of India. Garden soil (GS) was collected from top 15 cm from the Institute's garden at CSIR-CIMFR, Dhanbad located in Jharkhand state of India. Farmyard manure (FYM) was obtained from a nearby dairy farm. All the samples were air dried for a week, thoroughly mixed and grounded to pass through a 4 mm sieve.

Garden soil and farmyard manure were mixed with fly ash on weight basis, in proportions of 5%, 10% and 2%, 5% respectively. Treatment I was the control with no amendments while treatments IV (FA + 5% GS + 2% FYM), V (FA + 10% GS + 2% FYM), VI (FA + 5% GS + 5% FYM), and VII (FA + 10% GS + 5% FYM) had been prepared with different combinations of garden soil and farmyard manure. The resultant treatments had a pH near 7.00 at initial stage which usually doesn't pose any threat of metal contamination. The pots containing the mixtures were arranged for equilibration under natural light and temperature for 10 weeks. After equilibration period, first leachate (1L) collection was done while second (2L) and third leachates (3L) were collected at intervals of 2 months till final harvest of the plants near after 4 months. Leachate collection from the pots had been done, based on the design of Marseille *et al.* (2000). It was mainly done to investigate the contribution of amendments and vegetation on the amount of ions and metals leaching down which may contaminate the ground water. This was followed by plantation of 2–3 *Cymbopogon citratus* tillers per pot. Regular irrigation of the pots was done five – six times a week with equal amounts of water and recorded. The collected leachate samples were immediately taken to the laboratory for analysis. In an economical point of view *C. citratus* shoots are generally harvested after 3.5–4 months for oil extraction (Maiti, S. K., Maiti, D. 2015; Maiti, Prasad 2016). With this point in view, the plants in this study were harvested at maturity near after 4 months.

2.3. Analysis of samples

The root and shoot parts of the plants were segregated in the laboratory to determine their biomass. Average weight of all the plants from each site was calculated individually. They were repeatedly washed with tap water and deionized water to remove all the adhered fly ash particles followed by drying to a constant weight at 80 °C in oven and finally grinding to a fine powder for heavy metal analysis. Powdered samples of a known weight were reduced to ashes at 450 °C for 45 min and 850 °C for 1 h, followed by dissolution in 25 ml of 6 M HNO_3 and subsequent digestion. The solutions were filtered with Whatman 42 grade filter paper into 50 ml volumetric flasks (Rodriguez *et al.* 2011) followed by metal analysis through atomic absorption spectrophotometer (Thermo Scientific, Germany).

Fly ash samples were air-dried and ground to pass through a 2 mm sieve, for analysis of pH in 1:1 ratio (w/v) with distilled water (Jackson 1973), loss on ignition at 450 °C and 750 °C respectively in a muffle furnace for 1 hour each (Mishra, Das 2010; ASTM D 2974-87). Determination of available nitrogen was done by the method of Subbiah and Asija (1956) while water holding capacity was done through keen's box method (Piper 1942).

The leachate samples were analyzed for pH, EC (electrical conductivity), TH (total hardness), Ca H (calcium hardness), Cl (chloride), TDS (total dissolved solid), total alkalinity (Tot Alk), and SO_4 (sulphate) as per standard methods (APHA, AWWA, WEF 1992; IS-3025 2009).

Sample preparation was done before metal analysis. Approximately, 200 ml samples of each leachate was acidified using concentrated nitric acid and digested on a hot plate at 70 °C to a volume less than 50 ml. The digested and concentrated sample was filtered through Whatman cellulose filters, pore size 0.45 µm using sterile injectors into 50 ml volumetric flasks. The volumes were made up to 50 ml with distilled water. Analysis of each metal was done at respective sensitive wavelength i.e. at 324.7, 279.5, 217.0, 228.8, 213.9, 232.0 and 357.9 nm for Cu, Mn, Pb, Cd, Zn, Ni and Cr respectively. Limits of detection of various metals (Mn, Zn, Cu, Pb, Cd, Cr, and Ni) were in the range of 0.0005 to 0.01 mg L^{-1}. The calculations for metals in leachates were done by dividing the instrument reading for metals by concentration factor. Blanks and different dilution standards were used for calibration and quality assurance. The instrument was recalibrated after analysis of 15 samples. All the glassware used for analysis were acid washed and rinsed with distilled water, dried and kept in a clean dry environment prior to use. All reagents used were of analytical grade (Merck).

2.4. Statistical analysis

Statistical analyses were carried out to interpret the results by using SPSS (version 20.0 for Windows; SPSS Inc. Chicago, USA) and MS Excel 2010. Mean values along with their standard deviation (SD) has been computed and have been shown as error bars in the graphs. Descriptive statistics like one way analysis of variance was used to access the variance between the means of an analyzed parameter between the sampling sites and leachates from the pot study. Difference between individual means was carried out through DMRT (Duncan's multiple range test) model at 95% level of confidence. Hierarchical cluster analysis is performed on dataset of metal concentrations to group similar metal clusters using Eucledian distance and group linkage methods. Relation between general water quality measures and metal associations and their relations to pH have also been performed tested with two-tailed as well as one tailed Pearson correlation to get a matrix of correlations between them.

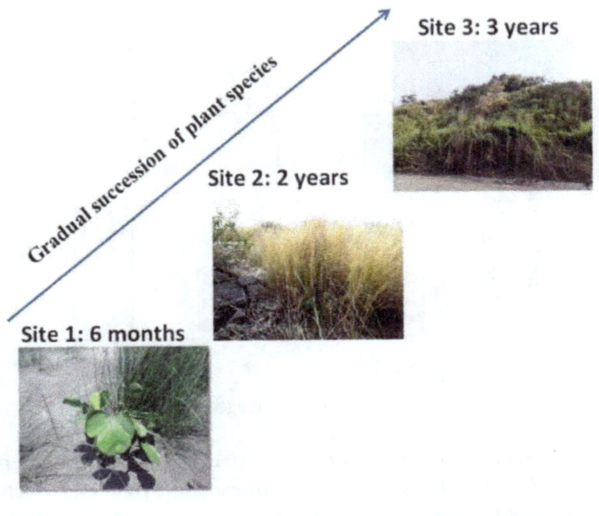

Fig. 1. Plant cover establishment in different fly ash dump sites

3. Result and discussion

3.1. Vegetation succession on fly ash disposal sites

Various field surveys done by us has revealed that natural colonisation of fly ash surfaces depend on the nutrient content of the substrate, efficiency of plant species to grow in adverse conditions and interactions between them (Belyaeva, Haynes 2012). The three field sites chosen by us varied in the age of the vegetation and were of 6 months, 2 years, and 3 years respectively (Fig. 1). Initial colonisation of the sites are dominated by weed herbs such as *Chromolaena odorata, Dysphania ambrosioides, Sida rhombifolia, Tephrosia purpurea* various shrubs and grasses as seen in site 1. At later stages hardy and tuft grasses cover the areas which were observed in site 2. A proper restoration program (such as topsoil blanketing, addition of amendments and seeding with native species) was followed in the 3rd field site (reference site) which showed the presence of more various plant species after 3 years of vegetation. The vegetation cover of more than a meter in site 3 was dominated by *Saccharum spontaneum, Ipomoea* sp., and *Pennisetum pedicellatum*. Presence of few number of tree species were also found on the dump slopes. It was recovered from the site authorities that during initiation of closure activity 2500 number of tree species were planted out of which a very few percent has survived till the present date. The slope length of the present reclaimed dump was 35–40 m with an angle of 55 degrees.

A noticeable fact as found in all the three sites was the dominance of *Saccharum spontaneum* and *Cynodon dactylon*. These types of results were also obtained by various other authors (Dwivedi *et al.* 2008; Pandey *et al.* 2012). These invasive grasses have tolerance to various adverse conditions and shows competitive behaviour towards other species. Narratively, their perennial nature, presence of light persistent seeds and rhizomatous roots imparts the advantage for their invasive nature (Graham *et al.* 2014). On the other hand they also stabilise the fly ash dump surface by holding on to the loose material through their strong roots (Kaith *et al.* 2010). Secondly it was also observed that later stages of reclamation of the dumps lacked an economic end use. A major strategy which can be focussed upon in this context is biomass production at later stages of restoration programmes. Few economically important trees which can be included are pulp and paper tree, biodiesel crops, firewood, timber wood and plywood trees.

3.2. Rhizospheric fly ash characteristics and plant biomass

Figure 2 shows the variation in physicochemical properties of the sites. Site 1 and 3 had near neutral pH and site 2 had an average pH of 5.6 significantly different from other sites may be due to the mixing of moderately acidic minesoil with fly ash. Previous studies on the groups of nearby opencast mines near site 2 also have revealed a mine soil pH value near 4.18–5.35 (Mukhopadhyay *et al.* 2013). Available nitrogen in the sites increased significantly with the age of the vegetation from 39–109 mg kg^{-1}. Approximately 64% of nitrogen increment was observed at the climax of vegetation establishment at site 3. In a study reported by Maiti, S. K., Maiti, D. (2015) a nitrogen level of 34–72 mg kg^{-1} was developed under grass legume vegetation while their reference site (nearby forest area) had approximately 100 mg kg^{-1} of available nitrogen. The presence of dense cover of *S. spontaneum* in site 2 incorporated fine root litter from its extensive root system which may be accountable for the increase in soil nitrogen after 2 years significantly similar to the restored site 3. Analogous to this the average ground cover biomass of the sites also had a significantly increasing trend (135–400 g per plant) with the age of vegetation (Fig. 3). Loss on ignition of the sites varied significantly from 3.7–18%. High value of loss on ignition in site 2 was probably due to mixing with mine soil. The proof of this statement relies on the fact that the loss on ignition value of the original fly ash (also used for pot scale study in this paper) used for filling the mine pit of site 2 is 5.55%. On the other hand the surface rhizospheric samples of site 1 and 2 had water holding capacity of nearly 64% while site 3 had a water holding capacity of 38%. The rhizospheric samples from site 3 were predominantly the topsoil which had been used to blanket the fly ash dump prior to the ecorestoration programme. High water holding capacity of fly ash compared to soil is attributed to the large surface area of the spherically shaped particles of fly ash which increases microporosity (Shaheen *et al.* 2014).

3.3. Metal accumulation in plants

Metal accumulation in the dominant plant species growing on two different aged fly ash dumps have been reported based on their yield in Table 2. The metal content in the plants were multiplied with the biomass data. Similar dominant plant species were chosen from the sites undergoing natural succession (site 1 and 2) to differentiate the metal accumulation under different conditions. *S. sponatneum* has been found to be more efficient and accumulated more amounts of metals than *C. dactylon*

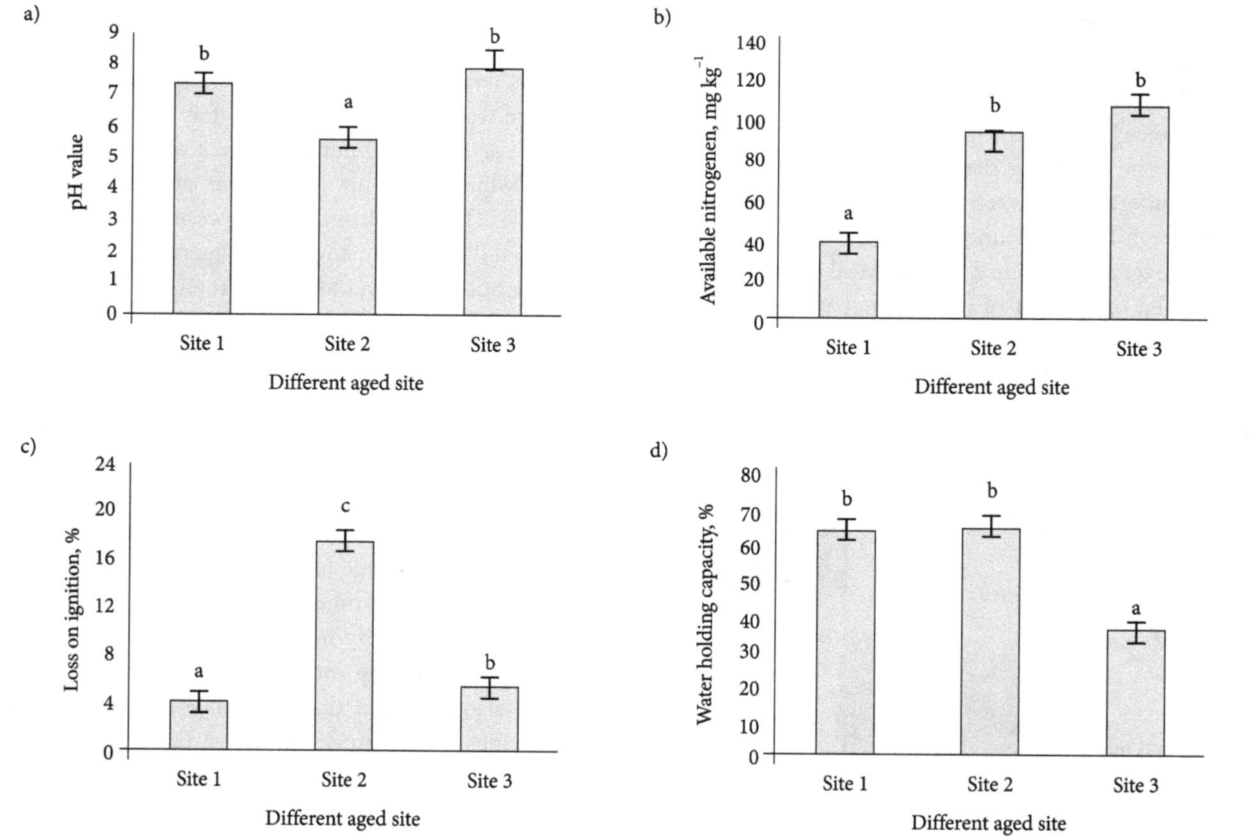

Fig. 2. Physicochemical parameters of rhizospheric samples from different aged fly ash dump sites such as pH (a); available nitrogen (b); loss on ignition (c); water holding capacity (d) for n = 7 replicates. The deviation between the data for each individual parameter for a site has been shown as error bars. Different alphabetical letters above each bar for a graph depict significant differences between the means at $p < 0.05$

due to its greater biomass. Bioaccumulation factor of Cr (4.21–77.27) and Pb (2.20) was found greater than 1 as indicated by bold letters in Table 2. Metal concentration in the plants increased with the age of the vegetation as also obvious from the higher values of metals in *S. spontaneum* from site 2 than in site 1. This is also evident from the data that toxic metals such as Cd, Ni, Pb, and Cr content varied from 0.01 to 0.05 mg/kg, 1.58 to 4.14 mg/kg, 0.30 to 0.58 mg/kg, 32.34 to 48.75 mg/kg in *S. sponatneum* which is approximately 80–90% more in site 2 than site 1. Even though, bioaccumulation factor of Cr in *C. dactylon* from site 1 was found highest (77.27) from site 2 (4.21) which are approximately 18 folds that of the later. Further, *S. spontaneum* also showed a twofold higher bioaccumulation factor for Cr in site 1 than site 2. Metal accumulation in plants also depends on pH, metal concentration in substrate, and differential response of each species to detoxify metals (Gupta *et al.* 2002).

Accumulation of metals in plants varied in the order Cr > Mn ~Zn > Cu ~ Ni > Pb > Cd and Mn ~ Cr > Zn > Ni > Cu > Pb > Cd respectively in site 1 and 2, which varied irrespective of the concentration of metals in fly ash. This was further proved by Figure 4 showing the dendrogram (between groups) obtained using hierarchical cluster analysis with pearson distance. It helps in showing the pattern of metal accumulation in plants from site 1 and 2. Here, Mn and Zn are in the same cluster linked with Cr showing that fly ash is contributing to elevated Cr levels in plants. Cr has also been found to be linked distantly with Ni whereas other metals such as Pb, Cd and Cu have been found in separate groups. Linkage of Cr and Ni are due to their nature of being transitional elements (Das *et al.* 2013). Metal accumulation by yield of Pb and Cd was found to be low in all the cases inspite of unit bioaccumulation factor of Pb in *C. dactylon*. The above results efficiently demonstrate that on virtue of its high biomass production potential *S. spontaneum* can be used for phytoextraction of some metals and subsequently phytoremediate metal contaminated sites.

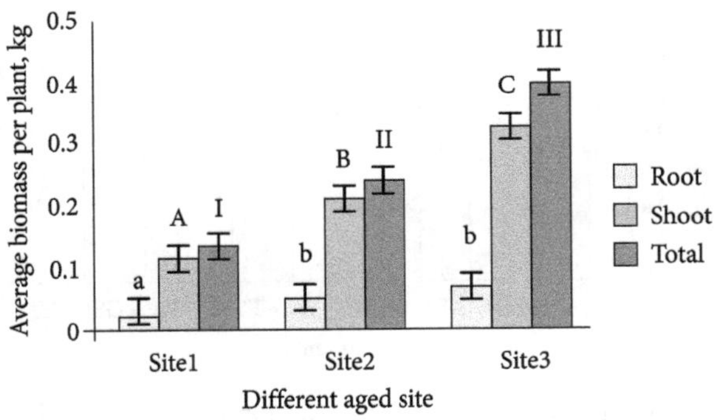

Fig. 3. Average shoot, root and total biomass per plant at different aged fly ash dump sites for $n = 7$ replicates. Different lowercase, uppercase and roman letters above each type of bar represent significant difference between the means for the parameter at $p < 0.05$

3.4. Growth of aromatic grass on amended fly ash with emphasis on leaching studies

In an another study we have found that lower rate of amendment application in fly ash such as 2–5% farmyard manure as well as 5–10% topsoil enhances the growth and biomass of perennial aromatic grass *Cymbopogon citratus*. *C. citratus*, also called lemon grass propagates through old tillers and was chosen for initial plantation in the pots as per literature studies. Maiti, S. K., Maiti, D. (2015) found it to be very promising on waste dumps of a sponge iron plant, in a field study. In this study, the grass grew to a height of more than 50 cm after 45 days of tiller plantation. It has deep fibrous root system which holds the fly ash particles and prevents its erosion (Fig. 7). A single plant produced approximately 30–40 tillers after 3–4 months of growth which usually follows a sigmoid-shaped curve.

Table 2. Metal accumulation in dominant plant species growing on different aged fly ash dump sites (mg kg^{-1}), bioaccumulation factors, and concentration in the rhizospheric fly ash samples (mg kg^{-1})

Parameters	Field site	Plant species	Metal						
			Mn	Cu	Zn	Cd	Cr	Ni	Pb
Accumulation in plants (mg kg^{-1})	1	*Cynodon dactylon*	4.60	0.53	5.35	0.01	12.17	1.04	0.46
	2	*Cynodon dactylon*	9.79	0.64	3.59	0.01	6.46	0.93	0.18
	1	*Saccharum spontaneum*	7.72	3.11	6.44	0.01	32.34	1.58	0.30
	2	*Saccharum spontaneum*	28.30	2.71	13.30	0.05	48.75	4.14	0.58
Bioaccumulation factor	1	*Cynodon dactylon*	0.12	0.29	0.54	0.28	77.27	0.39	2.20
	2	*Cynodon dactylon*	0.44	0.29	0.71	0.34	4.21	0.30	0.37
	1	*Saccharum spontaneum*	0.11	0.47	0.28	0.07	8.99	0.18	1.02
	2	*Saccharum spontaneum*	0.39	0.30	0.52	0.22	2.33	0.24	0.29
Average concentration in rhizosheric FA (mg kg^{-1})	1		255.51	15.69	71.04	0.43	4.93	22.35	1.52
	2		174.65	19.14	49.73	0.38	29.63	32.26	4.18

Commercially, multiple harvesting is generally done after 3.5 months for oil extraction (Akhila 2010). Thus the plants were harvested after 4 months and noticeably the root: shoot biomass ratios were within the range of 0.8–0.9 (Fig. 5). Highest ratio was observed in treatments with combination of amendments (both top soil and farm yard manure). Moreover, destructive harvest of lemon grass at the end of 4 months showed formation of a vigorous root mass in the shape of a basket around the fly ash as shown in Figure 7a.

3.5. Ions and metals in leachates obtained from fly ash treatments

Values obtained from analysis of the leachate samples from fly ash treatments had been compared to the standard drinking water specifications (BIS 10500: 1991). The change in these samples during the growth period of the grass has been represented in Figure 6a–n. Significant differences among the means of the parameters has been shown in Table 5 (supplementary section) to maintain clarity in the corresponding Figure 6a–n. It was observed that garden soil had a significant decreasing effect initially on pH (Fig. 6a) and alkalinity (Fig. 6e) due to its acidic nature. The pH of the control leachate was significantly higher than the other treatments all throughout the growth period may be partially due to high root growth of the grass in them (Maiti, Prasad 2016). Overally pH of 2nd leachate was higher than 1st leachate possibly due to consumption of protons during dissolution of glassy amorphous silicate minerals present in the fly ash (Seoane, Leiros 2001). The combined treatments having both fly ash and garden soil showed the lowest pH (6.97–7.2) in the later stages leachate again. Studies also suggest that grass roots promote weathering of fly ash and act to loosen the fly ash surface by excretion of saccharides (Neuschutz *et al.* 2010). Houben *et al.* (2012) stated that addition of manure increases leachate pH which was also found in our study, while Oste *et al.* (2002) concluded that addition of organic amendments reinforces alkalinity in leachates. Though in our study alkalinity was not alarming and was below the desirable limit. Leaching of ions and metals were observed at a significantly greater level in the first leachates as also supported by positive correlation among cations and anions with electrical conductivity which was primarily due to addition of farmyard manure (Fig. 6b). Electrical conductivity decreased to 275–544 and 197–327 μS cm^{-1} in the 2nd leachate and 3rd leachate respectively. Neuschutz *et al.* (2010) also observed decreasing electrical conductivity in pots with plant growth as they decrease the release of secondary minerals.

Control leachate had the least total hardness (TH) (96–214 mg L^{-1}) and total dissolved solid (TDS) (149–455 mg L^{-1}), chloride (26–320 mg L^{-1}) and sulphate (19–59 mg L^{-1}) in all the collection periods while treatment leachates had values marginally below the permissible limit in 1st leachates which went below desirable limit in 3rd leachate (decreased by 80%) (Fig. 6c, d, h, i). On comparison with the control the level of TDS was significantly higher in the treatment leachates and significantly positively correlated to other parameters (Table 3). Leaching of anions from treatments having high level of amendments was mainly due to the presence of high concentration of salts in them also observed by Shimozono *et al.* (2007). Mukhtar *et al.* (2003) in their study used high rate of application of farmyard manure as soil amendments which gave apparently high level of total solids in leachates and therefore they also recommended low rate application of farmyard manure to prevent groundwater contamination.

It is known that Ca is the most largely released cation from fly ash and this depends on its mode of occurrence

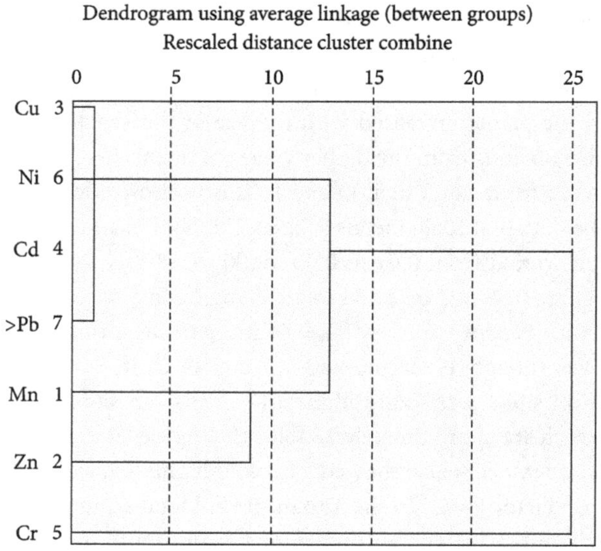

Fig. 4. Hierarchical dendrogram showing clustering of different metals in plants using Pearson correlation as distance and group linkage methods

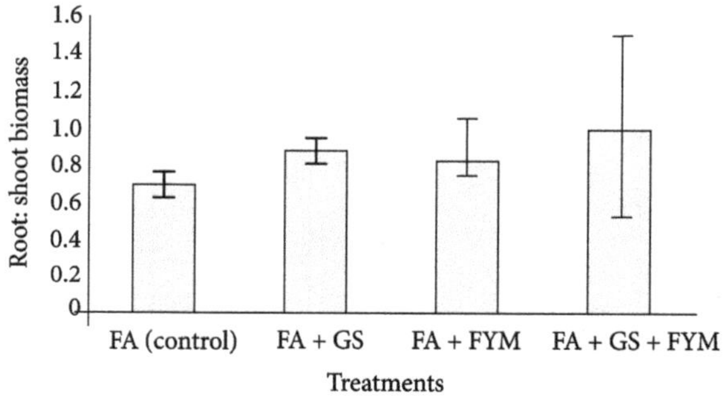

Fig. 5. Root is to shoot biomass ratio of *Cymbopogon citratus* grown of fly ash treatments compared to control having no amendments

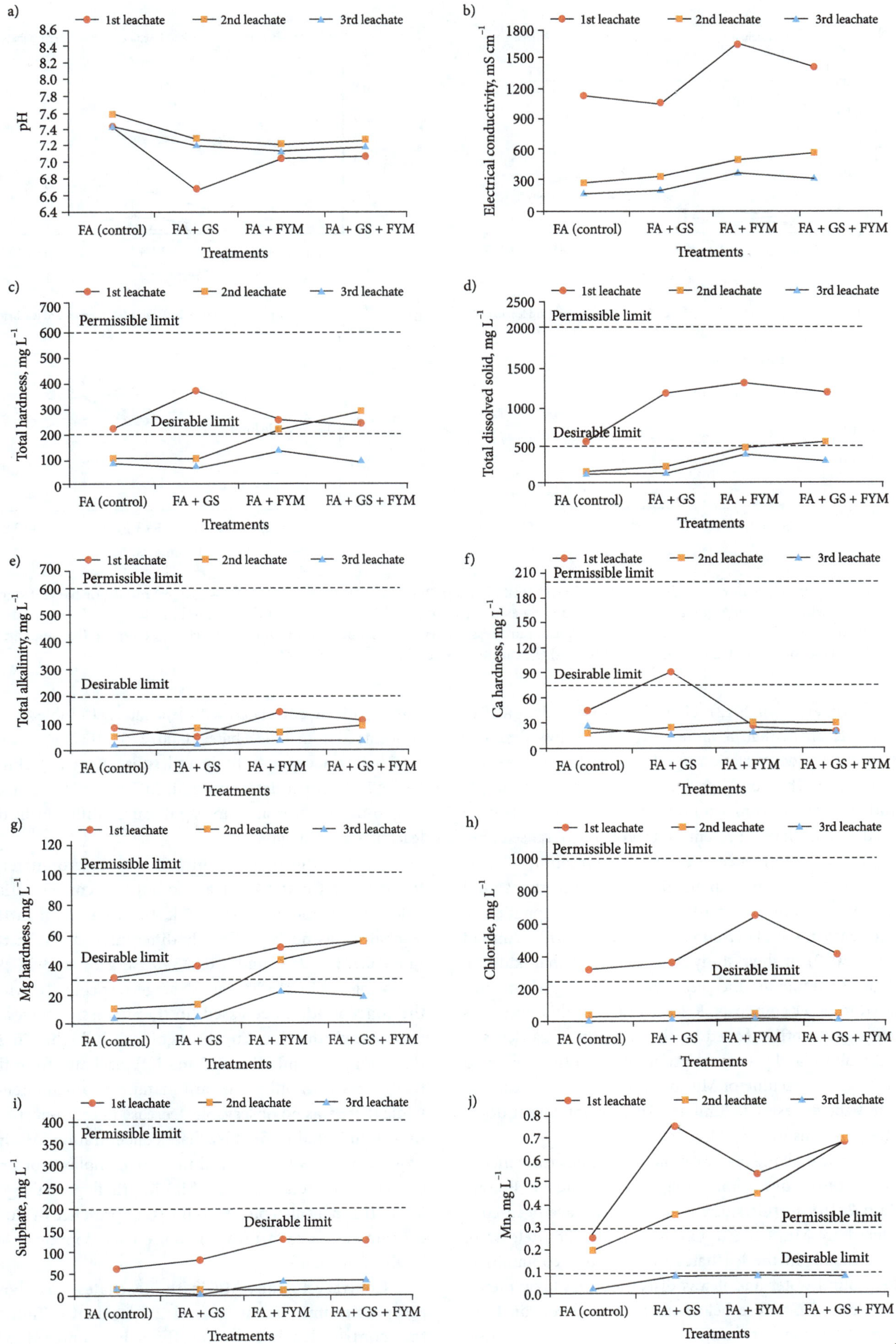
a)
1st leachate
2nd leachate
3rd leachate
pH
FA (control)
FA + GS
FA + FYM
FA + GS + FYM
Treatments
b)
Electrical conductivity, mS cm^{-1}
c)
Total hardness, mg L^{-1}
Permissible limit
Desirable limit
d)
Total dissolved solid, mg L^{-1}
e)
Total alkalinity, mg L^{-1}
f)
Ca hardness, mg L^{-1}
g)
Mg hardness, mg L^{-1}
h)
Chloride, mg L^{-1}
i)
Sulphate, mg L^{-1}
j)
Mn, mg L^{-1}

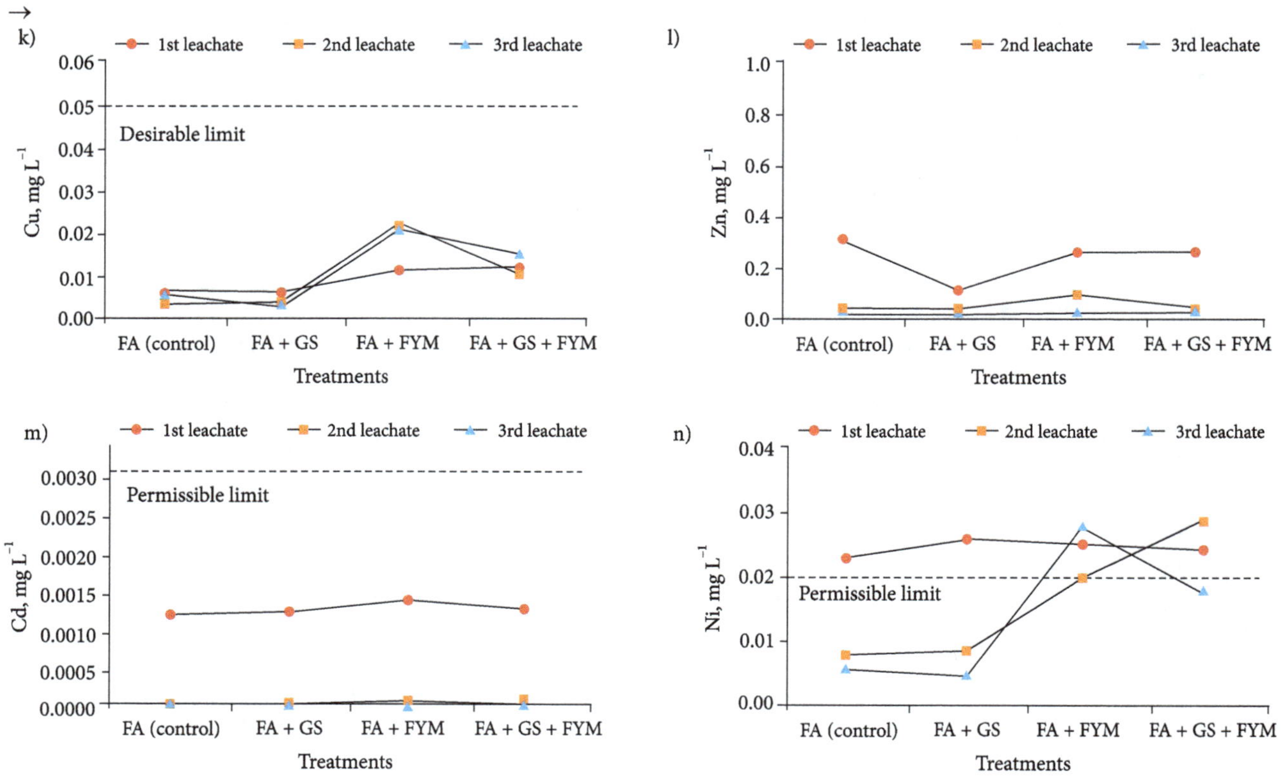

Fig. 6. Characteristics of pot leachates collected at 1st (initial stage), 2nd stage (3rd month), and 3rd stage (4th month or harvest stage) during growth of *Cymbopogon citratus* on amended fly ash; pH (a) electrical conductivity (EC) (b); total hardness (TH) (c), total dissolved solid (TDS) (d) total alkalinity (e), calcium hardness (Ca H) (f), magnesium hardness (Mg H) (g), chloride (h); sulphate (i) Mn (manganese) (j), Cu (copper) (k), zinc (Zn) (l), cadmium (Cd) (m), nickel (Ni) (n)

(Izquierdo, Querol 2012). Unlikely in this study, Ca H in all the leachates had a value marginally below desirable limit but a wider variation in release of Mg was observed (Fig. 6f, g). The pond ash samples collected for the pot study was taken from ash ponds which had already been weathered of the Ca present in it, probably because of being under the influence of water for a long time. Generally, Ca concentration in alkaline FA (like in this study) decreases in a matter of 10 weeks due to precipitation in secondary minerals (Neupane, Donahoe 2013). Ahmad *et al.* (2009) in their study had concluded that addition of farmyard manure increases leaching of Mg when compared to other composts. Release of Mg in this study was also significantly affected by the addition of garden soil and manure and was more pronounced in the combined treatments. Leaching of Mg continued till the 2nd leachate found above desirable limit and subsequently went down after 4 months in the 3rd leachate.

Amidst all the major cations and anions trace metals are most toxic due to their long term persistence (Prasad, Maiti 2016). In this study metals which have been considered are Mn, Cu, Zn, Cd, Ni, Cr, and Pb. Release of these metals in the leachates would finally contaminate the ground water and it was discerned that toxic metals like Pb and Cr were below instrument detection limit throughout the pot study. Further, metal concentration was negatively correlated with leachate pH (Table 4). Cd concentration in 1st leachates (0.0012–0.0013 mg L^{-1}) was also below acceptable limit and subsequently declined below instrument detection limit in 2nd and 3rd leachates (Fig. 6 m). Cd was also not significantly different in the leachates at 1st stage.

Among the metals which act as micronutrients for instance Cu and Zn, leached out in a concentration below acceptable limit (Fig. 6 k, l). Level of Zn in the leachates was not significantly different and was in the range 0.116–0.268 mg l^{-1}, 0.033–0.1 mg L^{-1} and 0.029–0.034 mg L^{-1} in 1st, 2nd and 3rd stage respectively, while the concentration decreased in the successive stages. It was noteworthy that Cu was significantly higher in 2nd (0.022 mg L^{-1}) and 3rd (0.023 mg L^{-1}) leachates from the treatments with only farmyard manure as amendments, while there was no observable decrease in the concentration from initial to final leachate concentration. Mixing of garden soil with farmyard manure complexes or precipitates the metals thus prohibiting their probable release. This may also be due to inherent presence of Cu in organic matter as farmyard manure itself contains high levels of nutrients.

Contrary to this, Mn in the leachates was above acceptable limit in the 1st and 2nd leachates. Though the control leachate (only fly ash) depicted Mn

(0.034–0.246 mg L^{-1}) above acceptable limit which indeed shows the intrinsic property of the fly ash to release loosely bound Mn, added to this the leachates obtained from amended treatments further had significantly higher concentration of Mn (0.087–0.753 mg L^{-1}) above permissible limit. This also exhibited that garden soil and farmyard manure contributed to the Mn release in the samples (Fig. 6j). Even among the amended treatments, pots with only garden soil showed significantly highest Mn leaching in 1st stage. Noticeably in the 3rd leachates Mn was found marginally below the acceptable

Table 3. Correlation between general parameters analyzed in leachate samples (n = 81)

	EC	TH	TDS	Ca H	Mg H	Cl	SO_4
EC	1						
TH	0.55**	1					
TDS	0.88**	0.61**	1				
Ca H	0.26*	0.79**	0.29**	1			
Mg H	0.64**	0.92**	0.70**	0.49**	1		
Cl	0.93**	0.44**	0.82**	0.26*	0.49**	1	
SO_4	0.89**	0.43**	0.83**	0.07	0.57**	0.85**	1

Notes: EC: electrical conductivity, TH: total hardness, TDS: total dissolved solid, Ca H: calcium hardness, Mg H: magnesium hardness, Cl: chloride, SO4: sulphate.
*Correlation is significant at the 0.05 level (2-tailed).
**Correlation is significant at the 0.01 level (2-tailed).

Table 4. Correlation between metals analyzed in leachate samples (n = 81)

	pH	Mn	Cu	Zn	Cd	Ni
pH	1					
Mn	−0.19	1				
Cu	−0.13	0.05	1			
Zn	−0.19	0.14	−0.09	1		
Cd	−0.38**	0.26*	−0.09	0.61**	1	
Ni	−0.15	0.65**	0.56**	0.13	0.15	1

Notes: *Correlation is significant at the 0.05 level (2-tailed). **Correlation is significant at the 0.01 level (2-tailed).

a) b)

Fig. 7. Growth of lemon grass after 4 months of plantation (a), massive root growth of lemon grass which stabilizes the fly ash (b)

limit (0.1 mg L^{-1}). Ni concentration also varied arbitrarily in the 3 stages and was significantly above permissible limit in the amended treatments when compared to the control (0.006–0.023 mg L^{-1}). More amount of the metal was observed in those treatments with only garden soil (0.004–0.026 mg L^{-1}) or farmyard manure (0.020–0.027 mg L^{-1}) which shows the presence of Ni in garden soil as well as farmyard manure. In fact, garden soil treatments showed a peak concentration of Ni in 1st leachates which lowered down in later stages whereas farmyard manure treatments gave constant elevated levels of Ni in all the stages. Yet treatments with both the amendments (garden soil and farmyard manure) depicted concentration below permissible level till the end exhibiting some precipitation of Ni. Metal levels in the leachates also showed some significant positive correlations between Mn, Ni; and Cu, Ni likewise observed by Edet and Offong (2002) in their study.

Moreover, at the time of harvest, leachates (3L) of all the treatments had homogeneous values for all parameters below or marginally around the desirable limit which is not an alarming issue regarding ground water contamination. Fly ash is known to have sufficient quantities of inorganic nutrients which can support plant growth. Therefore addition of lower percentage of easily available NPK supplements in the form of garden soil and farmyard manure can boost the growth of economically valuable aromatic grasses. Further, in previous studies workers have used high rate of amendments for ecorestoration of contaminated sites which can drastically affect the water quality beneath the ground. In this study the treatment with farmyard manure have proved their efficiency by enhancing the growth of *C. citratus* i.e high root: shoot biomass ratio. They can be successfully implemented in the field for revegetation of fly ash dumps without any possibilities of ground water contamination.

Conclusively, the perennial nature of *C. citratus* (Fig. 7b), unpalatability to grazing animals and high commercial end uses makes it an efficient plant species for restoration of fly ash disposal sites.

Conclusions

A reclamation strategy of fly ash disposal sites encompasses discussion on fly ash characteristics, followed by selection of type of amendment and its application rate as well as the plant species. Native and invasive species are an effective choice at initial stages. Apart from increment in fertility of the sites, fly ash surface stabilisation was adequately achieved which was the foremost goal in this context. This was proved from the present status of fly ash dumps in Damoda abandoned mines. Despite of less after care and maintenance species such as *Cynodon dactylon, and Saccharum spontaneum*, were found to colonize the dump creating a thick vegetation mat. Metal bioaccumulation studies proved that on virtue of high biomass potential of *Saccharum spontaneum* than *C. dactylon* it can also be used as phytoextractor of toxic metals and thus phytoremediate the metals in fly ash. Added to this, the pot scale study with amended fly ash showed that *Cymbopogon citratus* efficiently grew in low nutrient conditions and stabilized the fly ash volume through its extensive root system. It is another fruitful choice for revegetation of fly ash disposal sites. Initial leaching studies on the treatments have also supported the effectiveness of the dual amended treatments to be used in the field. The ephemeral effects on leachates due to addition of farm yard manure diminished in a matter of 4 months and acceptably the values were below permissible limit as per drinking water standards. A properly planned reclamation programme can thus be made by incorporating economically useful plant species such as *C. citratus* at later stages. Lastly a regular monitoring schedule is of prime importance in each restoration programme which will help in predicting long term change in fly ash properties due to vegetation and guide the future steps to improve the status of the sites.

Acknowledgements

Authors are grateful to Dr. Pradeep Kumar Singh, Director, CSIR – Central Institute of Mining and Fuel Research, Dhanbad, India for the continuous support and motivation during the investigation period. First author is thankful to (Department of Science and Technology, Government of India) DST for the (Innovation in Science Pursuit for Inspired Research) INSPIRE fellowship provided to her.

References

Ahmad, Z.; Honna, T.; Yamamoto, S.; Irshad, F. M.; Waleed, H. A. E. 2009. Effect of chloride and sulfate salinity on micronutrients release and uptake from different composts applied on total phosphorus basis, *Communication in Soil Science and Plant Analysis* 40: 1566–1589. http://dx.doi.org/10.1080/00103620902831644

Akhila, A. 2010. *Essential oil-bearing grasses: the genus Cymbopogon.* Medical and aromatic plants-industrial profile. Taylor and Francis Group: LLC.

APHA; AWWA; WEF. 1992. *Standard methods for the examination of water and wastewater.* 18th ed. Washington: DC.

ASTM D 2974-87:2000. *Standard test methods for moisture, ash, and organic matter of peat and other organic soils.* American Society for Testing and Materials, USA.

Babu, A. G.; Reddy, M. S. 2011. Dual inoculation of arbuscular mycorrhizal and phosphate solubilising fungi contributes in sustainable maintenance of plant health in fly ash ponds, *Water, Air, and Soil Pollution* 219: 3–10. http://dx.doi.org/10.1007/s11270-010-0679-3

Belyaeva, N. O.; Haynes, J. R. 2012. Comparison of the effects of conventional organic amendments and biochar on the chemical, physical and microbial properties of coal fly ash as a plant growth medium, *Environmental Earth Sciences* 66: 1987–1997. http://dx.doi.org/10.1007/s12665-011-1424-y

BIS IS 10500:1991. *Drinking water specifications.* First revision. Bureau of Indian Standards, India.

CEA. 2012. *Annual Report on Fly-ash utilization. Report on fly ash generation at coal/lignite based thermal power stations and its utilization in the country for the year 2011–12* [online], [cited 8 December 2015]. New Delhi. Available from Internet: http://cea.nic.in/reports/others/thermal/tcd/flyash_final_1415.pdf

Cheung, K. C.; Wong, J. P. K.; Zhang, Z. Q.; Wong, J. W. C.; Wong, M. H. 2000. Revegetation of lagoon ash using the legume species *Acacia auriculiformis* and *Leucaena leucocephala*, *Environmental Pollution* 109: 75–82. http://dx.doi.org/10.1016/S0269-7491(99)00235-3

Das, S.; Ram, S. S.; Sahu, H. K.; Rao, D. S.; Chakraborty, A.; Sudarshan, M.; Thatoi, H. N. 2013. A study on soil physicochemical, microbial and metal content in Sukinda chromite mine of Odisha, India, *Environment Earth Science* 69: 2487–2497. http://dx.doi.org/10.1007/s12665-012-2074-4

Dwivedi, S.; Srivastava, S.; Mishra, S.; Dixit, B.; Kumar, A.; Tripathi, R. D. 2008. Screening of native plants and algae growing on fly-ash affected areas near National Thermal Power Corporation, Tanda, Uttar Pradesh, India for accumulation of toxic heavy metals, *Journal of Hazardous Materials* 158: 359–365. http://dx.doi.org/10.1016/j.jhazmat.2008.01.081

Edet, A. E.; Offiong, O. E. 2002. Evaluation of water quality pollution indices for heavy metal contamination monitoring. A study case from Akpabuyo-Odukpani area, Lower Cross River Basin (southeastern Nigeria), *GeoJournal* 57: 295–304. http://dx.doi.org/10.1023/B:GEJO.0000007250.92458.de

Graham, D. B.; Josef, N. S. K.; Kristin, S. 2014. The reproductive biology of *Saccharum spontaneum* L.: implications for management of this invasive weed in Panama, *NeoBiota* 20: 61–79. http://dx.doi.org/10.3897/neobiota.20.6163

Gupta, A. K.; Sinha, S. 2008. Decontamination and/or revegetation of fly ash dykes through naturally growing plants, *Journal of Hazardous Materials* 153: 1078–1087. http://dx.doi.org/10.1016/j.jhazmat.2007.09.062

Gupta, D. K.; Rai, U. N.; Tripathi, R. D.; Inouhe, M. 2002. Impacts of fly ash on soil and plant responses, *Journal of Plant Research* 115: 401–409. http://dx.doi.org/10.1007/s10265-002-0057-3

Haynes, R. J. 2009. Reclamation and revegetation of fly ash disposal sites – challenges and research needs, *Journal of Environmental Management* 90: 43–53. http://dx.doi.org/10.1016/j.jenvman.2008.07.003

Houben, D.; Pircar, J.; Sonnet, P. 2012. Heavy metal immobilization by cost-effective amendments in a contaminated soil: effects on metal leaching and phytoavailability, *Journal of Geochemical Exploration* 123: 87–94. http://dx.doi.org/10.1016/j.gexplo.2011.10.004

Hrynkiewiez, K.; Baum, C.; Niedojadlo, J.; Dahm, H. 2009. Promotion of mycorrhiza formation and growth of willows by the bacterial strain *Sphingomonas* sp. 23 L on fly ash, *Biology and Fertility of Soils* 45: 385–394. http://dx.doi.org/10.1007/s00374-008-0346-7

IS-3025:2009. *Methods of sampling and test (physical and chemical) for water used in industry*. Bureau of Indian Standard, India.

Izquierdo, M.; Querol, X. 2012. Leaching behaviour of elements from coal combustion fly ash: an overview, *Intinternational Journal of Coal Geology* 94: 54–66. http://dx.doi.org/10.1016/j.coal.2011.10.006

Jackson, M. L. 1973. *Soil chemical analysis*. New Delhi: Prentice Hall of India.

Jain, K. A.; Gaggar, S. 2013. Present status of availability, utilization and other options to increase effective use of fly ash in India, in *3rd Annual International Summit: Fly Ash Utilisation*, 17–18 December 2013, New Delhi, India.

Jamil, S.; Abhilash, P. C.; Singh, N.; Sharma, P. N. 2009. *Jatropha curcas*: a potential crop for phytoremediation of coal fly ash, *Journal of Hazardous Materials* 172: 269–275. http://dx.doi.org/10.1016/j.jhazmat.2009.07.004

Kaith, B. S.; Jindal, R.; Jana, A. K.; Maiti, M. 2010. Development of corn starch based green composites reinforced with *Saccharum spontaneum* L. fiber and graft copolymers – Evaluation of thermal, physicochemical and mechanical properties, *Bioresource Technology* 101: 6843–6851. http://dx.doi.org/10.1016/j.biortech.2010.03.113

Krzaklewski, W.; Pietrzykowski, M.; Wos, B. 2012. Survival and growth of alders *Alnus glutinosa* (L.) Gaertn. and *Alnus incana* (L.) Moench on fly ash technosols at different substrate improvement, *Ecological Engineering* 49: 35–40. http://dx.doi.org/10.1016/j.ecoleng.2012.08.026

Maiti, D.; Prasad, B. 2016. Revegetation of fly ash–a review with emphasis on grass-legume plantation and bioaccumulation of metals, *Applied Ecology and Environmental Research* 14(2): 185–212. http://dx.doi.org/10.15666/aeer/1402_185212

Maiti, S. K.; Jaiswal, S. 2008. Bioaccumulation and translocation of metals in the natural vegetation growing on fly ash deposits: a field study from Santaldih thermal power plant, West Bengal, India, *Environment Monitoring Assessment* 136: 355–370. http://dx.doi.org/10.1007/s10661-007-9691-5

Maiti, S. K.; Maiti, D. 2015 Ecological restoration of waste dumps by topsoil blanketing, coir-matting and seeding with grass-legume mixture, *Ecological Engineering* 77: 74–84. http://dx.doi.org/10.1016/j.ecoleng.2015.01.003

Marseille, F.; Tiffreau, C.; Laboudigue, A.; Lecomte, P. 2000. Impact of vegetation on the mobility and bioavailability of trace elements in a dredged sediment deposit: a greenhouse study, *Agronomie* 20: 547–556. http://dx.doi.org/10.1051/agro:2000149

Mathur, R.; Chand, S.; Tezuka, T. 2003. Optimal use of coal for the power generation in India, *Energy Policy* 31: 319–331. http://dx.doi.org/10.1016/S0301-4215(02)00067-8

Mishra, D. P.; Das, S. K. 2010. A study of physico-chemical and mineralogical properties of Talcher coal fly ash for stowing in underground coal mines, *Materials Characterization* 61: 1252–1259. http://dx.doi.org/10.1016/j.matchar.2010.08.008

Mukhopadhyay, S.; Maiti, S. K.; Masto, R. E. 2013. Use of Reclaimed Mine Soil Index (RMSI) for screening of tree species for reclamation of coal mine degraded land, *Ecological Engineering* 57: 133–142. http://dx.doi.org/10.1016/j.ecoleng.2013.04.017

Mukhtar, S.; Kenimer, A. L.; Sadaka, S. S.; Mathis, J. G. 2003. Evaluation of bottom ash and composted manure blends as a soil amendment material, *Bioresource Technology* 89: 217–228. http://dx.doi.org/10.1016/S0960-8524(03)00085-3

Neupane, G.; Donahoe. R. J. 2013. Leachability of elements in alkaline and acidic coal fly ash samples during batch and column leaching tests, *Fuel* 104: 758–770. http://dx.doi.org/10.1016/j.fuel.2012.06.013

Neuschutz, C.; Bostrom, D.; Greger. M. 2010. Root growth into sealing layers of fly ash, *Journal of Plant Interaction* 5: 75–85. http://dx.doi.org/10.1080/17429141003592853

Oste, L. A.; Temminghoff, E. J. M.; Van Riemsdijk. W. H. 2002. Solid-solution partitioning of organic matter in soils as influenced by an increase in pH or Ca concentration, *Environment Science Technology* 36: 208–214. http://dx.doi.org/10.1021/es0100571

Page, A. L.; Elseewi, A. A.; Straughan, I. R. 1979. Physical and chemical properties of fly-ash from coal-fired power plants with reference to environmental impacts, *Residue Review* 71: 83–120. http://dx.doi.org/10.1007/978-1-4612-6185-8_2

Pandey, V. C. 2012. Invasive species based efficient green technology for phytoremediation of fly ash deposits, *Journal of Geochemical Exploration* 123: 3–18. http://dx.doi.org/10.1016/j.gexplo.2012.05.008

Pandey, V. C. 2013. Suitability of *Ricinus communis* L. Cultivation for phytoremediation of fly ash disposal sites, *Ecological Engineering* 57: 336–341. http://dx.doi.org/10.1016/j.ecoleng.2013.04.054

Pandey, V. C.; Singh, K.; Singh, R. P.; Singh, B. 2012. Naturally growing *Saccharum munja* L. on the fly ash lagoons: a potential ecological engineer for the revegetation and stabilization, *Ecological Engineering* 40: 95–99. http://dx.doi.org/10.1016/j.ecoleng.2011.12.019

Piper, C. S. 1942. *Soil and plant analysis*. The University of Adelaide, Adelaide, Australia. 368 p.

Prasad, B.; Maiti, D. 2016. Comparative study of metal uptake by *Eichhornia crassipes* growing in ponds from mining and nonmining areas – a field study, *Bioremediation Journal* 20: 144–152. http://dx.doi.org/10.1080/10889868.2015.1113924

Punshon, T.; Adriano, D. C.; Weber, J. T. 2002. Restoration of drastically eroded land using coal fly ash and poultry biosolid, *The Science of the Total Environment* 296: 209–225. http://dx.doi.org/10.1016/S0048-9697(02)00128-6

Ram, L. C.; Jha, S. K.; Tripathi, R. C.; Masto, R. E., Selvi, V. A. 2008. Remediation of fly ash landfills through plantation, *Remediation* 18: 71–90. http://dx.doi.org/10.1002/rem.20184

Seoane, S.; Leiros, M. C. 2001. Acidification–neutralization processes in a lignite mine spoil amended with fly ash or limestone, *Journal of Environment Quality* 30: 1420–1431. http://dx.doi.org/10.2134/jeq2001.3041420x

Shaheen, S. M.; Hooda, P. S.; Tsadilas, C. D. 2014. Opportunities and challenges in the use of coal fly ash for soil improvements – a review, *Journal of Environmental Management* 145: 249–267. http://dx.doi.org/10.1016/j.jenvman.2014.07.005

Shimozono, N.; Fukuyama, M.; Kawaguchi, M.; Iwaya Inoue, M.; Molla, A. H. 2007. Nutrient dynamics through leachate and turf grass growth in sands amended with food-waste compost in pots, *Communication in Soil Science and Plant Analysis* 39: 241–256. http://dx.doi.org/10.1080/00103620701759251

Subbiah, B. V.; Asija, G. L. 1956. A rapid procedure for the determination of available nitrogen in soils, *Current Science* 25: 259–260.

Zolnierz, L.; Weber, J.; Gilewska, M.; Strączynska, S.; Pruchniewicz, D. 2016. The spontaneous development of understory vegetation on reclaimed and afforested post-mine excavation filled with fly ash, *Catena* 136: 84–90. http://dx.doi.org/10.1016/j.catena.2015.07.013

Deblina MAITI. MSc, DST – INSPIRE Fellow (Senior Research Fellow), Academy of Scientific and Innovation Research, CSIR – Central Institute of Mining and Fuel Research, Dhanbad, Jharkhand, India. Research interests: Ecorestoration of fly ash landfills, Phytoremediation, Soil chemistry.

Bably PRASAD. PhD, Senior Principal Scientist, Natural Resources and Environmental Management, CSIR – Central Institute of Mining and Fuel Research, Dhanbad, Jharkhand, India. Research interests: Water pollution and treatment, Acid Mine drainage, Environmental Impact of Fly ash.

16

INVESTIGATION ON EFFECT OF MAGNETITE NANOFLUID ON PERFORMANCE AND EMISSION PATTERNS OF METHYL ESTERS OF BIO DIESEL

Yuvarajan D[a], Venkata Ramanan M[b]

[a] *Department of Mechanical Engineering, Jeppiaar Institute of Technology, Chennai, India*
[b] *Department of Mechanical Engineering, College of Engineering, Anna University, Chennai, India*

Abstract. The objective of this work is to reduce viscosity and other related problems associated with biodiesel by adding non polluting additives. Magnetite is chosen as an additive in this work. Significant reason for using magnetite as an additive is that the magnetic nano particles (magnetite) can be collected from exhaust using magnetic billets in the tail pipe. Performance and emission test is carried out in single cylinder compression ignition engine using base and modified fuel and is compared to diesel. Modified fuel consists of 98.5% base fuel (Rice bran oil methyl ester), 1.3% additives (magnetite) and 0.2% surfactant (tri methyl ammonium hydroxide) by volume. The experimental work confirmed that by adding magnetite to methyl esters of rice bran oil enhances the brake thermal efficiency by 4.27% with 5.17% reduction in SFC. In addition, 10.8%, 9.1% and 8.49% reduction in HC, CO and NOx emissions is observed respectively.

Keywords: air-fuel ratio, delay period, ignition, performance.

Introduction

Vegetable oil is found to be a potential alternative fuel owing to its properties very similar to Diesel. Vegetable oil is used as an alternative fuel due to its inbuilt advantages. The significant aspects of considering vegetable oil as a potential alternative fuel is that, it can be used in modern diesel engines with no modifications. Vegetable oils are used as direct fuel or blended with diesel fuel in diesel engine. It also emits a lesser amount of carbon monoxide and hydrocarbon emissions. In spite of numerous advantages, it also emits higher percentage of nitric oxide (NOx) emission (Aydin, Bayindir 2010; Ramadhas *et al.* 2005). The possible reason reported is due to loss in engine power as it increases NOx emissions (Ryu 2010; Karabektas 2009). NOx emissions in diesel engine fuelled with biodiesel are reduced by adding nano particles in liquid form. Nano particles in liquid form reduce the NO_X emissions as it acts as a catalyst and enhances the rate of combustion (Choi, Oh 2006). NO_X emissions are significantly related to the temperature in the combustion chamber. It has been experimentally found that by adding nano particles in liquid state lowers the thermal loading, heat flux and combustion chamber surface temperature (Galfetti *et al.* 2007). Further, it has been experimentally found that the brake thermal efficiency of biodiesel is improved significantly with addition of nanofluid. Chemical compositions of fuel are altered with inclusion of nanofluid (Jung *et al.* 2005). In addition nano fluid releases energy when diffused into higher temperature. Enhancement in oxidation reaction will take place with addition of nanofluid in the methyl ester (Selim, Elfeky 2001). Performance characteristics of methyl ester are improved due to its natural increase in density impulse of nanofluid when exposed to higher temperature in combustion chamber. Radiative heat transfer and mass diffusivity of methyl ester is improved by adding nanofluid to biodiesel (De Luca *et al.* 2005; Galfetti *et al.* 2007). Performance parameters of diesel engine fuelled with biodiesel is considerably increased with addition of nanofluid as it has high ratio of surface area to volume (Prasher *et al.* 2006). It is reported that the ignition probability of methyl ester is improved by incorporating nanofluid which results in increase in efficiency and reduction in NO_X emissions (Krishnamurthy *et al.* 2006).

In recent times, adding nano particles to liquid fuel is subjected to a lot of investigation. Arul Mozhi Selvan *et al.* (2014) studied the impact of cerium oxide nano particles of size 32 nm with a dosing level of 25 ppm using diesterol in diesel engine. They found a significant reduction in NO_X emissions. Shafii *et al.* (2011) investigated the effect of water based ferrofluid with a size of 10 nm and a concentration of

Corresponding author: Yuvarajan D
E-mail: dyuvarajan7@gmail.com

0%, 0.4%, and 0.8% by volume in diesel fuel to views its effects on performance and emissions characteristics. It was found that the magnetic particles improves the combustion characteristics and reduce NOX emissions. Further, magnetic nano particles increase in chaotic movements and reducing in settling velocity of the particles which aids combustions. However, the impact on addition of ferrofluid to rice bran oil methyl ester has never been experimented. The present experimental study is aimed to examine the effect of ferrofluid on the various properties of methyl esters of rice bran oil and also its influence on performance and emissions characteristics. Experiments were carried out with RBOME and RBOMEF and are compared with petroleum diesel. RBOMEF was prepared comprising 98.5% biodiesel (Rice bran oil methyl ester), 1.3% magnetite based ferrofluid and 0.2% surfactant by volume was used in a CI engine.

1. Experimental material & methods

1.1. Materials and reagents

1.1.1. Rice bran oil

Rice bran oil is best suitable for high-temperature cooking methods such as stir frying and deep frying. It is accepted as cooking oil in several Asian countries. It has a low-cost feed stock with high value added by- products. In addition, the chemical properties of rice bran are closer to that of petroleum diesel. The properties of transesterifed methyl esters of rice bran oil fall well within the acceptable limit of ASTM standards. Rice bran is obtained as a by-product during polishing of rice from its kernel. Rice bran is dried to remove the water present in it. It is then fed to oil extractor which extracts oil by crushing. The oil obtained is classified as defatted and crude rice bran oil. Defatted rice bran oil with 16–20% of protein content is used for edible purpose. Crude rice bran oil with 23–24% free fatty acid is used as the main source of biodiesel production process. Crude rice bran oil is converted into its methyl esters of rice bran oil by conventional transesterification process. The present work utilized refined Rice bran oil. The alcohol used in this work was methanol and KOH was used as catalyst for transesterifications.

1.1.2. Ferrofluid

Ferrofluid (Fe_3O_4) is prepared using the following Eq. (1).

$$FeCl_2 + 2FeCl_3 + 8NH_3 + 4H_2O \rightarrow Fe_3O_4 + 8NH_4Cl. \quad (1)$$

Diameter of ferrofluid was in the order of 10nm (Berger *et al.* 1999). Aqueous tri methyl ammonium hydroxide ((CH3)3NOH) solution is used as a surfactant to improve the bonding between ferrofluid to methyl ester. The most significant reason for using ferrofluid is that the magnetic nano particles can be collected from exhaust using magnetic billets. Furthermore, it can be easily diluted to biodiesel and as a result it can collect the benefits of water- biodiesel emulsions.

1.2. Apparatus & procedure

1.2.1. Base fuel preparation procedure

Batch transesterification process is followed during conventional transesterification. Molar ratio of vegetable oil and alcohol is 5.9:1. 0.51% of catalysts (wt/wt) to rice bran oil were used. Temperature is maintained at 60 °C during transesterification process. Vegetable oil, methanol and catalysts are then mixed and stirred at 280 rpm for 75 minutes. Further, the mixture is then allowed to cool. The top layer from the mixture is separated, washed and dried for further investigation. The ester obtained by conventional transesterification is used as a base fuel. Table 1 shows the fatty acids compositions of rice bran oil using Gas-liquid chromatography technique.

Table 1. Fatty acids compositions of rice bran oil

Fatty acids	Percentage
Myristic acid	0.9%
Palmitic acid	21.5%
Stearic acid	2.6%
Oleic acid	37.7%
Linoleic acid	35%
α-Linolenic acid	2.3%

1.2.2. Modified base fuel preparations procedure (RBOMEF)

0.3% of aqueous tri methyl ammonium hydroxide and 1% of magnetite by volume is used in the study. Surfactant, magnetite and methyl esters of rice bran oil are mixed using ultrasonic agitator. Stability test were performed to view its phase change quality. Test results in homogeneity of magnetite in methyl esters of rice bran oil and found stable. The ester achieved by adding magnetite is referred as RBOMEF.

1.3. Base and modified fuel properties

Fuel properties of methyl esters and Diesel are shown in Table 2. There is a significant change in all the properties with addition of ferrofluid. 3.15% enhancement in Kinematic viscosity is observed by adding magnetite to methyl esters of rice bran oil. This is caused by increase in resistance among fuel layer. 2 °C increase in flash point is observed by adding magnetite to methyl esters of rice bran oil. Higher flash point is beneficial for safety measures. 0.02% increase in water content was found by adding magnetite to methyl esters of rice bran oil. This is due to presence of water content in ferrofluid. Higher water content will reduce the NO_X emissions. Other property such as, density and calorific value of RBOMEF is enhanced by 2.96% and 2.61% respectively. This is due to positive effects of ferrofluid.

Table 2. Properties of test fuels

Properties	RBOMEF	RBOME	Diesel
Kinematic Viscosity [cst]	6.56	6.31	4.2
Flash point [°C]	143	137	50
Water content [%]	0.152	0.131	Nil
Density [gm/cc]	0.911	0.881	0.821
Calorific Value [kJ/kg]	39243	38223	42957
Cetane Index	54	52	45

1.4. Experimental set-up & testing procedure

Air cooled compression ignition engine is used for experimentation. Technical specification of engine is illustrated in Table 3. Engine exhaust pipe is linked to AVL444 di gas analyzer having a maximum response time of 8 sec for monitoring the engine emissions. The probe that is used to the exhaust gas analyzer was placed inside the exhaust pipe in which emissions such as HC, CO and NOx are measured. Exhaust gas temperature was measured using K-type thermocouple made of Chrome-alumel with digital indicating unit. U-tube manometer having one end connected to tank and other end left free to ambient is used to measure the air flow rate. Pressure inside the cylinder was measured with the help of an AVL pressure transducer with sensitivity of 16.04pc/bar, linearity of <± 0.3% and the crank angle were inferred using a TDC encoder. Both the pressure transducer and TDC encoder were linked to a computer for monitoring and recording the combustion parameters through AVL 617 indimeter of software version V2.00. Details of gas analyzer are shown in Table 4. Figure 1 shows the layout of engine and instrumentation set up.

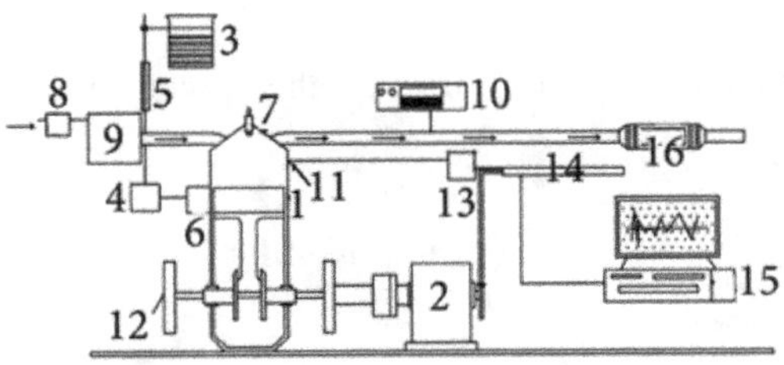

1. Kirloskar TV1 engine
2. Eddy current dynamometer
3. Fuel tank
4. Fuel filter
5. Fuel measurement setup
6. Fuel pump
7. Fuel injector
8. Air filter
9. Air stabilizing tank
10. Smoke meter and flue gas analyzer
11. Pressure transducer
12. TDC encoder
13. Charge amplifier
14. Indimeter
15. Monitor
16. Exhaust silencer

Fig. 1. Layout of experimental setup

Table 3. Specifications of the engine

Make	Kirloskar
Stroke	4
Number of Cylinders	1
Power rated	4.45 kW
Speed rated	1500 rpm
Bore diameter(D)	87.5 mm
Stroke length(L)	110mm
Compression ratio	17.5:1

Table 4. Details of gas analyzer

Parameters	Make	Accuracy	Relative Uncertainty
CO	AVL gas analyzer	0.01	±3%
HC		± 10 ppm	±2%
NOx		± 10 ppm	±2.5%

1.5. Error analysis

The error analysis for HC emission are calculated using the Equation (2)

$$\sigma_{HC} = \left[\frac{\sum_{i=1}^{n} \left((HC)_i - \left(\overline{HC} \right) \right)^2}{n-1} \right]^{\frac{1}{2}}. \quad (2)$$

The uncertainty in $\Delta(HC)$ in hydrocarbons is calculated by Equation (3)

$$\frac{\sigma_{HC}}{\sqrt{n}}, \quad (3)$$

where n = number of HC measurements,
$\overline{HC}$ = mean value of HC measurements,
σ_{HC} = unbiased standard deviation for HC measurements.

By adapting the above equation, the error analysis of other emissions such as CO, NOx and smoke were analysed.

2. Results and discussion

2.1. Effect of magnetite on emission parameters

2.1.1. Hydrocarbon emissions

Figure 2 shows the change in HC emissions for all the test fuels with respect to load. Average HC emission from diesel, RBOME and RBOMEF is 46.5 ppm, 44.7 ppm and 38.0 ppm at all loads respectively. Significant reduction in HC emissions were observed for RBOME and RBOMEF when compared to diesel. RBOME and RBOMEF have inbuilt oxygen content which helps in improved combustion and lesser HC emissions. Puhan *et al.* (2005) found that reduction in HC emission was about 63% for methyl ester compared with Diesel. Alam *et al.* (2004) observed significance reduction in HC emission for methyl ester comparing diesel citing ample availability of oxygen as the reason. By adding magnetite to RBOME, 9.1% reductions in HC emissions were found. Activation temperature of carbon combustion is lowered by adding magnetite which performs as oxidation catalysts. Similar trend was mentioned on the experimental work by (Sajith *et al.* 2010; Kannan *et al.* 2011).

2.1.2. Carbon monoxide emissions

Figure 3 shows the change in CO emissions for all the test fuels with respect to load. An average CO emission from diesel, RBOME and RBOMEF is 0.07425, 0.06875 and 0.066% volume correspondingly. CO emissions for RBOME and RBOMEF are significantly lower than diesel. This is due to enhanced oxidation reaction during combustion owing to higher content of oxygen of RBOME and RBOMEF. Puhan *et al.* (2005) found 70% reduction in CO emissions when fuelled with Maua oil methyl ester. For all the fuels, CO emission increases with increase in load. This is due necessity of rich mixture with increase in load to sustain steady power output. Ulusoy *et al.* (2009) found increase in CO emission with increase in load citing decrease in air fuel ratio at higher loads as the cause. By adding magnetite to RBOME, 6.34% reductions in HC emissions were observed. This is due to increased surface area to volume ratio of magnetite which improves the mixing rate of air and fuel as cited by Lenin *et al.* (2013). Further, RBOMEF has a lesser delay period due to the presence of magnetite which results in complete oxidation. The obtained result is justified with of similar work done by Selvan *et al.* (2004), Kim, Choi (2010).

2.1.3. Nitrogen oxide emissions

Figure 4 shows the change in NO_X emissions for all the test fuels with respect to load. Average NO_X emission from diesel, RBOME and RBOMEF is 149, 175 and 197 ppm correspondingly. NOX emissions for RBOME and RBOMEF are higher than diesel by 23.5% and 18.74%. Since the peak pressure of RBOME and RBOMEF is higher than diesel, higher NO_X emissions are obvious. By adding magnetite to RBOME, 8.49% reductions in NO_X emissions were observed. The possible reason is due to shorter ignition delay period for RBOMEF. By adding magnetite to methyl ester of rice bran oil the combustion rate is improved. Further, the water present in magnetite reduces the peak combustion temperature thereby reducing NO_X emissions for RBOMEF. Similar results were citied on the experimental work done by Shafii *et al.* (2011).

2.2. Effect of magnetite on performance parameters

2.2.1. Brake thermal efficiency

Figure 5 shows Brake thermal efficiency of RBOME, RBOMEF with respect to load. It is inferred that BTE for diesel is higher than RBOME and RBOMEF. This is attributed to higher viscosity of methyl esters when compared to diesel. It was reported from the past information that BTE decreases with the increase in cottonseed oil methyl ester in the blends comparing diesel due to its higher viscosity and lower heating value (Aydin, Bayindir 2010). Moreover the calorific value of diesel is higher as compared to RBOME and RBOMEF. Higher calorific value leads to lesser requirement of fuel for delivering any given rated power. BTE for RBOMEF is higher than RBOME by 4.27%. Since the density of RBOMEF is 2.97% higher than

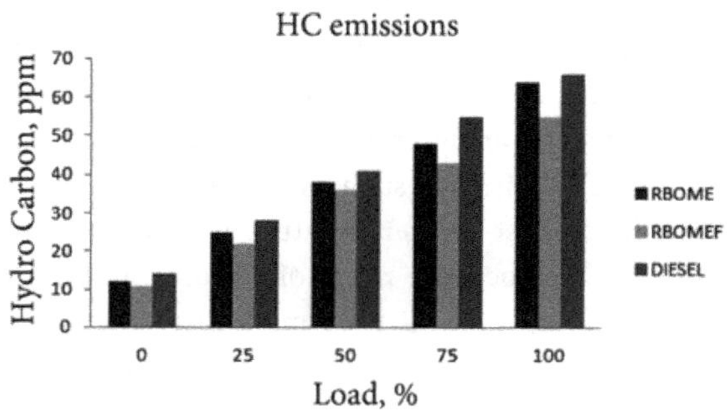

Fig. 2. Hydro carbon emissions with respect to load

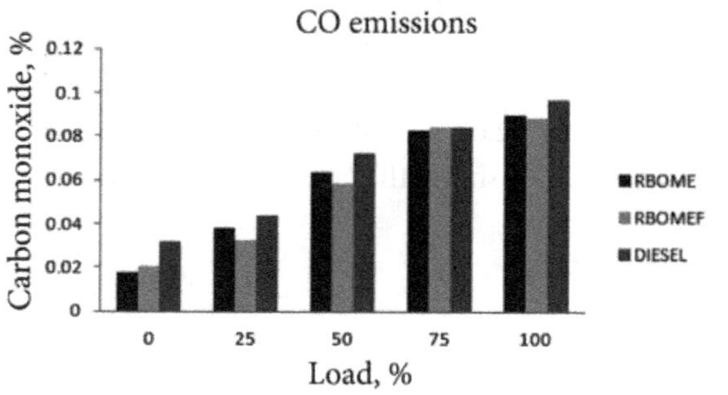

Fig. 3. Carbon monoxide emission with respect to load

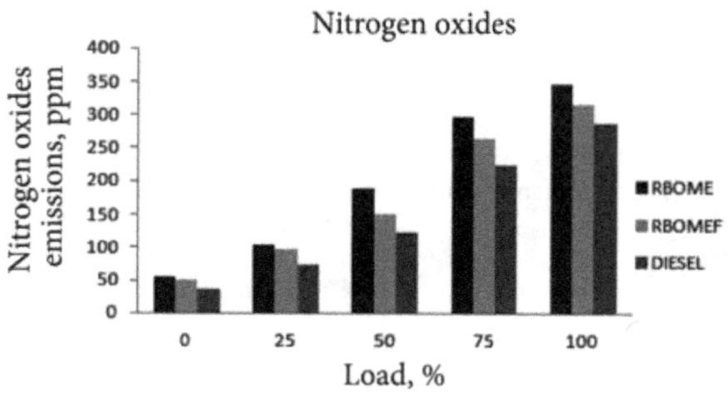

Fig. 4. NOx emission with respect to load

RBOME the mixture momentum and penetration depth in cylinder is controlled and result in shorter ignition delay and enhanced combustion. Furthermore, the fuel with higher density as in the case of RBOMEF restrains spray tip penetration, ensuing that the air fuel mixture is only created at the core region of the combustion chamber owing to improved combustion (Raheman, Phadatare 2004). The other reason is due to higher availability of oxygen which enhances air fuel mixing and increase efficiency Shafii *et al.* (2011)

2.2.2. Specific fuel consumption

SFC of RBOME, RBOMEF is shown in Figure 6. SFC for diesel is least during the trail. This is because of higher calorific value of diesel. Further, addition of ferrofluid decreases SFC. Average BSFC for RBOMEF is lesser than

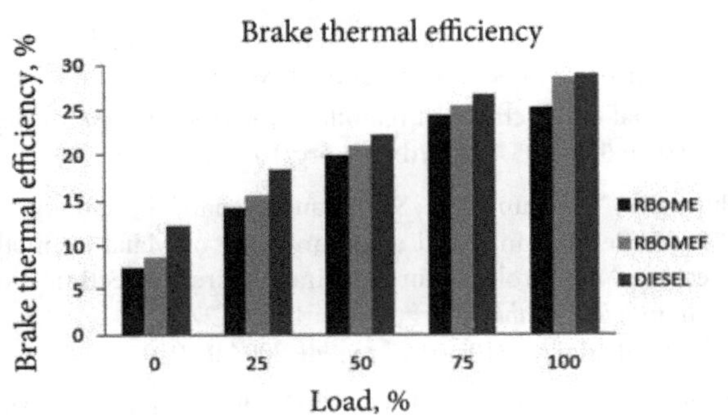

Fig. 5. Brake thermal efficiency with respect to load

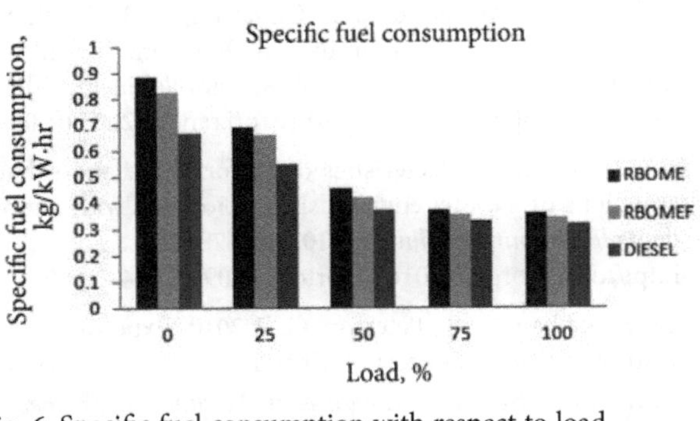

Fig. 6. Specific fuel consumption with respect to load

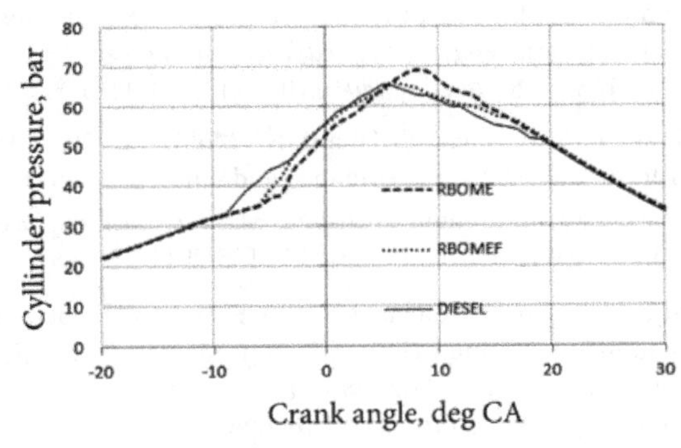

Fig. 7. Variation of pressure with crank angle

RBOME by 5.17%.This is due to positive effect of ferrofluid which reduces ignition delay thereby reducing fuel consumption. The other reason is due to surplus availability of oxygen in RBOMEF which promotes improved combustion and lesser consumption of fuel. Furthermore, nanosized particles can be dispersed into high-temperature zones for direct oxidation reaction leading shorter delay period thereby resulting in lesser consumption of fuel (Shafii *et al.* 2011).

2.2.3. Pressure vs crank angle

Figure 7 shows the variation in pressure with crank angle for RBOME, RBOMEF and Diesel at full load conditions. Peak pressure of RBOME is higher than RBOMEF and Diesel by 4.18% and 5.34%.Since the viscosity of RBOME is higher by 5.94% comparing RBOMEF atomization and mixing of fuel with air is not in uniform rate causing longer breakup length, lower dispersion rate and increased ignition delay. Due to longer delay period more measure of fuel is accumulated causing rapid increase in peak pressure. Similar results were cited on the experimental work by (Buyukkaya 2010).

It is also inferred that the peak pressure of Diesel is lesser than RBOME and RBOMEF. This is due to reduced fuel consumption of diesel in comparison with RBOME and RBOMESUM ensuring improved atomization and better combustion which result in shorter ignition delay and lower peak pressure. It is also inferred that by addition of ferrofluid the peak cylinder pressure is reduced by 3.35%. This is due to shorter delay period and improved ignition properties of nanoparticles which initiate the early combustion when compared to RBOME as cited by (Basha, Anand 2011). The other reason is as a result of higher cetane number of RBOMEF which initiates faster evaporation and improved combustion leading to shorter ignition delay. It is also observed that the occurrence of peak pressure is advanced with addition of ferrofluid due to shorter ignition delay. This is due to surplus oxygen in ferrofluid which promotes improved combustion. This is in agreement with similar work done by (Buyukkaya 2010).

Conclusion

The following conclusions were drawn by adding magnetite to methyl esters of rice bran oil in a constant speed diesel engine.

(1) 6.34% of CO emission is reduced by adding ferrofluid to methyl ester due to magnetite which undergoes catalytic oxidation reaction and causing complete oxidation during combustion.

(2) 9.1% of HC emission is reduced with addition of magnetite. HC oxidation reaction is improved due to the presence of ferrofluid which acts as an oxidation catalyst and lowers the carbon combustion activation temperature.

(3) 8.49% of NOX emission is achieved by doping magnetite to methyl ester. This is achieved as a result of latent heat of evaporation of water present in ferrofluid.

(4) Engine performance is enhanced by 4.27% along with 5.17% reduction in BSFC. This is due to higher availability of oxygen which enhances air fuel mixing during combustion. Ferro fluid also possesses desirable combustion characteristics such as high heats of combustion and fast energy release rates.

Adding magnetite to methyl esters of rice bran oil is promising technique to offset the drawbacks associated with it.

References

Alam, M.; Song, J.; Acharya, R.; Boehman, A.; Miller, K. 2004. Combustion and emissions performance of low sulfur, ultra low sulfur and biodiesel blends in a DI diesel engine, *SAE Technical Paper Series*. http://dx.doi.org/10.4271/2004-01-3024

Arul Mozhi Selvan, V.; Anand, R. B.; Udayakumar, M. 2014. Effect of cerium oxide nanoparticles and carbon nanotubes as fuel-borne additives in Diesterol blends on the performance, combustion and emission characteristics of a variable compression ratio engine, *Fuel* 130: 160–167. http://dx.doi.org/10.1016/j.fuel.2014.04.034

Aydin, H.; Bayindir, H. 2010. Performance and emission analysis of cottonseed oil methyl ester in a diesel engine, *Renewable Energy* 35(3): 588–592. http://dx.doi.org/10.1016/j.renene.2009.08.009

Basha, J. S.; Anand, R. B. 2011. An experimental study in a CI engine using nanoadditive blended water-diesel emulsion fuel, *International Journal of Green Energy* 8(3): 332–348. http://dx.doi.org/10.1080/15435075.2011.557844

Berger, P.; Adelman, N. B.; Beckman, K. J.; Campbell, D. J.; Ellis, A. B.; Lisensky, G. C. 1999. Preparation and properties of an aqueous ferrofluid, *Journal of Chemical Education* 76(7): 943–948. http://dx.doi.org/10.1021/ed076p943

Buyukkaya, E. 2010. Effects of biodiesel on a DI diesel engine performance, emission and combustion characteristics, *Fuel* 89(10): 3099–3105. http://dx.doi.org/10.1016/j.fuel.2010.05.034

Choi, S. H.; Oh, Y. 2006. The emission effects by the use of biodiesel fuel, *International Journal of Modern Physics B* 20: 4481–4486. http://dx.doi.org/10.1142/s0217979206041550

De Luca, L. T.; Galfetti, L.; Severini, F. 2005. Burning of nano-aluminized composite rocket propellants, Combustion, *Explosion and Shock Waves* 41(6): 680–692. http://dx.doi.org/10.1007/s10573-005-0080-5

Galfetti, L.; DeLuca, L. T.; Severini, F.; Meda, L.; Marra, G. 2007. Pre and post- burning analysis of nano-aluminized solid rocket propellants, *Aerospace Science and Technology* 11(1): 26–32. http://dx.doi.org/10.1016/j.ast.2006.08.005

Jung, H.; Kittelson, D. B.; Zachariah, M. R. 2005. The influence of a cerium additive on ultrafine diesel particulate emissions and kinetics of oxidation, *Combustion and Flame* 142(3): 276–278. http://dx.doi.org/10.1016/j.combustflame.2004.11.015

Kannan, G. R.; Karvembu, R.; Anand, R. 2011. Effect of metal based additive on performance emission and combustion characteristics of diesel engine fuelled with biodiesel, *Applied energy* 88(11): 3694–3703. http://dx.doi.org/10.1016/j.apenergy.2011.04.043

Karabektas, M. 2009. The effects of turbocharger on the performance and exhaust emissions of a diesel engine fuelled with biodiesel, *Renewable Energy* 34(4): 989–993. http://dx.doi.org/10.1016/j.renene.2008.08.010

Kim, H.; Choi, B. 2010. 2010. The effect of biodiesel and bioethanol blended diesel fuel on nanoparticles and exhaust emissions from CRDI diesel engine, *Renewable Energy* 35(1): 157–163. http://dx.doi.org/10.1016/j.renene.2009.04.008

Krishnamurthy, S.; Bhattacharya, P.; Phelan, P. E.; Prasher, R. S. 2006. Enhanced mass transport in nanofluids, *Nano Letters* 6(3): 419–423. http://dx.doi.org/10.1021/nl0522532

Lenin, M. A.; Swaminathan, M. R.; Kumaresan, G. 2013. Performance and emission characteristics of a DI diesel engine with a nanofuel additive, *Fuel* 109: 362–365. http://dx.doi.org/10.1016/j.fuel.2013.03.042

Prasher, R.; Bhattacharya, P.; Phelan, P. E. 2006. Brownian-motion-based convective-conductive model for the effective thermal conductivity of nanofluids, *Journal of Heat Transfer* 128(6): 588–595. http://dx.doi.org/10.1115/1.2188509

Puhan, S.; Vedaraman, N.; Sankaranarayanan, G.; Ram, B. V. B. 2005. Performance and emission study of Maua oil methyl ester in a 4-stroke natural aspirated direct injection diesel engine, *Renewable Energy* 30(8): 1269–1278. http://dx.doi.org/10.1016/j.renene.2004.09.010

Raheman, H.; Phadatare, A. G. 2004. Diesel engine emissions and performance from blends of karanja methyl ester and diesel, *Biomass and Bioenergy* 27(4): 393–397. http://dx.doi.org/10.1016/j.biombioe.2004.03.002

Ramadhas, A. S.; Muraleedharan, C.; Jayaraj, S. 2005. Performance, emission evaluation of a diesel engine fuelled with methyl esters of rubber seed oil, *Renewable Energy* 30(12): 1789–1800. http://dx.doi.org/10.1016/j.renene.2005.01.009

Ryu, K. 2010. The characteristics of performance and exhaust emissions of a diesel engine using a biodiesel with antioxidants, *Bioresource Technology* 101(1): S78–82. http://dx.doi.org/10.1016/j.biortech.2009.05.034

Sajith, V.; Sobhan, C. B.; Peterson, G. P. 2010. Experimental investigations on the effects of cerium oxide nanoparticle fuel additives on biodiesel, *Advances in Mechanical Engineering* 2(0): 581407. http://dx.doi.org/10.1155/2010/581407

Selim, M. Y. E.; Elfeky, S. M. S. 2001. Effects of diesel/water emulsion on heat flow and thermal loading in a pre combustion chamber diesel engine, *Applied Thermal Engineering* 21(15): 1565–1582. http://dx.doi.org/10.1016/s1359-4311(01)00019-9

Shafii, M. B.; Daneshvar, F.; Jahani, N.; Mobini, K. 2011. Effect of ferrofluid on the performance and emission patterns of a four stroke diesel engine, *Advances in Mechanical Engineering* 3(0): 529049. http://dx.doi.org/10.1155/2011/529049

Ulusoy, Y.; Arslan, R.; Kaplan, C. 2009. Emission characteristics of sunflower oil methyl ester, *Energy Sources, Part A Recovery, Utilization, and Environmental Effects* 31(11): 906–910. http://dx.doi.org/10.1080/15567030802087528

APPENDIX

Sample calculation to find Uncertainty for HC emission

The five measurements of HC for RBOME are 25 ppm, 38 ppm, 48 ppm and 64 ppm.
The standard deviation for the measurements is

$$\sigma_{HC} = 2.75 .$$

The uncertainty in HC is given by

$$\Delta\,(HC) = \frac{2.75}{\sqrt{5}} = 1.223 \text{ ppm} .$$

$$\text{Relative uncertainty} = \frac{1.223}{64} \times 100\% = 1.91 \approx 2\% .$$

In the same procedure, uncertainty of NO_X and CO are calculated.

YUVARAJAN D received his B.E degree in Mechanical engineering from Anna University, India in 2006 and M.E degree in Energy Engineering from College of Engineering, Anna University, India in 2009. He is currently an assistant professor in Jeppiaar Institute of Technology, India and correspondingly perusing his Ph.D. degree in Anna University, India. His research area includes performance, combustion and emissions characteristics in CI engines and fuel additives.

VENKATA RAMANAN M received his B.E degree in Mechanical engineering from University of Madras, India in 1998 and M. Tech degree in Energy conservation and Management from School of Energy, Tiruchirappalli, India in 2000. He received his Ph.D. in Biomass Gasification at Anna University, India in 2008. He is currently an Assistant Professor (Senior Grade) in Department of Mechanical Engineering, Anna University, India. He is an energy auditor and has consulted many industrial houses for energy conservation studies. His research area includes IC engines, gasification and energy optimization in process utilities. He has published more than 30 international journal papers on his area of research.

17

COPPER REMOVAL USING ACID ACTIVATED PEANUT HUSK FROM AQUEOUS SOLUTION

Pradnya K. INGLE, Attarkar KARISHMA, Virendra K. RATHOD

Department of Chemical Engineering,
Institute of Chemical Technology, Matunga (E), Mumbai-400019, India

Abstract. Peanut husk, an agricultural waste is a precursor biomass veteran for chemical activation using phosphoric acid. The work deals with activation of peanut husk using phosphoric acid followed by its application. The effect of various parameters such as impregnation ratio, temperature, particle size and agitation speed on copper adsorption were also studied. Physical characterization of the activated adsorbent was done using SEM, BET and FTIR to understand the modifications. The optimum parameter values of chemical activation influencing adsorption were 20% phosphoric acid concentration, 1:8 acid impregnation ratio, 60 °C activation temperature and 350 rpm stirring speed. The copper uptake was 14.3 mg/g for peanut husk whereas by chemical activation it was increased to 17.88 mg/g. Thus adsorption capacity of peanut husk can be successfully improved using phosphoric acid activation.

Keywords: peanut husk, phosphoric acid, chemical activation, bio adsorbent, copper adsorption, adsorption capacity.

Introduction

Due to rapid industrialization and urbanization heavy metals have entered into the ecological system in the form of water pollution (Sari, Tuzen 2008; Uluozlu *et al.* 2008; Donghee *et al.* 2010). These heavy metals get accumulated in living organisms, consequently leading to environmental hazards (Loutseti *et al.* 2009; Batley *et al.* 2004). Copper, one of the metals finds varied application in diverse fields such as electroplating, etching, rinsing, chemical and mechanical polishing, etc. which lead to accumulation of gallons of wastewater contributing to heavy metal pollution (Ingle *et al.* 2014). Moreover, copper being a micronutrient if consumed in the form of inhalation in higher dosage leads to be fatal, hence there is a need to remove copper before it is discharged (Ali *et al.* 2013). The heavy metals can be removed from the waste water by ion exchange (Gaikwad *et al.* 2010), sedimentation (Song *et al.* 2000), biological operations (Wang 2002), membrane processes (Valenzuela *et al.* 2005), chemical precipitation (Barakat 2011). But at the same time these conventional methods have drawbacks of high capital and operational cost (Sharma, Forster 1994). Moreover, these techniques fail to remove heavy metals when present at low concentration. From past few years researchers are interested in substituting costly methods of water treatment by low cost and agricultural by-product adsorbent (Namasivayam, Ranganathan 1995). Adsorption process has been proven one of the best water treatment technologies around the world and activated adsorbent is undoubtedly considered as a better adsorbent for the removal of diverse types of pollutants from water (Bhatnagar, Sillanpaa 2010). Adsorption using biological waste is an effective technique for the removal of heavy metals and the process is termed as bio sorption. This effective technique depends on the parameters like capacity, affinity, specificity of bio sorbent and the conditions in effluents. Bio sorption can be used for the treatment of wastewater containing low heavy metal content and found to be an inexpensive, simple and effective alternative to conventional methods. Bio sorption is the capability of active sites on the surface of biomaterials to bind and concentrate heavy metals from even the most dilute aqueous solutions (Wang, Chen 2006). Minimization of chemical sludge, recovery of the adsorbents and possibility of metal recovery are added advantages of these potential adsorbents. Among the agricultural waste reported, peanut husk is considered as one of the best precursors for copper removal as it is a cheaper, abundantly available. Although the bio sorbents are cheaper in cost, the low adsorption capacity is the major problem which can be improved by chemical activation. It has been reported that there is an increase in the adsorption of metal

Corresponding author: Virendra K. Rathod
E-mail: vk.rathod@ictmumbai.edu.in

ions after chemical activation of peanut hull (Wafwayo *et al.* 1999). Similar kind of behavior of higher adsorption of copper was obtained by other researchers (Charmarthy *et al.* 2001)

Activation of biosorbent consists of multiplying amount of pores of certain carbonaceous material to produce extremely porous structure. Activation can be performed by physical or/and chemical means. Physical activation involves carbonization of carbonaceous materials at high temperatures (500–900 °C) in an inert atmosphere followed by activation in the presence of activating agents such as CO_2 or steam resulting in activated carbon. In chemical activation, the precursor is impregnated with a chemical activator such as $ZnCl_2$, H_3PO_4, KOH, H_2SO_4 etc. This activated material can be used directly or the impregnated precursor is further heated in an inert atmosphere to convert it in activated carbon. Chemical activation is preferred over physical activation as it is carried at lower temperature with less activation time. Moreover, it develops better porous structure and also gives high adsorption capacity (Guo, Rockstraw 2007). It is revealed from the work of other researchers that phosphoric acid has the capability of creating physical and chemical modifications on the botanical structure by penetration, particle swelling, and partial dissolution of the biomass, bond cleavage and reformation of new polymeric structures resistant to thermal decomposition (Girgis *et al.* 2002).

The present work is focused on the preparation of activated adsorbent using phosphoric acid as an activating agent. The effect of various parameters affecting activation such as concentration of acid, impregnation ratio, temperature, particle size and stirring speed is studied. In order to understand the changes induced by the activation as well as to characterize the adsorbent, the acid activated materials before and after adsorption are analyzed and characterized using Scanning Electron Microscope (SEM, Jeol) and Fourier Trans-form Infrared Spectroscopy (FTIR, Bruker) and Specific surface area and porosity analyzer (BET, PMI Incorp. Ltd.).

1. Materials and methods

1.1. Materials

Peanut husk (APMC Market Vashi, Navi Mumbai) were washed extensively using tap water for 1–2 h to remove any mud and soil particles adhered to it. Further it was washed several times with deionized water and then dried in an oven at 80 °C for 48 h. The dried husk was grounded in a fine powder, sieved by ASTM standard sieve and stored in an airtight container for further use. Phosphoric acid was obtained from S.D. Fine Chem. Ltd., Mumbai, India. All the chemicals used were analytical grade and were used as received from the suppliers. The deionized water for the experiments was obtained from Sartorious stedimarium Water Purifier Unit.

1.2. Chemical activation of peanut husk

The chemical activation was performed in a 100 ml glass reactor provided with stirrer. Experiments were carried out in a reactor using 15 g of peanut husk and correspondingly phosphoric acid. The acid-adsorbent mixture with various impregnation ratios (1:4–1:12) was agitated at different stirring speeds (100–650 rpm) using mechanical stirrer within the temperature range 20–80 °C. The temperature is maintained throughout the experiment using a water bath. After predefined time, the activated samples were filtered and the pH was adjusted to 7 with 0.1 M NaOH and 0.1 M HCl using pH meter (Labtronic digital pH meter Model No. LT-10). Then the adsorbent was dried at 100 °C in the oven.

1.3. Copper adsorption experiment

In order to study the effect of activation on copper adsorption, 100 mL metal ion solution prepared in the laboratory of 100 mg/L copper concentration with initial pH 4.8 is equilibrated with 0.5 g of chemically activated adsorbent. Sorption process is carried out in a conical flask placed in Remi Orbital incubator shaker at room temperature (30±2 °C) for 12 h till equilibrium was attained. After equilibrium, 1 mL of sample was withdrawn to determine the metal ion concentration. The metal ion concentration was measured by using standard method for copper ion determination with PAN indicator (1-(2-Pyridylazo)-2-naphthol) using a UV-VIS Spectrophotometer. This method was used for all the experiments to determine the effect of chemical activation on adsorption capacity of the adsorbent. The metal uptake at equilibrium was evaluated from the given Eq. (1):

$$Metal\,uptake(q) = \frac{(C_i - C_e)}{M} \cdot V, \qquad (1)$$

where C_i is the initial metal ion concentration before sorption process in mg/L; C_e is the equilibrium metal ion concentration after sorption process in mg/L; V is the volume of metal ion solution in mL; M is the mass of bio sorbent taken in g.

1. 4. Characterization of acid treated peanut husk

In order to analyze the surface area and porosity change after phosphoric acid treatment the standard method of BET analysis (BET, PMI Incorporated Ltd.) was carried out. The surface area is evaluated by nitrogen multi-layer adsorption which is carried out at 77 K, measured as a function of relative pressure. Scanning Electron Microscope (SEM, Jeol) and Fourier Transform Infrared Spectroscopy (FTIR, Bruker) were done to examine the changes in the structure of acid activated peanut husk before and after adsorption.

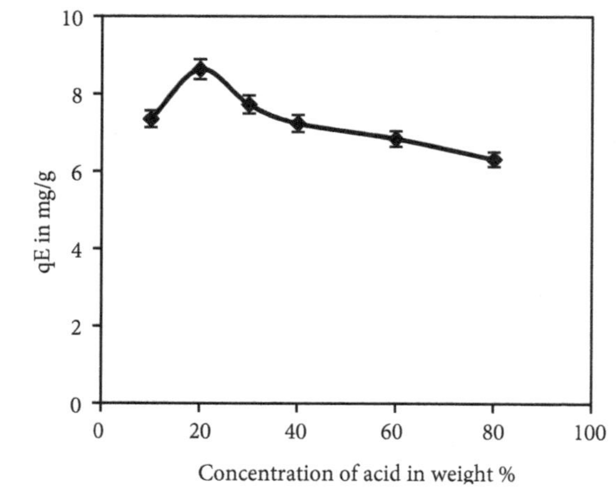

Fig. 1. Effect of acid concentration in weight % at 30 °C, impregnation ration 1:12 and speed of stirring 150 rpm

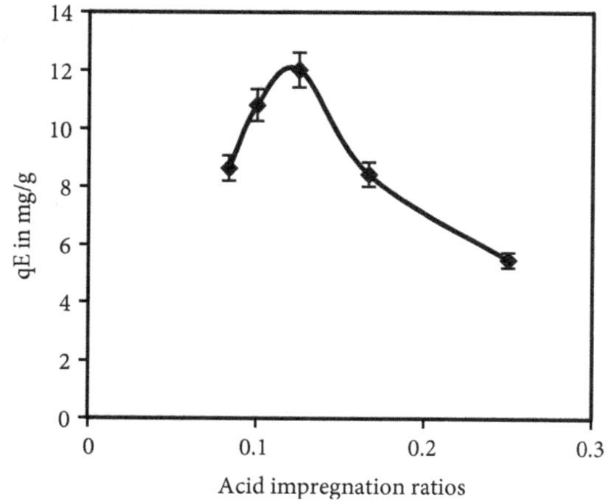

Fig. 2. Effect of acid impregnation ratios at 20% by weight acid concentration at 30 °C and speed of stirring 150 rpm

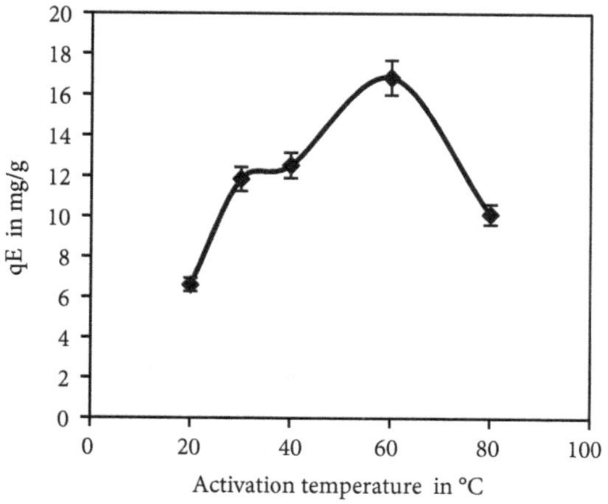

Fig. 3. Effect of activation temperature at 20% by weight acid concentration, impregnation ratio 1:8 and speed of stirring 150 rpm

2. Results and discussions

2.1. Effect of phosphoric acid concentration

The surface characteristics and porosity of the bio sorbent are mainly decided by the concentration of acid used for activation. The effect of acid concentration was studied by varying the concentration from 10 to 80%. The experiments were carried out at 30 °C with 1:12 impregnation ratio and stirring speed 150 rpm. It is known that phosphoric acid swells the botanical material and modifies the surface characteristics. The adsorption of copper after activation has increased at lower concentration of acid till 20% and then with further increase in concentration of acid the copper uptake has decreased considerably (Fig. 1). This may be due to the structural abrasions on the surface of the peanut husk. The copper ion uptake was 8.68 mg/g at 20% (weight) of phosphoric acid concentration. Similar trend is observed by (Rahman *et al.* 2005), where 20% (weight) of phosphoric acid concentration was used for activation of rice husk for the removal of malachite green dye.

2.2. Effect of acid impregnation ratios

Acid impregnation ratios are defined as weight of adsorbent to the weight of acid. This parameter decides the quantity of phosphoric acid required for activation. The acid impregnation ratios was studied using different ratio in the range of 1:4–1:12 by keeping other parameters constant at acid concentration 20%, temperature 30 °C and speed of stirring 150 rpm. With an increase in impregnation ratios the copper uptake increases till 1:8 impregnation ratios (Fig. 2).

This may be due to increase in size of active site of peanut husk with the interaction with phosphoric acid. The impregnation of acid might work through alteration in the surface functional group chemically rather than physical interaction (Kim *et al.* 2006). The copper uptake has increased up to the maximum value of 12.01 mg/g at optimum impregnation ratio 1:8 due to pore enlargement and surface modification. The pores may get ruined by the presence of excess acid as the impregnation ratios is decreased beyond the optimum value. So for the further studies the acid impregnation ratios is kept constant at 1:8.

2.3. Effect of temperature on chemical activation

In order to study the effect of temperature, the activation of peanut husk was carried out within the temperature range 20 to 80 °C. The other parameters like impregnation ratio 1:8, acid concentration 20% and the stirring speed of 150 rpm was kept constant throughout the study. It is observed that the copper uptake after activation of peanut husk was increased with elevation in temperature from 20 to 60 °C (Fig. 3). This is because an increase in temperature initiates the rate of diffusion of phosphoric acid into the pores of peanut husk. The diffusion results in the conversion of micro pores to mesopores and macro pores. The copper uptake after activation increased from 6.6 mg/g at 20 °C to 16.8 mg/g at 60 °C. The reduction in the adsorption capacity of copper beyond 60 °C may be consequence of increase in temperature, resulting into the pore destruction. The adsorption of copper using peanut husk as a precursor was also carried out and the copper uptake was found to be 9.6 mg/g whereas the phosphoric acid modified peanut husk has proved better as the copper uptake was 16.8 mg/g which is nearly doubled.

2.4. Effect of stirring speed on chemical activation

The stirring speed is an important parameter in the activation process, as it dictates the extent of mass transfer. The experiments were carried out at varied speed from 100--850 rpm. The experimental parameters of acid concentration, impregnation ratio, activation temperature were maintained at optimum values. It is seen that the copper uptake increased from 10.8 mg/g to 17.88 mg/g as the stirring speed is increased from 100 rpm to 350 rpm (Fig. 4). This is in relation with the fact that an increase in stirring speed reduces the film resistance created by the boundary layer surrounding the adsorbent particles (Aman *et al.* 2008). Due to low film resistance the phosphoric acid diffuses within the peanut husk, thereby modifying the pore structure. Thus increase in stirring speed increases the external film mass transfer coefficient leading to higher copper uptake. But beyond 350 rpm the surplus stirring may damage the pores resulting in decrease in copper uptake.

2.5. Effect of particle size on chemical activation

The particle size determines the time required by the phosphoric acid for transport within the pores which results in the formation of mesopores and micro pores (Ricordel *et al.* 2001). The particle size was varied from 150 microns to 720 microns to analyze the effect of particle size on chemical activation. The other experimental parameters were upheld at acid concentration 20%, impregnation ratio 1:8, activation temperature 60 °C and stirring speed 350 rpm. It is clear that with an increase in particle size the copper uptake decreases (Fig. 5).

The surface area of the particle increases with decrease in particle size. The phosphoric acid diffuses in a shorter period of time into the peanut husk which has smaller particle size as there is direct relation between equilibrium time and particle size. The adsorption of copper using peanut husk of 500 microns was carried out, which showed that copper uptake 9.6 mg/g whereas the phosphoric acid modified peanut husk has given copper uptake 12.3 mg/g.

2.6. Characterization of chemically activated peanut husk

To study the effect of chemical activation on the surface morphology and porosity, SEM analysis was carried out for the untreated peanut husk and the chemically activated peanut husk. In SEM micrograph of the untreated peanut husk there are some flaky structures and rudimentary pores while SEM micrograph of the chemically treated peanut husk has wide pores as a result of the treatment with phosphoric acid which is corrosive and leads to surface leaching (Fig. 6). This type of activation is dependent on the material under study and peanut husk is pectin rich, which undergoes swelling, during which the arrangement of the molecules in the direction of longitudinal axis is unaltered but the lateral bonds undergo drastic modification in terms of void formations likely due to the chemical activity of the acid (Smisek, Cerny 1970). Thus the SEM image clearly shows that the activation has played a major role in the enhancement of surface adsorptive capacity of the peanut husk. The SEM micrograph of the chemically activated peanut husk after the copper adsorption process

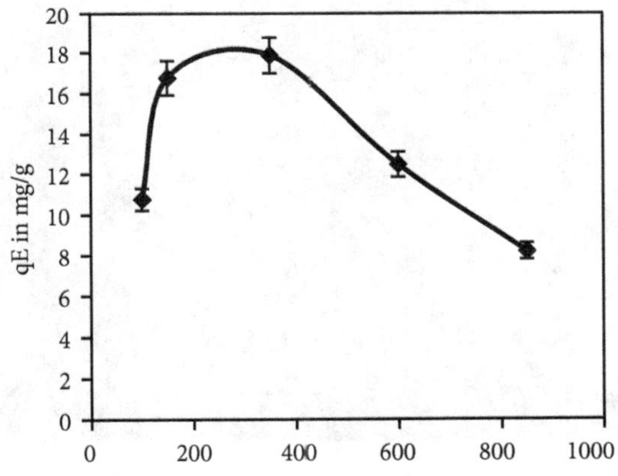

Fig. 4. Effect of stirring speed at 20% by weight acid concentration impregnation ratio 1:8 and temperature 60 °C

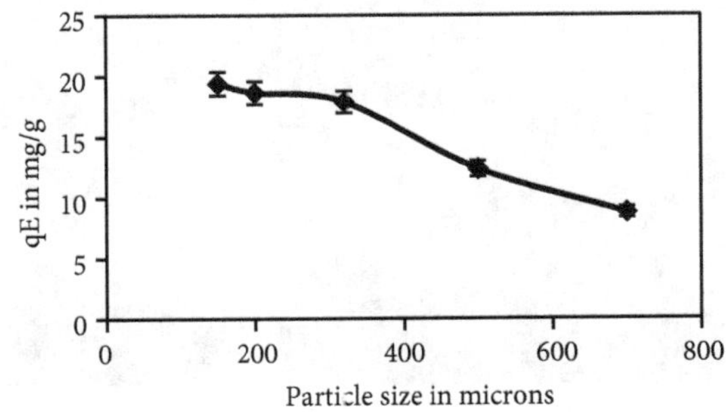

Fig. 5. Effect of particle size at 20% by weight acid concentration, impregnation ratio 1:8, temperature 60 °C and stirring speed 350 rpm

is shown in (Fig. 6c) which shows decreased void space and the surface charges as compared with SEM images (Fig. 6a and Fig. 6b) indicating the deposition of copper on the voids.

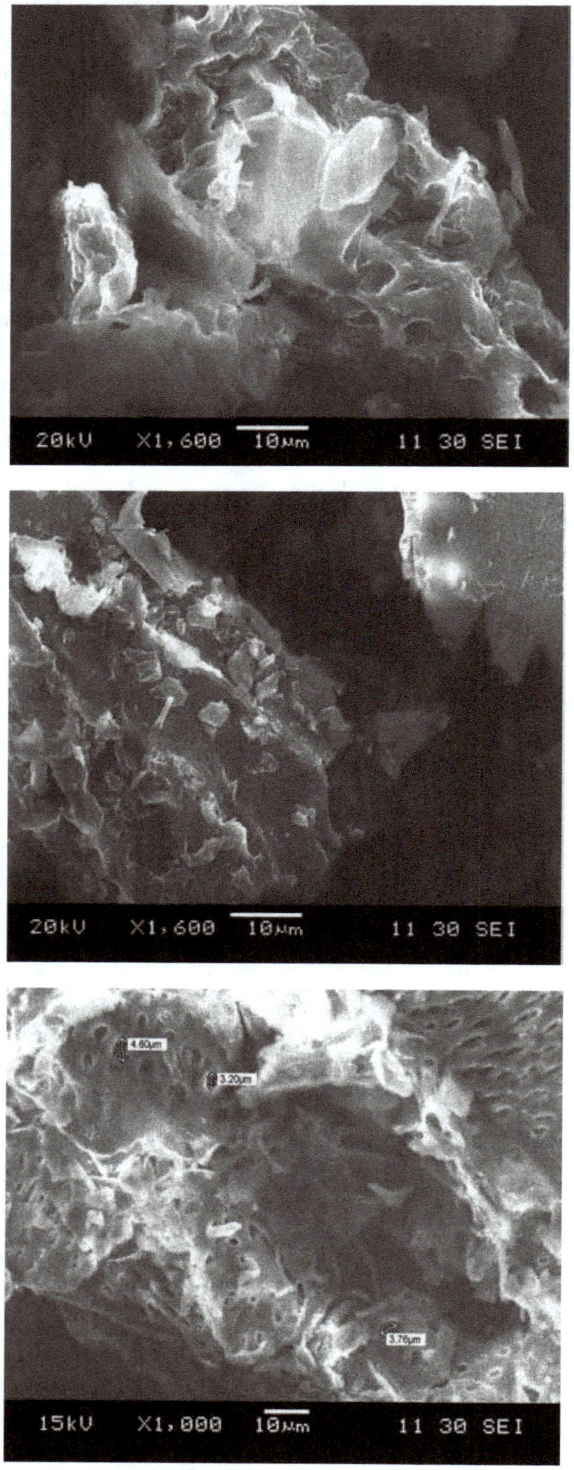

Fig. 6. SEM micrographs of untreated peanut husk, chemically activated peanut husk; and chemically activated peanut husk after adsorption: a) peanut husk; b) chemically activated peanut husk c) chemically activated peanut husk after adsorption

2.7. Fourier transform infrared spectroscopy

Infrared spectroscopy provides information on the chemical structure of the adsorbent material. Figure 7 shows FTIR spectra of the peanut husk and the acid modified peanut husk. The most of characteristic changes are observed with the range of 1700–1730 cm^{-1}. The band centred at 1731 cm^{-1} is ascribed to the stretching vibrations of carboxyl groups on the edges of layer planes or to conjugated carbonyl groups (C-O in carboxylic acid and lactones groups). The spectrum of the acid treated adsorbent shows a significant change in this band showing the effect of phosphoric acid on the husk. The broad band between 1300 and 900 cm^{-1} in acid treated peanut husk has a maximum at 1070–1080 cm^{-1}.

The absorption in this region is the characteristic for phosphorus and phosphor carbonaceous compounds. The peak at 1057–1033 cm^{-1} is ascribed to ionized linkage $P^{+}O^{-}$ in acid phosphate esters and to symmetrical vibration in a chain of P-O-P (polyphosphate). The spectra therefore suggest the formation of P-containing carbonaceous structures like acid phosphates and polyphosphates in phosphoric acid-activated peanut husk.

2.8. BET surface areas and pore structure

The BET surface areas and porous structures of samples were investigated based on nitrogen adsorption-desorption.

The BET specific surface area (SBET) of untreated peanut husk and acid-activated peanut husk are 18.088 and 31.140 m^2g^{-1}, respectively showing that there is a large increment in the surface area due to the acid treatment.

Figure 8 shows nitrogen adsorption-desorption isotherms for the two samples. The isotherms of all samples show type III isotherms, indicating the presence of macropores throughout the material. This type of isotherm is a characteristic of weak gas-solid interactions. This weakness causes uptake at low pressure to be small which is quite evident from the figure. But once the molecule is adsorbed, the adsorbate – adsorbate interactions will further promote the adsorption process. After the acid treatment

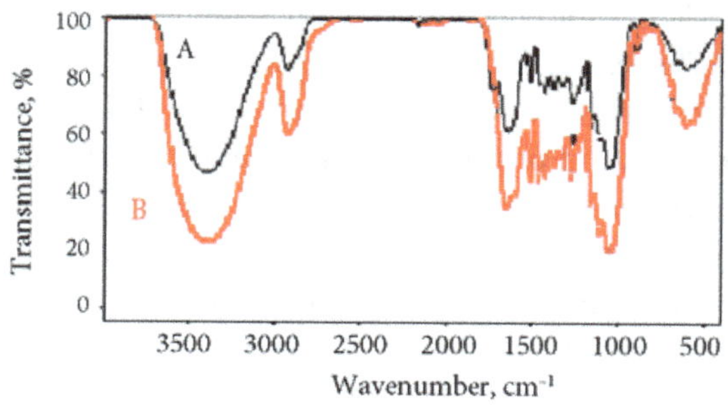

Fig. 7. FTIR of untreated peanut husk (A) and chemically treated peanut husk (B)

Table 1. Comparison of activating agents used for chemical activation of various adsorbent (Low *et al.* 1995; Acar, Eren 2006; Ngah, Hanafiah 2008; Baccara *et al.* 2009; Dubey, Gopal 2007; Li *et al.* 2007; Shukla, Pai 2005)

Sr. no	Agricultural waste as precursor	Chemical agent	Metal ions	Metal uptake
1	Banana pith Nitric acid	HNO_3	Cu^{2+}	13.46
2	Sawdust (Poplar tree)	H_2SO_4	Cu^{2+}	13.95
3	Sawdust (Poplar tree)	NaOH	Cu^{2+} Zn^{2+}	6.95 15.8
4	Carbon from Tunisian-olive waste	H_3PO_4	Cu^{2+}	35.5
5	Sawdust	Reactive Orange 13	Cu^{2+} Ni^{2+} Zn^{2+}	8.07 9.87 17.09
6	Groundnut husk	H_2SO_4 followed by silver impregnation	Cr^{6+}	11.4
7	Peanut husk	H_2SO_4	Cu^{2+}	10.15
8	Groundnut shells	Reactive Orange 13	Cu^{2+}	7.60
9	Jute fibers	Reactive Orange 13	Cu^{2+} Ni^{2+}	8.4 5.26
10	Peanut husk (present study)	H_3PO_4	Cu^{2+}	17.88

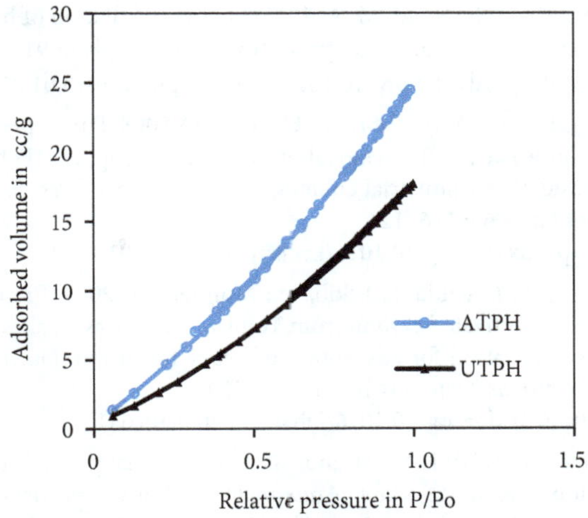

Fig. 8. N_2 adsorption – desorption isotherm of Acid treated peanut husk (ATPH) and Untreated peanut husk (UTPH)

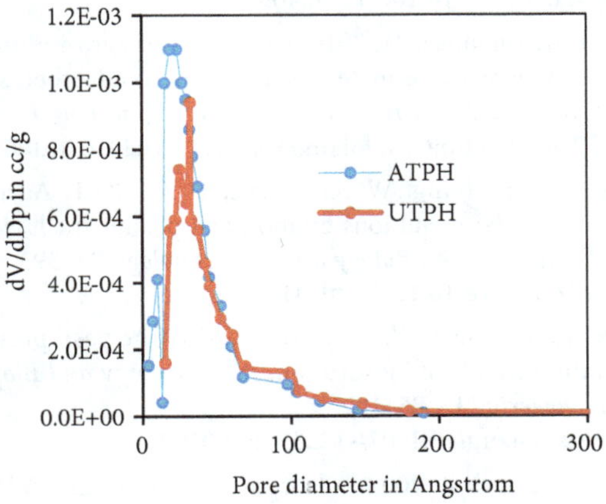

Fig. 9. Pore diameter distribution of acid treated peanut husk (ATPH) and untreated peanut husk (UTPH)

there was an increase in the surface leading to higher adsorption at higher relative pressures.

Also, from Figure 9, it can be concluded that the pore size distribution is quite broad in both the materials and the pore volume increased from 0.0214 cc/g to 0.03548 cc/g after the acid treatment. Thus, this type of results confirms the presence of mesoporous and macroporous nature of the adsorbent material.

Also, Table 1 depicts the various chemical agents used for activation and their adsorption capacities. From the table it can be concluded that the phosphoric acid is found to be the excellent chemical activating agent for enhancing the adsorption capacity.

Conclusion

1. The chemical activation of peanut husk by phosphoric acid has enhanced the copper adsorption capacity of peanut husk as compared to untreated peanut husk. Thus the activation of ligno-cellulosic material by phosphoric acid has improved the surface characteristics.
2. The optimum conditions obtained from the study of chemical activation are 20% phosphoric acid concentration, 1:8 acid impregnation ratio, 60 °C activation temperature and 350 rpm stirring speed.
3. This enhancement may be due to the surface modifications and pore widening as seen in SEM micrographs and FTIR images.
4. The results indicate that the chemical activation can be used as a technique to enhance the adsorption capacity of bio adsorbents.

Acknowledgement

The support by the UGC Networking Programme is gratefully acknowledged. Also, the analytical instrumental support by TEQIP and DST-FIST is highly acknowledged.

References

Acar, F. N.; Eren, Z. 2006. Removal of Cu (II) ions by activated poplar sawdust (Samsun clone) from aqueous solutions, *Journal of Hazardous Materials* 137: 909–914. http://dx.doi.org/10.1016/j.jhazmat.2006.03.014

Ali, H.; Khan, E.; Sajad, M. A. 2013. Phyto remediation of heavy metals - concepts and applications, *Chemosphere* 91: 869–881. http://dx.doi.org/10.1016/j.chemosphere.2013.01.075

Aman, T.; Kazi, A. A.; Sabri, M. U.; Bano, Q. 2008. Potato peels as solid waste for the removal of heavy metal copper (II) from waste water/industrial effluent, *Colloids and Surfaces. B, Biointerfaces* 63: 116–121. http://dx.doi.org/10.1016/j.colsurfb.2007.11.013

Baccara, R.; Bouzida, J.; Fekib, M.; Montiela, A. 2009. Preparation of activated carbon from Tunisian olive-waste cakes and its application for adsorption of heavy metal ion, *Journal of Hazardous Materials* 162: 1522–1529. http://dx.doi.org/10.1016/j.jhazmat.2008.06.041

Barakat, M. A. 2011. New trends in removing heavy metals from industrial waste water, *Arabian Journal of Chemistry* 4(4): 361–377. http://dx.doi.org/10.1016/j.arabjc.2010.07.019

Batley, G. E.; Apte, S. C.; Stauber, J. L. 2004. Speciation and bioavailability of trace metals in water: Progress since 1982, *Australian Journal of Chemistry* 57(10): 903–919. http://dx.doi.org/10.1071/CH04095

Bhatnagar, A; Sillanpaa, M. 2010. Utilization of agro-industrial and municipal waste materials as potential adsorbents for water treatment - a review, *Chemical Engineering Journal* 157: 277–296. http://dx.doi.org/10.1016/j.cej.2010.01.007

Charmarthy, S.; Chung, W. S.; Wayne, E. M. 2001. Adsorption of selected metal ions by modified peanut shells, *Journal of Chemical Technology and Biotechnology* 76: 593–597. http://dx.doi.org/10.1002/jctb.418

Donghee, P.; Yeoung, S. Y; Jong, M. P. 2010. The past, present, and future trends of biosorption, *Biotechnology and Bioprocess Engineering* 15: 86–102. http://dx.doi.org/10.1007/s12257-009-0199-4

Dubey, S. P.; Gopal, K. 2007. Adsorption of chromium (VI) on low cost adsorbents derived from agricultural waste material: a comparative study, *Journal of Hazardous Materials* 145: 465–470. http://dx.doi.org/10.1016/j.jhazmat.2006.11.041

Gaikwad, R. W.; Sapkal, R. S.; Sapkal, V. S. 2010. Removal of copper ions from acid mine drainage wastewater using ion exchange technique: factorial design analysis, *Journal of Water Resource and Protection* 2: 984–989. http://dx.doi.org/10.4236/jwarp.2010.211117

Girgis, B. S.; Hendawy, E. L.; Naseer, A. 2002. Porosity development in activated carbons obtained from date pits under chemical activation with phosphoric acid, *Microporous Mesoporous Materials* 52: 105–117. http://dx.doi.org/10.1016/S1387-1811(01)00481-4

Guo, Y.; Rockstraw, D. A. 2007. Physicochemical properties of carbons prepared from pecan shell by phosphoric acid activation, *Bioresource Technology* 98: 1513–1521. http://dx.doi.org/10.1016/j.biortech.2006.06.027

Ingle, P. K.; Gadipelly, C.; Rathod V. K. 2014. Sorption of copper (II) from aqueous solution onto Arachis hypogaea husk, *Desalination and Water Treatment*: 1–9.

Kim, K. J.; Kang, C. S.; You, Y. J.; Chung, M. C.; Woo, M. W.; Jeong, W. J.; Park, N. C.; Ahn, H. G. 2006. Adsorption-desorption characteristics of VOCs over impregnated activated carbons, *Catalysis Today* 111: 223–228. http://dx.doi.org/10.1016/j.cattod.2005.10.030

Li, Q.; Zhai, J.; Zhang, W.; Wang, M.; Zhou, J. 2007. Kinetic studies of adsorption of Pb (II), Cr (III) and Cu (II) from aqueous solution by sawdust and modified peanut husk, *Journal of Hazardous Materials* 141: 163–167. http://dx.doi.org/10.1016/j.jhazmat.2006.06.109

Loutseti, S.; Danielidis, D. B.; Economou-Amilli, A.; Katsaros, C; Santas, R.; Santas, P. 2009. The application of a micro-algal/bacterial biofilter for the detoxification of copper and cadmium metal waste, *Bioresource Technology* 100(20): 99–105. http://dx.doi.org/10.1016/j.biortech.2008.11.019

Low, K. S.; Lee, C. K.; Leo, A. C. 1995. Removal of metals from electroplating wastes using banana pith, *Bioresource Technology* 51: 227–231. http://dx.doi.org/10.1016/0960-8524(94)00123-I

Namasivayam, C.; Ranganathan, K. 1995. Removal of Pb(II), Cd(II) and Ni(II) and mixture of metal ions by adsorption onto waste Fe(III)/ Cr(III) hydroxide and fixed bed studies, *Environmental Technology* 16: 851–860.

Ngah, W. S. W.; Hanafiah, M. A. K. M. 2008. Removal of heavy metal ions from wastewater by chemically modified plant wastes as adsorbents: A review, *Bioresource Technology* 99: 3935–3948. http://dx.doi.org/10.1016/j.biortech.2007.06.011

Rahman, I. A.; Saad, B.; Shaidan, S.; Sya Rizal, E. S. 2005. Adsorption characteristics of malachite green on activated carbon derived from rice husks produced by chemical-thermal process, *Bioresource Technology* 96: 1578–1583. http://dx.doi.org/10.1016/j.biortech.2004.12.015

Ricordel, S.; Taha, S.; Cisse, I.; Dorange, G. 2001. Heavy metals removal by adsorption onto peanut husks carbon: characterization, kinetic study and modeling, *Separation and Purification Technology* 24: 389–401. http://dx.doi.org/10.1016/S1383-5866(01)00139-3

Sari, A.; Tuzen, M. 2008. Biosorption of total chromium from aqueous solution by red algae (*Ceramium virgatum*): equilibrium, kinetic and thermodynamic studies, *Journal of Hazardous Materials* 160: 349–355. http://dx.doi.org/10.1016/j.jhazmat.2008.03.005

Sharma, D. C.; Forster, C. F. 1994. A preliminary examination into the adsorption of hexavalent chromium using low-cost adsorbents, *Bioresource. Technology* 47: 257–264. http://dx.doi.org/10.1016/0960-8524(94)90189-9

Shukla, S. R.; Pai, R. S. 2005. Adsorption of Cu(II), Ni(II) and Zn(II) on dye loaded groundnut shells and sawdust, *Separation and Purification Technology* 43: 1–8. http://dx.doi.org/10.1016/j.seppur.2004.09.003

Shukla, S. R.; Pai, R. S. 2005. Adsorption of Cu(II), Ni(II) and Zn(II) on modified jute fibres, *Bioresource Technology* 96: 1430–1438. http://dx.doi.org/10.1016/j.biortech.2004.12.010

Smisek, M.; Cerny, S. 1997. *Active carbon manufacture properties and application*. Amsterdam: Elsevier Publishing Company Amsterdam.

Song, Z.; Williams, C. J.; Edyvean, R. G. J. 2000. Sedimentation of tannery waste water, *Water Resources* 34: 2171–2176.

Uluozlu, O. D.; Sari, A.; Tuzen, M.; Soylak, M. 2008. Biosorption of Pb(II) and Cr(III) from aqueous solution by lichen (*Parmelinatiliaceae*) biomass, *Bioresource Technology* 99: 2972–2980. http://dx.doi.org/10.1016/j.biortech.2007.06.052

Valenzuela, F.; Fonsecaa, C.; Basualtoa, O.; Correab, C.; Tapiaa, J.; Sapaga, C. 2005. Removal of copper ions from a waste mine water by a liquid emulsion membrane method, *Minerals Engineering* 18(1): 33–40. http://dx.doi.org/10.1016/j.mineng.2004.05.011

Wafwayo, W.; Seo, C. W.; Marshal, W. E. 1999. Utilization of peanut shells as adsorbent for selected metals, *Journal of Chemical Technology and Biotechnology* 74: 1117–1121. http://dx.doi.org/10.1002/(SICI)1097-4660(199911)74:11<1117::AID-JCTB151>3.3.CO;2-I

Wang, J.; Chen, C. 2006. Biosorption of heavy metals by *Saccharomyces cerevisiae*: a review, *Biotechnology Advances* 24: 427–451. http://dx.doi.org/10.1016/j.biotechadv.2006.03.001

Wang, J. 2002. Biosorption of copper (II) by chemically modified biomass of *Saccharomyce cerevisiae, Process Biochemistry* 37(8): 847–850. http://dx.doi.org/10.1016/S0032-9592(01)00284-9

Virendra K. RATHOD. Dr, Prof. in Chemical Engineering Department at Institute of Chemical Technology, Mumbai, India. His research interest includes extraction of natural ingredients and synthesis of perfumes and flavors, separation of biomolecules, enzyme-catalyzed reactions, biodiesel preparation and purification, separation processes, and wastewater treatment.
He is a Fellow of Maharashtra Academy of Sciences. Till 2015 has published around 80 papers in international peer reviewed journals.

Pradnya K. INGLE has completed her post-graduation (Masters in Chemical Engineering) from Institute of Chemical Technology, Mumbai in 2010. Currently she is pursuing her PhD from Institute of Chemical technology, Mumbai. Her research interests are biodiesel preparation and purification, and wastewater treatment.

Karishma ATTARKAR has completed her graduation from Institute of Chemical Technology, Mumbai. Her research interest is wastewater treatment.

18

EXPERIMENTAL RESEARCH INTO LEACHING OF METALS FROM IMMOBILIZED CIS SOLAR MODULE WASTE

Saulius VASAREVIČIUS[a], Gintautas SKRIPKIŪNAS[b], Vaidotas DANILA[c]

[a, c] *Department of Environmental Protection, Faculty of Environmental Engineering, Vilnius Gediminas Technical University, Saulėtekio al. 11, LT-10223 Vilnius, Lithuania*

[b] *Department of Building Materials, Faculty of Civil Engineering, Vilnius Gediminas Technical University, Saulėtekio al. 11, LT-10223 Vilnius, Lithuania*

Abstract. The aim of this research is to determine metal leaching from concrete specimens containing different quantities of waste recovered from copper indium selenide (CIS) solar module by replacing a certain share of sand aggregate. During the first stage of research the CIS solar module was shredded and leaching tests were performed on recovered waste by analysing six metals – Na, Mg, Fe, Cd, Cu and Zn. It has been determined that out of all metals analysed, the highest leaching was observed for sodium, while the highest leaching out of the heavy metals was found for zinc. In Phase II of the study concrete specimens with shredded CIS module waste were made and the physical properties of these specimens as well as the leaching of the same metals was determined. The results have shown that three metals, namely Fe, Cd, Zn, were successfully immobilised and did not leach from the specimens.

Keywords: solar module waste, concrete, leaching, water absorption, porosity, immobilization.

Introduction

Copper indium selenide (CIS) solar module is an attractive and clean technology for electricity production (Zhang, Ciftja 2008). Since there are only a few second generation solar module production facilities of commercial purpose in the world (Alsema *et al.* 2006), no large amounts of waste of these solar modules have been produced so far. As production of CIS solar modules is growing (Granata *et al.* 2014); therefore, large amounts of this photovoltaic module waste will be produced in the future. Since the expected active time of solar modules is 20 to 25 years, this means that at the end of their life cycle solar modules have to be disassembled and handled in a certain way or otherwise they will be disposed in landfills (Fthenakis 2000; Study on… 2011).

Unusable CIS solar modules are electronic waste containing several potentially hazardous elements such as heavy metals (Cd, Cu, Zn), which can have an adverse impact on the environment (Fernández *et al.* 2011). Human health can be endangered due to substance leaching from broken photovoltaic modules. Leaching from cracked or broken modules can occur after landfilling CIS solar modules (Malaiškienė *et al.* 2014). The landfilling of modules in large quantities can increase possible risks to humans and biota (Ladwig, Hope 2003). Chemical substances leached from these landfilled modules can pollute local ground and surface water (Özbaş, Balkaya 2014).

An effective method for managing used module waste can alleviate an adverse impact on the environment (Klugmann-Radziemska 2012). The waste management method involving immobilization of materials in cement matrixes is widely used all over the world for handling different types of waste (Gougar *et al.* 1996). It is a relatively inexpensive technology used for treating hazardous waste containing heavy metals. Mechanisms for contaminant immobilization cover chemical stabilisation and physical encapsulation (Lampris *et al.* 2011; Falciglia *et al.* 2014). A significant role in the process of immobilization is attributed to the C-S-H (calcium silicate hydrates) phase which results from the hydration reaction of concrete and thus heavy metals are both chemically fixed in the lattice of hydration products and physically encapsulated in the matrix. The C-S-H phase is the main component of hardened cement paste, which is responsible for high absorbability and low permeability of liquids thus preventing their migration and the formation of micro-pores. These factors have an influence on high waste immobilization

Corresponding author: Vaidotas Danila
E-mail: vaidotas.danila@vgtu.lt

potential in concrete. The possibility of hazardous component leaching from waste immobilised in concrete is significantly reduced. Cement restricts the mobility of heavy metals due to high pH and its capability to precipitate metals in an insoluble form (Giergiczny, Król 2008; Gupta, Surwade 2007).

Giergiczny and Król researched into the possibility of immobilising heavy metals in the form of salts in mineral additions containing concrete composites and found that these metals were effectively immobilised (up to 99%). Concrete containing immobilised waste heavy metals is eco-friendly and can be used for road building (Giergiczny, Król 2008). During concrete lifetime immobilized heavy metals precipitate in insoluble salts that results in the reduction of metals mobility and toxicity. After the useful lifetime of the construction, concrete which contain immobilized waste, can be recycled again and used for the new production of building constructions.

The aim of the research is to determine the possibility of immobilising metals (Na, Mg, Fe, Cd, Cu, Zn) present in CIS solar module waste into concrete and evaluate their leaching from concrete specimens with different waste content.

1. Methodology

1.1. Preparation of CIS solar module waste

A CIS solar module, whose structure is shown in Figure 1, was used in research.

The CIS solar module was shredded in order to prepare waste with particle size below 4 mm and use it for

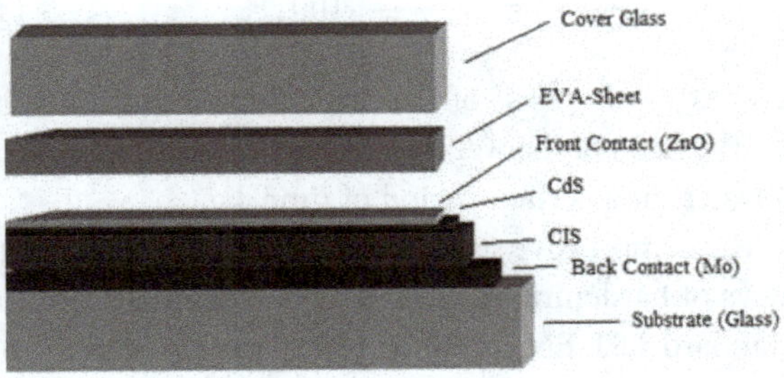

Fig. 1. Structure of the CIS solar module (Raugei *et al.* 2007)

replacing a certain quantity of sand aggregate in concrete. The solar module was physically disintegrated by delamination. The frame, the junction box and connecting cables were removed from the module. The Ethylene Vinyl Acetate (EVA) film was also physically removed from the coating glass. The remaining part was shredded and milled. The microstructure images of solar module waste particles were obtained using a Carl Zeiss microscope Evo LS 25.

As it can be seen from the images CIS solar module waste particles are not spherical and have irregular shape and sharp edges. One of the surfaces of the back side glass contains the active solar cell layers.

CIS solar module particle size distribution was determined according to standard EN 933-1 *Tests for geometrical properties of aggregates Part 1: Determination of particle size distribution.* Sieving method. Sieves used in waste particle size analysis: 4 mm; 2 mm; 1 mm; 0.5 mm; 0.125 mm.

1.2. Batch test

A leaching test is used to describe the leaching of contaminants from solid to liquid phase (Broberg 2009). A two-stage batch leaching test according to standard EN 12457-3 was used to test the leaching of metals (Na, Mg, Fe, Cu, Cd, Zn) from shredded CIS solar module waste. Six metals were chosen considering the composition of CIS solar module and their amount in it. Glass makes the largest part of CIS solar module, about 25 kg for 1 m^2 during manufacturing (Raugei *et al.* 2007). The soda-lime glass used in CIS solar module production contains highly soluble metals in forms of Na_2O (14.3%), MgO (4.3%), and Fe_2O_3 (0.03%) (Rudmann 2004). The total quantity of heavy metals used in CIS solar module production is low, 70 g for 1 m^2 (Raugei *et al.* 2007). Cd was chosen due to its high toxicity, and Cu and Zn have relatively high solubility in water compared to other heavy metals used in CIS solar module.

Five shredded CIS solar module waste samples (particle size ≤ 4 mm), of 175 g each, were prepared for batch leaching tests. Two liquid to solid ratios were used (L/S = 2

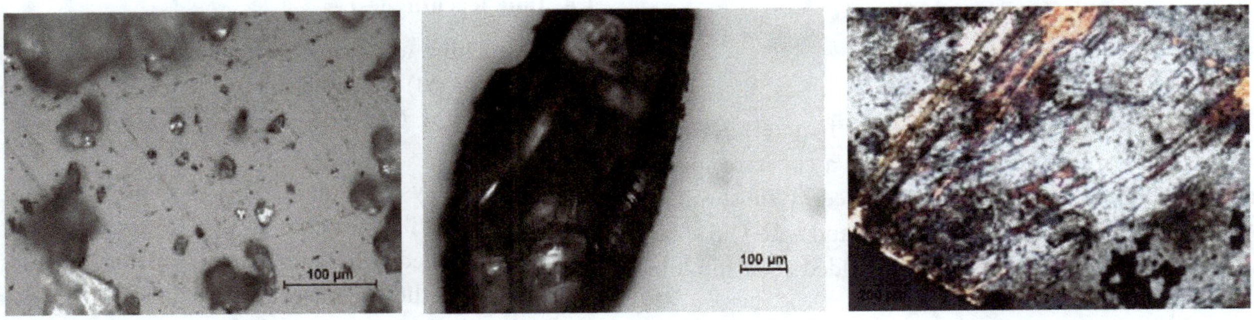

Fig. 2. The images of crushed CIS solar module waste particles used in concrete

in Stage 1, L/S = 8 in Stage 2). In this test the batching was done in two stages and distilled water was used as leachant. In Stage 1, a glass bottle of 0.5 l with the solution was batched for 6 ± 0.5 h at L/S = 2. In stage 2, a glass bottle of 2 l with the same test portion was batched for 18 ± 0.5 h at L/S = 8. The cumulative L/S ratio obtained in both stages is = 10.

Eluates obtained in both stages were left to settle and later filtered through membrane filters (0.45 μm). Then the pH, temperature and electrical conductivity of these eluates were measured (EN 12457-3:2002).

The obtained eluates were analysed in order to determine the amounts of metals leached from the shredded CIS solar module waste. The final results were expressed as the amount of certain metal relatively leached from the total mass of the sample, mg/kg of dry matter.

The amount of metals leached from shredded waste in batching Stage 1 at L/S = 2 was calculated according to the formula:

$$A_2 = C_2\left(\frac{L_2}{M_D}\right), \qquad (1)$$

where: A_2 is the release of metal at L/S = 2, mg/kg; C_2 is the concentration of a particular metal in the eluate, mg/l; L_2 is the leachant volume, l (0.35 l); M_D is the dry mass of the test portion, kg (0.175 kg).

The cumulative release (L/S = 10) of metals after Stage 2 of the batching test at L/S = 8 was calculated according to the formula:

$$A_{2-10} = C_2\left(VE_1 / M_D\right) + C_8\left[\left(L_2 + L_8 - VE_1\right) / M_D\right], \qquad (2)$$

where: A_{2-10} – cumulative release of metal at L/S = 10, mg/kg; C_8 – the concentration of a particular metal in the eluate from the second extraction, mg/l; L_8 – the volume of the leachant applied in the second extraction, l (1.4 l); VE_1 – the volume of the eluate recovered from the first extraction, l (EN 12457-3:2002).

1.3. Preparation of concrete specimens and determination of their physical properties

Concrete specimens (100×100×100 mm) were made in the laboratory. Materials used for concrete specimens:

1. coarse aggregate (gravel), 4/16 mm fraction;
2. fine aggregate (sand), 0/4 mm fraction;
3. cement;
4. water;
5. shredded solar module waste.

Firstly, the required quantities of materials were weighted and mixed in order to make a concrete specimen. Concrete specimens were prepared by substituting 0%, 2.5%, 5%, 10%, 20%, and 40% of sand with CIS solar module waste. Concrete without shredded solar module waste was used as the control specimen.

Table 1. Concrete mix composition

Waste content, %	Concrete mix composition				
	Cement, kg/m³	Gravel, kg/m³	Sand, kg/m³	Waste, kg/m³	Water, kg/m³
0	277	1002	927.00	0	158
5	277	1002	880.65	46.35	158
10	277	1002	834.30	92.70	158
20	277	1002	741.60	185.40	158
30	277	1002	648.90	278.10	158
40	277	1002	556.20	370.80	158

The prepared concrete mixes were placed in oiled moulds. After 24 hours concrete specimens were removed from the moulds and placed for hardening in water.

After 27 days of hardening in water the capillary porosity of concrete specimens was determined in accordance with Annex G to standard LST EN 13369:2004 Common rules for precast concrete products. The concrete specimens with different content of waste were divided in two parts and dried in the oven at 105 °C and then weighed.

The capillary porosity of concrete specimens was determined according to the formula:

$$P = W\frac{\rho_c}{\rho_w}, \%, \qquad (3)$$

where: W – absorption of water of the concrete specimen, %; ρ_c – the density of the concrete specimen, kg/m³; ρ_w – water density, kg/m³.

Absorption of water was calculated using the following formula:

$$W = \frac{W_2 - W_1}{W_1} \cdot 100, \%, \qquad (4)$$

where: W_1 – the mass of the dry concrete specimen, kg; W_2 – the mass of the wet concrete specimen after soaking it in water for a certain period of time, kg.

The compressive strength of concrete specimens after 27 days of hardening in water was determined according to standard LST EN 12390-3:2009 *Testing hardened concrete – Part 3: Compressive strength of test specimens.* Loading speed of used machine was 0.5 MPa/s.

1.4. Tank leaching test

A tank leaching test according to the modified standard NEN 7375 (2004) was used to test the leaching of metals (Na, Mg, Fe, Cu, Cd, Zn) from hardened concrete specimens.

After 28 days of hardening the specimen cubes were immersed in distilled water at V_L/V_S = 1.5, where V_L is the volume of distilled water and V_S is the volume of distilled water and V_S is the volume of the specimen. The leachant was replenished after 1, 4, 11, 25 and 64 days.

The pH of the leaching per fraction was measured with a pH meter. Each leachate sample was filtered via a membrane filter, 0.45 µm.

For each analysed component, the measured leaching per fraction was calculated according to the formula:

$$E_i^* = \frac{c_i V}{fA}, \qquad (5)$$

where: E_i^* – measured leaching of a component in the fraction i, mg/m²; c_i – concentration of a component in the fraction i, µg/l; V – volume of eluate, l; f – conversion factor: 1000 µg/mg; A – surface area of the specimen, m².

The concentration c_i indicated in the formula is the concentration of the metal originally present in the eluate.

For each analysed component the cumulative leaching was calculated according to the formula:

$$\varepsilon_n^* = \sum_{i=1}^{n} E_i^* \text{ (n = 1 to N)}, \qquad (6)$$

where: ε_n^* – measured cumulative leaching of a component for the period n comprising fractions from i = 1 to n, mg/m²; E_i^* – measured leaching of a component in the fraction i, mg/m²; N – the number of leachant replenishment times.

The concentration of metals (Na, Mg, Fe, Cd, Cu, Zn) in all leachates was determined by the atomic absorption spectrometry method using flame atomic absorption spectrometer Buck Scientific 210 VGP. In this method specific wavelengths of light absorbed by analysed metal atoms were used. Wavelengths correspond to the energies, which are needed that electrons could move to higher energy level (Atomic absorption… 2012). The absorbance value was used to determine the concentration of an analysed metal in the leachate.

2. Results and discussion

2.1. Analysis of solar module waste particles

Table 2 shows the results of sieving obtained after shredding solar module waste particles. Data given in the table show the distribution of the mass of residues of waste particles of a certain size on each sieve.

Table 2. Solar module waste particle size distribution

Size of a sieve, mm	Retained mass on each sieve, kg	Percentage of waste sample retained on each sieve, %
4	0.251	2.41
2	4.293	41.22
1	2.130	20.46
0.5	1.553	18.91
0.125	1.770	17.00
Pan	0.416	4.00

Figure 3 shows the cumulative percent of waste particles passed through each sieve.

The data provided in Table 2 and the distribution function of crushed waste particles show that the highest percentage of the obtained particles was between 2 mm and 4 mm (41.22%).

2.2. Analysis of batch leaching test

Six metals, namely sodium (Na), magnesium (Mg), iron (Fe), cadmium (Cd), copper (Cu) and zinc (Zn) were analysed in the filtered eluates after the first and second stages of the batch leaching test. The values of metal quantities leached from the shredded module depended on the content of metals in solar module waste and their solubility in neutral water.

The quantities of metals (mg/kg) leached from solar module waste after the first stage of batch (at liquid to solid ratio of 2 l/kg) are presented in Figure 4.

As the figure shows, of all six metals analysed, Na had the highest release from CIS solar module waste – 132.67 mg/kg. Compared to magnesium and iron release, sodium leaching from solar module waste was 64 times and 310 times higher, respectively. This could be attributed to a higher content of sodium in soda-lime glass of solar modules: the content of Na_2O was 14.3%, MgO was 4.3%, and Fe_2O_3 was 0.03% (Rudmann 2004).

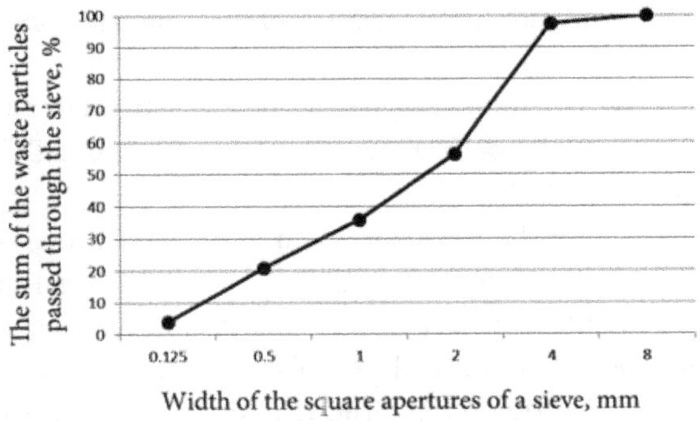

Fig. 3. Particle size distribution of crushed CIS solar module waste

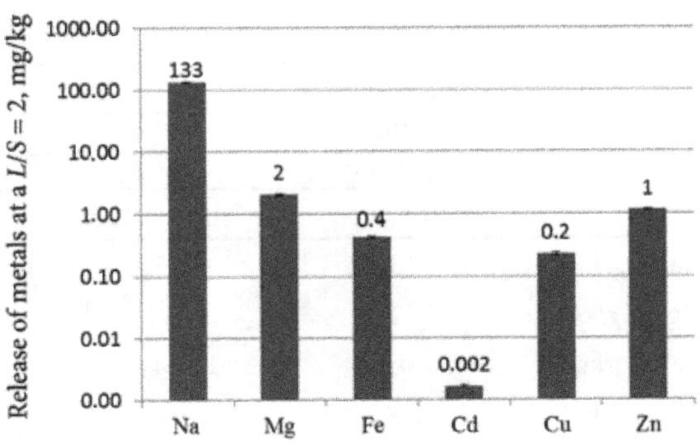

Fig. 4. The quantities of metals leached from shredded CIS solar module waste during the first test stage at L/S = 2 (agitation time: 6±0.5 h)

The comparison of leached heavy metal quantities showed that Zn had the highest release – 1.149 mg/kg and it was 5 times above the release of copper and 570 times above the release of cadmium. The lowest release of cadmium from solar module waste could be explained by the fact that cadmium in the solar module is in the form of cadmium sulphide (CdS) compound, which is almost insoluble in neutral water (CdS is better soluble in acid). Also, the buffer layer of CIS solar cells containing cadmium sulphide is thinner than the layers of the absorber containing copper, indium, gallium and selenium metals, and of the front contact containing zinc oxide.

Figure 5 shows the cumulative (when liquid to solid ratio is equal to 10 *l*/kg) quantities of metals released from solar module waste.

The leaching potential of metals at $L/S = 10$ showed a similar pattern to the previous on in Stage 1 of the leaching test at $L/S = 2$. The obtained cumulative release of Na was by 1.77 times higher than in Stage 1 of the leaching, which shows that the major part of sodium was leached out from the solar module in Stage 1 of the leaching because this metal is well soluble in water. The cumulative releases of magnesium, iron and zinc are about 5 ÷ 6 times above the respective values obtained in Stage 1 of the leaching, and the cumulative releases of cadmium and iron in Stage 2 were about 10 times higher than in Stage 1 of the leaching test. This could be attributed to lower solubility of these metals in water and, therefore, more time was necessary for higher releases of these metals from solar module waste.

Table 3 shows comparison of the leaching results with limit values of the Council decision 2003/33/EC (Directive 2003/33/EC).

As it can be seen, the leaching values of analysed heavy metals doesn't exceed limits for non-hazardous waste landfill. Considering only analysed heavy metals, it would be possible to send CIS solar module waste to non-hazardous waste landfill. The release of Zn from waste at $L/S = 10$ was 5 mg/kg, exceeding the leaching limit value of 4 mg/kg for inert waste landfill. Considering Zn leaching results CIS solar module waste cannot be sent to inert waste landfill.

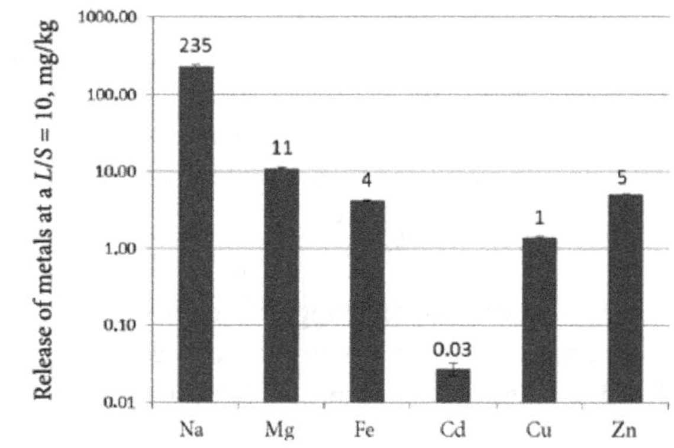

Fig. 5. Cumulative quantities of metals leached from shredded CIS solar module waste at $L/S = 10$ (agitation time: 18 ± 0.5 h)

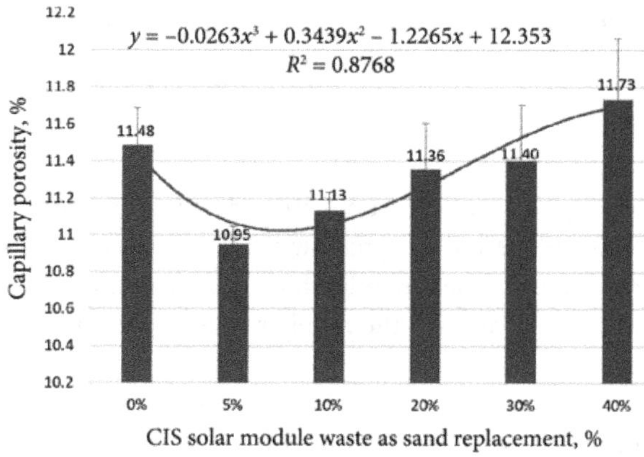

Fig. 6. Dependence of the capillary porosity of concrete specimens (%) on the quantity of solar module waste used

2.3. Capillary porosity of concrete

Porosity is an important parameter since the number of reduced pores improves the physico-mechanical and environmental properties of concrete (Bouzalakos *et al.* 2008).

Figure 6 shows calculated capillary porosity values for concrete specimens with different quantities of solar module waste used.

Table 3. Metals contents in water extracts from CIS solar module waste

Metal	Concentrations of leached elements		European landfill directive			
			Inert		Non-hazardous	
	$L/S = 2$	$L/S = 10$	$L/S = 2$	$L/S = 10$	$L/S = 2$	$L/S = 10$
Na (mg/kg)	133	235	–	–	–	–
Mg (mg/kg)	2	11	–	–	–	–
Fe (mg/kg)	0.4	4	–	–	–	–
Cd (mg/kg)	0.002	0.03	0.03	0.04	0.6	1
Cu (mg/kg)	0.2	1	0.9	2	25	50
Zn (mg/kg)	1	5*	2	4	25	50

Note: * Higher than the limit value of the decision 2003/33/EC for inert waste)

From the figure it can be seen that the capillary porosity of concrete containing 5% solar module waste as sand replacement is the lowest. Then further increase of the amount of solar module waste in concrete resulted in the constant increase of the capillary porosity of concrete.

The lowest capillary porosity, which was in concrete where 5% of sand was replaced with solar module waste, was resulted by fine (>0.2 mm) solar module waste particles which lowered the volume of voids. Also, the reaction products of calcium hydroxide – calcium hydrosilicates, formed during cement hydration, and amorphous silicon dioxide, which is present solar module waste glass, filled capillary pores and therefore decreased absorption and porosity of the concrete. The increasing amount of waste in concrete resulted in a gradual decline of this effect as all amount of CaOH is used in the reaction. As the waste particles have smooth surface so the increase of waste particles in the concrete resulted adhesion deterioration and it increased water absorption and porosity.

Similar tendencies towards a decrease in water absorption and capillary porosity, compared to control specimens, have been also observed by other authors that conducted experiments on concrete specimens in which glass waste was used (Omoniyi *et al.* 2014).

2.4. Analysis of compressive strength of concrete specimens

The results of testing the compressive strength of concrete specimens, having different content of solar module waste particles, after 28 days of curing, are presented in Figure 7.

At first, an average increase in the compressive strength of concrete specimens was observed, and the maximum average strength was obtained in the concrete where 5% of sand aggregate had been replaced by waste particles and was equal to 39.3 MPa. This value is 17% higher than the average compressive strength of the concrete specimens containing no waste. Next, an increase in the content of waste in concrete resulted in a constant decrease of the compressive strength of concrete specimens. Within the interval between 5 to 20%, the compressive strength of concrete specimens was higher than that of the control concrete specimens. When 10% of sand aggregate was replaced by waste, compressive strength was higher by 12%, and when 20% of sand aggregate was replaced by waste, compressive strength was higher by 3% than the average compressive strength of concrete specimens with no waste. An increase in compressive strength compared with control specimens have been influenced by the formation of a denser microstructure of concrete. The average compressive strength of the specimens where 30% and 40% of sand have been replaced with solar module waste particles was respectively 4% and 8% lower than that of control samples. A decrease in concrete strength has been affected by the granulometric composition of the waste particles.

2.5. Tank leaching results

As the leaching results showed, three metals – Na, Mg, Cu – leached from concrete samples, while the concentrations of other metals, Fe, Cd and Zn, were below detection values in the analysed eluates.

The cumulative release of Na from concrete specimens with different content of solar module waste replacing sand aggregate is presented in Figure 8.

It has been determined that the highest release of Na, 1077.75 mg/m^2, was from concrete specimens in which 40% of sand was replaced by shredded solar module waste, which is, respectively, 11% and 22% higher compared to the specimens in which 30% and 20% of sand was replaced by waste. The lowest release of Na, 646.08 mg/m^2, was in the specimens in which 5% of sand was replaced by solar module waste and that is, respectively, 13% and 16% less compared to the specimens with 10% of sand replaced by waste and the specimens without waste. Lower capillary porosity of the specimens in which 5% and 10% of sand were replaced by waste is responsible for a lower release of Na.

The cumulative release of Mg from concrete specimens with different content of solar module waste used instead of sand aggregate is presented in Figure 9.

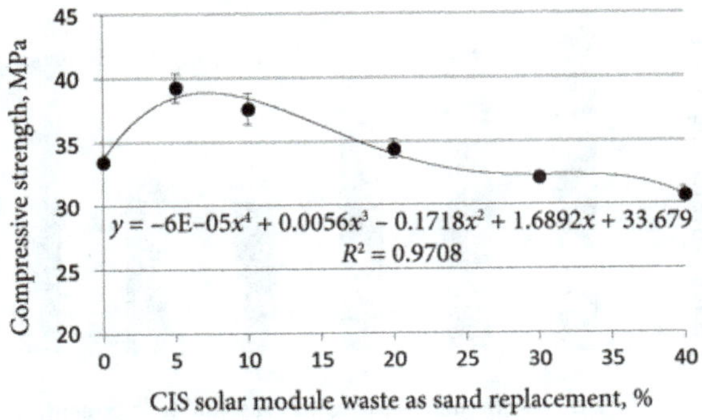

Fig. 7. The dependence of the compressive strength of concrete specimens on the amount of solar module waste

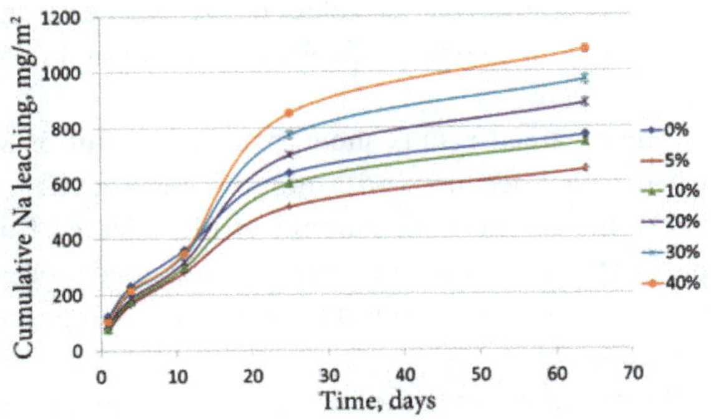

Fig. 8. Na release from concrete specimens over time

As the graph shows, Mg release had a different pattern comparing to Na release. Since the content of Mg in solar module waste is smaller, the capillary porosity of concrete in this case is more important for leaching.

It has been found that the highest magnesium release, 18.72 mg/m², was from concrete specimens with no waste, while the lowest release, 7.31 mg/m², was from the specimen in which 20% of waste was used instead of sand.

The cumulative release of Cu from concrete specimens with different content of solar module waste instead of sand aggregate is presented in Figure 10.

It was determined that the released amounts of Cu from concrete specimens were very low compared to the release of Na and Mg. The highest cumulative leaching of Cu, 2.95 mg/m², was from concrete specimens in which 20% of sand was replaced by waste. The lowest cumulative leaching of Cu, 0.91 mg/m², was from concrete specimens in which 5% of sand was replaced by waste.

2.6. Analysis of cumulative metal leaching

The quantity of metals leached from concrete specimens depends on the four main factors:

1. Content of waste in concrete specimens;
2. Percentage of a respective metal in a solar module;
3. Capillary porosity of concrete specimens;
4. Waste particle distribution in a mix.

Figure 11 presents the values of Na leaching results depending on the content of solar module waste in concrete specimens.

Since, of the six metals analysed, sodium had the largest content in solar module waste, it is evident from the obtained results that this factor had a bigger influence on the obtained leaching values than the capillary porosity of specimens. Compared to the control specimen, the leaching of Na from concrete specimens in which 5% of sand was replaced by waste decreased. Since these specimens contained a small quantity of solar module waste, the decreased capillary porosity also determined the decreased leaching of sodium, which improved the environmental properties of concrete. When the content of waste in concrete specimens further increased, the leaching of sodium increased steadily.

Figure 12 presents the values of Mg leaching results depending on the content of solar module waste in concrete specimens.

The obtained results show that the amount of Mg leached from specimens with immobilised waste in all cases is smaller than the amount of Mg released from concrete specimens without waste. The decreased capillary porosity of concrete improved the environmental properties of concrete.

Figure 13 presents the values of Cu leaching results depending on the content of shredded solar module waste in concrete specimens.

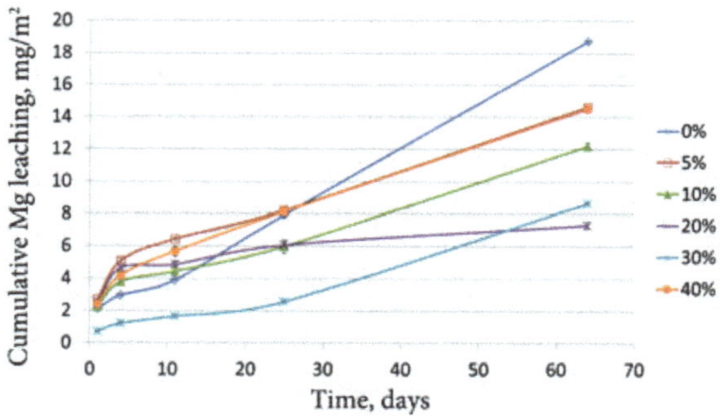

Fig. 9. Mg release from concrete specimens over time

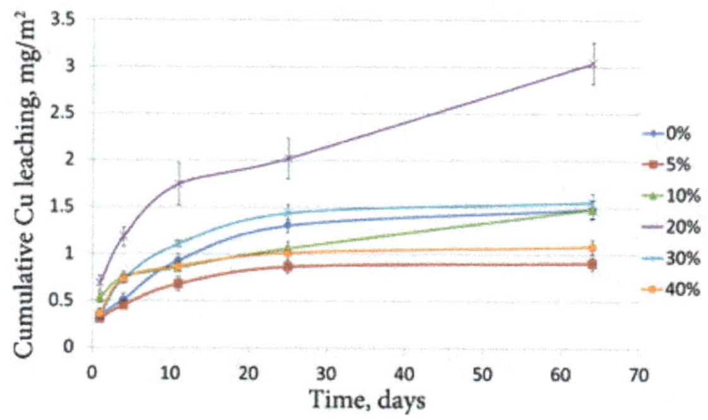

Fig. 10. Cu release from concrete specimens over time

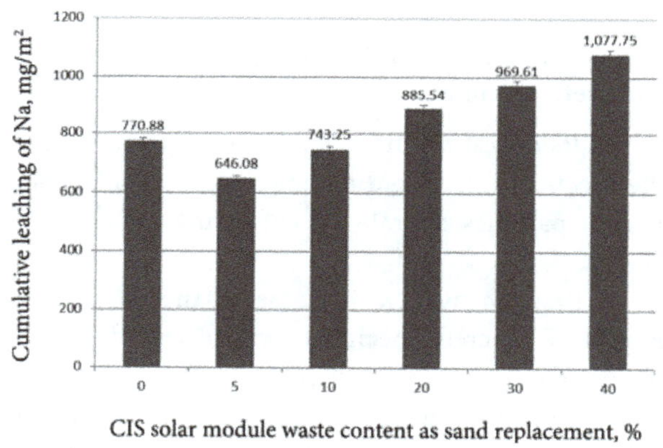

Fig. 11. Cumulative leaching of Na depending on the quantity of waste used

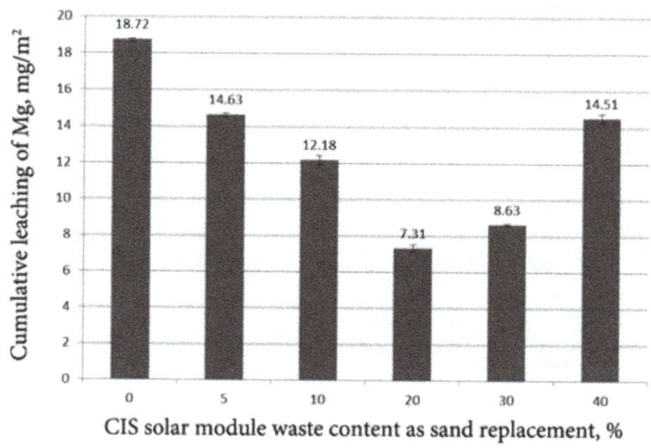

Fig. 12. Cumulative leaching of Mg depending on the quantity of waste used

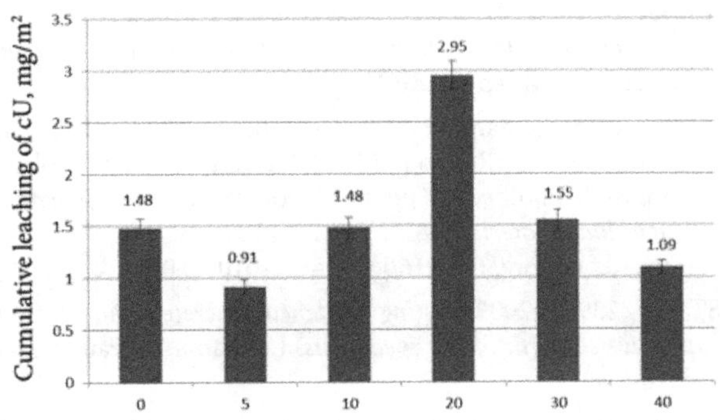

Fig. 13. Cumulative leaching of Cu depending on the amount of waste used

Reduced leaching of Cu from the concrete specimens with immobilised waste compared to the specimens with no waste was observed in the specimens in which 5% and 40% of solar module waste was used instead of sand.

The decreased leaching of copper from concrete specimens in which 40% of solar module waste was used instead of sand could be attributed to uneven distribution of waste particles in a concrete mix. The largest release of Cu was in concrete specimens in which 20% of sand was replaced by solar module waste.

Conclusions

1. The results of the batch leaching test showed that the largest release from shredded CIS solar module waste, 132.67 mg/kg, was that of Na. The analysis of heavy metals (copper, cadmium and zinc) in eluates was performed. According to the leaching test results, the leaching of Zn was the highest, comparing to other heavy metals. The concentrations of other heavy metals in eluates were much lower due to a thin layer of solar cells in which these metals are present and low solubility in water thereof.

2. Research into the capillary porosity of concrete specimens with immobilised shredded solar module waste has shown that higher waste content (replacing up to 40% of sand aggregate in concrete) was responsible for the reduced capillary porosity of the specimens. Thus, the properties and durability of concrete improved. The lowest capillary porosity of concrete, about 11%, was determined in the specimens in which sand replacement by solar module waste accounted for 5%.

4. The metal leaching test for concrete specimens with immobilised shredded solar module waste showed that three metals, Fe, Cd, Zn, were successfully immobilised and were not released from the tested specimens.

5. Of all metals contained in concrete specimens, the highest release was found for Na. Compared to the control specimens, the leaching of Na from specimens in which sand replacement by waste represented 5% and 10% decreased due to reduced capillary porosity of these specimens and a small quantity of used waste. The cumulative leaching of Mg and Cu from concrete specimens mainly depended on the capillary porosity of concrete, which in most cases reduced the release of these metals. In all cases the specimens with immobilised waste had lower release of Mg compared to the specimens with no waste, whereas the release of Cu from the specimens with immobilised waste, compared to the control specimens, was lower in the specimens in which the replacement of sand by waste was 5% and 40%.

6. In summary of the results of the analysis of metals, Na, Mg, Cu, leaching from concrete specimens we can conclude that the best option is to replace 5% to 10% of sand aggregate by solar module waste, as it best reduces the leaching of these metals from concrete specimens.

7. Generally, research into concrete specimens with different content of solar module waste has shown that the concrete production replacing 5% to 10% of sand aggregate by solar module waste can be successfully applied for immobilization of metals Na, Mg, Fe, Cd, Cu, Zn. Since the compressive strength of concrete specimens, where 5% to 10% of sand aggregate in concrete was replaced with waste particles, is higher than the compressive strength of control concrete specimens such concrete can be used for making constructions.

Acknowledgements

The authors are sincerely grateful to the Lithuanian company JSC Reenpro for assistance in research by providing the CIS solar module and cooperation on the issues of solar module waste handling.

Disclosure statement

There are not any competing financial, professional, or personal interests from other parties.

References

Alsema, E. A.; de Wild-Scholten, M. J.; Fthenakis, V. M. 2006. Environmental impacts of PV electricity generation-a critical comparison of energy supply options, in *21st European photovoltaic solar energy conference*, 4–8 September 2006, Dresden, Germany.

Atomic absorption spectrometry. 2012 [online], [cited 3 April 2016]. Available from Internet: https://s3-ap-southeast-1.amazonaws.com/erbuc/files/cefeb931-4829-4b7e-af2c-d8b-27f00a2ee.pdf

Bouzalakos, S.; Dudeney, A. W. L.; Cheeseman, C. R. 2008. Controlled low-strength materials containing waste precipitates from mineral processing, *Minerals Engineering* 21(4): 252–263. http://dx.doi.org/10.1016/j.mineng.2007.09.006

Directive 2003/33/EC of the Council decision of 19 December 2002 on establishing criteria and procedures for the acceptance of waste at landfills pursuant to Article 16 of and Annex II to Directive 1999/31/EC [online], [cited 3 April 2016]. Available from Internet: http://eur-lex.europa.eu/leg

EN 12457-3. 2002. *Characterisation of waste. Leaching compliance test for leaching of granular waste materials and sludges. Part 3: Two stage batch test at a liquid to solid ratio of 2 l/kg and 8 l/kg for materials with high solid content and with particle size below 4 mm*. European Committee for Standarization.

EN 933-1:1997. *Tests For Geometrical Properties Of Aggregates - Part 1: Determination Of Particle Size Distribution – Sieving Method*. European Committee for Standarization.

Falciglia, P. P.; Al-Tabbaa, A.; Vagliasindi, F. G. 2014. Development of a performance threshold approach for identifying the management options for stabilisation/solidification of lead polluted soils, *Journal of Environmental Engineering and Landscape Management* 22(2): 85–95. http://dx.doi.org/10.3846/16486897.2013.821070

Fernández, J. L.; Ferrer, R.; Aponte, F. D.; Fernández, P. 2011. Recycling silicon solar cell waste in cement-based systems, *Solar Energy Materials and Solar Cells* 95(7): 1701–1706. http://dx.doi.org/10.1016/j.solmat.2011.01.033

Fthenakis, V. M. 2000. End-of-life management and recycling of PV modules, *Energy Policy* 28(14): 1051–1058. http://dx.doi.org/10.1016/S0301-4215(00)00091-4

Giergiczny, Z.; Król, A. 2008. Immobilization of heavy metals (Pb, Cu, Cr, Zn, Cd, Mn) in the mineral additions containing concrete composites, *Journal of Hazardous Materials* 160(2): 247–255. http://dx.doi.org/10.1016/j.jhazmat.2008.03.007

Gougar, M. L. D.; Scheetz, B. E.; Roy, D. M. 1996. Ettringite and C- S- H Portland cement phases for waste ion immobilization: a review, *Waste Management* 16(4): 295–303. http://dx.doi.org/10.1016/S0956-053X(96)00072-4

Granata, G.; Pagnanelli, F.; Moscardini, E.; Havlik, T.; Toro, L. 2014. Recycling of photovoltaic panels by physical operations, *Solar Energy Materials and Solar Cells* 123: 239–248. http://dx.doi.org/10.1016/j.solmat.2014.01.012

Gupta, S. K.; Surwade, M. T. 2007. Immobilization of heavy metals from steel plating industry sludge using cement as binder at different pH, *in Conference on Moving Forward Wastewater Biosolids Sustainability: Technical, Managerial, and Public Synergy*, 24–27 June 2007, Moncton, New Brunswick, Canada.

Klugmann-Radziemska, E. 2012. Current trends in recycling of photovoltaic solar cells and modules waste, *Chemistry-Didactics-Ecology-Metrology* 17(1–2): 89–95. http://dx.doi.org/10.2478/cdem-2013-0008

Ladwig, K.; Hope, L. 2003. *Potential health and environmental impacts associated with the manufacture and use of photovoltaic cells*. Final report. EPRI.

Lampris, C.; Stegemann, J. A.; Pellizon-Birelli, M.; Fowler, G. D.; Cheeseman, C. R. 2011. Metal leaching from monolithic stabilised/solidified air pollution control residues, *Journal of Hazardous Materials* 185(2): 1115–1123. http://dx.doi.org/10.1016/j.jhazmat.2010.10.021

LST EN 12390-3:2009. *Testing hardened concrete – Part 3: Compressive strength of test specimens*. Lithuania Bureau of Standards.

Malaiškienė, J.; Mačiulaitis, R.; Mikalauskaitė, R. 2014. Possibilities to recycle auto glass waste in building ceramics, *Journal of Environmental Engineering and Landscape Management* 22(1): 21–29. http://dx.doi.org/10.3846/16486897.2013.867863

NEN 7375. 2004. *Leaching characteristics of moulded or monolithic building and waste materials. Determination of leaching of inorganic components with the diffusion test*. Netherlands Normalisation Institute Standard.

Omoniyi, T. E.; Akinyemi B. A.; Fowowe, A. O. 2014. Effects of waste glass powder as pozzolanic material in saw dust cement brick, *Scholars Journal of Engineering and Technology* 2(4A): 517–522.

Özbaş, E. E.; Balkaya, N. 2014. Removal of heavy metals (Cu, Ni, Zn, Pb, Cd) from compost by molasses hydrolysate, *Journal of Environmental Engineering and Landscape Management* 22(4): 301–310. http://dx.doi.org/10.3846/16486897.2014.919922

Raugei, M.; Bargigli, S.; Ulgiati, S. 2007. Life cycle assessment and energy pay-back time of advanced photovoltaic modules: CdTe and CIS compared to poly-Si, *Energy* 32(8): 1310–1318.

Rudmann, D. 2004. *Effects on sodium on growth and properties of Cu(In, Ga)Se2 thin films and solar cells*: a dissertation. Swiss Federal Institute of Technology (ETH), Zurich. http://dx.doi.org/10.3929/ethz-a-004796411

Study on photovoltaic panels supplementing the impact assessment for a recast of the WEEE directive. Final report 2011 [online], [cited 24 November 2015]. Available from Internet: http://ec.europa.eu/environment/waste/weee/pdf/Study%20on%20PVs%20Bio%20final.pdf

Zhang, L.; Ciftja, A. 2008. Recycling of solar cell silicon scraps through filtration, Part I: Experimental investigation, *Solar Energy Materials and Solar Cells* 92(11): 1450–1461. http://dx.doi.org/10.1016/j.solmat.2008.06.006

Saulius VASAREVIČIUS. Dr Assoc. Prof. (since 1999), Head of Environmental Protection Department, Vilnius Gediminas Technical University. Doctor of Science (Environmental engineering), VTU (now VGTU), 1995. Master of Science, VTU, 1991. First degree in Civil Engineering and Management, Vilnius Civil Engineering Institute (VISI, now VGTU), 1989. Publications: author of more than 50 research papers. Probation in Germany. Research interests: environmental management, air pollution, waste management, soil treatment.

Gintautas SKRIPKIŪNAS. Civil Engineer, Dr.sc.ing., Prof., Vilnius Gediminas Technical University, Faculty of Civil Engineering, Department of Building Materials Publications: ~ 50 scientific and methodological works. Research interests include material science, concrete technology, rheological, mechanical, structural properties of concrete, durability of concrete, concrete modification.

Vaidotas DANILA. Probationer, VGTU, Faculty of Environmental Engineering, Department of Environmental Protection. Master of Science (environmental management and cleaner production), VGTU, 2015. Bachelor of Science (environmental engineering), Vilnius Gediminas Technical University, 2013. Research interests: environmental management, air pollution, waste management.

19

STAKEHOLDER PERCEPTIONS OF THE BENEFITS AND BARRIERS OF IMPLEMENTING ENVIRONMENTAL MANAGEMENT SYSTEMS IN THE NIGERIAN CONSTRUCTION INDUSTRY

Victoria O. OWOLANA[1], Colin A. BOOTH[2]

[1]*Department of Architecture and the Built Environment, Faculty of Environment and Technology, University of the West of England, Bristol BS16 1QY, UK*

[2]*Department of Architecture and the Built Environment, Faculty of Environment and Technology, University of the West of England, Bristol BS16 1QY, UK*

Abstract. This study investigates stakeholder opinions of the major benefits and barriers of Environmental Management Systems (EMS) to the Nigerian construction industry, and the perceived issues to EMS adoption among organisations in the industry. The study highlights the environment as an important stakeholder in the industry because it affects and is affected by construction activities on a regular basis. It identifies the importance of ISO 14001 in ensuring adequate consideration for the environment is maintained on construction projects. The research adopts a quantitative approach by analysing responses from an online survey among construction industry professionals in Nigeria. The questions on the survey were drawn from a similar study carried out in Asia and the results were analysed using the Weighted Average and Standard Deviation statistical approach. Results reveal that the major benefits of EMS to the Nigerian construction industry were improved efficiency in waste management and environmental protection, as well as an overall increase in employee motivation due to better opportunities for training and development. Lack of technological support in organisations and the high cost of implementing EMS were viewed as the major barriers towards its uptake in construction companies. The findings also indicate that a feasible EMS implementation strategy must not ignore the unique nature of the Nigerian construction industry, which comprises mostly small and medium enterprises. The study concludes by recommending the use of a waste management plan based on the Reuse-Reduce-Recycle-Recover model and an employee training plan to ensure continuous improvement in the organisation's environmental management strategy.

Keywords: ISO 14001, corporate responsibility, sustainability, environmental management, construction, Nigeria.

Introduction

Sustainability has become an important factor in today's world, and huge importance is increasingly attached to the environment as a stakeholder in major construction industries (Haigh, Griffiths 2009; Hammond, Booth 2010; Medineckiene *et al.* 2010). Many construction projects have included sustainability as one of their key performance indicators (Fewings 2013); productivity and progress are often based on the ability to meet several environmental criteria. According to Chan and Chan (2004), it appears that the industry is no longer just focusing on maximising value for money for the client. Preserving the environment has undoubtedly become part of the social responsibility of organisations towards the global community (Castka, Balzarova 2008).

An Environmental Management System (EMS) is a tool that helps organisations to manage and positively improve their level of impact on the environment (Christini *et al.* 2004; Oke 2004). An EMS helps a company measure its environmental performance and provides the framework for the integration of sustainable development goals within the organisation's corporate plan (Ilinitch *et al.* 1998; Jolevski 2013). Typically, an EMS is made up of policies, goals, plans, regulatory requirements and is usually reflected in the company's annual reports. Christini *et al.* (2004) gives the following basic characteristics for an organisation's EMS: (i) goals, methods and a timeline for

Corresponding author: Victoria O. Owolana
E-mail: victoria.owolana@gmail.com

achieving environmental criteria; (ii) procedures for maintaining a paper trail in relation to those goals; (iii) a defined structure and a matrix of responsibilities, as well as allocated resources; (iv) corrective actions and emergency procedures; (v) employee training plan; and (vi) a plan for monitoring and auditing the organisation's performance in achieving the EMS goals.

The earliest set of standard EMS was released by the British Standards Institution in 1992 as the standard BS7750 'Specification for Environmental Management Systems'. BS7750 was designed to complement the European regulation, Eco-management and Audit Scheme, which had also been launched by the EU in the mid-90s (Bohoris, O'Mahony 1994; Wenk 2004). One of the current, most widely used and recognised standards for environmental management is the ISO 14001. This is a specification document that forms part of the ISO 14000 series designed for the purposes of audit and specification by the International Organisation for Standardisation (ISO) (Zhang *et al.* 2000). Christini *et al.* (2004) compared ISO 14001 to an integrated management system that streamlines the policies, documentation, data collection, auditing of quality, environmental, and health and safety management systems. In essence, ISO 14001 provides guidelines that can improve the management portfolio of any type of organization in any country and not just their environmental performance (Morrow, Rondinelli 2002).

As evidence of the importance of EMS in large construction related organisations: (i) Rio Tinto Plc demonstrates a commitment to maintaining their responsibility to their stakeholders by ensuring that risks related to social wellbeing, environmental stewardship and economic prosperity are understood and managed (Rio Tinto 2008). This commitment has formed the foundation of their business conduct and their sustainable development framework is named: 'The way we work'. The framework forms part of their EMS, which governs their overarching sustainability strategy (Rio Tinto 2008); (ii) Carillion Plc is one of the international construction companies that have fully integrated their EMS into their wider corporate risk management processes (Carillion Plc 2014). The majority of their establishments in the UK, Middle East and North Africa operate under ISO 14001 (Carillion Plc 2014); and (iii) The British Standards Institution (BSI) has also published a case study of Costain Plc, a leading British engineering and construction firm, which shows that the implementation of ISO 14001 has helped in reducing the amount of waste generated by 53% and 93% has been diverted from landfill in the last four years (BSI Group 2014).

Industrial sectors have been slow to embrace EMS in Africa and, moreover, most construction companies have simply failed to engage with EMS. Therefore, the purpose of this study is to explore the opinions of construction professionals towards their perceptions of the benefits and barriers of implementing EMS in the Nigerian construction industry, so as to identify a strategy for improved organisational ownership of sustainable issues.

1. Environmental management systems in the African construction industry

The paradigm shift towards the implementation of environmental management in Africa is gradually gaining significance (Aminu *et al.* 2010). This has materialised out of the need for an environmentally conscious approach within the construction industry, which accounts for a large percentage of development in Africa. Adebayo (2002) pointed out that the driving force was best explained by the Amsterdam Treaty in 1997, which takes on an all-encompassing approach to sustainable development and encourages an integrated approach in managing and implementing environmental policies that affect the economy, leading to parallel progress in all the relevant fields. In other words, the underlying idea of sustainable construction in Africa is the fuel that drives development within the construction industry, which is one of the most productive sectors in Africa (Ofori 2006; Medineckiene *et al.* 2010). It could, therefore, be said that the environment is a primary stakeholder in the economic development of Africa, especially in the area of construction (Marzouk *et al.* 2013).

Haigh and Griffiths (2009) justified the case for considering the environment as a stakeholder by approaching this from a strategic viewpoint because the environment is not only affected by an organisation but affects it, hence solidifying Freeman's (1984) criteria. However, when the influence of the environment is examined with respect to climate change, it is seen as possessing powerful, legitimate, urgent and proximate properties, all of which qualify it as a primary stakeholder (Haigh, Griffiths 2009). In this sense, the natural environment has a strong economic stake in the affairs of businesses, organisations and governments (Dixon *et al.* 2007; Medineckiene *et al.* 2010).

There are multiple benefits to be reaped through the implementation of EMS in African construction organisations. Ofori (1998) suggested that one major benefit is the ease with which effective construction waste management can be achieved using EMS frameworks and standards. This will amount in notable cost savings and reduce many health, safety and environmental risks that may arise during the course of any project (Ofori 1998; Jolevski 2013). Shen and Tam (2002) also noted that effective waste management would lead to reduced hazardous waste generated from both onsite and offsite construction processes.

Adebayo (2002) explained that in spite of obvious benefits of environmental management and the international legislation and guidelines that are available, few African countries have succeeded in making remarkable strides in this area. There is still a consistent gap between theory and practice in most African nations and between legislation and implementation (Ofori 2006). Many of the construction companies have failed to implement the necessary adjustments and progress is still remarkably slow compared to the rest of the world (Adebayo 2002).

Hill and Bowen (1997) noted that the initial implementation of the environmental management framework in South Africa was voluntary because it enabled the solutions to the issues of implementing sustainability in construction to be tested in the African environment before consequently being assimilated into a legal agenda. However, Adebayo (2002) explained that in African countries, like Nigeria, a number of construction project stakeholders in Africa chose to neglect the more important aspects of protection and quality of the environment and focus more on the economic angle. As such, there is still a slow rate of progress in terms of the adoption of EMS in Nigeria, even though it is one of the largest economies in Africa, alongside South Africa (BBC News 2014). Only 44 companies have been equipped with ISO 14001 certifications (ISO 2012) and a large percentage of these are multinational corporations (Windapo, Jegede 2013). Moreover, many of these organisations view this certification as a diploma earned after an examination, rather than as a process that needs to be implemented and improved upon (Boiral, Sala 1998).

Du Plessis (2005) suggested that one reason for this lapse in sustainable development in Africa is that many policies do not take into account the unique nature of the African environment and the differences in the way the construction industry is structured across various parts of the continent. The appropriateness of many EMS measures does not take into account the local economic conditions especially the reliability of key infrastructure services. Moreover, the EMS framework needs to be embedded into organisational culture and government policy (Balzarova *et al.* 2004).

Windapo and Jegede (2013) explained that the majority of the construction companies in Nigeria are small and medium enterprises. Many of the EMS standards are tailor made to be implemented by large and well established organisations and it is difficult to apply them in smaller companies (Sjostrom, Bakens 1999). This may be due to the high initial cost implication of implementing the environmental management framework (Oke 2004; Ervin *et al.* 2013), which was identified as a major barrier to EMS implementation in a similar study in Ghana (Famiyeh *et al.* 2014).

Adebayo (2002) argued that in order to bridge this conflict between theory and practice, policies need to be tailored to suit the African context. Current research in the African built environment is now shifting in this direction. Adegbite (2013) demonstrated that various aspects of established systems from other economies can be scrutinised and analysed to determine best practice for a developing economy such as Nigeria.

Nigeria, officially the Federal Republic of Nigeria is often referred to as "the Giant of Africa", due to its large population and economy (Holmes 1985). Nigeria has been identified as a regional power in Africa, with its economy (Gross Domestic Production (GDP)) becoming the largest in Africa with more than $500 billion, overtaking South Africa and becoming the world's 26th largest economy (BBC News 2014). The Nigerian construction industry contributes a substantial amount to the GDP of the economy and is mostly made up of small and medium indigenous companies, while a majority of the larger companies are managed by international organisations (Idoro 2012).

2. Methods

A survey method was chosen to explore the opinions of construction professionals towards their perceptions of the benefits and barriers of implementing EMS in the Nigerian construction industry. The questions were adapted from a similar study carried out by Shen and Tam (2002). The questionnaire was compiled and administered using an online distribution technique with the aid of Survey Monkey software (www.surveymonkey.com) to 500 construction industry professionals in Nigeria via email. Members of professional bodies were targeted via an online directory of Nigerian Industry professionals (www.nigerianprofessional.com). Respondents were also sourced from tertiary institutions and members of various organisations who have the data of their staff published publicly on their websites.

Forty completed responses were received (an 8% response rate). Although it is considered a low response rate, it is typical of online surveys carried out in developing countries (Business Day Online 2013). Furthermore, the quality of the Nigerian internet service is known to lead to partial or no response from participants, as explained by Czaja and Blair (2005).

The data was analysed using one of the statistical measurements of central tendency: Mean or Weighted Average. The Weighted Average formula shown in Equation (1) is taken from the same form of analysis performed by Shen and Tam (2002). WAS_i denotes the Weighted Average Score for each given factor i, α_j denotes the numerical value given to each of the ranking categories, where one is allocated to the lowest rank, and five is allocated to the highest rank, n_{ij} denotes the number of respondents for factor i with respect to the ranking category j, and N denotes the total number of respondents for the whole question.

$$WAS_i = \frac{\sum_{j=1}^{5}(\alpha_j n_{ij})}{N}. \quad (1)$$

Shen and Tam (2002) noted that this formula alone is not sufficient to calculate the ranking of each of the factors. This analysis, therefore, employed the use of a coefficient of variation, which is determined by dividing the WAS_i by the standard deviation, which is a recognised measure of dispersion (Naoum 2013). This coefficient was then added to the WAS_i to give an adjusted score on which the final ranking was then based. The Adjusted Score formula is shown in Equation (2), where WAS_i denotes the score calculated for each factor i and δ_i is the standard deviation for each factor i, calculated using the weighted average score.

$$\text{Adjusted Score} = WAS_i + \frac{WAS_i}{\delta_i}. \quad (2)$$

3. Results

3.1 Benefits of environmental management systems

The study presented a set of nine factors, which are considered the benefits of implementing EMS in the construction industry. These factors were selected based on a synthesis of relevant literature and were presented to the respondents as follows:

- BF-a: Cost saving due to the reduction of fines associated with convictions
- BF-b: Improving corporate image in environmental performance
- BF-c: Contribution to the improvement of public environmental standards
- BF-d: Contribution to environmental protection
- BF-e: Increasing overall business competitiveness
- BF-f: Reduction of environmental complaints
- BF-g: Improving staff work environment, thus increasing their morale
- BF-h: Reduction of environment-related sickness and injuries
- BF-i: Reduction of environmental risks – polluted air, land and water

Respondents were asked to rank these factors into the following categories: Large Extent, Moderate Extent, Some Extent, Minimal Extent and No Extent. Table 1 gives a summary of the participant responses and the frequency distribution of this ranking. In order to determine the overall highest and lowest ranking factors, the weighted average of each factor was calculated using the formula in Equation (1). The results of this calculation are shown in Table 2, under Average Significant Score (ASS).

The ASS could be used to determine the ranking of each factor. However, the weighted average does not take into account the variation of the individual responses (Shen, Tam 2002) and gives an inaccurate representation of the highest and lowest ranking factors. Therefore, an adjusted score was calculated using the coefficient of variation for each factor, which was calculated with the formula in the figure above. This adjusted score was tagged Beneficial Index Value (BIV). The final ranking of each factor is shown under the column BIV^R in Table 2.

The final ranking of the factors is, therefore, shown as follows:

- BF-h: Reduction of environment-related sickness and injuries
- BF-d: Contribution to environmental protection
- BF-g: Improving staff work environment, thus increasing their morale
- BF-i: Reduction of environmental risks - polluted air, land and water
- BF-c: Contribution to the improvement of public environmental standards
- BF-b: Improving corporate image in environmental performance
- BF-e: Increasing overall business competitiveness
- BF-f: Reduction of environmental complaints
- BF-a: Cost saving due to the reduction of fines associated with convictions

Table 1. Frequency distribution of participant responses to the benefits of implementing EMS in the Nigerian construction industry

Beneficial Factor	Large Extent	Moderate Extent	Some Extent	Minimal Extent	No Extent
BF-a	4	7	9	12	6
BF-b	12	9	9	7	1
BF-c	18	7	5	6	2
BF-d	20	7	5	4	2
BF-e	8	9	10	10	1
BF-f	10	9	6	11	2
BF-g	8	13	10	7	0
BF-h	16	10	7	4	1
BF-i	21	4	7	3	3

Table 2. Calculated Average Significant Score (ASS) and Beneficial Index Value (BIV) for participant responses to benefits of implementing EMS in the Nigerian construction industry

BF	ASS	δ	BIV	BIV^R
BF-a	2.76	1.22	5.02	9
BF-b	3.63	1.18	6.71	6
BF-c	3.87	1.30	6.84	5
BF-d	4.03	1.25	7.26	2
BF-e	3.34	1.15	6.24	7
BF-f	3.37	1.29	5.99	8
BF-g	3.58	1.02	7.10	3
BF-h	3.95	1.12	7.46	1
BF-i	3.97	1.33	6.97	4

Based on the results presented it can be seen that factors BF-h (Reduction of environment-related sickness and injuries), BF-d (Contribution to environmental protection) and BF-g (Improving staff work environment, thus increasing their morale) are deemed the top three benefits of EMS ranked by respondents. The lowest ranking benefits were BF-e (Increasing overall business competitiveness) BF-f (Reduction of environmental complaints) and BF-a (Cost saving due to the reduction of fines associated with convictions) in that order. Figure 1 is a graphical representation of the final ranking.

3.2 Barriers to environmental management systems

A list of barriers to implementing EMS in the construction industry was presented to the respondents as follows:

EB-a: Lack of government legal enforcement
EB-b: Increase in management and operation costs
EB-c: Lack of trained staff and expertise
EB-d: Lack of client support
EB-e: Lack of sub-contractor cooperation
EB-f: Lack of supplier co-operation
EB-g: Difficult co-ordination of environmental performance among multi-tier subcontractors
EB-h: Lack of working staff support
EB-i: Time-consuming for improving environmental performance
EB-j: Change of existing practice of company structure and policy
EB-k: Increase in documentation workload
EB-l: Lack of tailor-made training on environmental management
EB-m: Lack of technological support within organisation

Respondents were again asked to rank these factors into the following categories: Large Extent, Moderate Extent, Some Extent, Minimal Extent and No Extent. Table 3 provides a summary of the distribution of the participant responses.

In order to determine the overall highest and lowest ranking factors, the weighted average of each factor was calculated. The results of this calculation are shown in Table 4, under Average Barrier-effect Score (ABS). As in the previous section, an adjusted score was calculated using the coefficient of variation for each factor. This adjusted score was tagged Environmental Barrier-effect Index Value (EBIV). The final ranking of each factor is shown under the column $EBIV^R$ in Table 4.

The final ranking of the factors is, therefore, shown as follows:

EB-a: Lack of government legal enforcement
EB-m: Lack of technological support within organisation
EB-l: Lack of tailor-made training on environmental management
EB-c: Lack of trained staff and expertise
EB-b: Increase in management and operation costs
EB-d: Lack of client support
EB-j: Change of existing practice of company structure and policy
EB-h: Lack of working staff support

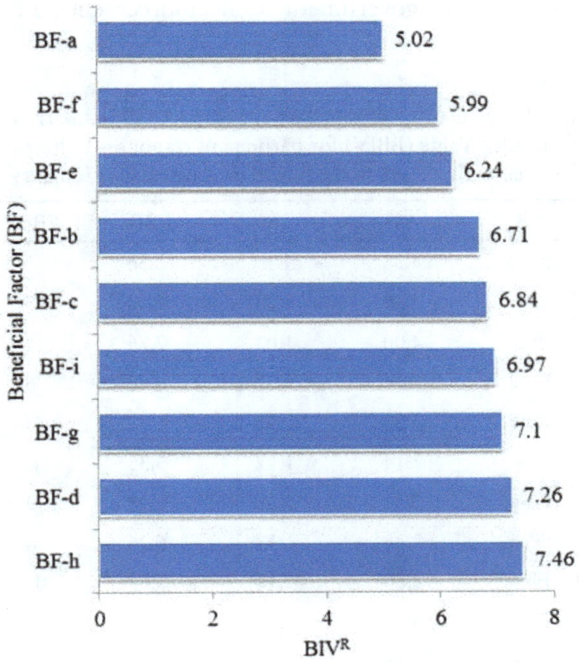

Fig. 1. Ranking profile of the beneficial factors perceived by Nigerian construction professionals

Table 3. Frequency distribution of participant responses to the barriers of implementing EMS in the Nigerian construction industry

Environmental Barrier-effect Factor	Large Extent	Moderate Extent	Some Extent	Minimal Extent	No Extent
EB-a	25	6	3	1	2
EB-b	13	11	7	6	0
EB-c	16	9	9	2	1
EB-d	16	3	14	2	2
EB-e	7	13	7	4	6
EB-f	8	9	7	6	7
EB-g	10	9	8	6	4
EB-h	8	9	11	7	2
EB-i	3	10	8	11	5
EB-j	6	12	11	7	1
EB-k	8	8	7	11	3
EB-l	20	6	6	5	0
EB-m	15	13	5	4	0

EB-g: Difficult co-ordination of environmental performance among multi-tier subcontractors
EB-e: Lack of sub-contractor cooperation
EB-k: Increase in documentation workload
EB-f: Lack of supplier co-operation
EB-i: Time-consuming for improving environmental performance

Based on the results presented above, the factors EB-a (Lack of government legal enforcement), EB-m (Lack of technological support within organisation) and EB-l (Lack of tailor-made training on environmental management) emerged as the top three barriers to achieve benefits of environmental management practice ranked by the survey respondents. The lowest ranking barriers were EB-k (Increase in documentation workload), EB-f (Lack of supplier co-operation) and EB-i (Time-consuming for improving environmental performance). Figure 2 is a graphical representation of the final ranking.

Table 4. Calculated Average Barrier Score (ABS) and Barrier-effect Index Value (EBIV) for participant responses to barriers of implementing EMS in the Nigerian construction industry

EB	ASS	δ	EBIV	EBIV^R
EB-a	4.38	1.10	8.36	1
EB-b	3.84	1.08	7.40	5
EB-c	4.00	1.07	7.75	4
EB-d	3.76	1.21	6.86	6
EB-e	3.30	1.33	5.77	10
EB-f	3.14	1.42	5.35	12
EB-g	3.41	1.32	5.98	9
EB-h	3.38	1.17	6.26	8
EB-i	2.86	1.19	5.27	13
EB-j	3.41	1.05	6.64	7
EB-k	3.19	1.29	5.66	11
EB-l	4.11	1.11	7.81	3
EB-m	4.05	0.98	8.17	2

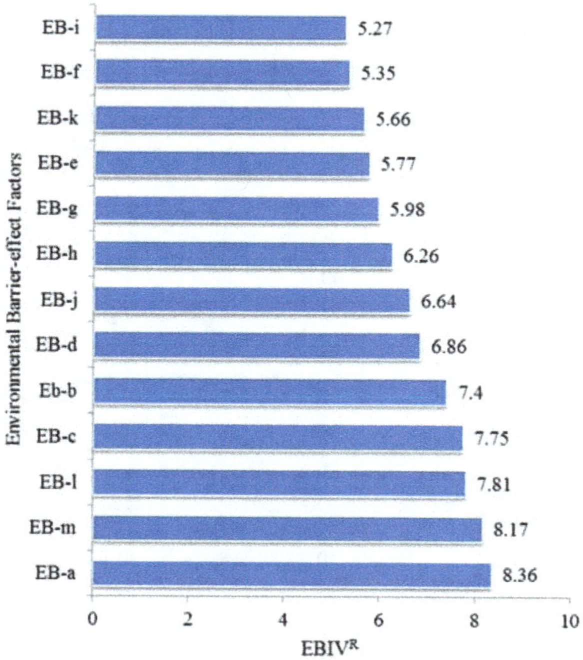

Fig. 2. Ranking profile of the barrier factors perceived by Nigerian construction professionals

3.3 Implementation of environmental management systems

A list of strategies for implementing EMS in the construction industry, which could contribute to the growth in the adoption of EMS within Nigeria, was presented to the respondents as follows:

EM-a: Legal requirements on environmental protection
EM-b: Reduction, reuse and recycling of construction and demolition wastes
EM-c: Imposing responsibilities of protecting environment on managerial staff
EM-d: Applying environmentally friendly technology on site
EM-e: Providing in-house training on environmental management
EM-f: Establishing waste management plan
EM-g: Continuous efforts in improving environmental management
EM-h: Inclusion of environmental management in tendering requirements
EM-i: Effective communication on environmental issue between all layers of subcontractors
EM-j: Close supervision at site level

Again respondents were asked to rank these factors into the following categories: Large Extent, Moderate Extent, Some Extent, Minimal Extent and No Extent. Table 5 provides a summary of the distribution of the participant responses.

To determine the overall highest and lowest ranking factors, the weighted average of each factor was calculated. The results of this calculation are shown in Table 6, Average Effectiveness Score (AES). By adopting the same analytical method employed in previous sections, an adjusted score was calculated using the coefficient of variation for each factor. This adjusted score was tagged Effective Index Value (EIV). The final ranking of each factor is shown under the column EIV^R in Table 6.

The final ranking of the factors is, therefore, shown as follows:

EM-i: Effective communication on environmental issue between all layers of subcontractors
EM-j: Close supervision at site level

Table 5. Frequency distribution of participant responses to the strategies of implementing EMS in the Nigerian construction industry

Environmental Management Measures	Large Extent	Moderate Extent	Some Extent	Minimal Extent	No Extent
EM-a	24	5	6	1	2
EM-b	14	12	6	5	1
EM-c	11	12	11	3	1
EM-d	18	9	7	3	1
EM-e	16	10	8	4	0
EM-f	18	10	7	2	1
EM-g	14	16	4	3	1
EM-h	19	9	5	4	1
EM-i	13	13	12	0	0
EM-j	21	12	3	0	2

Table 6. Calculated Average Effectiveness Score (ASS) and Effectiveness Index Value (EIV) for participant responses to strategies of implementing EMS in the Nigerian construction industry

EB	ASS	δ	EIV	EIV^R
EB-a	4.26	1.14	8.00	4
EB-b	3.87	1.13	7.30	10
EB-c	3.76	1.04	7.39	9
EB-d	4.05	1.10	7.74	7
EB-e	4.00	1.03	7.90	6
EB-f	4.11	1.05	8.03	3
EB-g	4.03	1.01	8.00	4
EB-h	4.08	1.13	7.68	8
EB-i	4.03	0.81	8.99	1
EB-j	4.32	1.00	8.62	2

EM-f: Establishing waste management plan
EM-a: Legal requirements on environmental protection
EM-g: Continuous efforts in improving environmental management
EM-e: Providing in-house training on environmental management
EM-d: Applying environmentally friendly technology on site
EM-h: Inclusion of environmental management in tendering requirements
EM-c: Imposing responsibilities of protecting environment on managerial staff
EM-b: Reduction, reuse and recycling of construction and demolition wastes

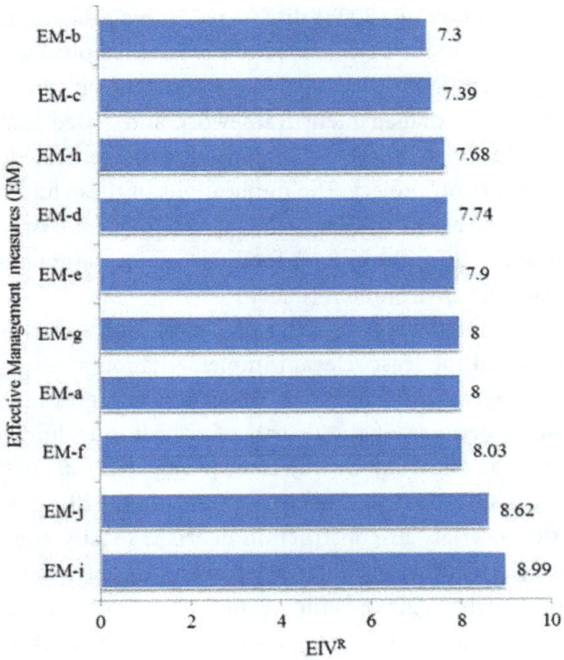

Fig. 3. Ranking profile of the strategic implementation measures perceived by Nigerian construction professionals

Based on the results presented above, the factors EM-i (Effective communication on environmental issue between all layers of subcontractors), EM-j (Close supervision at site level) and EM-f (Establishing waste management plan) emerged as the top three benefits of environmental management practice ranked by the survey respondents. The lowest ranking benefits were EM-h (Inclusion of environmental management in tendering requirements), EM-c (Imposing responsibilities of protecting environment on managerial staff) and EM-b (Reduction, reuse and recycling of construction and demolition wastes) in that order. Figure 3 is a graphical representation of the final ranking.

4. Discussion

The study has revealed that environmental protection and waste management are the perceived benefits of implementing EMS in the Nigerian construction industry. This is supported by Zhang *et al.* (2000) who explored the role of EMS in ensuring the conservation of renewable and non-renewable resources. Previous research has shown that the construction industry in Nigeria has impacted the environment negatively in a number of ways. Aminu *et al.* (2010) gives a detailed breakdown of the percentage contribution of the building industry in Nigeria to the overall carbon emission, stating that the country is in a vulnerable state.

This study also revealed that the provision of an effective waste management plan is essential to meeting the needs of waste management under the EMS framework, just as Poon *et al.* (2004) discovered in his evaluation of the management of construction waste in housing developments. Poon *et al.* (2004) also pointed out the ISO 14001 waste management framework and stated that it also helps in reducing the amount of waste generated on a construction project. The implications of these findings include the ability of EMS to increase significant cost savings and reduce the carbon footprint and hazardous impact on the environment (Ofori 1998).

The reuse and recycling of demolition wastes was considered the least relevant implementation strategy to managing waste by Nigerian construction professionals, perhaps because of the dearth of recycling technology within the Nigerian environment (Awopetu *et al.* 2013). There is still a heavy reliance of landfills in the disposal of waste materials, as construction debris is mostly seen as waste for disposal rather than resources for recycling and reuse (Adebayo 2002; Simion *et al.* 2013). Thus, the waste management plan may often be based on effective disposal of waste, instead of reuse or recycling of materials (Tannahill, Booth 2012). Aminu *et al.* (2010) outlined a proposal for the adoption of the Reuse-Reduce-Recycle-Recover model as a means of reversing this trend.

In a similar study carried out by Morrow and Rondinelli (2002), employee motivation was listed as one of the major reasons that organisations implement EMS and this study demonstrated that this factor was considered to have considerable benefit to the implementation of environmental management in Nigeria. One of the possible indicators of this is the amount of training and development that is invested in employees whose organisations implement EMS (Morrow, Rondinelli 2002). This benefit appeals to the employees need for self-actualisation, which is the highest motivating factor based on Maslow's hierarchy of needs (Martin 2005). Many employees are motivated by the nature of the work itself (Martin 2005) and the study suggests that this is the case in many Nigerian construction companies. Morrow and Rondinelli (2002) discuss the implication of this as leading to a higher level of productivity (Martin 2005) within that organisation.

One major barrier unearthed by the study is the limited access to training as a result of lack of technological support within the organisation. Many construction companies in Nigeria are small and medium enterprises (Windapo, Jegede 2013) and, therefore, may be unable to afford the initial capital investment into training materials and software required to implement EMS in the organisation. Sjostrom and Bakens (1999) attribute the high cost of implementation of EMS as one of the reasons for poor implementation of EMS in smaller organisations. As a result, the training might prove irrelevant to the organisation, especially if it is too expensive to apply. Du Plessis (2005) stated that this barrier may pose a demotivating factor to managers and employees and recommends that rigid systems be modified to suit the unique African climate.

Many organisations are reluctant to adopt EMS for the purpose of increasing their overall business competitiveness due to its high cost of implementation (Ervin *et al.* 2013). There are many reasons for this gap in the Nigerian construction industry, one of which is the aim of clients and contractors to maximise delivery costs by keeping projects under-budget (Windapo, Jegede 2013). This may also lead to corruption within the industry, which could jeopardise the quality of the eventual project deliverables (Windapo, Jegede 2013). In a similar study, Dada (2012) identified some of the factors that directly influenced the choice of procurement methods on Nigerian construction projects, many of which were based on the need for cost certainty, cost reduction and improved project cost performance. Hence, competitive advantage is defined by which organisation can supply the cheapest design option, in sharp contrast to the criteria used in most developed countries like the UK, where value for money ranks equally alongside environmental performance among the key performance indicators on most projects (Chan, A. P. C., Chan A. P. L. 2004).

The lack of adequate technology can be demotivating to organisations' attempts to adopt EMS, and as the study shows, this is the case in Nigeria, based on the high relevance of the lack of technology as a barrier to EMS implementation. Du Plessis (2005) points out the inability of Africa to compete with most continents in the area of technological advancement and the regular supply of basic infrastructure and Nigeria is no exception. As a result of this, many companies are unable to maximise the potential of the technology that is part of the EMS package. This is also responsible for poor investment of these companies in EMS training and development programmes, as earlier explained.

The difficulty in implementing EMS without appropriate technology is more vivid when examined against the complexity of the environmental management process. Christini *et al.* (2004) outlines the factors that need to be taken into account when implementing an EMS including proper documentation, which is difficult to achieve without the right infrastructure. However, the study concluded that an increase in documentation workload is considered one of the least relevant barriers to the adoption of EMS by Nigerian construction firms. It also concluded that the time consuming nature of the environmental improvement process had the least relevance to the decision to adopt EMS in organisations. This is in contrast to the view presented by Windapo and Jegede (2013) with regards to the priority placed on the value of time by managers of small businesses.

Effective communication on environmental issues between all layers of sub-contractors emerged as the most effective environmental management measure in Nigeria based on the findings from the study. This is validated by the case study of Beers Skanka, a USA construction firm, which was discussed by Christini *et al.* (2004) and which highlighted the importance of communication in EMS policy formation. Effective communication is also the foundation to the Plan-Do-Check-Act Model adapted into an EMS framework by Zhang *et al.* (2000) as it is one of the tools by which continuous improvement can be achieved (Gupta 1995). Christini *et al.* (2004) recommended the development of an employee training plan to further enhance the training and development of the supply chain by employing tools such as site inductions and documentation of best practice from previous projects. The essence of this is to encourage the collation of a set of internal standards by which the organisation can improve its environmental management processes. This strategy is useful in developing informal frameworks which may then be tested and passed on to legal institutions for policy formation, as Hill and Bowen (1997) recommend in their analysis of South African implementation of EMS. The strategy is, therefore, relevant in dealing with the lack of formal environmental frameworks highlighted by Aminu *et al.* (2010).

The need to closely monitor activities on site was highlighted in the study as an important factor in ensuring that suppliers, sub-contractors, and other project stakeholders are aware of the organisation's environmental goals and adhere to the relevant EMS. Christini *et al.* (2004) explain the success of any approach to environmental management is dependent on the effectiveness of the management strategy employed in coordinating the activities of all relevant stakeholders. The study showed that the challenge with supply chain coordination was seen as a minor barrier to EMS. This implies that supply chain management in Nigeria is not a pressing challenge to environmental performance.

Conclusion

The construction industry is a major contributor to the GDP of many developing countries and is pivotal to the growth and development of the Nigerian economy. The study has pointed out that the implementation of EMS in the construction industry will contribute to productivity within that sector, as the environment has become a more relevant industry stakeholder. Furthermore, it is important that construction companies are more proactive in dealing with sustainability issues, by ensuring that their activities are more eco-friendly and comply with recognised environmental standards. The findings from the study survey administered among construction professionals in Nigeria has given insight into the major benefits and barriers of EMS and how these insights could be used to generate a more effective implementation strategy within construction organisations.

The study showed that the need for project managers to keep projects under budget and the inability of many organisations to afford the high cost of EMS personnel training are some of the major barriers to EMS adoption. When matched against the identified importance of employee training to the adoption of EMS and its role in increasing employee morale, the need for a strategy that will be less financially cumbersome for small and medium enterprises becomes apparent. The study identified that further research into a framework that accommodates the priority that project managers and clients place on project cost-savings in Nigeria may be more effective in ensuring an increase in the implementation of EMS. In addition, the importance of an employee training plan was highlighted in this study as an effective tool in ensuring continuous development and improvement on environmental management legislation.

The study supported previous research on the effectiveness of ISO 14001 in managing construction waste, which is a major environmental challenge in Nigeria. It is therefore recommended that further studies on a waste management plan based on the Reuse-Reduce-Recycle-Recover model in place of the current waste disposal model be carried out. Effective communication in order to effect continuous improvement, and a close monitoring of site activities to ensure that all stakeholders are aware of the importance of environmental management were also identified as enablers to ensure stakeholder adoption of an organisation's environmental strategy.

This study has highlighted a number of benefits that can be reaped from the successful adoption of the EMS framework in the Nigerian construction industry. It is recommended that further research be carried out in this area as an increase in adoption of EMS will not only be of benefit to more organisations in the construction industry, but also to other industry sectors in Nigeria. A well-defined EMS implementation framework in Nigeria may also serve as the basis for developing an environmental management framework for developing countries in Sub-Saharan Africa and other parts of the world.

References

Adebayo, A. A. 2002. *Sustainable construction in Africa: agenda 21 for sustainable development in developing countries. African Position papers.*

Adegbite, B. O. 2013. Assessment and adaptation of an appropriate green building rating system for Nigeria, *Journal of Environment and Earth Science* 3(1): 1–10.

Aminu, D. Y.; Kandar, M. Z.; Ossen, D. R. 2010. Evoking the green-shift in the building industry for sustainable development in Nigeria, in S. Laryea, R. Leiringer, W. Hughes (Eds.). *Proceedings of West Africa Built Environment Research*

(WABER) Conference, 27–28 July 2010, Accra, Ghana, 333–343.

Awopetu, M. S.; Coker, A. O.; Awopetu, R. G.; Awopetu, S. O.; Booth, C. A.; Fullen, M. A.; Hammond, F. N.; Tannahill, K. 2013. Reduction, Reuse and recycling of solid waste in the Makurdi Metropolitan area of Nigeria: public opinions and perceptions, *International Journal of Education & Research* 1(11): 1–12.

Balzarova, M. A.; Castka, P.; Bamba, C. J.; Sharp, J. M. 2004. How organisational culture impacts on the implementation of ISO 14001:1996 – a UK multiple-case view, *Journal of Manufacturing Technology Management* 17(1): 89–103. http://dx.doi.org/10.1108/17410380610639524

BBC News. 2014. *Nigeria becomes Africa's biggest economy* [online], [cited 15 April 2014]. Available from Internet: http://www.bbc.co.uk/news/business-26913497

Bohoris, G.; O'Mahony, E. 1994. BS 7750, BS 5750 and the EC's eco management and audit scheme, *Industrial Management & Data Systems* 94(2): 3–6. http://dx.doi.org/10.1108/02635579410055819

Boiral, O.; Sala, J. 1998. Environmental management: should industry adopt IS0 14001, *Business Horizons* 41(1): 57–64. http://dx.doi.org/10.1016/S0007-6813(98)90065-9

BSI Group. 2014. *BSI case study: Costain group* [online], [cited 25 April 2014]. Available from Internet: http://www.bsigroup.com/Documents/iso-22301/case-studies/Costain-case-study-UK-EN.pdf

Business Day Online. 2013. *How Nigerians use social media* [online], [cited 25 April 2014]. Available from Internet: http://businessdayonline.com/2013/08/how-nigerians-use-social-media/#.U2EEmVfiiSo

Carillion Plc. 2014. *2013 Sustainability Report* [online], [cited 25 April 2014]. Available from Internet: http://www.carillionplc.com/media/196204/carillion-sr2013-summary.pdf

Castka, P.; Balzarova, M. A. 2008. The impact of ISO 9000 and ISO 14000 on standardisation of social responsibility – an inside perspective, *International Journal of Production Economics* 113(1): 74–87. http://dx.doi.org/10.1016/j.ijpe.2007.02.048

Chan, A. P. C.; Chan, A. P. L. 2004. Key performance indicators for measuring construction success, *Benchmarking: An International Journal* 11(2): 203–221.

Christini, G.; Fetsko, M.; Hendrickson, C. 2004. Environmental management systems and ISO 14001 certification for construction firms, *Journal of Construction Engineering and Management* 130(3): 330–336. http://dx.doi.org/10.1061/(ASCE)0733-9364(2004)130:3(330)

Czaja, R. F.; Blair, E. A. 2013. *Designing surveys: a guide to decisions and procedures.* SAGE Publications Incorporated.

Dada, M. O. 2012. Predictors of procurement selection: an investigation of traditional and integrated methods in Nigeria, *Journal of Construction in Developing Countries* 17(1): 69–83.

Dixon, T.; Colantonio, A.; Shiers, D.; Gallimore, P.; Reed, R.; Wilkinson, S.; Keeping, M. 2007. *A green profession? An audit of sustainability tools, techniques and information for RICS members.* RICS Report.

Du Plessis, C. 2005. Action for sustainability: preparing an African plan for sustainable building and construction, *Building Research & Information* 33(5): 405–415. http://dx.doi.org/10.1080/09613210500218974

Ervin, D.; Wu, J.; Khanna, M.; Jones, C.; Wirkkala, T. 2013. Motivations and barriers to corporate environmental management, *Business Strategy and the Environment* 22(6): 390–409. http://dx.doi.org/10.1002/bse.1752

Famiyeh, S.; Kuttu, S.; Anarfo, E. B. 2014. Factors influencing the implementation of environmental management systems in Ghanaian firms, *Environmental Management and Sustainable Development* 3(2): 18–30. http://dx.doi.org/10.5296/emsd.v3i2.5893

Fewings, P. 2013. *Construction project management: an integrated approach.* 2nd ed. London: Routledge.

Freeman, R. E. 1984. *Strategic management: a stakeholder approach.* Boston: Pitman.

Gupta, M. C. 1995. Environmental management and its impact on the operations function, *International Journal of Operations & Production Management* 15(8): 34–51. http://dx.doi.org/10.1108/01443579510094071

Haigh, N.; Griffiths, A. 2009. The natural environment as a primary stakeholder: the case of climate change, *Business Strategy and the Environment* 18(6): 347–359. http://dx.doi.org/10.1002/bse.602

Hammond, F. N.; Booth, C. A. 2010. Environmental stakeholder management, in E. Chinyio, P. Olomolaiye (Eds.). *Construction stakeholder management.* Oxford: Wiley-Blackwell, 317–337.

Hill, R. C.; Bowen, P. A. 1997. Sustainable construction: principles and a framework for attainment, *Construction Management & Economics* 15(3): 223–239. http://dx.doi.org/10.1080/014461997372971

Holmes, P. 1985. *Nigeria: giant of Africa. Nigeria: National Oil and Chemical Marketing Company of Nigeria.*

Idoro, G. I. 2012. The influence of project documents on the outcome of construction projects procured by traditional contracts in Nigeria, *Journal of Construction in Developing Countries* 17(1): 1–19.

Ilinitch, A. Y.; Soderstrom, N. S.; Thomas, T. E. 1998. Measuring corporate environmental performance, *Journal of Accounting and Public Policy* 17(4): 383–408.

ISO 2012. The ISO survey of management system standard certifications – survey data.xls [online], [cited 06 July 2014]. Available from Internet: http://www.iso.org/iso/iso-survey.zip

Jolevski, A. 2013. Benefits of ISO14001 certification in the Republic of Macedonia, *International Journal of Environment and Sustainable Development* 12(2): 168–183. http://dx.doi.org/10.1504/IJESD.2013.052983

Martin, J. 2005. *Organizational behaviour and management.* Cengage Learning EMEA.

Marzouk, M.; Abdelhamid, M.; Elsheikh, M. 2013. Selecting sustainable building materials using system dynamics and ant colony optimization, *Journal of Environmental Engineering and Landscape Management* 21(4): 237–247. http://dx.doi.org/10.3846/16486897.2013.788506

Medineckiene, M.; Turskis, Z.; Zavadskas, E. K. 2010. Sustainable construction taking into account the building impact on the environment, *Journal of Environmental Engineering and Landscape Management* 18(2): 118–127. http://dx.doi.org/10.3846/jeelm.2010.14

Morrow, D.; Rondinelli, D. 2002. Adopting corporate environmental management systems: motivations and results of ISO 14001 and EMAS certification, *European Management Journal* 20(2): 159–171. http://dx.doi.org/10.1016/S0263-2373(02)00026-9

Ofori, G. 1998. Sustainable construction: principles and a framework for attainment-comment, *Construction Management & Economics* 16(2): 141–145. http://dx.doi.org/10.1080/014461998372448

Ofori, G. 2006. Revaluing construction in developing countries: a research agenda, *Journal of Construction in Developing Countries* 11(1): 1–16.

Oke, S. A. 2004. A review on environmental research, *Journal of Environmental Engineering and Landscape Management* 12(4): 153–157.

Poon, C.; Yu, A. T.; Jaillon, L. 2004. Reducing building waste at construction sites in Hong Kong, *Construction Management and Economics* 22(5): 461–470. http://dx.doi.org/10.1080/0144619042000202816

Rio Tinto. 2008. *2008 Annual Report* [online], [cited 27 April 2014]. Available from Internet: http://www.riotinto.com/annualreport2008/pdfs/AnnualReport2008.pdf

Shen, L.; Tam, V. W. 2002. Implementation of environmental management in the Hong Kong construction industry, *International Journal of Project Management* 20(7): 535–543.

Simion, I. M.; Fortuna, M. E.; Bonoli, A.; Gavrilescu, M. 2013. Comparing environmental impacts of natural inert and recycled construction and demolition waste processing using LCA, *Journal of Environmental Engineering and Landscape Management* 21(4): 273–287. http://dx.doi.org/10.3846/16486897.2013.852558

Sjostrom, C.; Bakens, W. 1999. CIB Agenda 21 for sustainable construction: why, how and what, *Building Research & Information* 27(6): 347–353. http://dx.doi.org/10.1080/096132199369174

Tannahill, K.; Booth, C. A. 2012. Linkages of waste management strategies and climate change issues, in C. A. Booth, F. Hammond, J. E. Lamond, D. G. Proverbs (Eds.). *Solutions to climate change challenges in the built environment.* Oxford: Wiley-Blackwell, 207–221.

Wenk, M. S. 2004. EU's eco-management and audit scheme, *Environmental Quality Management* 14(1): 59–70. http://dx.doi.org/10.1002/tqem.20026

Windapo, A. O.; Jegede, O. P. 2013. A study of health, safety and environment (HSE) practices of Nigerian construction companies, *The Professional Builder* 4(1): 92–103.

Zhang, Z. H.; Shen, L. Y.; Love, P. E.; Treloar, G. 2000. A framework for implementing ISO 14000 in construction, *Environmental Management and Health* 11(2): 139–149. http://dx.doi.org/10.1108/09566160010321541

Victoria O. OWOLANA is an Assistant Project Manager at Courtney Michael Partnership, Nigeria and a former postgraduate student at the University of the West of England, Bristol (UK). Research interests: architecture, project management, construction law and procurement, sustainability, and environmental management.

Colin A. BOOTH Dr., Associate Professor of Sustainability, is Associate Head of Research and Scholarship in the Faculty of Environment and Technology at the University of the West of England, Bristol (UK). Publications: author/co-author of several books and ~140 scientific papers and chapters. Research interests: sustainability, environmental management, climate change mitigation and adaptation strategies, water resources management, and built environment studies.

20

ANALYSIS ON TRENDING ELECTROMAGNETIC EXPOSURE LEVELS AT HOMES AND PROXIMITY NEXT TO BASE STATIONS ALONG THREE YEARS IN A CITY

Osman ÇEREZCİ, Baha KANBEROĞLU, Şuayb Çağrı YENER

Department of Electric-Electronic Engineering, Faculty of Engineering, Sakarya University, Esentepe Campus, 54055, Serdivan, Sakarya, Turkey

Abstract. Electromagnetic field exposure levels of people living in the closest houses to the GSM transmitting antennas were investigated in a city between 2010 and 2012. At the end of 3-year period, trend of the electromagnetic exposure levels was determined especially for indoor/outdoor environments near the base station antennas. Because of increasing number of base stations by years and changing of the technology, it is determined that average electromagnetic exposure values in the city increased in a certain extent each year. Total and frequency selective measurements were performed in indoor/outdoor places. The results were compared by International ICNIRP limits. In addition, possibility of the compliance with some European countries which have applied low limit values are discussed to minimize involuntary exposure to electromagnetic fields at indoor/outdoor environments. Consequently, a suggestion is made and discussed for obtaining relatively homogeneous distribution of electromagnetic field exposure at indoor/outdoor environments near base station antennas to eliminate extent values.

Keywords: electromagnetic field exposure, base station, indoor environment, measurement.

Introduction

In recent years, with continuous development of mobile communication, electromagnetic exposure from GSM base station antennas has changed due to the increasing number of base stations placed in close of residential areas, schools and streets. In order to control and limit the electromagnetic field exposure levels of people, international guidelines and standards were established (ICNIRP 1998; EN 50413: 2008). Also, national regulations based on these guidelines and standards are prepared and taken into account (BTK 2011; Karpowicz *et al.* 2006; FOEN 1999). A number of researchers studied on public and occupational exposure from base station antennas with performing field measurements (Baltrenas *et al.* 2012; Wu *et al.* 2012; Alanko *et al.* 2008). Meanwhile, theoretical analysis of electromagnetic exposure levels has been also studied.

The level of human exposure to electromagnetic radiation from base station antennas can be calculated theoretically (Poljak *et al.* 2004). A numerical and experimental analysis has been performed with human exposure in close of the base station antenna via phantom model (Gosselin *et al.* 2009). With reflecting walls and buildings, human exposure studies are expanded to a urban environment model (Bernardi *et al.* 2000). Existence of an exclusion zone is important to keep people away from excessive exposure in the vicinity of base station antennas (Singh 2012). The electric fields radiated from base station antennas are evaluated and compared with the measurements and permissible exposure limits to determine the safety zone for electromagnetic exposure of base station antenna (Chen *et al.* 2009). On the other hand, another way to determine safety distance for electromagnetic exposure is to calculate Specific Absorption Rate (SAR) and compare with basic restrictions (Joseph, Martens 2005). By means of SAR calculations, minimum safety distance is to be calculated for a person in the near field of a base station for single frequency (Cooper *et al.* 2002). The actual situation is highlighted via measuring the GSM frequency-based electromagnetic environment parameters as electric field, magnetic field or electromagnetic power density which are spread to the house and street of the cities (Wu *et al.* 2012). Also, measurements are necessary to compare them

Corresponding author: Osman Çerezci
E-mail: cerezci@sakarya.edu.tr

with standards to verify whether the electromagnetic exposure levels are appropriate for limits (Neskovic *et al.* 2011). Electromagnetic field distribution measurements can be used to determine compliance with exposure limits (Higashiyama, Tarusawa 2011). The number and distribution of base stations are more densely in populated areas than in non-populated areas (Viel *et al.* 2009). Base stations operating in densely populated cities in GSM 900, 1800 and 2100 MHz frequency bands are continuously evolving due to the conditions of competition and the need of better response to increasing customer demands. This evolution is carried out by replacing base antennas with new high-gain antennas or establishing a new base station in the targeted area. Naturally, electromagnetic field density of environment has also changed in parallel to this evolution process. Because of changing process, people living near base stations, especially in indoor places, may be exposed to several times more electromagnetic field than those living farther away from base stations. Excessive exposure of people to electromagnetic fields can be avoided by positioning the base antennas appropriately (Alanko *et al.* 2008).

Intensity of electromagnetic radiation emitted from base station antennas changes according to parameters such as being in the main lobe of antenna, distance from the antenna and exposure duration (Baltrenas *et al.* 2012). Unnoticed continuous exposure situation occurs for a human being who lives against or near the base station antennas or spends a considerable time there. Therefore, monitoring the trend of increasing electromagnetic field in GSM frequencies by performing control measurements at indoor places in parallel with the evolution of base stations in cities are important to determine the involuntary exposure to electromagnetic radiation. Information derived from these studies is used to control the levels and estimate the future values of electromagnetic radiation in cities and specify the minimum electromagnetic exposure possibility (Çerezci, Şeker 2010).

In this study, GSM frequency-based electromagnetic radiation was measured especially in living areas closest to the base antennas and change of electromagnetic exposure over the years investigated during a 3-year period starting from 2010. Measurements were performed in Turkey's fourth largest metropolis Bursa's 300000-populated Nilüfer district. 86 and 98 measurements were performed at indoor locations in 2010 and 2011, respectively. In 2012, with increasing number of base stations, 228 GSM frequency-based electromagnetic radiation measurements were done at indoor and outdoor locations. 61 measurements were performed at outdoor locations like bus stops and frequently used, crowded points in front of the main lobe of base station antenna that are mounted on the street lighting poles. Briefly, during 3-year period, 412 measurements were performed at 250 locations (189 indoor and 61 outdoor) in the distance 15–40 meters from the base stations. 64 of 250 locations investigated over 3 years (2010 to 2012), 34 investigated twice(2010–2011 or 2011–2012) and 152 investigated only one time (14 in 2010 and 138 in 2012).

The measurement results were compared with ICNIRP and national exposure limits. Also, the increase trends in graphics were interpreted by years. The aim of the regulations of EMF's are to protect people from discomforting effects. Therefore, maximum potential fields need to be determined at so called "sensitive areas" in principle rooms which are used by people regularly for continuous periods.

This study is an updated version of previous studies (Baltrenas *et al.* 2012; Bernardi *et al.* 2000; Viel *et al.* 2009; Wu *et al.* 2012) that had been published. The measurement period and the continuity of the study are the main differences. This study showed continuation and measurements were repeated in same places every coming year from 2010. Meanwhile, Nilüfer municipality had announced the results of the measurements as an annual report to increase public awareness on this subject and help people to monitor electromagnetic exposure level in Nilüfer city. During 3-year period, with repeated and continuous measurements, involuntary exposure to electromagnetic fields at indoor/outdoor environments, especially at sensitive areas, was controlled and analyzed. These measurements are helpful to obtain relatively homogeneous distribution of electromagnetic field exposure at indoor/outdoor environments near base station antennas and to minimize the exposure to electromagnetic fields. Also, the presented measurement data, in this study, is expected to reflect the general pattern of electromagnetic exposure level in the distance 15–40 meters from standardized base station antennas in Turkey and also in other countries because of the use of similar infrastructure worldwide. But detailed analysis of electromagnetic exposure in particular locations needs individual assessment, because of the significant variations of exposure level between particular locations presented in this study.

1. Electromagnetic safety regulations

Safety requirements for limiting exposure to time varying electric and magnetic fields are enforced by regulations. At international level, safety guidelines for electromagnetic exposure of workers/general public and controlled/uncontrolled environments have been issued by the International Commission on Non-Ionizing Radiation Protection (ICNIRP 1998), American National Standards Institute (ANSI) and Federal Communication Commission (FCC) and some of the other regularly boards.

The ICNIRP guidance's for EMF exposure which GSM base stations must satisfy changes between 41 V/m and 61 V/m depending on the operating frequency. Some countries such as Switzerland, Italy and Poland have applied lower limits in sensitive areas from ICNIRP levels. The limits for the total EMF's from GSM base station antennas are 5 V/m in Switzerland and 7 V/m for Poland. The regulations for GSM frequency range have been generated with reference to the ICNIRP regulations by BTK (Information and Communication Technologies Authority) in Turkey. Total GSM exposure limits generated by BTK are equal to ICNIRP reference levels.

Maximum permissible exposure limits for single GSM and total environment at GSM frequencies presented in BTK guidelines are given in Table 1 and 2. Single GSM exposure limit for electric field is 1/4 of total GSM exposure limit of BTK and ICNIRP reference level. Also, single GSM exposure limit for power density is 1/16 of total GSM exposure limit as given in Table 2.

Due to the signals emitted from base station antennas at different frequencies, the electromagnetic field exposure study was focused on measuring the total of all GSM frequencies at any point of the city.

A comparison table among the reference level of Turkey, Switzerland, Italy and Poland are given in Table 3. As it can be seen from the table that the total limit levels of Turkey for electric field is seven-eight times greater than other countries and at least 50 times greater for power density.

Table 1. Maximum permissible exposure limits at GSM frequencies presented in BTK guidelines (electric field)

Frequency	GSM Band Electric Field Strength (V/m) (Single GSM)	GSM Band Electric Field Strength (V/m) (Total GSM)
900 MHz	10.23	41.25
1800 MHz	14.47	58.34
2100 MHz	15	61

Table 2. Maximum permissible exposure limits at GSM frequencies presented in BTK guidelines (power density)

Frequency	GSM Band Power Density (W/m^2) (Single GSM)	GSM Band Power Density (W/m^2) (Total GSM)
900 MHz	0.28	4.5
1800 MHz	0.56	9
2100 MHz	0.66	10.5

Table 3. Maximum permissible total exposure limits of Turkey, Italy, Switzerland and Poland at GSM frequencies

Country (Total Environment)	GSM Band Electric Field Strength (V/m)	GSM Band Power Density (W/m^2)
Turkey	41–61	4.5–10.5
Italy	6	0.1
Switzerland	5	0.042–0.095
Poland	7	0.1

2. Measurement and method

In this study, with the partnership of Bursa-Nilüfer municipality public health department and Sakarya University, 412 electromagnetic field measurements were performed at sensitive places of residents and streets near base stations between 2010 and 2012. At the first step of the project, houses to be measured were decided at the locations where active base station existed. The choice of locations was made in cooperation with Bursa-Nilüfer municipality public health department. The parameters as whether the location of measurement point was in front of the main lobe of the base antenna or not and distance to the base antenna used for evaluation of the measurement results were taken into account during the decision process. Then, permissions required for the measurement program were taken from the people living in the houses. During 3-year period, the decision and permission steps repeated at the beginning of every year.

In 2010 and 2011, base stations in Nilüfer district were located on the roof of the buildings. Due to this situation, measurements were performed in the rooms of houses at the upper floor of the neighbour buildings to determine the total maximum exposure. In 2012, it was noticed that in addition to the base station antennas placed on the roof of the buildings, new base stations antennas were mounted on the street lighting poles at a height of 4–5 meters located through the street. The base stations on the lighting poles were located frequently on the streets and the number of base stations increased by 25% in the city due to this application. Finally, the measurements in the last year were performed at locations such as bus stops located in front of base antennas or on the way of people used for going to their homes and works where base station antennas were placed on lighting poles. Street measurements that consist can give an idea about determining the electromagnetic field levels at people's walking and transition areas placed opposite to the increasing base station placement application on low altitude street lighting poles.

All spot and continuous measurements were done at far field of base antennas by using PMM 8053 broadband field meter with EP33M isotropic E-field Probe 700 MHz-3GHz, Narda NBM 550 broadband field meter with EF0691 isotropic E-field Probe and Narda SRM 3006 Selective radiation meter with 27 MHz-3 GHz isotropic E-field Probe. The broadband field meter measures instantaneous field strength of all signals presented in a large frequency band. Therefore, broadband field meter cannot distinguish between signals coming from different transmitting sites. This does not expose a great problem when measuring field levels are in regard to limits recommended by ICNIRP regulations which range up to 61 V/m. Electric field values at other frequency bands except GSM frequency were found to be lower than 0.01 V/m via using spectrum analyzer at the measurement points. Thus, signal level outside GSM frequency bands was recognized ineffective and total in addition to the base station antennas placed on the roof of the buildings. Electromagnetic field measurements were considered at 900/1800/2011 MHz frequencies.

Measuring techniques used in the study were based on recommendations specified in standards. Measurements were performed at the far field of the base station antennas. According to several guidelines, the exposure to radiofrequency electromagnetic fields is quantified in terms of measured electric field (V/m) and estimated power density (W/m^2) that are the main physical parameters which have been assessed in this study.

All measurements were performed between 11:00 and 16:30 via a field meter placed on a tripod. Electric field measurements at each selected points were performed for different heights from the floor as 1 m, 1.5 m and 1.7 m for 6 minutes durations. All measurements were repeated at different hours of measurement period and a difference of 25–30% was noticed at some measurement results. Due to these differences, the highest field strength value was recorded as the measurement value of selected point. Distance between measurement points and base station antennas varied from 15 m to 40 m. The diversity of distances resulted from physical conditions of measurement points. Vertical difference between measurement point and base station antenna was measured from 0 to 25 m and horizontal difference was from 12 to 35 m. All distances were measured with Leica Disto D8 laser meter (0.2–200 m range).

The study was repeated in 2011 and 2012. In the second and third years of program, a great attention was paid for doing measurements at the same points and conditions of previous years (time period, weekday, weather condition). New-built settlements, ascending number of base stations, base stations that undergo revisions (direction or tilt change, replaced antennas) caused an increase at the number of measurements by years. A few base stations removed completely or moved to another places. So the change of the electromagnetic environment was investigated at the same locations and conditions over the previous year.

During measurements, it is aimed to prevent measurements from the effects of the operator and suppress the uncertainty of measurements by using the same measurement method, same measurement setup and performing the measurements at same conditions. But many other factors affect the measurements like intensity of communication traffic, weather, physical conditions or device and probe based uncertainties. Uncertainties of probes and devices, used in this study, can be determined from the calibration certificate documents of measurement devices and listed in Table 4.

Table 4. Uncertainties of E-field measurement devices calibration

NBM550 EF Probe		SRM 3006 E Probe	
Frequency (MHz)	Uncertainty (dB)	Frequency (MHz)	Uncertainty (dB)
0.1–100	0.5	26–300	1
200–300	0.8	433–1600	1.5
500–1000	1.5	1900–3000	1
1800–6000	0.6	–	–

Measurements were also repeated in same conditions to confirm the measurement results and to overcome the effect of uncertainties like communication traffic, weather etc. Some different results were determined in terms of the overall effect of the uncertainty of measurement as (25–30)% on measurement results and the highest results were taken into account as measurement results.

At the end of the project, electromagnetic environment at homes and streets near the base stations in Nilüfer district was analyzed and change of electromagnetic environment was observed during three years. The electromagnetic fields radiated from transmitters of base stations measured between frequencies from 700 MHz to 3 GHz. 900 MHz, 1800 MHz and 2100 MHz GSM frequencies were especially focused on. Frequency level measurement was realized with spectrum analyzer when needed. The distribution of the measured values, analysis of statistical situation and review in terms of national and international limits were given with the graphs.

3. Results

In this study, the electromagnetic field measurement results obtained between 2010 and 2012 in Bursa/Nilüfer are presented. 86, 98 and 228 measurements were performed in 2010, 2011 and 2012, respectively. Electromagnetic exposure was undertaken at indoor and outdoor locations at vicinity of the base station antennas over three years in the city. An assessment was made by comparing the measured electric field levels with guidelines on limiting people exposure published by ICNIRP and other regulations which have been applying exposure of the precaution limits. The measurement values for all points and levels according to the limits are given graphically. In addition, the numerical distributions of electric field measurements are given in graphs. Also, the changing level of the results according to years are presented in the study. The total electromagnetic exposure with GSM band frequencies quotients due to all GSM signals frequencies with 900/1800/2100 MHz measured ranged from 0.3% to 27% of the ICNIRP reference levels for exposure of the general public(for 900 MHz). The results of this work may be useful to gain information on public exposure to total electromagnetic fields from GSM antennas. Also uptrending rate of the electromagnetic power density determined in the environment of sensitive areas with respect to year by year.

Table 5. Electric field measurement results – maximum value and ratios by reference levels

Year	Maximum Electric Field Strength (V/m)	Ratio by ICNIRP Reference Level (%)	Ratio by BTK Reference Level (for Single GSM) (%)
2010	4.7	12	46
2011	11	27	108
2012	10	24	98

Table 6. Electric field measurement results – averaged values and ratios by reference levels

Year	Average Electric Field Strength (V/m)	Ratio by ICNIRP Reference Level(%)	Ratio by BTK Reference Level (for Single GSM) (%)
2010	0.91	2	9
2011	1.38	3	13
2012	1.54	4	15

Table 7. Power density estimation – maximum values and ratios by reference levels

Year	Maximum Power Density Strength (mW/m^2)	Ratio by ICNIRP Reference Level (%)	Ratio by BTK Reference Level (for Single GSM) (%)
2010	59.8	1	21
2011	321	7	114
2012	267	6	95

Table 8. Power density estimation – averaged values and ratios by reference levels

Year	Average Power Density Strength (mW/m^2)	Ratio by ICNIRP Reference Level (%)	Ratio by BTK Reference Level (for Single GSM) (%)
2010	2.2	0.05	0.8
2011	5.1	0.11	1.8
2012	6.3	0.14	2.3

The measurement results are presented with the graphs by years. The information contained is given both in graph titles and figure captions. The measurement values for all measurement locations are given in Figures 1, 4 and 7 for 2010, 2011 and 2012 respectively. Limit levels of Turkey (for single GSM and environment), Switzerland and Poland are also given in graphs. Numerical distributions of electric field measurements are given in Figures 2, 5 and 8 by years, respectively. The numerical distribution of all measurements performed during 3-year period is given in Figure 10. In addition, distribution of electric field measurements by distance are given in Figures 3, 6 and 9 for 2010, 2011 and 2012 respectively.

Maximum and average values of electric field measurements are given with ratios by limit values of ICNIRP and BTK at Tables 5 and 6 for 2010, 2011 and 2012, separately.

It is clear from Table 5 that maximum electric field measurement results are quite below the ICNIRP limits (max. 27%). However, 2011 maximum value exceeds the permissible exposure limits for single GSM and also 2012 maximum value is nearly equal to the limit value. According to Table 6, average electric field values of each measurement year are below the ICNIRP reference level and BTK reference level for single GSM but there is a notable increasing trend of average electric field values by years.

Maximum and average values of power density are given with ratios by limit values of ICNIRP and BTK at Tables 7 and 8 for 2010, 2011 and 2012, respectively.

In 2010, 86 electric field measurements were performed at indoor places in homes located in far field of the GSM base station antennas and values of measurements are shown in Figure 1. All of the measurement results were below the national reference level for single GSM

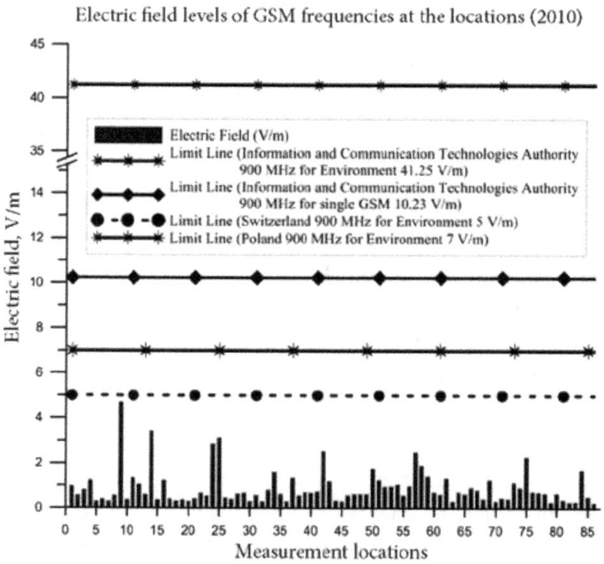

Fig. 1. Electric field measurement levels of GSM frequencies at the locations in 2010

and ICNIRP limits. The maximum electric field strength reached 4.7 V/m and below the reference levels. The maximum exposure was 12% of ICNIRP limits and 46% of national reference level for single GSM as shown at Table 5.

In 2010, power density values at measurement locations were in range from 0.17 to 59.8 mW/m². The

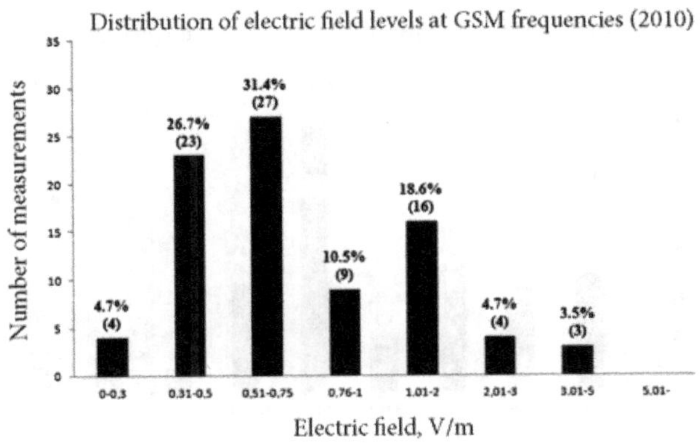

Fig. 2. Distribution of electric field levels at GSM frequencies (2010)

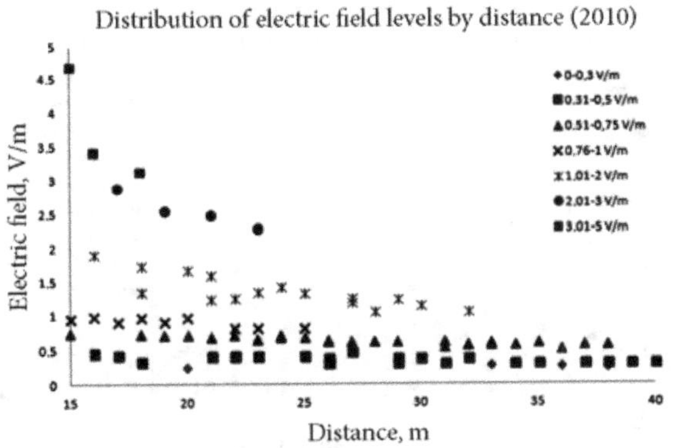

Fig. 3. Distribution of electric field levels by distance (2010)

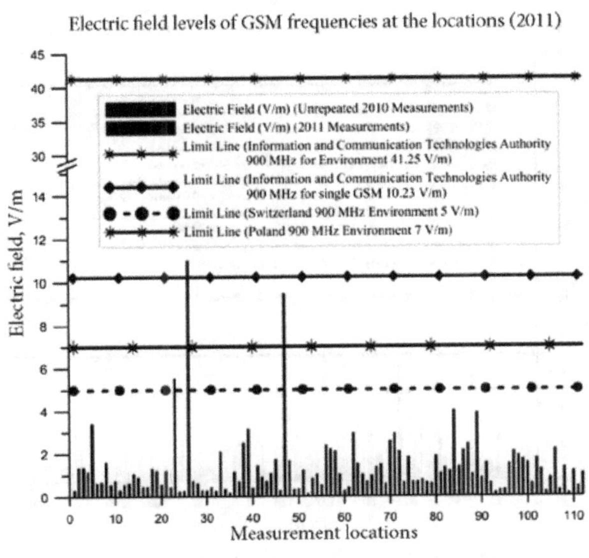

Fig. 4. Electric field measurement levels of GSM frequencies at the locations in 2011

maximum power density value is 21% of BTK single GSM reference level and also 1% of ICNIRP limit. Average power density of locations in 2010 was 2.20 mW/m².

Also, numeric distribution of 2010s electric field measurements is given in Figure 2. It is noticed that 97% (83 of 86) of the measurements are lower than 3 V/m and 73% (63 of 86) are lower than 1 V/m. Due to these electric field values, average value of measurements was calculated 0.91 V/m as given at Table 6. Average value of the measurements was 11 times lower than national reference level for single GSM and 2% of ICNIRP limits (for 900 MHz).

Distribution of electric field measurements by distance are given in Figure 3. It is clear that higher measurement results are located close to base station antennas (mostly located from 15 m to 25 m). The highest electric field value was measured at one of the closest locations (4.7 V/m at 15 m distance). 23 of 86 measurement results were greater than 1 V/m. 9 of measurement results were located between 15 m to 20 m and 8 of them between 20 m to 25 m. With increasing distance, measurement values were decreasing and mostly 0.5 V/m and smaller values were measured from 25 m and further distances. This graphic also showed and proved the inverse relation between distance and electric field strength.

In 2011, 98 measurements were performed at indoor places at homes located in the far field of GSM base station antennas and values are shown in Figure 4. 72 of 98 measurements were repeated measurements at same locations of previous year and 26 new measurement locations added to these locations. All except one of the measurement results were below the national reference level for single GSM. The maximum electric field strength was 11 V/m and 27% of ICNIRP limits. Also, it was noticed that maximum electric field measurement value was higher than national reference level for single GSM and equal to 108% of national reference level.

In addition, different electric field values are noticed at some measurement points compared to previous year during analyzing the repeated measurements. 34 of 72 measurement values increased and 12 decreased. Measurement results in 26 locations were identified as equal in both years. Uncertainty of measurement devices are also taken into account during analyzing process.

Power density values ranged from 0.06 to 321 mW/m². The maximum power density value is 114% of BTK reference level for single GSM and 7% of ICNIRP and BTK total reference level. Average power density was 5.1 mW/m².

In Figure 5, distribution of electric field levels are shown by numerical ranges of measurement results. Due to Figure 4, 3% of measurement values (3 of 98) are higher than 5 V/m and only 1% greater than 10 V/m. Electric field values are below 3 V/m at 94% of measurement points (92 of 98) and below 1 V/m at 48% (47 of 98).

Average value of measurements was 1.38 V/m in

2011. It was 13% of BTK national reference level for 900 MHz and 30 times smaller than ICNIRP reference level as shown at Table 6. Average electric field strength increased 51% against 2010 average value of measurements. According to the comparison of 2010 and 2011 measurements, it was observed that the number of measurements above 1 V/m increased 121% and raised from 23 to 51. Also, 100% increase was noticed at number of measurements above 3 V/m (from 3 to 6).

In Figure 6, distribution of electric field measurements by distance in 2011 is given. 50% of measurement locations are between 15 m to 25 m. 30 of 51 electric field measurements greater than 1 V/m are located between 15–25 m scale. The highest electric field value was measured at 16 m away from the related base station. With increasing distance, measurement values were decreasing and mostly 0.75 V/m and smaller values were measured from 25 m and further distances. 23 of 30 measurement values are smaller than 1 V/m between 30–40 m scale.

Electric field measurement levels of 2012 are shown in Figure 7. Differ from 2010 and 2011, number of measurements increased and 228 measurements were performed. 90 of 228 measurements were repeated measurements at same locations of previous year and 138 new measurement locations added to these locations. The main reason of increasing number of measurements is new base stations antennas mounted on the street lighting poles at a height of 4–5 meters located through the street. Also, new settlements and base stations that undergo revisions are the other reasons of increase.

The maximum electric field strength was measured as 10 V/m and was equal to 98% of national reference level for single GSM. Also, it was four times lower than ICNIRP limits and the measurement location was previous year's highest value measured point.

90 measurements performed in 2011 were repeated in 2012. Compared to 2011 measurement values, 27 of 90 measurement values increased and 7 measurement values decreased. Measurement results at 56 locations were identified as equal in 2011 and 2012.

In 2012, Power density values ranged from 0.08 to 267 mW/m². The maximum power density is 95% of BTK reference level for single GSM and 6% of ICNIRP and BTK total reference level. Average power density was increased to 6.3 mW/m².

Distribution of measurement values by numerical ranges are given in Figure 8. 3% of all measurement values (7 of 228) are greater than 5 V/m. It is noticed 51% of measurements (117 of 228) are below 1 V/m and 90% of measurements (205 of 228) are less than 3 V/m.

In 2012, average value of measurements was 1.54 V/m and continued increasing. It was 4% of ICNIRP limits and 15% of national reference value for single GSM as shown at Table 6. Average electric field strength increased 12% against 2011 average value of measurements and 69% against 2010. According to the comparison of 2011 and 2012 measurements, it was observed that the number of measurements higher than 1 V/m increased 217% and raised from 51 to 111. Also, number of measurements higher than 3 V/m nearly quadrupled and

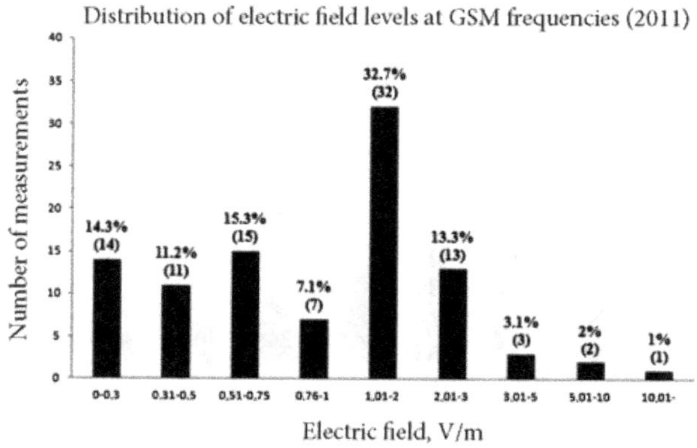

Fig. 5. Distribution of electric field levels at GSM frequencies (2011)

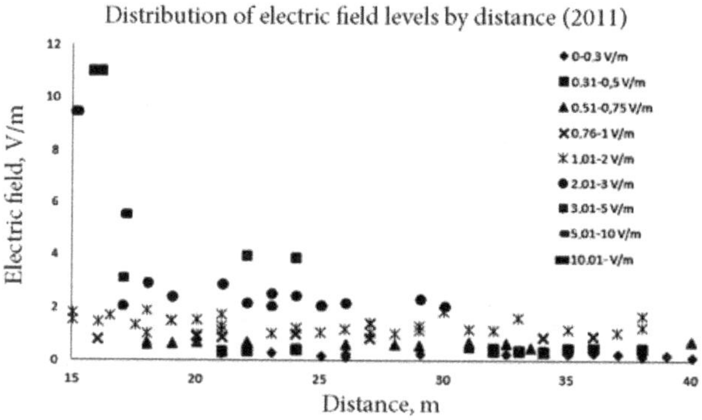

Fig. 6. Distribution of electric field levels by distance (2011)

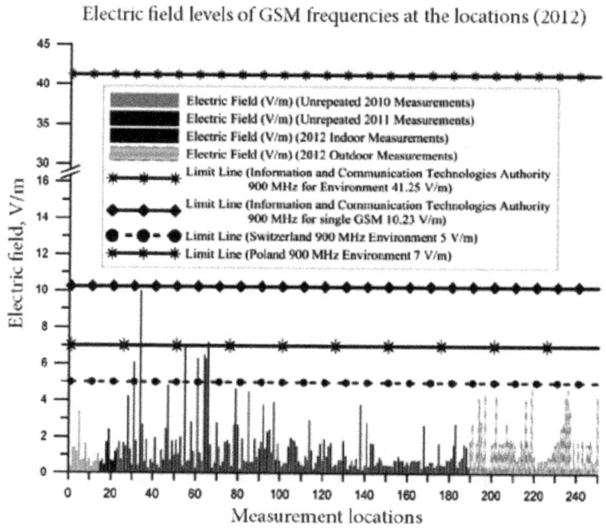

Fig. 7. Electric field measurement levels of GSM frequencies at the locations in 2012

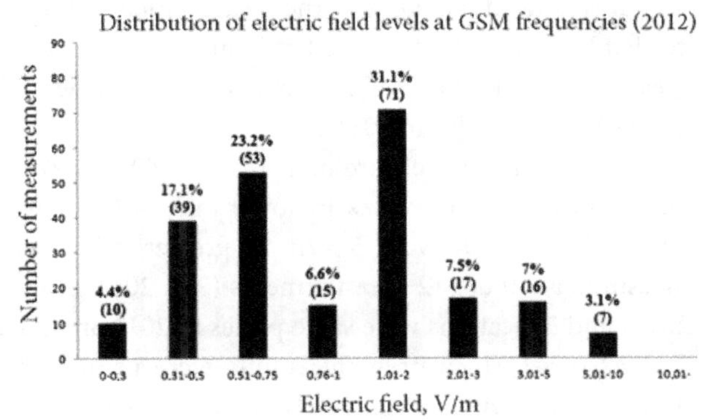

Fig. 8. Distribution of electric field levels at GSM frequencies (2012)

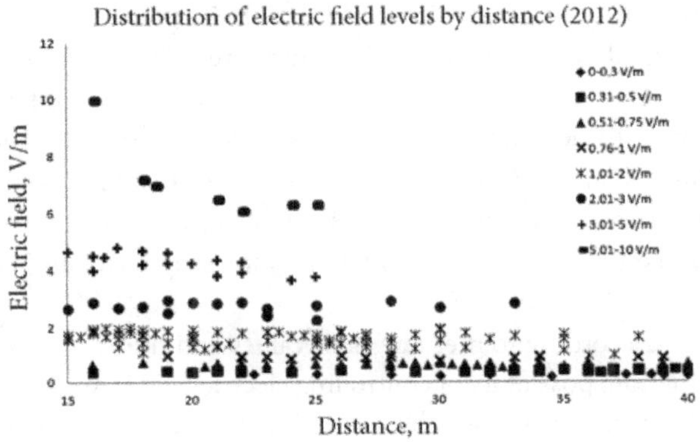

Fig. 9. Distribution of electric field levels by distance (2012)

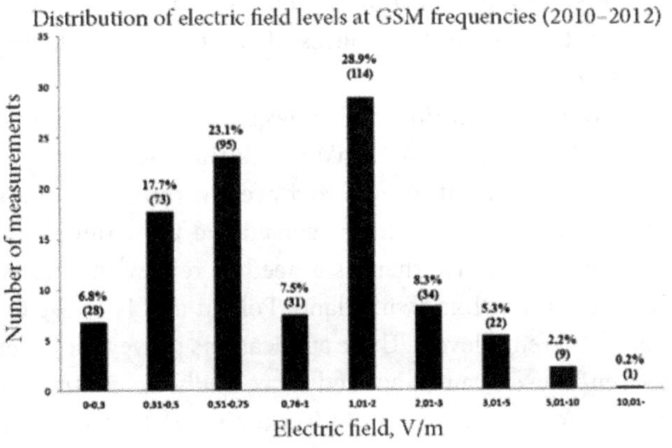

Fig. 10. Distribution of electric field levels at GSM frequencies (2010–2012)

raised to 23 from 6. The main reason of the increment at number of measurements was 61 base station mounted on street lightning poles.

Distribution of electric field measurements by distance in 2012 are given in Figure 9. The highest measurement values are clearly close to base station antennas and located between 15 m to 25 m. All of the measurement values greater than 3 V/m are located closer than or equal to 25 m. 111 measurement results were greater than 1 V/m. 48 of measurement results were located between 15–20 m and 32 of between 20–25 m. With increasing distance, measurement values were decreasing and mostly 0.75 V/m and smaller values were measured from 25 m and further distances. Some values are greater than 0.75 V/m located in this scale.

In 2012, number of measurement locations were increased due to newly installed base stations mounted on street lightning poles. These base antennas mounted at a height of 8–10 m from the ground and distances between base antenna and measurement locations are generally 15 to 25 m. Due to this situation, differ from Figures 3 and 6, number of measurements between 15 to 25 m are increased as shown in Figure 9. Compared to Figures 3, 6 and 9, the increase of measurement results between 1–2 V/m scale mostly depended on these new installations.

With this detailed and long-term work, large numbers of measurements were realized. Many significant results have been obtained with analyzing and evaluating the result of measurements. First of all, it can be easily seen that the limit values are much greater than the real environment results. All except one of the electric field measurement values are lower than single GSM base station limit value (10.23V/m for 900 MHz frequency) and all of measurement results are lower than ICNIRP and BTK total GSM exposure reference levels. 97% of measurement values are below 5 V/m and 84% are below 2 V/m as shown in Figure 10. In this context, total environment limit values (41–61 V/m) are pointed out an extremely high level and current limit values should be revised and lowered. If electromagnetic fields in sensitive places is desired to be constituted homogeneous distribution, the regulation limits can be lower to 5 V/m from 41–61 V/m.

Countries have used different limit values for electric field strength. Some countries such as Switzerland, Italy and Poland have applied lower limits than ICNIRP levels in sensitive areas. Also countries like France, Germany, Spain, Sweden, Cyprus, Finland, Austria, Ireland and UK have set ICNIRP limits (41–61 V/m) for electric field exposure (Stam 2011).

During 3-year period, it was observed that all of the measurement results higher than 3 V/m were located between 15 to 25 m away from base station antennas. 20 of 32 measurement results were 15 to 20 m and the rest were 20 to 25 m. This situation mostly resulted from selection of emission points of the base stations close and at same height to the homes or crowded outdoor places. The combination of this situation with high electromagnetic exposure limits applied in Turkey can be caused high measurement results. Therefore, placement of base station antennas was also an important factor to be considered during the planning phase.

412 measurements were performed at 250 locations (189 indoor and 61 outdoor) from stations 2010 to 2012.

64 of 250 locations investigated over 3 years (2010 to 2012). Measurement results at 12 locations were identified as equal during measurement period and measurement results changed at 52 locations. Electromagnetic exposure level at 19 of 64 locations increased over 3 years. Also, at 3 of 64 locations, electromagnetic exposure level decreased continuously during measurement period.

Average measurement results show that the electromagnetic exposure at residential areas increased continuously during 3-year period as shown at Tables 6 and 8. Due to increasing population and demand, new base stations installed and revisions made at some base stations (change of tilt or/and direction, replace base antennas with powerful antennas, relocated or removed base stations). Also, new-built settlements around base stations and base station mounted on street lightning poles caused the increment of electromagnetic exposure at indoor and outdoor places at the vicinity of base stations.

Average electric field values increased from 0.91 V/m to 1.54 V/m. 69% increase determined from 2010 to 2012. The highest average value was 1.54 V/m in 2012 and 7 times lower than BTK single GSM reference level. Also, power density values showed the same characteristics and increased from 2.2 mW/m^2 to 6.3 mW/m^2. About 200% increase noticed from 2010 to 2012 at average power density values. The maximum average value that obtained in 2012 was 2.3% of BTK reference level for single GSM (900 MHz).

Average exposure levels obtained from the study are quite below compared to the limits. But increase of average values during 3-year period is remarkable and evolution of electromagnetic exposure must be monitored carefully at coming years.

Conclusions

In this study, electromagnetic exposure levels were determined especially for the indoor and outdoor environments near base station antennas at Bursa/Nilüfer district via 412 electromagnetic field measurements from 2010 to 2012. Bursa/Nilüfer district is a growing city with its 300000 population. It can be stated that there are variations between measurements values by years. The main reason of variations are new established or removed base stations, revisions of local old base stations and new built settlements in the city.

It was understood from the repeated measurements at same points that evolving GSM base antennas, application of ICNIRP limits at the highest levels and increasing number of the base stations based on the rapid progress at communication technology caused electromagnetic exposure at indoor places ranging from 0.3 V/m to 11 V/m for 15 m to 40 m distances between measurement points and base station antennas in Nilüfer/Bursa.

It is noticed that none of the measurement values exceeded the ICNIRP reference level and only one value is greater than national BTK reference level for single GSM at 900 MHz (11 V/m in 2011).

According to measurement results, 97%, 92% and 84% of measurement values are lower than 5 V/m, 3 V/m and 2 V/m, respectively. 5 V/m or greater values were measured at 10 of 412 measurements (3 in 2011 and 7 in 2012) and 3 locations were same places in 2011 and 2012.

32 of 412 measurement results were greater than 3 V/m. 20 of 32 measurement results were located between 15 to 20 m and the rest were 20 to 25 m. Selection and placement of base station antennas is also an important factor to be considered during the planning phase.

Average electromagnetic exposure levels in sensitive places has risen up by 3-year period. In 2010, average electric field value was 0.91 V/m and increased to 1.38 V/m and 1.54 V/m in 2011 and 2012, respectively. During 3-year period, an increase of 69% was determined and this instant increase has to be monitored in coming years and some precautions taken into account. This increase can contain data's that will be basis for long term studies, such as 5–10 years.

According to measurement values and numerical distribution, possibility of controlling electric field levels below 5 V/m can be questionable to minimize the electromagnetic exposure radiated from GSM antennas in living areas. Also, this study shows that it may be reasonable to apply precaution limit values used in Switzerland, Italy and Poland.

Electromagnetic power density limits of ICNIRP were not exceed indoor places of the residences during measurement period.

Average electromagnetic power density has increased from 2.20 mW/m^2 to 6.29 mW/m^2 during 3-year period.

It is important to note that present threshold limits described by the ICNIRP are considered to be rather too generous and hence there is a need to review the limits. Some countries like Switzerland, Poland and Italy use lowered reference levels. These applications prove that lower reference levels can be applied successfully. To control the electromagnetic exposure in Turkey restricting ICNIRP reference limits as given European countries can be recommended as more accurate method.

This research shows an increasing trend emerging in the electromagnetic field by years of exposure values and it can be reference to discuss possible situations will be caused by the rapid development of communication technology that concern every country on the world. As a result, it is clear that keeping the electromagnetic exposure limits as low as possible is the most efficient and correct application to reduce and control the continuous involuntary electromagnetic field exposure.

Safety concern regarding the biological effects of electromagnetic radiation including low level of exposure have been raised. Long time exposure to radiation of GSM transmission antennas in rooms near the base station antennas is a fact to be considered. And the potential biological effects of exposure to these sources are to take into account.

Acknowledgments

The authors would like to thank to Bursa/Nilüfer municipality for their support and opportunities provided to do this work.

References

Alanko, T.; Hietanen, M.; von Nandelstadh, P. 2008. Occupational exposure to RF fields from base station antennas on rooftops, *Annals of Telecommunications* 63: 125–132. http://dx.doi.org/10.1007/s12243-007-0001-6

Baltrenas, P.; Buckus, R.; Vasarevicius, S. 2012. Research and evaluation of the intensity parameters of electromagnetic fields produced by mobile communication antennas, *Journal of Environmental Engineering and Landscape Management* 20(4): 273–284. http://dx.doi.org/10.3846/16486897.2012.738680

Bernardi, P.; Cavagnaro, M.; Pisa, S.; Piuzzi, E. 2000. Human exposure to radio base-station antennas in urban environment, *IEEE Transactions on Microwave Theory and Techniques* 48(11): 1996–2001. http://dx.doi.org/10.1109/22.884188

Chen, H. Y.; Su, T. Y.; Chuang, C. Y. 2009. Electric fields radiated from a cellular phone base station antenna constructed on a roof-top, *Electromagnetics* 29(3): 235–249. http://dx.doi.org/10.1080/02726340902718468

Cooper, J.; Marx, B.; Buhl, J.; Hombach, V. 2002. Determination of safety distance limits for a human near a cellular base station antenna, adopting the IEEE standards or ICNIRP guidelines, *Bioelectromagnetics* 23(6): 429–443. http://dx.doi.org/10.1002/bem.10037

Çerezci, O.; Şeker, S. 2010. Determining of electromagnetic pollution in Bursa Nilufer district and recommending a sample model to decrease exposure levels, in *National Conference on Electrical, Electronics and Computer Engineering ELECO' 2010* [online], 2–5 December 2010, 1–5 [cited 10 February 2013]. Available from Internet: http://ieeexplore.ieee.org/xpls/abs_all.jsp?arnumber=5698224

EN 50413: 2008. Basic Standard on Measurement and Calculation Procedures for Human Exposure to Electric, Magnetic and Electromagnetic Fields (0 Hz–300 GHz). European Committee for Electrotechnical Standardization (CENELEC). 59 p.

Gosselin, M. C.; Christ, A.; Kühn, S.; Kuster, N. 2009. Dependence of the occupational exposure to mobile phone base stations on the properties of the antenna and the human body, *IEEE Transactions on Electromagnetic Compatibility* 51(2): 227–235. http://dx.doi.org/10.1109/TEMC.2009.2013717

Higashiyama, J.; Tarusawa, Y. 2011. RF exposure compliance assessment for radio base station with built-in antenna, in *Proceedings of 10th International Symposium of Electromagnetic Compatibility EMC Europe 2011* [online], 26–30 September 2011, York, UK, 90–93 [cited 12 March 2013]. Available from Internet: http://ieeexplore.ieee.org/xpl/articleDetails.jsp?tp=&arnumber=6078580

International Commission on Non-Ionizing Radiation Protection (ICNIRP). 1998. ICNIRP guidelines for limiting exposure to time-varying electric, magnetic and electromagnetic fields (up to 300 GHz), *Health Physics* [online] 74(4): 494–522 [cited 12 March 2013]. Available from Internet: http://www.icnirp.org/cms/upload/publications/ICNIRPemfgdl.pdf

Information and Communication Technologies Authority (BTK). 2011. *Regulation on Electromagnetic Exposure the limit values of the Electromagnetic Field force from the fixed telecommunication devices operating within the frequency range of 10 kHz–60 GHz.* 7 p. [online], [cited 12 March 2013]. Available from Internet: http://tk.gov.tr/mevzuat/yonetmelikler/dosyalar/ehc.pdf

Joseph, W.; Martens, L. 2005. Comparison of safety distances based on the electromagnetic field and based on the SAR for occupational exposure of a 900-MHz base station antenna, *IEEE Transactions on Electromagnetic Compatibility* 47(4): 977–985. http://dx.doi.org/ 10.1109/temc.2005.854100

Karpowicz, J.; Hietanen, M.; Gryz, K. 2006. EU directive, ICNIRP guidelines and Polish legislation on electromagnetic fields, *International Journal of Occupational Safety and Ergonomics* [online] 12: 125–136 [cited 12 March 2013]. Available from Internet: http://www.ciop.pl/17469

Neskovic, N.; Koprivica, M.; Neskovic, A.; Paunovic, G. 2011. Improving the efficiency of measurement procedures for assessing human exposure in the vicinity of mobile phone (GSM/DCS/UMTS) base stations, *Radiation Protection Dosimetry* 149(3): 238–244. http://dx.doi.org/10.1093/rpd/ncr248

The Federal Office for the Environment (FOEN). 1999. *Ordinance relating to Protection from Non Ionizing Radiation (ONIR).* 20 p. [online], [cited 12 March 2013]. Available from Internet: http://www.bafu.admin.ch/elektrosmog/01079/index.html?lang=en

Poljak, D.; Kovac, N. 2004. The boundary element electromagnetic–thermal analysis of human exposure to base station antennas radiation, *Engineering Analysis with Boundary Elements* 28(7):763–770. http://dx.doi.org/10.1016/j.enganabound.2004.02.004

Singh, R. K. 2012. Assessment of electromagnetic radiation from base station antennas, *Indian Journal of Radio & Space Physics* [online] 41: 557–565 [cited 28 February 2013]. Available from Internet: http://nopr.niscair.res.in

Stam, R. 2011. *Comparison of international policies on electromagnetic fields (power frequency and radiofrequency fields)* [online]. Netherlands National Institute for Public Health and the Environment. 13 p. [cited 15 February 2014]. Available from Internet: http://www.rivm.nl

Viel, J. T.; Clerc, S.; Barrera, C.; Rymzhanova, R.; Moissonnier, M.; Hours, M.; Cardis, E. 2009. Residential exposure to radiofrequency fields from mobile phone base stations and broadcast transmitters: a population-based survey with personal meter, *Occupational & Environmental Medicine* 66: 550–556.

Wu, T.; Shao, Q.; Yang, L.; Qi, D.; Lin, J.; Lin, X.; Yu, Z. 2012. A large-scale measurement of electromagnetic fields near GSM base stations in Guangxi, China for risk communication, *Radiation Protection Dosimetry* 155(1): 25–31. http://dx.doi.org/10.1093/rpd/ncs309

Osman ÇEREZCİ. Dr, Prof. (since 1999), Department of Electric-Electronic Engineering, Sakarya University. Doctor of Science (Electronic Engineering), Uludag University, 1985. First degree in Physics, Istanbul University, 1976. Employment: Professor (1999), Associate Professor (1994), Assistant Professor (1986), Sakarya University Electric-Electronic Engineering Department. Publications: author of 6 books, over 70 research papers. Research interests: biological effects of electromagnetic fields, electromagnetic compatibility and electromagnetic wave propagation.

Baha KANBEROĞLU. Master, PhD student (since 2005) and Research Assistant (since 2002), Department of Electric-Electronic Engineering, Sakarya University. Master of Science (Electronic Engineering), Sakarya University, 2005. First degree in Electric-Electronic Engineering, Sakarya University, 2002. Publications: author of 14 research papers. Research interests: electromagnetic fields and theory, electromagnetic radiation, antenna theory.

Şuayb Çağrı YENER. Dr, Research Assistant (since 2004), Department of Electric-Electronic Engineering, Sakarya University. Doctor of Science (Electronics Engineering), Department of Electronics and Communication Engineering, Istanbul Technical University, 2014. Master of Science (Electronics and Communication Engineering), Istanbul Technical University, 2007. First degree in Electrical-Electronics Engineering, Sakarya University, 2004. Publications: author of 25 research papers and proceedings. Research interests: CMOS analog circuits, memristor based circuits, electromagnetic compatibility, microwave and antenna technique.

RADIOLOGICAL ASSESSMENT OF IONIZING RADIATION IMPACT ON THE TERRESTRIAL NON-HUMAN BIOTA IN LITHUANIA

Marina KONSTANTINOVA[a], Nina PROKOPČIUK[a], Arūnas GUDELIS[a], Donatas BUTKUS[b]

[a]*State research institute Center for Physical Sciences and Technology, Institute of Physics, Savanorių pr. 231, LT-02300 Vilnius, Lithuania*

[b]*Vilnius Gediminas Technical University, Saulėtekio al. 11, LT-10223 Vilnius, Lithuania*

Abstract. The quantitative assessment of radionuclides transfer to non-human biota using their activity concentration ratios is required for models of predictive doses of ionizing radiation. Based on long-term data regarding activity concentration of radionuclides in the top soil layer of the entire territory of Lithuania, and with the help of ERICA Assessment Tool – a software application that calculates dose rates to selected biota, we estimated the radiological impact on the terrestrial non-human biota with special emphasis on the protected areas located in the vicinity of Ignalina Nuclear Power Plant (INPP). Estimated total dose rates of artificial radionuclides – after-Chernobyl ^{137}Cs and ^{90}Sr as well as discharged by INPP – and natural radionuclides, such as ^{238}U and ^{232}Th, were found to be less than ERICA screening value of 10 μGy h^{-1}.

Keywords: ERICA Tool, terrestrial non-human biota, ^{137}Cs, ^{90}Sr, ^{238}U, ^{232}Th, evaluated total dose rate.

Introduction

Radiological impact on non-human biota in regulatory context is now of interest to a number of international and national bodies and is under scientific investigation. In order to develop predictive models, the quantification of transfer of radionuclides to biota using their activity concentration ratios is highly needed. An ability to predict the radionuclide activity concentrations in biota is an essential component of any approach assessing exposure of non-human biota (Beresford *et al.* 2008; Beresford *et al.* 2010; Hosseini *et al.* 2008; ICRP 2008; Fesenko *et al.* 2011; Wilson *et al.* 2010).

Up to the present time, great attention was paid to the human radiation protection in Lithuania: normative acts regulating the human exposure at national levels were validated. However, at present there are no guidelines for non-human biota exposure levels in Lithuania.

Radioactive waste disposal became especially important after the decommissioning of Ignalina Nuclear Power Plant (INPP). The Interim Spent Fuel Storage Facility (ISFSF) is located directly nearby the current INPP facilities. It is important to investigate the present ionizing radiation background in Lithuania. Presented data provide a possibility to put an assessment of the nuclear sites into context and compare the present terrestrial non-human biota exposure due to discharged anthropogenic radionuclides with that of background radiation considering the possible future exposure in the environment by the radioactive waste from the ISFSF.

The aim of this work is to estimate the radiological impact on the terrestrial non-human biota based on long-term data of radionuclide activity concentration in the top soil layer in the whole territory of Lithuania emphasizing the designated area for the ISFSF.

1. Subject of investigation and methods

INPP is situated in the north-eastern part of Lithuania close to the borders with Latvia and Belarus on the shore of Lake Druksiai (Fig. 1a). It has two RBMK reactors belonging to the thermal neutron reactor category with an electric capacity of 1500 MW. Unit 1 was shut down on 31 December, 2004 and Unit 2 – on 31 December, 2009.

The ISFSF and near-surface repository for redundant materials and waste will be located directly near the current INPP facilities. INPP uses Lake Druksiai as a cooling pond. However, the territory of Lake Druksiai and its

Corresponding author: Marina Konstantinova
E-mail: marina.konstantinova@ftmc.lt

neighbourhood are protected areas now. These territories are part of the NATURA network (EU Council Directive 1992), a network of protected areas of European Community. Documents related to projects on decommissioning of the INPP were evaluated and agreed (IAEA 2006; Interim storage... 2007). Lake Smalvos is at about 10 km from the ISFSF while Lake Dysnai is at about 12 km from the ISFSF, and also are protected areas (Fig. 1b).

There were two most important periods of the Lithuanian territory contamination with ^{90}Sr and ^{137}Cs: the first one covered nuclear weapon tests in the atmosphere in 1945–1980, and the second one started after the Chernobyl Nuclear Power Plant (ChNPP) accident in 1986. Before the Chernobyl accident, data of the ^{137}Cs and ^{90}Sr pollution were derived from measurements taken during 1979–1984 and samples were collected over the whole territory of Lithuania. Data of ^{137}Cs and ^{90}Sr pollution were determined by different research institutes and laboratories in 1987–2000 (Butkus, Konstantinova 2003; Butkus, Konstantinova 2006; Paskauskas, Mazeika 1997). In 1987, the ^{137}Cs distribution over the territory of Lithuania was determined by airborne spectrometry (Butkus *et al.* 2001).

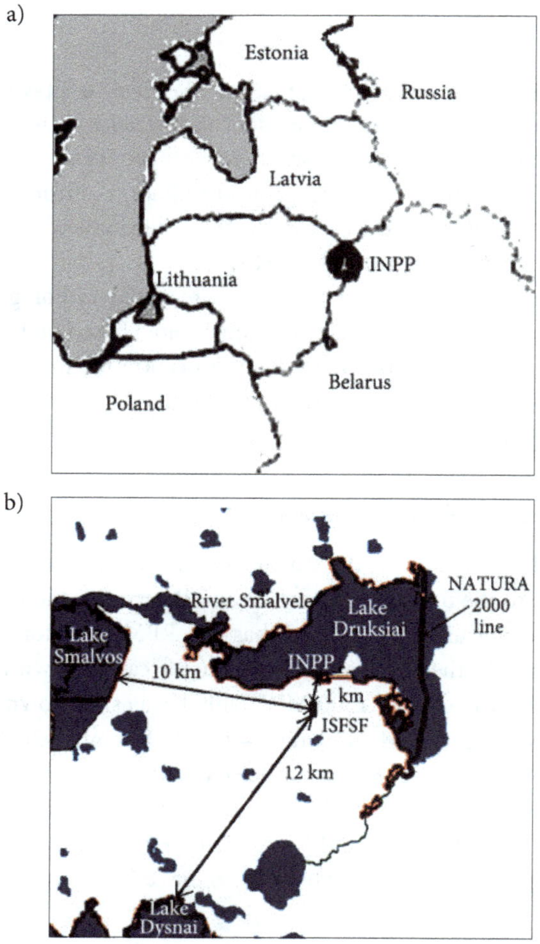

Fig. 1. (a) INPP geographical location, (b) The regional level map with marked INPP, ISFSF locations and outlined NATURA 2000 border (EU Council Directive 1992)

Predominant investigated natural series radionuclides are ^{232}Th and ^{238}U. Distribution of activity concentration of natural gamma emitters in topsoil was estimated at about the same time over the whole territory of Lithuania. Methods of sampling and measurement were described in (Butkus, Konstantinova 2003).

The major evaluated anthropogenic radionuclides discharged by the INPP are ^{54}Mn, ^{60}Co, $^{134,137}Cs$, ^{90}Sr and ^{238}Pu. Data on the state of terrestrial ecosystem prior to the operation of the INPP were presented in (Zukauskas *et al.* 1992). In addition, the results of large-scale radioecological studies had been carried out since 1979 and during the operation of the INPP and they were discussed in (Paskauskas, Mazeika 1997; Butkus *et al.* 2001; Mazeika 2002; Nedveckaite 2004). These data were then used as input into the model simulations to assess radionuclide transport and the associated radiological doses to non-human biota.

Based on these data the exposure dose rates to terrestrial reference organisms were evaluated using ERICA Assessment Tool (Beresford *et al.* 2008; Howard, Larsson 2008; Brown *et al.* 2008) version of 2011. ERICA (Environmental Risk from Ionizing Contaminants: Assessment and Management) is a software application that calculates weighted external and internal dose rates to selected biota. Monte-Carlo probabilistic simulation uses distributions of available input data. The result of such stimulation is the probabilistic distribution of the dose rate that facilitates evaluation of the most and the least probable (but possible) distribution values. The value of dose rate not only depends on the reference organism species but also on the dose conversion factor and the ratio of radionuclide concentrations in soil (in our case) and in organisms. Values of the parameters were estimated as indicated in the FASSET (FASSET 2003) database in case of soil.

The databases of ERICA use a number of reference organisms. Each reference organism is representative of a contaminated environment: terrestrial, freshwater or marine ecosystems, and shows the radiological impact on non-human biota. ERICA applies a screening dose rate value of 10 $\mu Gy\ h^{-1}$ for all types of organisms. The screening dose rate is a proposed generic (across all species and ecosystems) predicted no effects screening dose rate for application in conservative assessments of screening dose rate. In our work, we investigate terrestrial biota only.

2. Results and discussions

In order to compare exposure of terrestrial reference organism due to natural background radionuclides with that of anthropogenic radionuclides the available data on the levels of natural radionuclides were used (Table 1). Special emphasis was placed on measurements of ^{238}U and ^{232}Th top soil activity in previous studies (Butkus, Konstantinova

2003). ^{238}U and ^{232}Th concentrations in soil were determined by the radiation of their decay products: ^{226}Ra, ^{214}Pb, ^{214}Bi for ^{238}U and ^{208}Tl for ^{232}Th (Fig. 2).

Table 1. Examples of estimated total dose rates to terrestrial reference organisms attributed to natural background radionuclides

Terrestrial reference organisms	^{238}U mean total dose rates, (μGy h^{-1})	^{232}Th mean total dose rates, (μGy h^{-1})
Soil invertebrate	$1.44{\cdot}10^{-1}$ ($1.70{\cdot}10^{-1}$)*	$4.42{\cdot}10^{-3}$ ($7.80{\cdot}10^{-3}$)
Lichen & bryophytes	$3.18{\cdot}10^{-1}$ ($3.76{\cdot}10^{-1}$)	$5.17{\cdot}10^{-2}$ ($9.13{\cdot}10^{-2}$)
Grasses & herbs	$6.22{\cdot}10^{-2}$ ($7.36{\cdot}10^{-2}$)	$2.19{\cdot}10^{-2}$ ($3.86{\cdot}10^{-2}$)
Shrub	$3.92{\cdot}10^{-2}$ ($4.63{\cdot}10^{-2}$)	$8.02{\cdot}10^{-3}$ ($1.41{\cdot}10^{-2}$)
Tree	$3.98{\cdot}10^{-3}$ ($4.70{\cdot}10^{-3}$)	$5.41{\cdot}10^{-4}$ ($9.55{\cdot}10^{-4}$)
Mammal (Deer)	$4.22{\cdot}10^{-2}$ ($4.99{\cdot}10^{-2}$)	$6.11{\cdot}10^{-5}$ ($1.08{\cdot}10^{-4}$)
Mammal (Rat)	$4.83{\cdot}10^{-2}$ ($5.71{\cdot}10^{-2}$)	$6.34{\cdot}10^{-5}$ ($1.12{\cdot}10^{-4}$)
Bird	$6.02{\cdot}10^{-2}$ ($7.12{\cdot}10^{-2}$)	$1.95{\cdot}10^{-4}$ ($3.44{\cdot}10^{-4}$)
Amphibians	$5.69{\cdot}10^{-2}$ ($6.73{\cdot}10^{-2}$)	$1.95{\cdot}10^{-4}$ ($3.44{\cdot}10^{-4}$)

**Note*: in brackets – 95th Percentile

The maximum radiation impact on the organism is caused by ^{238}U radiation (Fig. 3), and the total dose rate does not exceed 2.42 μGy h^{-1}. Exposure dose rates of reference organisms reach 0.14 μGy h^{-1} for soil invertebrates due to the presence of ^{238}U in soil, and 0.05 μGy h^{-1} for lichens and bryophytes due to the presence of ^{232}Th in soil.

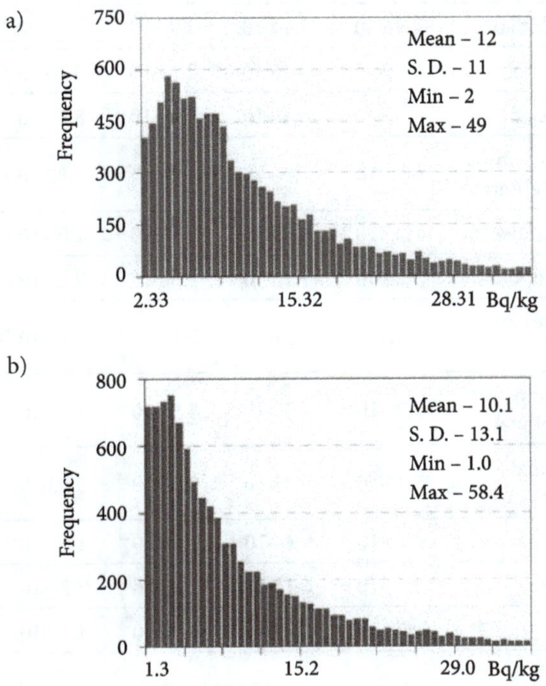

Fig. 2. Frequency histograms of ^{238}U (a) and ^{232}Th (b) activity concentration (d.w.) in Lithuanian soils

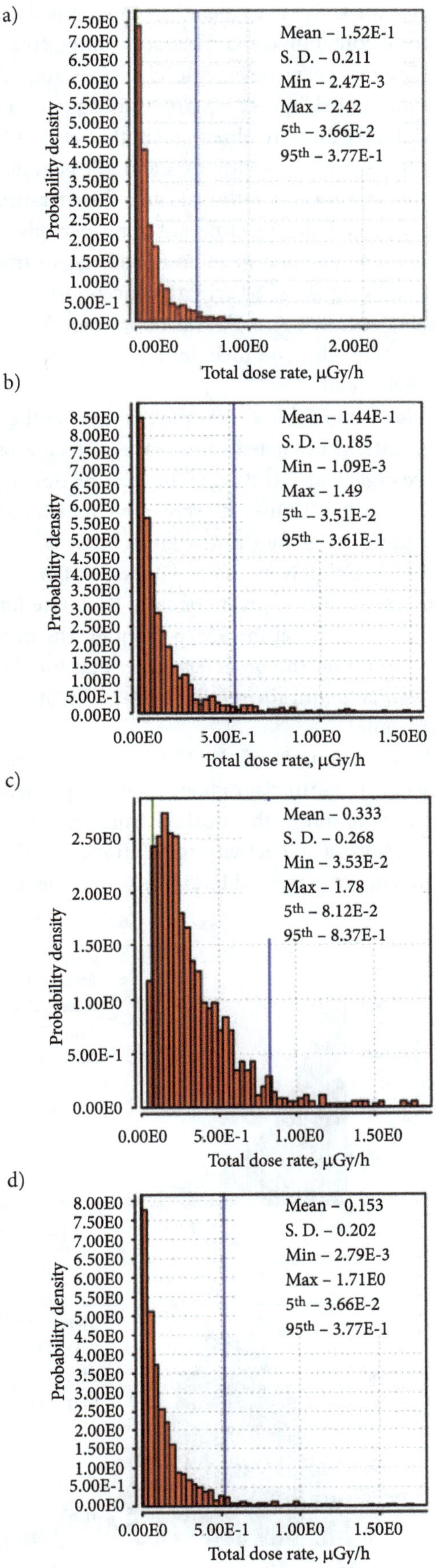

Fig. 3. Examples of estimated total dose rates to terrestrial reference organisms due to ^{238}U only: a) detritivorous invertebrate, b) flying insects, c) lichen & bryophytes, and d) soil invertebrate (worm)

After the ChNPP accident Lithuania was directly on the path of contaminated air masses coming from Chernobyl. The regions of elevated level of ^{137}Cs and ^{90}Sr contamination, called "spots", ranging from square meters to square kilometres, were observed (Butkus *et al.* 1992). At present, contamination with ^{137}Cs and ^{90}Sr is evaluated in the whole territory of Lithuania but the measurements were performed during different time periods. Nowadays ^{137}Cs activity concentration in soil ranges from 6 to 160 Bq kg^{-1} (Fig. 4a). ^{90}Sr activity concentration after the Chernobyl accident nowadays is on average 5 Bq kg^{-1} in Lithuania (Fig. 4b), and includes both the bomb and the Chernobyl fallout.

Figure 5 and Table 2 show the examples of the corresponding ERICA evaluated values of terrestrial ecosystem reference organisms. As it might be seen, all dose rates are far below ERICA 10 μGy h^{-1} screening value, and reach the maximum value for deer ($1.90 \cdot 10^{-2}$ μGy h^{-1} on average, and $7.43 \cdot 10^{-2}$ μGy h^{-1} at the 95th percentile). Lichen & bryophytes are also of great concern, dose rate for them is $4.44 \cdot 10^{-2}$ μGy h^{-1} at the 95th percentile. The dose rates due to ^{90}Sr are on the order less than that for ^{137}Cs and reach the maximum for reptiles ($9.3 \cdot 10^{-2}$ μGy h^{-1} at the 95th percentile).

The environment of the INPP was investigated in more detail. The artificial radionuclides, originated in the reactors, are found in the local environment. The investigation results of soil activity concentration in the INPP environment are presented in Figure 6, and the data were

a)

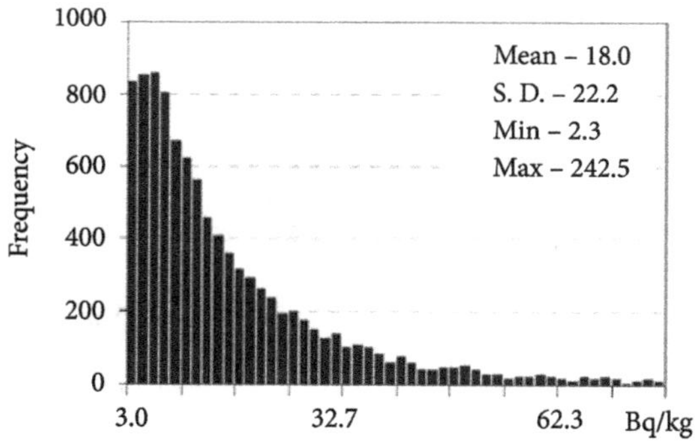

b)

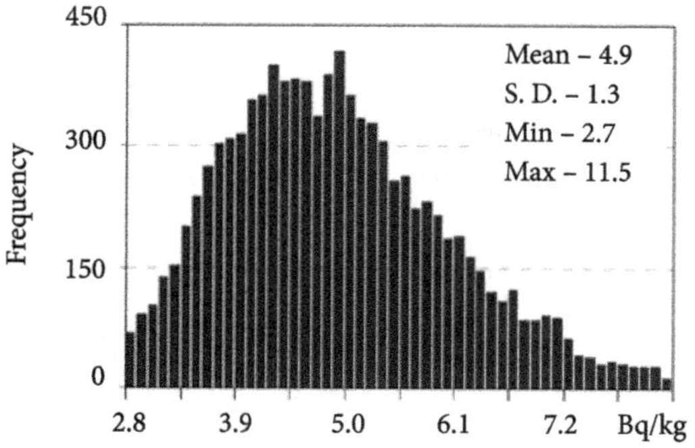

Fig. 4. Frequency histograms of ^{137}Cs (a) and ^{90}Sr (b) activity concentrations after the Chernobyl accident in Lithuanian top soils

Table 2. Estimated total dose rates (μGy h^{-1}) to terrestrial reference organisms attributed to artificial radionuclides after the Chernobyl accident

Terrestrial reference organisms	Mean	SD	Min.	Max.
		^{137}Cs		
Amphibian	$3.62 \cdot 10^{-3}$	$4.60 \cdot 10^{-3}$	$2.79 \cdot 10^{-4}$	$4.78 \cdot 10^{-2}$
Bird	$4.79 \cdot 10^{-3}$	$7.26 \cdot 10^{-3}$	$2.99 \cdot 10^{-4}$	$6.52 \cdot 10^{-2}$
Bird egg	$2.19 \cdot 10^{-3}$	$2.25 \cdot 10^{-3}$	$2.61 \cdot 10^{-4}$	$2.71 \cdot 10^{-2}$
Detritivorous invertebrate	$6.23 \cdot 10^{-3}$	$6.59 \cdot 10^{-3}$	$7.36 \cdot 10^{-4}$	$7.27 \cdot 10^{-2}$
Flying insects	$2.43 \cdot 10^{-3}$	$2.49 \cdot 10^{-3}$	$2.89 \cdot 10^{-4}$	$2.70 \cdot 10^{-2}$
Gastropod	$2.41 \cdot 10^{-3}$	$2.48 \cdot 10^{-3}$	$2.94 \cdot 10^{-4}$	$2.78 \cdot 10^{-2}$
Grasses & Herbs	$4.06 \cdot 10^{-3}$	$6.29 \cdot 10^{-3}$	$2.91 \cdot 10^{-4}$	$1.04 \cdot 10^{-1}$
Lichen & bryophytes	$1.16 \cdot 10^{-3}$	$1.76 \cdot 10^{-3}$	$3.34 \cdot 10^{-4}$	$1.88 \cdot 10^{-1}$
Mammal (Deer)	$1.90 \cdot 10^{-2}$	$3.69 \cdot 10^{-3}$	$3.07 \cdot 10^{-4}$	$4.02 \cdot 10^{-1}$
Mammal (Rat)	$1.47 \cdot 10^{-3}$	$2.58 \cdot 10^{-3}$	$7.87 \cdot 10^{-4}$	$4.46 \cdot 10^{-1}$
Reptile	$1.34 \cdot 10^{-3}$	$3.83 \cdot 10^{-3}$	$3.07 \cdot 10^{-4}$	$7.16 \cdot 10^{-1}$
Shrub	$1.28 \cdot 10^{-3}$	$2.30 \cdot 10^{-3}$	$4.05 \cdot 10^{-4}$	$4.39 \cdot 10^{-1}$
Soil Invertebrate (worm)	$5.97 \cdot 10^{-3}$	$6.18 \cdot 10^{-3}$	$7.22 \cdot 10^{-4}$	$6.68 \cdot 10^{-2}$
Tree	$2.65 \cdot 10^{-3}$	$3.08 \cdot 10^{-3}$	$2.32 \cdot 10^{-4}$	$4.26 \cdot 10^{-2}$
		^{90}Sr		
Amphibian	$2.46 \cdot 10^{-3}$	$3.74 \cdot 10^{-3}$	$2.96 \cdot 10^{-5}$	$5.38 \cdot 10^{-2}$
Bird	$1.79 \cdot 10^{-3}$	$3.16 \cdot 10^{-3}$	$2.45 \cdot 10^{-5}$	$3.76 \cdot 10^{-2}$
Bird egg	$4.09 \cdot 10^{-3}$	$6.48 \cdot 10^{-3}$	$3.08 \cdot 10^{-5}$	$8.46 \cdot 10^{-2}$
Detritivorous invertebrate	$6.55 \cdot 10^{-4}$	$1.96 \cdot 10^{-3}$	$9.46 \cdot 10^{-7}$	$3.32 \cdot 10^{-2}$
Flying insects	$1.33 \cdot 10^{-4}$	$1.39 \cdot 10^{-4}$	$5.31 \cdot 10^{-8}$	$1.05 \cdot 10^{-3}$
Gastropod	$2.24 \cdot 10^{-4}$	$9.75 \cdot 10^{-5}$	$5.88 \cdot 10^{-5}$	$9.53 \cdot 10^{-4}$
Grasses & Herbs	$6.22 \cdot 10^{-4}$	$4.51 \cdot 10^{-3}$	$2.84 \cdot 10^{-8}$	$1.22 \cdot 10^{-1}$
Lichen & bryophytes	$1.24 \cdot 10^{-2}$	$1.22 \cdot 10^{-2}$	$4.53 \cdot 10^{-4}$	$1.51 \cdot 10^{-1}$
Mammal (Deer)	$5.46 \cdot 10^{-3}$	$6.85 \cdot 10^{-3}$	$9.79 \cdot 10^{-5}$	$5.93 \cdot 10^{-2}$
Mammal (Rat)	$5.60 \cdot 10^{-3}$	$9.44 \cdot 10^{-3}$	$1.05 \cdot 10^{-4}$	$1.59 \cdot 10^{-1}$
Reptile	$3.47 \cdot 10^{-2}$	$6.11 \cdot 10^{-2}$	$2.47 \cdot 10^{-4}$	$7.25 \cdot 10^{-1}$
Shrub	$1.27 \cdot 10^{-4}$	$1.29 \cdot 10^{-4}$	$4.70 \cdot 10^{-6}$	$1.18 \cdot 10^{-3}$
Soil Invertebrate (worm)	$2.34 \cdot 10^{-5}$	$2.41 \cdot 10^{-5}$	$7.14 \cdot 10^{-8}$	$1.32 \cdot 10^{-4}$
Tree	$1.59 \cdot 10^{-3}$	$6.21 \cdot 10^{-4}$	$5.11 \cdot 10^{-4}$	$4.74 \cdot 10^{-3}$

taken from (Paskauskas, Mazeika 1997; Mazeika 2002). Table 3 shows the examples of corresponding ERICA evaluated exposure values of terrestrial reference organisms at the INPP operation sites. Presented data also are far below ERICA 10 μGy h^{-1} screening value and reach the maximal value of 0.8 μGy h^{-1} for deer due to ^{137}Cs (2.37 μGy h^{-1} at the 95th percentile).

Table 3. Examples of ERICA evaluated exposure values of terrestrial ecosystem reference organisms at INPP operation sites

Terrestrial reference organisms	Mean total dose rates, (μGy h^{-1})		
	^{60}Co	^{134}Cs	^{137}Cs
Soil invertebrate	5.69·10^{-1} (3.64) *	4.46·10^{-2} (1.82·10^{-1})	2.54·10^{-1} (8.25·10^{-1})
Grasses & herbs	2.10·10^{-1} (1.35)	2.01·10^{-2} (8.21·10^{-2})	1.68·10^{-1} (5.47·10^{-1})
Shrub	2.21·10^{-1} (1.41)	3.65·10^{-2} (1.49·10^{-1})	5.40·10^{-1} (1.76)
Tree	1.76·10^{-1} (1.13)	1.83·10^{-2} (7.46·10^{-2})	1.15·10^{-1} (3.75·10^{-1})
Mammal (Deer)	2.24·10^{-1} (1.43)	1.05·10^{-1} (4.26·10^{-1})	8.38·10^{-1} (2.73)
Mammal (Rat)	5.47·10^{-1} (3.5)	6.74·10^{-2} (2.75·10^{-1})	6.24·10^{-1} (2.03)
	^{90}Sr	^{54}Mn	^{238}Pu
Soil invertebrate	1.12·10^{-4} (2.45·10^{-4})	6.31·10^{-2} (2.33·10^{-1})	5.26·10^{-5} (9.08·10^{-5})
Grasses & herbs	2.54·10^{-3} (5.54·10^{-3})	2.33·10^{-2} (8.62·10^{-2})	2.62·10^{-5} (4.52·10^{-5})
Shrub	6.08·10^{-4} (1.33·10^{-3})	2.52·10^{-2} (9.33·10^{-2})	5.72·10^{-5} (9.87·10^{-5})
Tree	7.64·10^{-3} (1.67·10^{-2})	2.01·10^{-2} (7.45·10^{-2})	5.72·10^{-5} (9.87·10^{-5})
Mammal (Deer)	2.72·10^{-2} (5.94·10^{-2})	1.21·10^{-2} (4.49·10^{-2})	4.25·10^{-5} (7.33·10^{-5})
Mammal (Rat)	2.59·10^{-2} (5.66·10^{-2})	6.02·10^{-2} (2.23·10^{-1})	4.25·10^{-5} (7.33·10^{-5})
Soil invertebrate	1.12·10^{-4} (2.45·10^{-4})	6.31·10^{-2} (2.33·10^{-1})	5.26·10^{-5} (9.08·10^{-5})

**Note*: in brackets – 95th Percentile

Conclusions

Radionuclide activity concentrations in the top soil layer have been used as the basis for a case study in assessing radiological impacts on terrestrial ecosystem non-human biota. The radiation of organisms has the maximum impact due to ^{238}U. Exposure dose rates to reference organisms reach 1.70·10^{-1} μGy h^{-1} at the 95th percentile for soil invertebrates due to the presence of ^{238}U in soil, and 9.13·10^{-2} μGy h^{-1} at the 95th percentile for lichens and

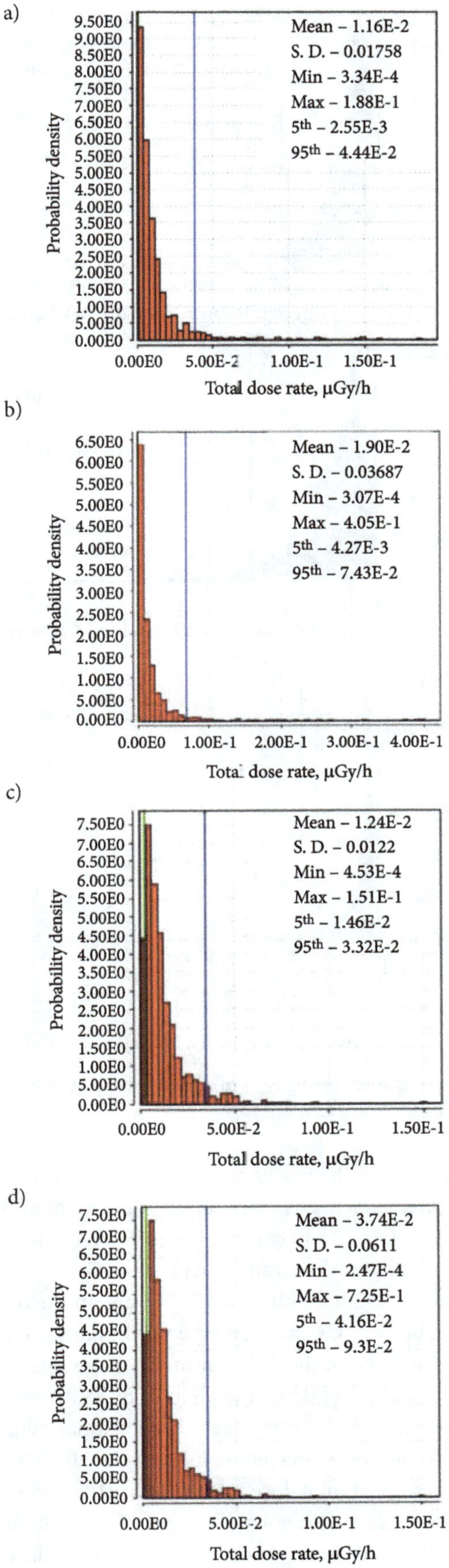

Fig. 5. Examples of estimated total dose rates to terrestrial reference organisms attributed to artificial radionuclides after the Chernobyl accident: lichen & bryophytes, ^{137}Cs (a); mammal (deer), ^{137}Cs (b); lichen & bryophytes, ^{90}Sr (c); and reptile ^{90}Sr (d)

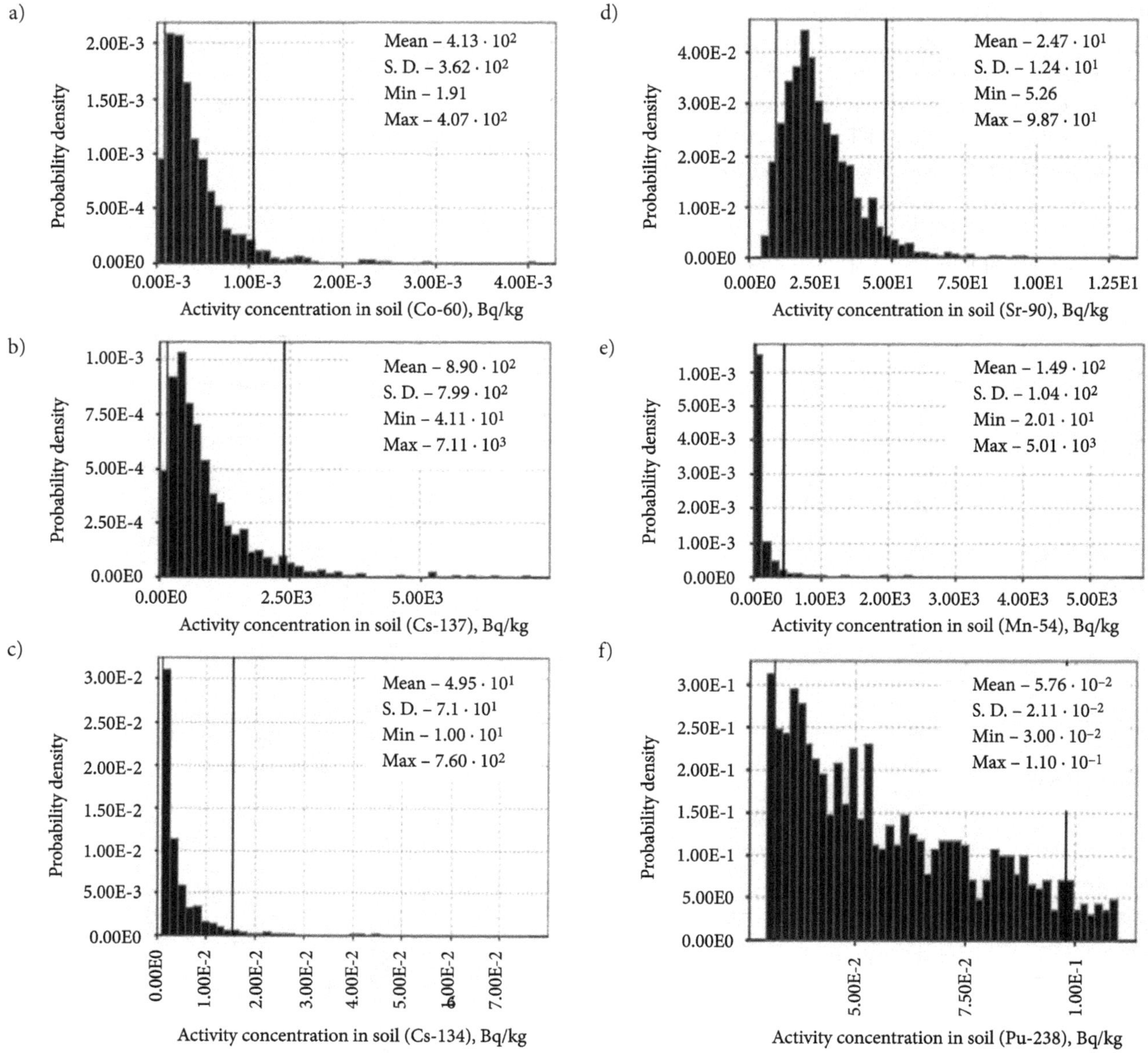

Fig. 6. The frequency histograms of anthropogenic radionuclide activity concentration in soils of INPP operation sites

bryophytes due to the presence of ^{232}Th in soil. As for artificial radionuclides, dose rates reach the maximum value for deer ($7.43 \cdot 10^{-2}$ μGy h^{-1} at the 95th percentile). Lichen & bryophytes dose rate is 4.44 10^{-2} μGy h^{-1} at the 95th percentile and dose rates due to ^{90}Sr are on the order less than that for ^{137}Cs and reach the maximum for reptiles ($9.3 \cdot 10^{-2}$μGy h^{-1} at the 95th percentile). Dose rates due to artificial radionuclide in the vicinity of INPP exceed unity in whole Lithuania, and reach the maximal value of 2.37 μGy h^{-1} at the 95th percentile for deer due to ^{137}Cs.

The evaluated total dose rates due to artificial radionuclides, both after-Chernobyl ^{137}Cs and ^{90}Sr, and discharged by INPP, were found to be less than a boundary of 10 μGy h^{-1}, as presented in ERICA approach, therefore, there is no adverse effect on the integrity of non-human biota. The present day total dose rates of up to the 95th percentile for the worst-affected organisms in the regions for nuclear and NATURA 2000 sites are less than 10 μGy h^{-1}. Presented data provide a possibility for the comparison of primary non-human biota exposure in context with the environment of radioactive waste and Interim Spent Fuel Storage Facility assumed to be located in this part of Lithuania.

References

Beresford, N. A.; Barnett, C. L.; Howard, B. J.; Scott, A.; Brown, J. E.; Copplestone, D. 2008. Derivation of transfer parameters for use within the ERICA Tool and the default concentration ratios for terrestrial biota, *Journal of Environmental Radioactivity* 99: 1393–1407. http://dx.doi.org/10.1016/j.jenvrad.2008.01.020

Beresford, N. A.; Barnett, C. L.; Brown, J. E.; Cheng, J.-J.; Copplestone, D.; Gaschak, S.; Hosseini, A.; Howard, B. J.; Kamboj, S.; Nedveckaite, T.; Olyslaegers, G.; Smith, J. T. Vives i Batlle, J.; Yu, C. 2010. An international comparison of approaches to assess the radiation exposure of wildlife to radioactivity: terrestrial species in the Chernobyl exclusion zone, *Journal of Radiological Protection* 30: 341–346. http://dx.doi.org/10.1088/0952-4746/30/2/S07

Brown, J. E.; Alfonso, B.; Avila, R.; Beresford, N. A.; Copplestone, D.; Prohl, G.; Ulanovsky, A. 2008. The ERICA Tool, *Journal of Environmental Radioactivity* 99: 1371–1383. http://dx.doi.org/10.1016/j.jenvrad.2008.01.008

Butkus, D.; Lebedyte, M.; Lubyte, G.; Matusevicius, K.; Mazeika, J. 2001. ^{137}Cs and ^{90}Sr in the soils of Lithuania, *Geochemistry International* 39: 719–24.

Butkus, D.; Konstantinova, M. 2003. On long-term ^{137}Cs migration in the upper layer of Lithuanian soils Lithuanian, *Journal of Physics* 43: 403–408.

Butkus, D.; Konstantinova, M. 2006. Estimation of capacity of an equivalent dose in the ground air, *Vestnik MANEB* 11: 36–43 (in Russian).

Butkus, D.; Krenevicius, R.; Morkunas, G.; Kandrotas, G.; Korkutis, P. 1992. *Investigations of Cs-137 pollution in the territory of Lithuania. Radionuclide pollution in Lithuania and its effects*: Workshop papers. Open Society Fund-Lithuania. Vilnius: Academia Press, 21–26.

EU Council Directive 92/43/EEC of 21 May 1992 on the conservation of natural habitats and wild fauna and flora.

FASSET 2003. Deliverable 5. Handbook for assessment of the exposure of biota to ionizing radiation from radionuclides in the environment. Appendix 1: transfer factors and dose coefficient look-up tables. European Commission, Fifth Framework, Contract No. FIGE-CT-2000-00102 [online]. 2003. [cited 12 January 2010]. Available from Internet: https://wiki.ceh.ac.uk/display/rpemain/FASSET+reports

Hosseini, A.; Thørring, H.; Brown, J. E.; Saxe´n, R.; Ilus, E. 2008. Transfer of radionuclides in aquatic ecosystems–default concentration ratios for aquatic biota in the ERICA Tool, *Journal of Environmental Radioactivity* 99: 1408–29.

Howard, B. J.; Larsson, C.-M. 2008. The ERICA integrated approach and its contribution to protection of the environment from ionizing radiation, *Journal of Environmental Radioactivity* 99: 1361–1363. http://dx.doi.org/10.1016/j jenvrad.2008.04.013

IAEA. 2006. *An international peer review of the programme for evaluating sites for near surface disposal of radioactive waste in Lithuania report of the IAEA International Review Team.* Vienna.

ICRP. 2008. Environmental protection: the concept and use of reference animals and plants, *ICRP Publication 108. Ann ICRP* 38: 4–6.

Fesenko, S.; Fesenko, J.; Sanzharova, N.; Karpenko, E.; Titov, I. 2011. Radionuclide transfer to freshwater biota species: review of Russian language studies, *Journal of Environmental Radioactivity* 102: 8–25. http://dx.doi.org/10.1016/j.jenvrad.2010.09.006

Interim Storage of RBMK Spent Nuclear Fuel from Ignalina NPP Units 1 and 2 Consortium GNS – NUKEM, LEI – Nuclear Engineering Laboratory. *EIA Report. Revision 4.* 24 October 2007. LEI Report S/14-658.5.9/EIA-R-04

Mazeika, J. 2002. *Radionuclides in geoenvironment of Lithuania.* Vilnius: Academia Press.

Nedveckaite, T. 2004. *Radiation protection in Lithuania.* Vilnius: Kriventa.

Paskauskas, R.; Mazeika, J. (Eds.) 1997. *Atomic energy and environment (Lithuanian National scientific programme). The collection of scientific reports.* Vilnius: Academia Press.

Wilson, R. C.; Vives i Batlle, J.; Watts, S. J.; McDonald, P.; Jones, S. R.; Craze, A. 2010. An approach for the assessment of risk from chronic radiation to populations of phytoplankton and zooplankton, *Radiation Environment Biophysics* 49: 87–95. http://dx.doi.org/10.1007/s00411-009-0254-8

Zukauskas, A.; Jankevicius, K.; Lasinskas, M.; Tamosaitis, J.; Virbickas, J. (Eds). 1992. *Thermal power generation and environment* (Lithuanian National scientific programme).

Marina KONSTANTINOVA. Dr., State research institute Center for Physical Sciences and Technology, Institute of Physics. Doctor of Science, physics, Institute of Physics, 2002. Publications: co-author of 28 research papers. Research interests: gamma spectrometry, contamination of environment by artificial radionuclides, parameters of 137Cs migration in soil, 137Cs soil-plant transfer, 137Cs and 3H release from radioactive waste repository and migration with ground water, 222Rn in atmosphere and soil, human and non-human exposure due to artificial and natural radionuclides, accelerator mass spectrometry, 14C measurements, graphite sample preparation features.

Nina PROKOPČIUK. Dr., State Research Institute Center for Physical Sciences and Technology. Institute of Physics. Doctor of Science, Physics, Institute of Physics, 2013. Co-author of 8 research papers. Research interests: radioecology, radiation protection, atmospheric aerosol particle dynamics and particle optical properties, aerosol particles and human health.

Arūnas GUDELIS. Dr., State research institute Center for Physical Sciences and Technology, Metrology Department, Head of Ionizing Radiation Metrology laboratory. Doctor of Science, Physics, Institute of Physics, 1998. Publications: author and co-author of 64 research papers. Research interests: radionuclide metrology, radionuclides and dosimetry in health, industry and environment, radioactive waste management.

Donatas BUTKUS. Dr Habil, Prof., Dept of Environment Protection, Vilnius Gediminas Technical University (VGTU). Doctor Habil of Science (Environmental Engineering) VGTU, 1999. A corresponding member of International Academy of Ecology and Life Protection. Publications: author of more than 270 research paper, co-author of the monographs "Geophysical problems of atmospheric krypton-85" and "Radionuclides in vegetation". Research interests: accumulation of radioactive noble gases, self-cleaning in the atmosphere, influence of ionizing radiation on geophysical processes, investigation of the consequences of the Chernobyl accident in Lithuania, soil-to-plant transfer of radionuclides.

22

LABORATORY SIMULATIONS OF WAVE ATTENUATION BY AN EMERGENT VEGETATION OF ARTIFICIAL PHRAGMITES AUSTRALIS: AN EXPERIMENTAL STUDY OF AN OPEN-CHANNEL WAVE FLUME

Li YIPING[a, b], Desmond Ofosu ANIM[a, b], Ying WANG[c], Chunyang TANG[b], Wei DU[b], Ni LIXIAO[b], Zhongbo YU[d, e], Kumud ACHARYA[f], Li CHEN[e, f]

[a]*Key Laboratory of Integrated Regulation and Resource Development on Shallow Lakes, Ministry of Education, Hohai University, Nanjing 210098, China*
[b]*College of Environment, Hohai University, Nanjing 210098, China*
[c]*Fujian Provincial Investigation, Design and Research Institute of Water Conservancy and Hydropower, Fuzhou, 350001, China*
[d]*Department of Geoscience, University of Nevada, Las Vegas, NV 89119, USA*
[e]*State Key Laboratory of Hydrology Water Resources and Hydraulic Engineering, Hohai University, Nanjing 210098, China*
[f]*Division of Hydrologic Sciences, Desert Research Institute, Las Vegas, NV, 89119, USA*

Abstract. This paper presents a well-controlled laboratory experimental study to evaluate wave attenuation by artificial emergent plants (*Phragmites australis*) under different wave conditions and plant stem densities. Results showed substantial wave damping under investigated regular and irregular wave conditions and also the different rates of wave height and within canopy wave-induced flows as they travelled through the vegetated field under all tested conditions. The wave height decreased by 6%–25% at the insertion of the vegetation field and towards the downstream at a mean of 0.2 cm and 0.32 cm for regular and irregular waves respectively. The significant wave height along the vegetation field ranged from 0.89–1.76 cm and 0.8–1.28 cm with time mean height of 1.38 cm and 1.11 cm respectively for regular and irregular waves. This patterns as affected by plant density and also location from the leading edge of vegetation is investigated in the study. The wave energy attenuated by plant induced friction was predicted in terms of energy dissipation factor (fe) by Nielsen's (1992) empirical model. Shear stress as a driving force of particle resuspension and the implication of the wave attenuation on near shore protection from erosion and sedimentation was discussed. The results and findings in this study will advance our understanding of wave attenuation by an emergent vegetation of Phragmites australis, in water system engineering like near shore and bank protection and restoration projects and also be employed for management purposes to reduce resuspension and erosion in shallow lakes.

Keywords: wave attenuation, aquatic emergent vegetation, open-channel, regular wave, irregular wave, shear stress.

Introduction

Coastal regions, estuaries and other natural water systems are susceptible to inundation by extreme waves and storm damping which results in flooding and events of shoreline erosion. These current situations in aquatic environment engineering and management have brought the idea of using aquatic vegetation by water resources managers in shoreline protection. Aquatic vegetation plays an important role in maintaining water level, wave and heavy storm damping and also helps maintain habitat diversity. The presence of aquatic vegetation can reduce shear stress at the flow-sediment interface and dampen waves, thus contributing to particle deposition, reducing sediment erosion, enhancing sediment stability, and controlling sediment suspension (Koch *et al.* 2006; Gruber, Kemp 2010). Also, aquatic vegetation can substantially affect hydrodynamic processes at the water systems depending on fluvial morphology, vegetation spatial distribution, position and stem density (Augustin *et al.* 2009). Moreover, since hydrodynamics is a significant contributing factor to

Corresponding authors: Desmond Ofosu Anim; Li Chen
E-mails: desofosa@gmail.com; Li.Chen@dri.edu

the ecosystem, the interaction between water waves and aquatic vegetation need to be quantified. Although bottom characteristics are usually a main factor in determining wave dissipation, wetlands ecosystem is particularly important as its effect on hydrodynamics is not only at the bottom but throughout the water column (Augustin *et al.* 2009). Hence understanding the vegetation affected hydrodynamic processes is key in understanding and maintaining the ecosystem. One important aspect is the attenuation of wave energy by aquatic vegetation (Suzuki *et al.* 2011). In large shallow lakes like Lake Taihu, wetlands and other natural water systems, aquatic emergent and submerged vegetation are expected to reduce bed sediments resuspension by dissipating wave energy (Houwing *et al.* 2002; Mendez, Losada 2004; Möller 2006).

For decades, many studies have been carried out to explore the wave and aquatic vegetation interaction focusing on vegetation induced wave attenuation (e.g. Dalrymple *et al.* 1984; Kobayashi *et al.* 1993; Mendez *et al.* 1999; Lovas 2000; Suzuki *et al.* 2011). The initial theoretical mode of wave damping was given by Dalrymple *et al.* (1984) who assumed plants as rigid cylinders exerting drag force on the regular waves. Kobayashi *et al.* (1993) proposed the idea adopting momentum and continuity equations, results of which showed exponential decay of wave height. Mendez *et al.* (1999) continued the work of Dalrymple *et al.* (1984) by considering irregular waves (Mendez, Losada, 2004). Other researchers (e.g. Dean, Bender 2006; Krauss *et al.* 2009) applied linear wave theory to demonstrate that the wave attenuation as a result of momentum transfer from breaking wave to the water column is minimized by two-thirds of incoming wave height as it travels through the vegetation. Augustin *et al.* (2009) also used linear wave theory to laboratory experiments to quantify bulk drag force exerted by vegetation. The study of aquatic vegetation attenuating wave is investigated and quantified in several laboratory studies with both natural and artificial vegetation (e.g. Dubi, Tørum 1996; Tschirky *et al.* 2000; Augustin *et al.* 2009; Cavallaro *et al.* 2010) and also field measurements in various different environments (Möller, Spencer 2002, Möller 2006; Quartel *et al.* 2007; Bradley, Houser 2009; Lövstedt, Larson 2010). In addition, modeling studies were also performed to investigate wave-vegetation interactions (Dalrymple *et al.* 1984; Asano *et al.* 1992; Mendez *et al.* 1999; Mendez and Losada 2004; Mullarney, Henderson 2010).

The above mentioned studies have reported changing rate of wave attenuation regarding the vegetation type and the study environment. Knutson *et al.* (1982) in an early study reported a decrease in wave energy in salt marsh vegetation up to 26%. Mork (1996) conducted experiments on hyperborea kelp beds in the coast of Norway that showed attenuation of more than 60% of wave height. Tschirky *et al.* (2000) in Lake Ontario recorded a 40% reduction of wave height in emergent wetland plants. Cooper(2005) found that wave damping over salt marshes was two times higher than over mudflats. In coastal mangrove, wave damping was shown to be five times more than that due to only bed friction (Quartel *et al.* 2007). Fonseca and Calahan (1992) also studied the wave energy reduction by seagrass and recorded that the effect is negligible provided the plants leaf height is less than half of the water depth. However, Elwany *et al.* (1995) reported that wave energy is not attenuated significantly by the vegetation field which moves concurrently with the flow and also with no substantial drag effect. Large variability exists among studies of wave damping and it is very difficult to describe a general behaviour of vegetation induced wave attenuation (Mendez, Losada 2004). The complexity shows that the dynamics depend largely on the vegetation characteristics such as vegetation type, density, stiffness, spatial distribution, geometry and also the hydrodynamic conditions such as incoming wave height, wave period and water depth. The vegetation field is exposed to varying wave forcing and water flow that changes with time as plant stem characteristics changes thus making wave-vegetation interaction highly dynamic (Bradley, Houser 2009; Anderson *et al.* 2011).

Even though there are existing studies that describe wave-vegetation damping patterns, successful applications are still rare. Since aquatic vegetation shows large variability across the world, seeking an inferred generalized method to quantify this behaviour is difficult. The attenuation of flow within vegetation canopies is a significant part of vegetation flows which has not fully been integrated especially in prior models of dissipation by rough surfaces (vegetation) which this study will aim to explore. This study also aims to advance our understanding knowledge or description of wave-vegetation behaviour by advancing our understanding of the related processes. Also, most previous studies investigated the wave-vegetation flow into sparsely arranged vegetation with some few spaced measuring points of investigation which to the best of our knowledge do not give a detailed profile of the interactions of the in-canopy flow with individual vegetation canopy elements. We hypothesise that the dynamics affecting the dissipation of flow inside the vegetation canopies will eventually control the rate at which wave energy is attenuated which this study aims to provide a detailed wave-vegetation with various observation points. Also, as the emergent macrophytes common Reed (Phragmites australis) is becoming a predominantly species in our water ecosystem especially in eastern China (Zhang 2009; Yan *et al.* 2011; Li *et al.* 2014), its use in this study will help to understand the extent by which these species could change the habitat modifying characteristics of the ecosystem in terms of wave dissipation. The investigations about the ecosystem engineering impacts of these species is very significant from a management perspective. This is achieved by investigating and quantifying the wave-vegetation interaction, focusing on wave damping and transmission, and energy dissipation. The effects of plant stem density, wave conditions are discussed. Also the vertical and horizontal shear stress variations as the wave travels through the vegetation field were investigated.

In this paper, wave attenuation was investigated in well-controlled laboratory experiments with artificial emergent vegetation field under different wave conditions and vegetation stem densities. In addition, the wave energy dissipation expressed as wave energy dissipation factor and wave orbital amplitude was studied and also investigating if friction by emergent plants relates can be described by existing empirical models.

1. Materials and method

1.1. Theory

1.1.1. Wave height and energy decay

Generally, wave attenuation by aquatic vegetation is a result of the energy loss to the plant stems (Dalrymple *et al.* 1984; Mork 1996; Bradley, Houser 2009). Theories for wave energy and height decay through vegetation field have been proposed in various studies and, in the end, a general solution was proposed irrespective of the type of vegetation (Mendez *et al.* 1999; Möller *et al.* 1999; Augustin *et al.* 2009; Bradley, Houser 2009). The resulting general equation for wave height decay is given as:

$$Q_c = \frac{H}{H_o}, \tag{1}$$

where Q_c is the wave decay coefficient, H is the incoming wave height, H_o is the local wave height. For irregular waves, H and H_o becomes the root mean square wave height (i.e. ***Hrms*** and ***$H_o$rms*** respectively).

Dalrymple *et al.* (1984) proposed an algebraic equation using the linear wave theory and wave energy conservation which was later modified by Mendez and Losada (2004):

$$\frac{H}{H_o} = \frac{1}{1+\beta x}, \tag{2}$$

where β is parameter related to the plant and wave characteristics given as

$$\beta = \frac{1}{3\pi} C_d N d_v H \left[\frac{sin^3 kl_v + 3 sinhkl_v}{(sinh2kh + 2kh) sinhkh} \right], \tag{3}$$

where k is the wave number (m^{-1}), l_v is the vegetation height, h is water depth. Generally, the dissipation of wave energy depends on the characteristics of the vegetated field (configuration, density, size) and morphology of plants.

Wave energy dissipation factor above roughness elements (in the presence of vegetation), f_e or the wave friction factors, f_w at the bed are usually a function of the wave orbital amplitude and the hydraulic roughness, r (Nielsen 1992; Soulsby 1997). Based on a large number of laboratory studies, Nielsen (1992) derived empirically the wave friction equation that reads

$$f_w = exp\left(5.5\left(\frac{r}{A}\right)^{0.2} - 6.3\right), \tag{4}$$

where A is the wave orbital amplitude calculated by

$$A = UT_p / 2\pi, \tag{5}$$

where T_p is peak wave period. This equation has been widely and successfully applied over rough surfaces (Manca *et al.* 2010; Lowe *et al.* 2005). In the present study, the wave energy loss as a result of the presence of vegetation was estimated by the vegetation induced friction and calculated using the commonly used dimensionless parameter f_e modified by Manca *et al.* (2010). This parameter is determined from the wave energy loss in the vegetation field assuming exponential wave height decay. This friction factor was applying the formulae of Jonsson (1966) as

$$f_e = \left(\frac{3}{2\pi}\frac{\varepsilon_f}{\rho U_*}\right), \tag{6}$$

where ρ is the water density, U_* is the horizontal wave velocity above canopy, ε_f is defined as the rate of wave energy dissipation per unit area as a result of induced friction. These two friction-based coefficients (f_w and f_e) are different in definitions, Nielsen (1992) proved that, in spite of the disparity of these two parameters shown in various experimental results, they can be assumed equal for practical applications. Manca *et al.* (2010) also made the similar conclusion. If the linear wave theory is applicable, ε_f can be calculated assuming the waves are normal to the vegetation field as (Dalrymple *et al.* 1984)

$$\varepsilon_f = \frac{\partial EC_g}{\partial x}, \tag{7}$$

where $E = \frac{1}{8}\rho g H^2 C_g$ is the wave energy density, g is the acceleration due to gravity, and $C_g = \frac{1}{2}c\left(1 + \frac{2kh}{\sin(2kh)}\right)$, where C_g and c are group velocity and wave phase speed respectively (Infantes *et al.* 2012).

1.1.2. Vegetative roughness and drag coefficient

For the parameters mentioned above, it is shown that the mean drag coefficient, C_d, of the vegetation field is one of the most significant parameters for determining attenuation coefficients. However, it is very difficult to estimate C_d. For flows through vegetation fields, both the Darcy-Weisbach coefficient and the Manning's n are used to represent the resistance coefficient. For a glass-walled wave flume like the one used in this study, the side wall is considered smooth and thus the sidewall resistance is negligible. The flume roughness resistance was then dominated by the vegetative roughness. The wave energy is greatly attenuated by the vegetative resistance in the wave flume. Kouwen (1996) established that vegetative roughness dominates the bed resistance instead of the surface resistance. To estimate the vegetative drag induced by the vegetal plant elements, the hydrodynamic forces (F) that waves exerts on the vegetation is examined as $F = (F_x, 0, F_z)$ where F_z normally assumed to be negligible likened to F_x especially for shallow flow water depth. Generally, the horizontal force (F_h) is given as $F_h = F_d + F_i$ (Morison *et al.* 1950) where F_d is the vegetative drag force and F_i is the inertial force. However, defining a generalized value for vegetation-induced drag is impossible since the C_d is highly dependent on hydrodynamic and plant properties. However, C_d is commonly regarded as a function of some non-dimensional flow parameters (Kobayashi *et al.* 1993; Mendez, Losada 2004), in which the most relevant one is the stem Reynolds number, *Re*, (Kobayashi *et al.* 1993; Augustin *et al.* 2009) which is related to C_d as

$$C_d = a + \left(\frac{\beta}{Re}\right)^{\zeta}, \tag{8}$$

where a, β, ζ are coefficients depending on the plant properties and *Re* is defined as

$$Re = \frac{Ud}{\nu}, \tag{9}$$

where ν is the kinematic viscosity of water (= 10^{-6}m^2/s at 24 °C) and u is the maximum horizontal characteristic velocity acting on the plants at the measurement point calculated from the linear wave theory as

$$U = \frac{\pi H}{T}\frac{\cosh(kl_v)}{\sinh(kh)}. \tag{10}$$

Kobayashi *et al.* (1993) calibrated C_d values for Asano *et al.* (1988) experiments and plotted against the corresponding stem *Re* and established a relationship as

$$C_d = 0.08 + \left(\frac{2200}{Re}\right)^{2.4}, \tag{11}$$

Property	Value
	Emerged
Plant total height	65 cm
Approximate height	
Leaf section	50 cm
Stem section	15 cm
Average diameter	
Leaf section	2.6 cm
Stem section	0.4 cm
Average width	
Leaf section	8 cm
Average surface area	
Leaf section	0.025 m^2

Fig. 1. Image of the artificial emergent plant used for the experiment and a summary of its physical properties

for *Re* range of 2200 < *Re* < 18000. However, Mendez *et al.* (1999) through laboratory studies extended Kobayashi *et al.* (1993) work and derived also an empirical formula for C_d varying with *Re* as

$$C_d = 0.08 + \left(\frac{2200}{Re}\right)^{2.2}, \tag{12}$$

for *Re* range of 200 < *Re* < 15500. This formula does not account for vegetation swaying (see Mendez *et al.* 1999) where it is assumed that the horizontal velocity of the vegetation is substantially smaller than the orbital velocity thus the vegetation can be considered to be rigid. Mendez, Losada (2004) however investigated that Keuglan Carpenter (KC) number relates appropriately the C_d with the stem length and wave parameters expressed as

$$KC = \frac{U_{\max}T}{d_v}. \tag{13}$$

However, for emergent vegetation field conditions, C_d was found to relate better with *Re* (Augustin *et al.* 2009; Bradley, Houser 2009; Cavallaro *et al.* 2010).

1.1.3. Bottom roughness and shear stress

Generally, in the field and laboratory studies, Manning and Darcy-Weisbach friction factors have been successfully applied to flows through aquatic vegetation which has also been related to the wall shear stress (Galema 2009; Augustin *et al.* 2009). However, it has been established in literature that for a glass-walled flume, the wave energy is more attenuated by the vegetative resistance, thus vegetative roughness dominates the bed resistance (Fathi-Maghadam, Kouwen 1997). Various studies have applied different approaches in estimating the mean bottom shear stress in a flume for wave-vegetation flow interaction (Mazda *et al.* 1997; Cheng *et al.* 1999; Mendez *et al.* 2004; Rashid 2010; Thompson *et al.* 2004). Thompson *et al.* (2004) mentioned that the Quadratic Stress Law is suitable for estimating the mean shear stress for all kinds of fluid flow and vegetation interaction. The shear stress, τ with the Quadratic Stress Law is defined as

$$\tau = \rho C_d U_\infty^2, \tag{14}$$

where U_∞ is the depth averaged longitudinal flow velocity (m/s).

1.2. Experimental setup

The experiments were conducted in an open wave flume constructed in the Coastal, Harbor and Offshore Engineering Hydraulics Laboratory at Hohai University, China. A series of experimental runs were performed for waves propagating over an array of plastic artificial emergent plants (see Fig. 1).

The laboratory wave flume is a 30×0.5 m glass-walled rectangular flume with a height of 1.0 m. A piston wave maker controlled by a computer was used to generate waves at the upstream of the wave flume. The incoming water flow was controlled by an inlet turning valve at the upstream end and a triangular adjustable weir at the downstream end, which was also used to control the water level. At the downstream of the wave flume was a rubber wave absorber built to minimize wave reflection effects. A 7 m long vegetation section was constructed 15 m downstream from the wave maker. Artificial plastic plants were used in this study to simulate aquatic vegetation. These plants mimicked real species of common reed (*Phragmites Australis*). They are produced from material to generate appropriate density and flexural rigidity to approximate the real plants as possible. The plants have biomechanical and morphological resemblance to the real species they mimic and were thus suitable to be used in our studies. Fiala (1976) described the vegetative morphology and mechanical properties of *Phragmites australis* plant stems. This approach of using artificial plant is similar to that used by many other researchers (e.g. Nepf, Vivoni 2000; Järvelä 2005; Chen *et al.* 2011; Li *et al.* 2014) who investigated scale models of vegetation and have found it is important to use simplified form plants to the real vegetation. Each individual artificial plant has a total height of 0.65 m (0.5 m stem section and 0.15 m leaf section) with a mean plant width of 0.08 m. It has an average diameter of 0.026 m and 0.04 m at the leaf and stem section respectively. The plant were built on a 7 m × 0.5 m × 0.01 m (L×W×H) board with holes and placed on the flume bed to develop the vegetation section. The schematic view of the experimental setup is shown in Figure 2. Figure 3 shows an image of the flume with the planted vegetation. The water depth in the flume was kept approximately at 0.5 m for all the experimental runs.

1.3. Instrumentation and experimental runs conditions

To investigate the effect of plant stem density and varied wave conditions on attenuation under emergent conditions, twelve scenarios were examined. Table 1 shows the conditions of these experimental runs. In addition, four vegetation densities (stems/m^2) ranging from 30 to 90 stems/m^2 which represent a typical aquatic configuration in most rivers in the eastern china (CVEC 1980; Ma *et al.* 2008; Zhang 2009) and six wave conditions for both regular (*R*) and irregular (IR) waves were selected to represent different wave heights and periods (see Table 1). Each experiment was assigned a specific code name in this paper based on the plant stem density and wave condition (see Table 1). For example, WIE90 indicates experimental run 1 under 90 stems/m^2 plant density and a regular wave of 3 cm wave height (*H*) and 1.0s wave period (*T*). Also, W4E90 indicated experimental run 4 under 90 stems/m^2 plant density and an irregular wave of H = 4.24 cm and T = 1.11 s. Code names with W1, W2 and W3 represent regular waves and the rest for irregular waves. The wave transformation along the flume was monitored by 10 wave gauges which measured the free surface oscillations time series (wave height and period). These were distributed at one meter interval beginning 1 m upstream of the vegetation section. Figure 4 shows the locations (Locations 1 to 10) of the wave gauges along the vegetation section in the flume. Locations 1 and 10 were located at upstream and downstream outside the vegetation section respectively whereas Locations 2 to 9 were located within the vegetation section.

Instantaneous velocity measurements were taken by a 3D macro acoustic Doppler velocimeter (ADV) (SonTek, San Diego, CA, USA) at a sampling frequency of 20 Hz at various positions (Positions 1 to 5) along the vertical direction with a 30 s sampling time. A total of 600 data points were collected at each position. A mean velocity value at each measuring point was obtained by using Win-ADV post-processing software. This velocity data were to

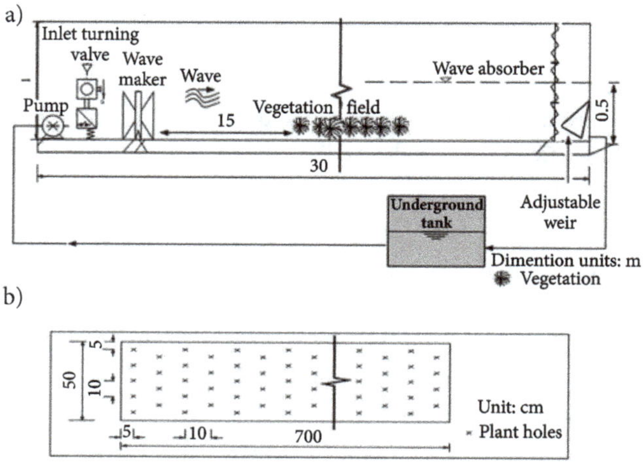

Fig. 2. a) Schematic view of the experimental flume setup; b) Plan view of the boards in which the artificial plants were arrange to create the vegetation field

Fig. 3. Image of the flume with the planted vegetation

be used later in estimating the vertical and longitudinal shear stresses in the flume. Figure 4 shows the ADV measuring positions. Positions 1 and 5 are located outside the vegetated regions whereas Positions 2, 3 and 4 are within the vegetated region. Velocity measurements were taken at fourteen points along the vertical direction starting from 5 cm above the flume bed. Measurements were taken at 2.5 cm and 5 cm intervals spaces near the bed and towards the surface respectively (i.e. 2.5 cm increments from 5 to 20 cm, and then 5 cm increments from 20 to 40 cm). Thus, there were a total of 14 measuring points vertically.

1.4. Wave data analysis

For the various experimental runs, the data were analyzed to quantify the wave decay by the vegetation field. The wave attenuation patterns as well as the estimation of the drag properties are of interest. For each of the 10 wave gauge points along the vegetation field, the root mean-square height (H_{rms}) and the mean wave period were determined from each water oscillation time series. This was obtained by employing a wave-by-wave analysis using post processing wave software which applies a zero-crossing method (Mizuguchi 1982; Augustin *et al.* 2009; Zhang *et al.* 2011) to calculate the significant wave height (H_s) and the peak wave period (T). The zero-crossing method represents average water surface fluctuations most accurately (Möller *et al.* 1999). Wave heights were obtained from the wave spectral for each wave gauge points for irregular waves. For irregular waves, H_s were converted to H_{rms} using the relation ($H_{rms} = H_s / 1.416$). The wave height decay coefficient (Q_c) was estimated from equation 1.

Table 1. The twelve experimental runs tested and their conditions. R and IR represent regular and irregular wave respectively

Expt. run	Vegetation density (stem/m^2)	Wave type	T (s)	H (cm)	Code name
1	90	R	1	3	W1E90
2	90	R	1.2	3	W2E90
3	90	R	1	4.5	W3E90
4	90	IR	1.11	4.24	W4E90
5	90	IR	1.33	4.24	W5E90
6	90	IR	1.11	6.36	W6E90
7	70	R	1	4.5	W3E70
8	70	IR	1.11	6.36	W6E70
9	50	R	1	4.5	W3E50
10	50	IR	1.11	6.36	W6E50
11	30	R	1	4.5	W3E30
12	30	IR	1.11	6.36	W6E30

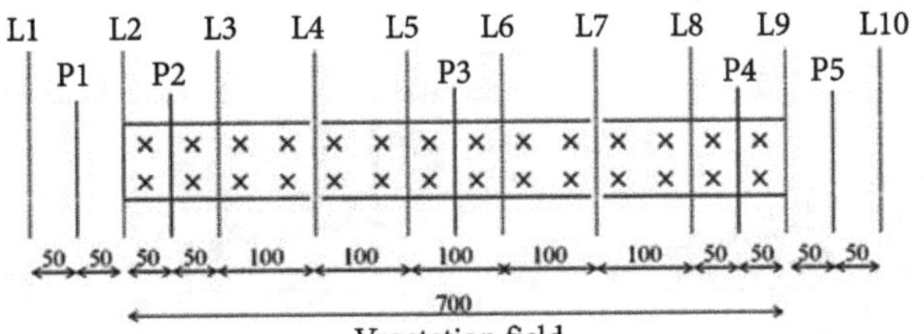

Fig. 4. Measuring points within the flume for both ADV and wave gauges. Wave gauge locations are shown by L1 to L10 and ADV velocity measuring points are shown by P1 to P5 where P and L represents Positions and Locations respectively.

2. Results and discussions

2.1. Wave heights and energy attenuation

The experimental data indicated that the wave height greatly decayed whiles the wave travelled through the vegetation field (see Fig. 5). For emergent conditions, the whole velocity profile is obstructed by the plant stems. Figure 5 shows the trend of wave height decay coefficient changing with distance through the vegetation section for all runs.

Generally, for emergent conditions, irrespective of the plant stem density or wave condition, the wave height decreased when the wave propagated through the vegetation section. The greater part of wave height attenuated over the vegetation field mostly occurred along the first few meters of the vegetation field (from Locations 1 to 3) (see Fig. 5) which was observed for both regular and irregular waves runs. This was similar to the findings observed by previous researchers (Kobayashi *et al.* 1993; Mendez, Losada 2004; Augustin *et al.* 2009). This obvious great effect is due to the fact that when wave travelling through emergent vegetation, the drag effects from the plant are distributed throughout the water depth where the velocity profile is obstructed. The decay however decreased with distance along the vegetation field.

Also in Fig. 5, the theoretical wave decay coefficient as calculated by Equation 2 was plotted against the experimental results of Q_c . It was observed that the theoretical equation provides a good fit to the experimental data. For all experimental runs, the experimental wave height data follows the proposed theoretical equation for wave height decay by Dalrymple *et al.* (1984) and Mendez and Losada (2004) which describe exponential wave height decays through vegetation zones.

The wave height was observed to decrease beyond gauge measuring point 3 at Location 3. Generally, the H_s decrease by a mean 0.2 cm for regular waves experimental runs and 0.32 cm for irregular waves runs over the vegetation filed between Locations 2 to 9. The H_s decreased by ~25% between Location 2 and 5 at a distance of 3 m and then averagely ~6% from Locations 5 to 9. The H_s turns to increase at Location 2 when it first contacts the vegetation field and also from Locations 9 to 10 after leaving the vegetation field. The significant wave height along the vegetation field ranged from 0.89–1.76 cm with a time mean H_s of 1.38 cm for regular waves and 0.8–1.28 cm with a time averaged H_s of 1.11 cm for irregular waves. This study suggest that for an incident regular wave of 3 cm < H < 4.5 cm and 1 s < T < 1.2 s moving over a constant depth of 0.5 m, the percentage wave height attenuation was of values of 13.6% decrease for the first 3 m within the vegetation section at position 5 and 6.4% decrease for the next 3 m till the end of the vegetation section and increase slightly just after at position 10. Similar trend occurred for an incident irregular wave of 4.24 cm < H < 6.36 cm and 1.11 s < T < 1.33 s with an attenuation of 7.6% decrease for first 3 m and a further mean 2.4% decrease. Qualitatively, the results in this present study correspond to that of Mendez and Losada (2004) and Augustin *et al.* (2009), even though the magnitudes of wave height decay are not the same. Hence, such deviation could be as a result of different condition pertained to the depth of water, different stem densities and plants characteristics. For most experimental runs, the wave height increased at the edge of the vegetation field before decreasing for both regular and irregular waves. This observation could be as due to the frictional dissipation in some of the experimental runs. This could also be explained as a result of the local wave interaction caused by the immediate impediment and change in bottom configuration experienced by the waves. This behavior was also reported in regular waves results by Bradley and Houser (2009) and Stratigaki *et al.* (2011).

The experimental runs with the larger canopy density showed the greatest wave height decay. The higher stem densities resulted in greater attenuation irrespective of the wave condition but were most prominent in the regular wave experiment runs. The results also exhibit a rapid increase in wave decay with large incident wave height for regular wave experimental runs and less wave attenuation with increasing wave height for irregular wave experimental runs. For the same plant stem density, this trend also existed. The wave height decay reduction with the accumulation of the water depth and wave length ratio could be explained as the results of the very less penetration of the wave propagation within the water column as it travels along the vegetation section which results in a less wave induced flow and plants interactions.

The wave height decay (Q_c) which occur exponentially ranged from 0.2 to 0.58 with a mean coefficient of 0.3 for regular wave runs whereas that of irregular waves runs ranged from 0.12 to 0.26 with a mean coefficient of 0.16. The fluctuation of the Q_c turns to depend on the *Re* where the characteristic velocity was the velocity related with the peak spectral wave. The highest Q_c was observe for the largest *Re* observed (25000 < *Re* < 45000) and the smallest Q_c observed for the lowest *Re* observed (*Re* < 20 000). Figure 6 shows the rate of wave energy attenuation as it passes through the vegetation field. The result shows a significant wave energy reduction as a result

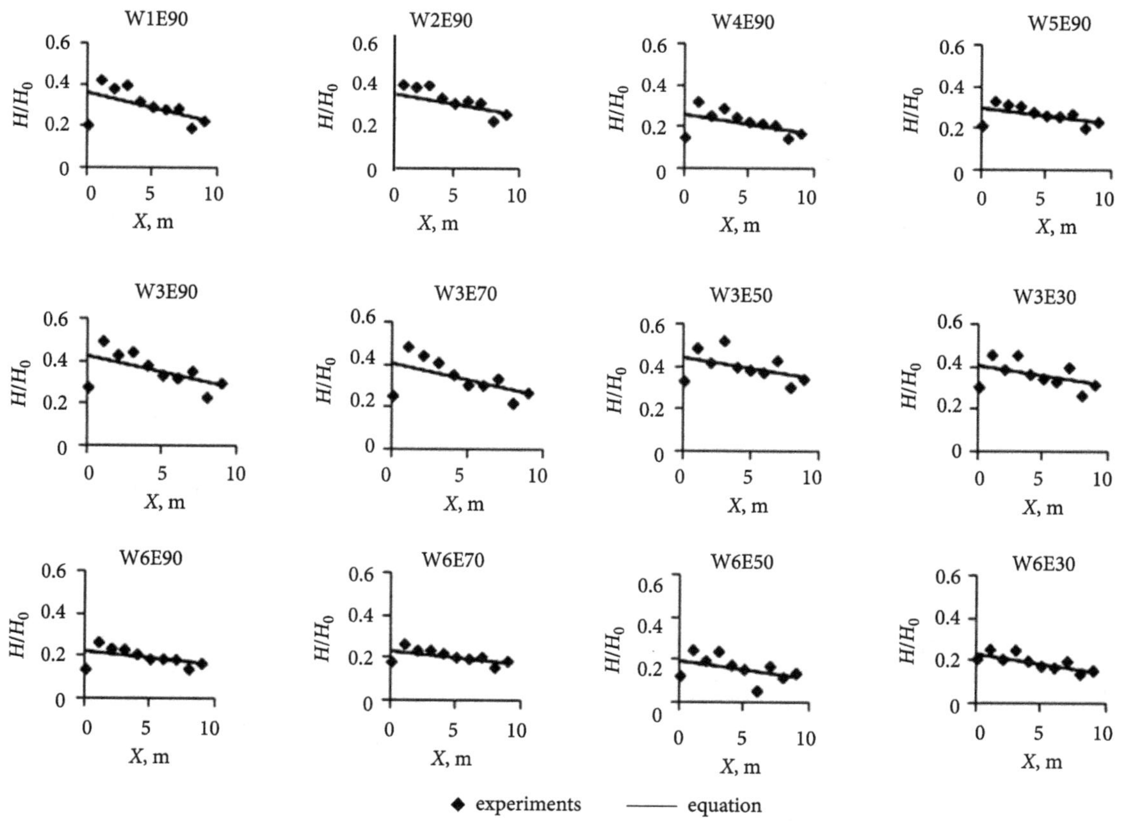

Fig. 5. Experimental and numerical results of wave height decay coefficient ($Q_{C=}H_o/H$) variation over the emergent artificial vegetation field for all the experimental runs tested.

of the interaction between the plants and wave induced flows. The wave energy dissipation was found to be maximum at the first two transects within the vegetation. For the same plant stem density (see Fig. 6a), the rate of dissipation trend seems not to change after travelling through the first 2 m of the vegetation section. The rate of dissipation for a high incident wave period and wave height was minimum and decrease gradually as the wave travels through the vegetation section towards a near-zero value. Similarly, for short wave period and small incident wave height, the rate of dissipation was high (with a mean 10% dissipation rate every 1 m) and decreased rapidly as the wave travels. However, a higher amount of wave energy was dissipated about 3 m within vegetation field. From Figure 6c, it was could be said that irregular wave energy was largely dissipated compared to regular wave experimental runs. This corresponded to the drag coefficient which was higher for irregular wave run cases than the regular wave run cases. Considering the vegetation density, it was observed that the small plant stem density induced higher rate of energy dissipation especially at the edge of the vegetation field which decreased as the wave travels

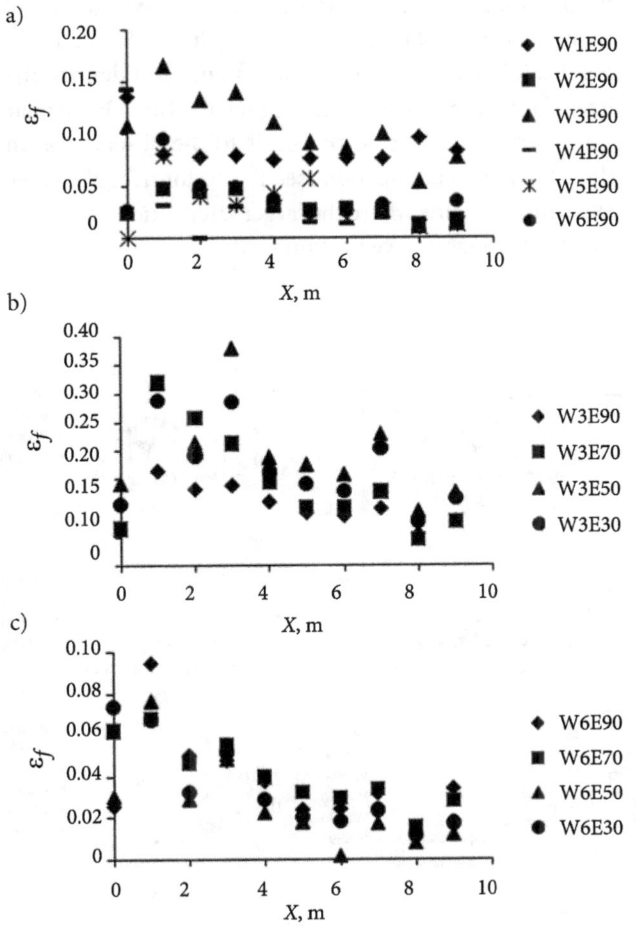

Fig. 6. The rate of wave energy attenuation variation over the emergent artificial vegetation field for all the experimental runs tested: a) for same plant stem density; b) regular wave experimental runs; c) irregular wave experimental runs

through the vegetation section. However, the rate of energy dissipation caused by the higher plants stem density seems to small and turns to be stable with little change as the wave travels through the vegetation section.

2.2. Wave energy dissipation factor

In this study, the wave energy attenuation as influenced by the friction induced by the presence of the vegetation field was estimated in terms of f_e. This has often been expressed as a function of the wave orbital amplitude (Manca *et al.* 2012). The f_e decayed with increasing *A*. For each experimental runs, the corresponding wave energy dissipation factor (f_e) were computed using Equation (5) for regular wave (f_e^r) and irregular wave runs (f_e^{ir}). Figure 7 shows wave energy dissipation factors (f_e) plotted against the wave orbital amplitude (*A*). There were not much difference of f_e values for low and high plant stem density. For low plant stem density, the f_e values were lower (0.007–0.014) and 0.006–0.09 for high plant stem density. The friction produced by vegetation field under irregular wave was higher than regular wave.

Augustin *et al.* (2009) modeled f_e and had values between 0.05–0.19 for low density artificial Salt-marsh under emergent conditions and Möller *et al.* (1999) obtained values between 0.07–0.28 in a natural Salt-marsh for plant stem density between 150 gm^{-2} and 800 gm^{-2}. These f_e values were in the higher part of range of that found in this study. This could be due to the difference in vegetation characteristics. The ability for emergent vegetation to attenuate wave energy is known to change with vegetation characteristics (stiffness, morphology) and wave conditions (Bouma *et al.* 2005; Koch *et al.* 2009; Paul, Amos 2011; Manca *et al.* 2012). Thus, to have meaningful comparisons of results becomes quite difficult. Generally, this study shows a clear decrease in f_e with an increasing *A* over the emergent vegetation field (see Fig. 7). This could be explained that the friction induced by the vegetation field is not the same across all wave amplitudes. The decreasing trend demonstrates the dependency of the f_e on the wave periods. This is however fairly scattered for irregular wave runs which is most likely to be due to f_ε dependence on other local flow conditions (i.e. *Re*). The f_e expectedly changed as a function of *A* in agreement to the Nielsen (1992) empirical model (see Eq. (4)). With this, a strong correlation coefficient (0.95, $n = 11$, $P < 0.05$) was produced for a hydraulic roughness ($r = 0.03$), where *n* is the sample size for each experimental run and *P* is the probability value. Higher peak wave period and incident wave height experience gradual decrease dissipation rate than short wave period and small incident wave height which experience high dissipation and decreases rapidly as the wave travels especially within the vegetated region.

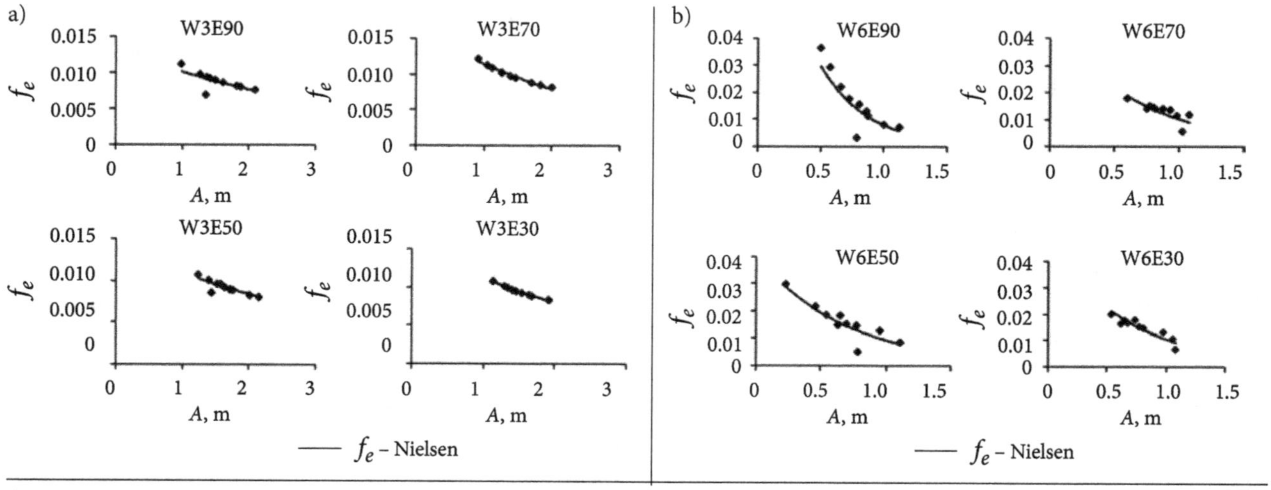

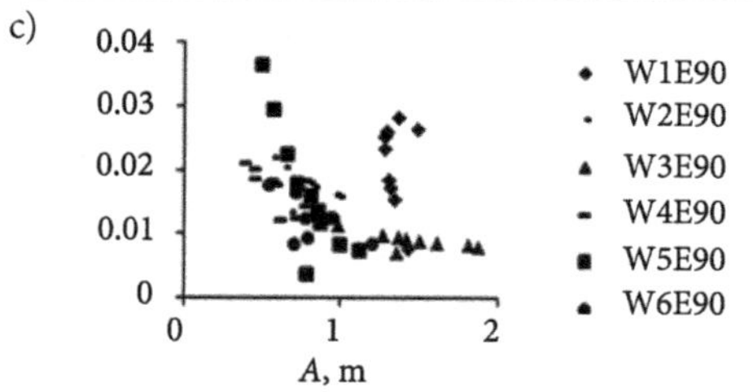

Fig. 7. Wave energy dissipation factors (f_e) plotted against the wave orbital amplitude (A): a) regular and b) irregular wave experimental runs. A fit of Nielsen (1992) model (Eq. (4)) was overlaid with data from test represented by solid line; c) for the same plant density

2.3. Vegetation roughness and drag coefficient

Using Equation 2 and the observed wave energy decay, the plant stem drag coefficient was calculated for each experimental runs. It is important to note that the representative effective drag coefficient was computed under the presumption of rigid vegetation as the reflection measured from the plants was less than 8% and therefore neglected. The C_d changes considerably similar to the wave decay coefficient and dissipation which ranging from 0.05 to 0.8 with a mean C_d of 0.12 for regular waves conditions and from 0.1 to 1.5 with a mean C_d of 0.6 for irregular waves. Fig. 8 shows a plot of the C_d against *Re* and *KC* following Kobayashi *et al.* (1993) and Mendez *et al.* (1999). It shows that the drag coefficient decreased as the waves decay under emergent conditions which resulted in reduction of the attenuation rate. It was observed that the C_d decreased with increasing *Re*. For very large *Re*, the C_d was to an order of 0.01. For the irregular waves experimental runs, the C_d produced a more scattering phenomena which resulted in a weak correlation coefficients. However, the correlation between C_d and *Re* for regular wave experimental runs yielded an improvement in the correlation coefficient results with less scatter. Similar observation was made for the correlation coefficient between C_d and *KC*. This indicated a weak dependency between C_d and wave period. However, the correlation here yielded is very weak. This confirms the claims made by various researchers that C_d depends directs on the *Re* (e.g. Mendez and Losada, 2004; Augustin *et al.* 2009; Bradley and Houser 2009). Considering relatively the impact of the plant stem density on C_d, it is generally observed that the lower stem density (30 and 50 stems/m^2) produced lower stem drag coefficient compared to the higher stem density (70 and 90 stems/m^2). Table 2 gives the mean C_d for each run. From Fig. 8a, it is seen that for the same stem plant density, the irregular wave run cases yield higher C_d than the regular wave run cases. This corresponds to the observation in the wave height attenuation (see Fig. 5) for irregular wave experimental runs where the larger attenuation was experienced through the vegetation field.

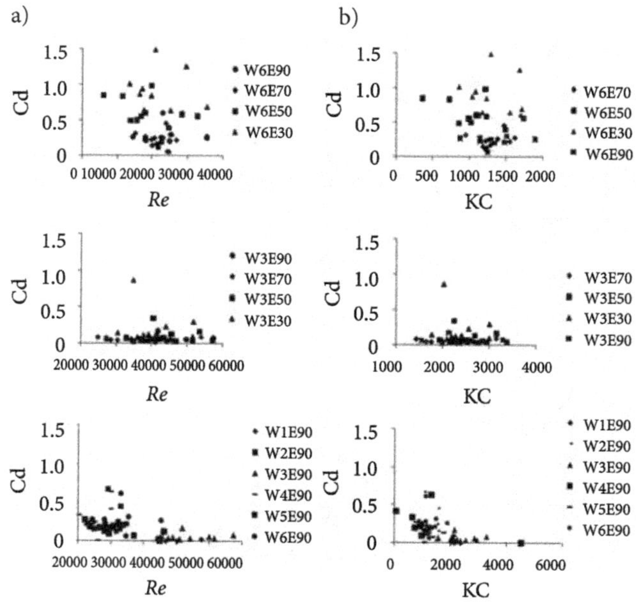

Fig. 8. Relationship between the plant stem drag coefficient (C_d) and the Reynolds number (*Re*) (a) and Keuglan-Carpenter (KC) (b) for regular (W3) and irregular (W6) and the same plant stem density experimental runs calculated locally at each measurement location.

Table 2. The vegetation drag coefficient values for the twelve experimental runs tested. R and IR represent regular and irregular wave respectively

Code name	Wave type	C_d	Code name	Wave type	C_d
W1E90	R	0.205	W3E70	R	0.061
W2E90	R	0.245	W6E70	IR	0.295
W3E90	R	0.058	W3E50	R	0.104
W4E90	IR	0.202	W6E50	IR	0.238
W5E90	IR	0.149	W3E30	R	0.220
W6E90	IR	0.253	W6E30	IR	0.301

2.4. Effect of plant density

Figure 9 shows the effect of plant stem density (N) on the wave height attenuation for a constant wave height and wave period (see Fig. 9 a, b). The experimental data shows that the stem density greatly impacts the wave attenuation under emergent conditions. This effect is obvious and mostly within the first two wavelengths of wave propagation within the vegetation field. It was found that immediately after the entering the vegetation section, a small local wave height increase occurred particularly for higher N which could be to be as a result of the sudden presence of the vegetation section. It was also found that the wave height increases at the last two wavelengths within the vegetation section towards the end with increasing N. This may be due to the fact that as wave induced flow is more and more impeded by the plants as it travels, part of wave energy is reflected and the wave height increases. The influence of the stem density on the wave attenuation was also characterized in this study as a function of the effective drag coefficient representing the average drag force per unit area across the wave propagation. This was obtained from theories in the literature as proposed by Dalrymple *et al.* (1984) and improved by Mendez and Losada (2004) related to coefficients which depend on the plant properties like the stem density and often related to the *Re*. Generally, in some cases the lower N experimental runs for both wave types resulted in higher C_d values compare to higher N runs. This resulted in higher wave decay coefficients which suggest that the analytical decayed wave height as given in Equation (2) will be higher. The higher C_d by lower N in these cases was explained by Augustin *et al.* (2009) as a result of an artifact of sheltering among individual plant stem such that the effective velocity acting on an individual plant at higher plant stem densities is reasonably low compared to that of an unsheltered stem. This could also be explained by the lower *Re*. For the same wave conditions, wave height decay depending on the N appears to be increased with increasing plant density.

2.5. Effect of wave period

Figure 10 shows the effect of wave period on the wave attenuation in the vegetation section. Here, Q_c is plotted for the same plant stem density (90 stems/m^2) and different wave conditions. Increases of the wave period and wave length resulted in a higher reduction of wave height as wave travelled along the vegetation zone, even though the difference observed between the tested wave periods was not very substantial for both the regular (W1E90 and W2E90) and irregular (W4E90 and W5E90). The high incident wave height are normally attenuated by the vegetation section with maximum 12% wave height attenuation observed even for $T = 1$ and 1.11 s and dense plant stem. The wave decay rate rises with the wave period for the small values of T to a point and reduces gradually for the large values of T to an almost asymptotic

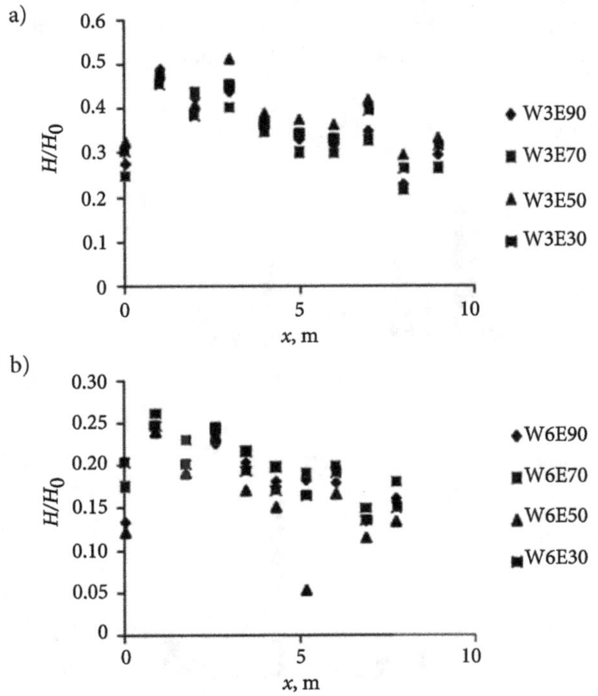

Fig. 9. Effect of plant stem density (N, stem/m^2 on the wave height decay coefficient ($QC = H_0/H$) variation over the emergent artificial vegetation field: a) eregular wave experimental runs; b) irregular wave experimental runs

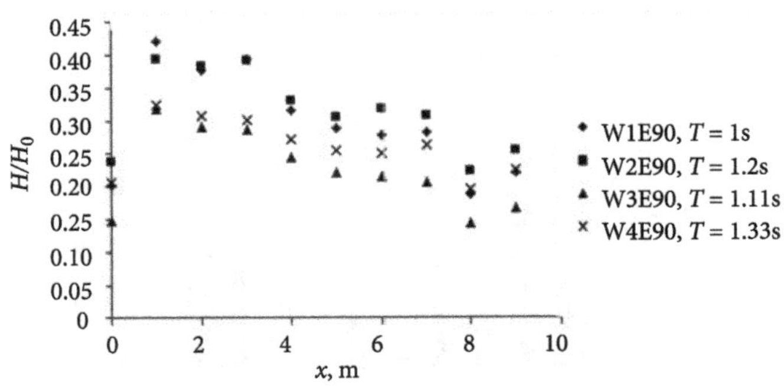

Fig. 10. Effect of wave period (T) on the wave height decay coefficient ($QC = H_0/H$) variation over the emergent artificial vegetation field

value as it travels through the vegetation field. The experimental wave decay coefficients changes more widely for short peak wave periods and nearly converge for longer *T*. It could be said considering the various peak wave periods, the theoretical wave decay coefficient is nearly independent of the peak periods regardless of the slight decrease. This is in agreement with Huang *et al.* (2011) investigations which reported that the wave transmission coefficients mostly depend on the vegetation density and incident wave height.

2.6. Shear stress distribution under wave-vegetation interaction

Figure 11 shows profiles of the longitudinal and vertical distribution of shear stress for the different plant stem density and wave conditions. The results show that the vertical distribution could be divided into two parts: (a) the lower layer ($z \leq 15$ cm), from the bottom bed which is also the stem section of the plant and (b) the upper layer which is from $z \geq 15$ cm to the canopy of the plant towards the free water surface. This shows that the shear stress profile is complex and keeps changing within the flow depth. The shear stress the lower layer (z = 2–10 cm) was nearly zero for most flow conditions. In general, the shear stress within the vegetated region (Position 2 to 4) was maximum at z = 0–2cm and decreased gradually at the lower layer (2 cm $\leq z \leq$ 10cm) and then experienced a rapid increase at z =10–15 cm. It then decreased gradually towards the free water surface especially at Positions 3 and 4 within the vegetated region. For regular wave runs, maximum shear occurred at the lower layer and decreased towards free surface. Also, outside the vegetated region, shear stress was higher which explains that emergent plant stem greatly reduced the shear. Within the vegetated region, shear stress increased slightly at z = 25–30 cm at Positions 3 and 4. Longitudinally, in the direction of the flow, it is observed that τ decreased as the wave travelled within the vegetated field especially at the lower layer where there is a lot of impediment for emergent conditions.

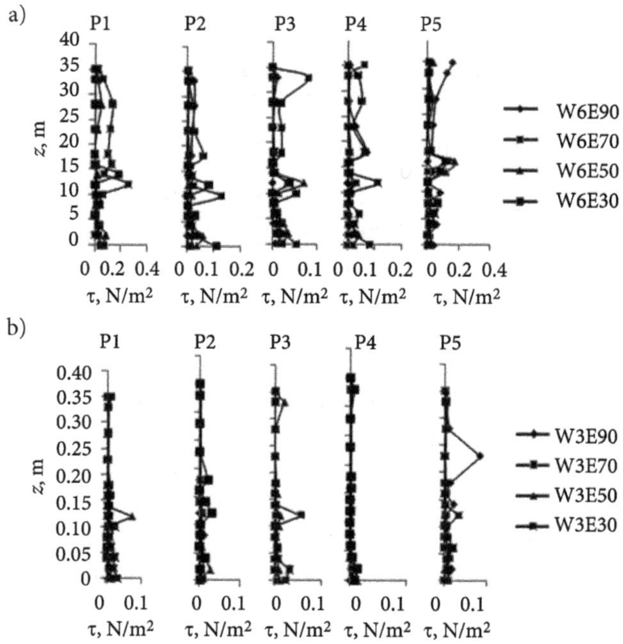

Fig. 11. The vertical and longitudinal distribution profiles of shear stress over the vegetation field. (a) Irregular wave experimental runs and (b) Regular wave experimental runs. z represents the flow depth; P1, P2, P3, P4, and P5 represent Position 1 to Position 5 as described in Figure 4

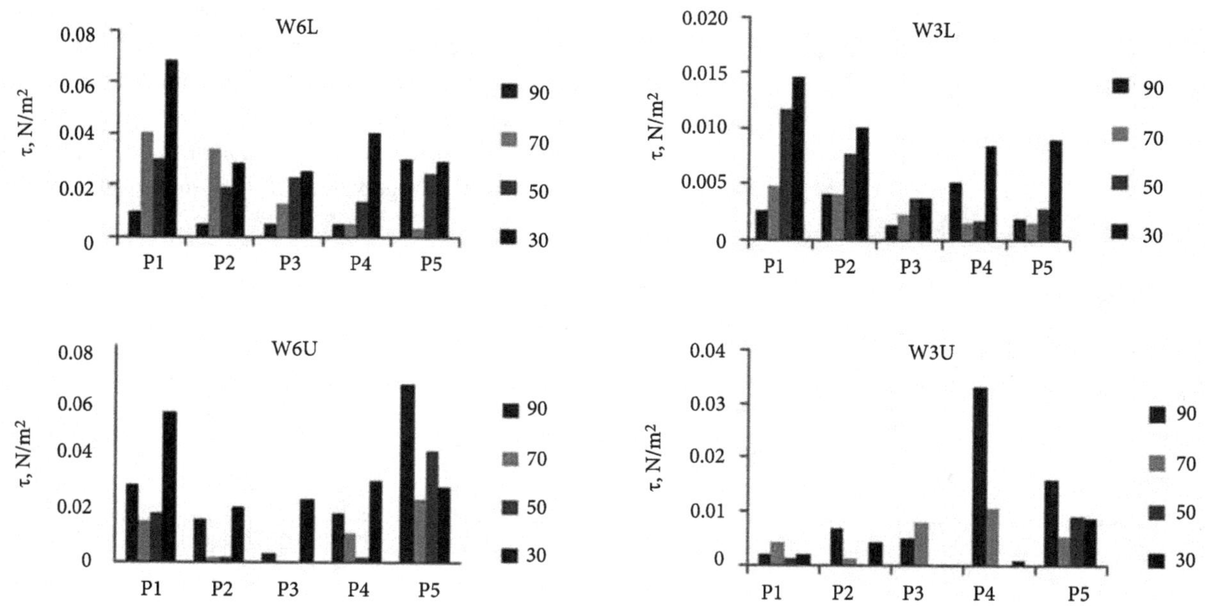

Fig. 12. The longitudinal profiles of shear stress induced by different plant stem densities under different wave conditions. W3 and W6 represent regular and irregular waves runs respectively. L and U represent the lower ($z \leq 0.15$ m) and upper ($z \geq 0.15$ m) layer within the flow depth respectively. P1 to P5 are the various positions within the vegetation field (Fig. 4)

To investigate the impact of plant stem density inducing shear stress reduction, a mean τ was estimated for the lower and upper layer respectively. Figure 12 shows the impact of plant stem density on τ as the waves travelled through the vegetation field at the various positions (1 to 5) both at the lower and upper layer. It is observed that, averagely, the τ was maximum at the upper layer for all plant stem density compared to the lower layer. The τ was smallest for the highest plant stem density at the upper layer especially within the vegetated region from Positions 2 to 4. Maximum τ reduction both at the upper and lower layer were with the lowest plant stem density (30 stem/m^2) followed by 50 stem/m^2. The shear stress was nearly linear distributed in the longitudinal direction in the downstream of vegetated region. This shows that τ variation turned to be stable within the water depth from the middle (Position 3) of the vegetated region to downstream. By comparing the near bed shear stress downstream outside of the vegetation area with that upstream before the vegetation area, we found that the near bed shear stress reduction rate by vegetation field was larger with smaller vegetation density but this trend diminishes in the upper layer towards the free water surface. This correspond with the rate of wave energy dissipation where the lower plant stem density seems to cause more energy loss as the wave travels through the vegetation field (Fig. 6). It was found that the effect of plant stem density in inducing τ decrease as the wave travels through the vegetation field compared to outside the vegetated zones (Positions 1 and 5).

2.7. Implication of wave attenuation on nearshore protection and sedimentation dynamics

Aquatic emergent vegetation has the ability to protect near shore and river banks from erosion and also to mitigate riverbed sediment resuspension (Kothyari *et al.* 2009; Anderson *et al.* 2009; Harvey *et al.* 2011). Research has proven that aquatic plants generally acting as transition zones between dry land and open water could serve as buffer and reduce storm surge and travelling waves significantly before they reach near shore or banks. Also, it is known that aquatic plants can reduce shear stress especially close to the bottom bed near sediment interface by damping wave activity effect hence contributing to reducing the rate of particle resuspension and sediment erosion reduction and also enhance sediment stability (Larsen *et al.* 2009; Gruber, Kemp 2010). In this study, it has been established that aquatic emergent plants has the ability to create rough surface impacting the travelling waves which resulted in substantial wave energy decay especially for small-orbital amplitude waves. The wave damping rate increases under emergent conditions where wave height decreased significantly as it travels through the vegetation field either for a small or long values of peak wave period. Enough work was performed by the plant stems resulting in reduction of energy and thus smaller wave heights affecting the wave-induce flows significantly especially for long wavelengths. These findings suggest that emergent plants are capable to protect the water bed and also provide protection for near shore and banks from erosion. The wave decay is a function of plants characteristics such as density and also wave conditions such as incident wave height and period. In this study it was established that shear stress which is a driving force behind sediment resuspension and erosion (Hamilton, Mitchell 1997) within the flow depth is substantially reduced under emergent conditions as the wave-induced flow is reduced. It was observed that the near bed shear reduction rate by emergent plants was high even with smaller plant stem density. The small plant stem density will be able to produce locally sheltered conditions as they were effective to minimize wave-induce flow at the bed thereby reducing shear stresses. The shear stress near the lower layer was nearly reduced zero for most flow conditions. This was in agreement with James *et al.* (2004) who reported that the bottom shear stress was very small even at high wind speeds (i.e. >30 km/h). It is reported by Dieter (1990) that in shallow marsh lakes, sediment resuspension rate in areas protected by emergent plants was only one-third of that in the open water. It was also found that, the wave-induced flow reduction hence wave height and energy decay and also shear stress reduction rate decreases as the wave travels through the vegetation field. This indicates that vegetation field size should be considered especially in restoration projects when assessing the capacity of the vegetation to stabilize the bottom bed and also protect nearshore and bank erosion. Manca *et al.* (2012) mentioned that fragmentation of the vegetation field to a series of aquatic plants will decrease its potential to minimize wave energy and stabilize the bed.

Conclusions

The present experimental study has focused on wave attenuation and transformation around an artificial emergent vegetation field under different conditions in an open-channel wave flume. The experimental results indicate that artificial emergent plants (*Phragmitis australis*) are effective in attenuating wave energy under both regular and irregular waves. The wave height measured along the vegetated region indicated wave damping along the emergent vegetated plants. An increases in plant density in the experimental runs tested induced high wave damping, heightened wave-induced flow deceleration. This trend was confirmed under both regular and irregular waves. The wave height decay had the same pattern for all the experimental run conditions and seemingly depends on the incident wave height and the plant stem density where greater decay occurred along the first few meters from the leading edge of the vegetation field and decrease as it travels through the vegetated field. The experimental wave decay coefficient changes more widely for short wave peak periods and nearly converge for longer wave periods whereas theoretical wave decay coefficient is nearly independent of the peak periods regardless of the slight decrease. A decrease of the wave period reduces the wave height along the vegetation for higher plant stem density. The wave energy attenuation as influenced by the friction induced by the emergent plants could be predicted in terms of energy dissipation factor (f_e) by an empirical formula [Nielsen (1992) model]. There was a clear decrease in f_e with an increasing A over the emergent vegetation field. However the energy decay was not uniform for all A. Within the vegetated region, the emergent plants are efficient at reducing wave-induced flows near the bed thereby reducing local momentum exchange towards the bottom bed which is good for reducing local sediment transport and resuspension. The near bed shear reduction rate by emergent plants was high even with smaller plant stem density. The results from the shear stress distribution analysis under wave-vegetation interaction shows this trend which indicates that they will be able to promote greater sediment stabilization compared to un-vegetated bed. The shear results show that conditions at the upper layer at the edge of canopy are more turbulent and wave-induced flows are less reduced than within the lower layer at the stem section where more uniform and stable hydrodynamics conditions exists. These findings would prove useful to help further our understanding of wave attenuation by emergent aquatic plants and in water system engineering such as near shore and bank protection and in restoration projects.

Acknowledgements

The research was supported by Chinese National Science Foundation (51579071, 51379061, 41323001, 51539003), and Jiangsu Province National Science Foundation (BK20131370), and National Science Funds for Creative Research Groups of China (No. 51421006); the program of Dual Innovative Talents Plan and Innovative Research Team in Jiangsu Province, the Special Fund of State Key Laboratory of Hydrology-Water Resources and Hydraulic Engineering and the Fundamental Research Funds for the Central Universities The authors are indebted to the anonymous reviewers for their constructive comments and thorough reviews of the paper.

References

Anderson, M. E.; Smith, J. M.; Mckay, S. K. 2011. Wave dissipation by vegetation, *ERDC/CHL Coastal and Hydraulics Engineering Technical Note* 1(82): 1–22.

Asano, T.; Deguchi, H.; Kobayashi, N. 1992. Interactions between water waves and vegetation, in *Proceedings of the 23rd International Conference on Coastal Engineering* ASCE, 4–9 October 1992, Venice, Italy, 2710–2723.

Augustin, L. N.; Irish, J. L.; Lynett, P. 2009. Laboratory and numerical studies of wave damping by emergent and near-emergent wetland vegetation, *Coastal Engineering* 56: 332–340. http://dx.doi.org/10.1016/j.coastaleng.2008.09.004

Bradley, K.; Houser, C. 2009. Relative velocity of seagrass blades: implications for wave attenuation in low-energy environments, *Journal of Geo-physical Research* 114: 1–14. http://dx.doi.org/10.1029/2007JF000951

Bouma, T. J.; De Vries, M. B.; Low, E.; Peralta, G., *et al.* 2005. Trade-offs related to ecosystem engineering: a case study on stiffness of emerging macrophytes, *Ecology* 86(8): 2187–2199. http://dx.doi.org/10.1890/04-1588

Cavallaro, L.; Re, C. L.; Paratore, G.; Viviano, A.; Foti, E. 2010. Response of *Posidonia oceanica* to wave motion in shallow-waters: preliminary experimental results, in *Proceedings of the 32nd International Conference on Coastal Engineering, Coastal Engineering Research Council*, 30 June – 5 July 2010, Shanghai, China, 1–10.

Cheng, R. T.; Ling, C.; Gartner, J. W. 1999. Estimates of bottom roughness length and bottom shear stress in San Francisco Bay, California, *Journal of Geophysical Research* 104: 7715–7728. http://dx.doi.org/10.1029/1998JC900126

Chen, S. C.; Kuo, Y. M.; Li, Y. H. 2011. Flow characteristics within different configurations of submerged flexible vegetation, *Journal of Hydrology* 398: 124–134. http://dx.doi.org/10.1016/j.jhydrol.2010.12.018

Cooper, N. J. 2005. Wave dissipation across intertidal surfaces in the Wash Tidal inlet, Eastern England, *Journal of Coastal Research* 21(1): 28–40. http://dx.doi.org/10.2112/01002.1

Chinese Vegetation Editional Council (CVEC). 1980. *Vegetation in China*. Beijing: Science Press (in Chinese).

Dalrymple, R. A.; Kirby, J. T.; Hwang, P. A. 1984. Wave diffraction due to areas of energy dissipation, *Journal of Waterway, Port, Coastal, and Ocean Engineering* 110(1): 67–79. http://dx.doi.org/10.1061/(ASCE)0733-950X(1984)110:1(67)

Dean, R. G.; Bender, C. J. 2006. Static wave setup with emphasis on damping effects by vegetation and bottom friction, *Coastal Engineering* 53: 149–156. http://dx.doi.org/10.1016/j.coastaleng.2005.10.005

Dieter, C. D. 1990. The importance of emergent vegetation in reducing sediment resuspension in wetlands, *Journal of Freshwater Ecology* 5(4): 467–473. http://dx.doi.org/10.1080/02705060.1990.9665263

Dubi, A.; Tørum, A. 1996. Wave energy dissipation in kelp vegetation, in *Proceedings of the 25th International Conference on Coastal Engineering,* ASCE, 2–6 September 1996, Orlando, USA, 2626–2639.

Elwany, M. H. S.; O'Reilly, W. C.; Guza, R. T.; Flick, R. E. 1995. Effects of Southern California kelp beds on waves, *Journal of Waterway Port Coastal Ocean Engineering* 121(2): 143–150. http://dx.doi.org/10.1061/(ASCE)0733-950X(1995)121:2(143)

Fathi-Maghadam, M.; Kouwen, N. 1997. Nonrigid, nonsubmerged, vegetative roughness on floodplains, *Journal of Hydraulic Engineering* 123(1): 51–57. http://dx.doi.org/10.1061/(ASCE)0733-9429(1997)123:1(51)

Fiala, K. 1976. Underground organs of *Phragmites communis*, their growth, biomass and net production, *Folia Geobotanica et Phytotaxonomica* 11(3): 225–259. http://dx.doi.org/ 10.1007/BF02909471

Fonseca, M. S.; Cahalan, J. A. 1992. A preliminary evaluation of wave attenuation by four species of seagrass, *Estuarine, Coastal, and Shelf Science* 35(6): 565–576. http://dx.doi.org/10.1016/S0272-7714(05)80039-3

Galema, A. 2009. *Evaluation of vegetation resistance descriptors for flood management*: Master Thesis. University of Twente, Enschede, Netherlands.

Gruber, R. K.; Kemp, W. M. 2010. Feedback effects in a coastal canopy-forming submerged plant bed, *Limnology and Oceanography* 55(6): 2285–2298. http://dx.doi.org/10.4319/lo.2010.55.6.2285

Hamilton, D. P.; Mitchell, S. F. 1997. Wave-induced shear stresses, plant nutrients and chlorophyll in seven shallow lakes, *Freshwater Biology* 38: 159–168. http://dx.doi.org/10.1046/j.1365-2427.1997.00202.x

Harvey, J. W.; Noe, G. B.; Larsen, L. G., *et al.* 2011. Field flume reveals aquatic vegetation's role in sediment and particulate phosphorus transport in a shallow aquatic ecosystem, *Geomorphology* 126(3–4): 297–313. http://dx.doi.org/10.1016/j.geomorph.2010.03.028

Houwing, E. J.; Tanczos, I. C.; Kroon, A.; de Vries, M. B. 2002. Interaction of submerged vegetation, hydrodynamics and turbidity; analysis of field and laboratory studies, *Proceedings in Marine Science* 5: 441–453. http://dx.doi.org/10.1016/S1568-2692(02)80032-8

Huang, Z.; Yao, Y.; Sim, S. Y.; Yao, Y. 2011. Interaction of solitary waves with emergent, rigid vegetation, *Ocean Engineering* 38: 1080–1088. http://dx.doi.org/10.1016/j.oceaneng.2011.03.003

Infantes, E.; Orfila, A.; Simarro, G., *et al.* 2012. Effect of a seagrass (*Posidonia oceanica*) meadow on wave propagation, *Marine Ecology Progress Series* 456: 63–72. http://dx.doi.org/10.3354/meps09754

James, W. F.; Barko, J. W.; Butler, M. G. 2004. Shear stress and sediment resuspension in relation to submersed macrophyte biomass, *Hydrobiologia* 515: 181–191. http://dx.doi.org/10.1023/B:HYDR.0000027329.67391.c6

Järvelä, J. 2005. Effect of submerged flexible vegetation on flow structure and resistance, *Journal of Hydrology* 307: 233–241. http://dx.doi.org/10.1016/j.jhydrol.2004.10.013

Jonsson, I. G. 1966. Wave boundary layers and friction factors, in *Proceedings of 10th International Conference on Coastal Engineering*, September 1966, Tokyo, Japan, 127–148.

Knutson, P. L.; Bronchu, R. A.; Seelig, W. N. 1982. Wave damping in *Spartina alternifora* marshes, *Wetlands* 2(1): 87–104. http://dx.doi.org/10.1007/BF03160548

Kobayashi, N.; Raichle, A. W.; Asano, T. 1993. Wave attenuation by vegetation, *Journal of Waterway, Port, Coastal, and Ocean Engineering* 119(1): 30–48. http://dx.doi.org/10.1061/(ASCE)0733-950X(1993)119:1(30)

Koch, E. W.; Barbier, E. B.; Silliman, B. R., *et al.* 2009. Non-linearity in ecosystem services: temporal and spatial variability in coastal protection, *Frontiers in Ecology and the Environment* 7: 29–37. http://dx.doi.org/10.1890/080126

Koch, E. W.; Sanford, L. P.; Chen, S-N.; Shafer, D. J.; Smith, J. M. 2006. Waves in seagrass systems: review and technical recommendations. United States Army Corps of Engineers.

Kothyari, U. C.; Hashimoto, H.; Hayashi, K. 2009. Effect of tall vegetation on sediment transport by channel flows, *Journal of Hydraulic Research* 47: 700–710. http://dx.doi.org/10.3826/jhr.2009.3317

Kouwen, N.; Fathi-Moghadam, M. 1996. Friction factors for vegetation, in *The Proceedings of 2nd International Symposium on Habitat Hydraulics, Ecohydraulics*, A251-A267, June 1996, IAHR/INRS-EAU, Quebec, Canada.

Krauss, K. W.; Doyle, T. W.; Doyle, T. J., *et al.* 2009. Water level observations in mangrove swamps during two hurricanes in Florida, *Wetlands* 29(1): 142–249. http://dx.doi.org/10.1672/C7-232.1

Larsen, L. G.; Harvey, J. W.; Crimaldi, J. P. 2009. Predicting bed shear stress and its role in sediment dynamics and restoration potential of the Everglades and other vegetated flow systems, *Ecological Engineering* 35(12): 1773–1785. http://dx.doi.org/10.1016/j.ecoleng.2009.09.002

Li, Y.; Anim, D. O.; Ying, W.; Chunyang, T.; Wei, D.; Yu, Z.; Kumud, A. 2014. An open channel flume study of flow characteristics through combined layer of submerged and emerged flexible vegetation, *Ecohydrology* 7: 633–647. http://dx.doi.org/10.1002/eco.1384

Lovas, S. M. 2000. Hydro-physical condition in kelp forests and effect on wave damping and dune erosion: a case study on *Laminaria Hyperborea*: PhD Thesis, The Norwegian University of Science and Technology, Trondheim, Norway.

Lövstedt, C. B.; Larson, M. 2010. Wave damping in reed: Field measurements and mathematical modelling, *Journal of Hydraulic Engineering* 136(4): 222–233. http://dx.doi.org/10.1061/(ASCE)HY.1943-7900.0000167

Lowe, R. J.; Falter, J. L.; Bandet, M. D.; Pawlak, G.; Atkinson, M. J.; Monismith, S. G.; Koseff, J. R. 2005. Spectral wave dissipation over a barrier reef, *Journal of Geophysical Research: Oceans* 110(C04001): 1–16.

Ma, R.; Duan, H.; Gu, X.; Zhang, S. 2008. Detecting aquatic vegetation changes in Taihu Lake, China using multi-temporal satellite imagery, *Sensors* 8(5): 3988–4005. http://dx.doi.org/10.3390/s8063988

Manca, E.; Caceres, I.; Alsina, J. M., *et al.* 2012. Wave energy and wave-induced flow reduction by full-scale model *Posidonia oceanica* seagrass, *Continental Shelf Research* 50–51: 100–116. http://dx.doi.org/10.1016/j.csr.2012.10.008

Manca, E.; Stratigaki, V.; Prinos, P. 2010. Large scale experiments on spectral wave propagation over *Posidonia oceanica* seagrass, in *Proceedings of 6th International Symposium on Environmental Hydraulics, 23–25 June 2010*, Athens, Greece, 463–468. http://dx.doi.org/10.1201/b10553-74

Mazda, Y.; Magi, M.; Kogo, M.; Hong, P. N. 1997. Mangroves a coastal protection from waves in the Tong King delta, Vietnam, *Mangroves and Salt Marshes* 1: 127–135. http://dx.doi.org/10.1023/A:1009928003700

Mendez, F. J.; Losada, I. J. 2004. An empirical model to estimate the propagation of random breaking and nonbreaking waves over vegetation fields, *Coastal Engineering* 53: 103–118. http://dx.doi.org/10.1016/j.coastaleng.2003.11.003

Mendez, F. J.; Losada, I. J., Losada, M. A. 1999. Hydrodynamics induced by wind waves in a vegetation field, *Journal of Geophysical Research* 104(C8): 18383–18396. http://dx.doi.org/10.1029/1999JC900119

Mizuguchi, M. 1982. Individual wave analysis of irregular wave deformation in the nearshore zone, *in Proceedings of 18th*

Coastal Engineering Conference, 14–19 November 1982, Cape Town, South Africa, 485–504.

Möller, I. 2006. Quantifying saltmarsh vegetation and its effect on wave dissipation. Results from the UK East Coast saltmarsh, *Estuarine, Coastal, and Shelf Science* 69: 337–351. http://dx.doi.org/10.1016/j.ecss.2006.05.003

Möller, I.; Spencer, T. 2002. Wave dissipation over macrotidal saltmarshes: Effects of marsh edge typology and vegetation change, *Journal of Coastal Research* S136: 506–521.

Möller, I.; Spencer, T.; French, J. R.; Leggett, D. J.; Dixon, M. 1999. Wave transformation over salt marshes: a field and numerical modeling study from North Norfolk, England, *Estuarine, Coastal, and Shelf Science* 49(3): 411–426. http://dx.doi.org/10.1006/ecss.1999.0509

Morison, J. R.; O'Brien, M. P.; Johnson, J. W.; Schaaf, S. A. 1950. The force exerted by surface waves on piles, *Journal of Petroleum Technology* 2(5): 149–154. http://dx.doi.org/10.2118/950149-G

Mork, M. 1996. The effect of kelp in wave damping, *Sarsia* 80: 323–327.

Mullarney, J. C.; Henderson, S. M. 2010. Wave-forced motion of submerged single-stem vegetation, *Journal of Geophysical Research* 115(C12). http://dx.doi.org/10.1029/2010JC006448

Harvey, J. W.; Noe, G. B.; Larsen, L. G.; Nowacki, D. J.; McPhillips, L. E. 2011. Field flume reveals aquatic vegetation's role in sediment and particulate phosphorus transport in a shallow aquatic ecosystem, *Geomorphology* 126: 297–313. http://dx.doi.org/10.1016/j.geomorph.2010.03.028

Nepf, H. M.; Vivoni, E. R. 2000. Flow structure in depth-limited vegetated flow, *Journal of Geophysical Research* 105: 547–557. http://dx.doi.org/10.1029/2000JC900145

Nielsen, P. 1992. Coastal bottom boundary layers and sediment transport, in *Advanced Series on Ocean Engineering* 4. Singapore: World Science.

Paul, M.; Amos, C. L. 2011. Spatial and seasonal variation in wave attenuation over *Zostera noltii*, *Journal of Geophysical Research* 116 (C08019): 1–16.

Quartel, S.; Kroon, A.; Augustinus, P. G. E. F.; Van San-ten, P.; Tri, N. H. 2007. Wave attenuation in coastal mangroves in the Red River Delta, Vietnam, *Journal of Asian Earth Sciences* 29(4): 576–584. http://dx.doi.org/10.1016/j.jseaes.2006.05.008

Rashid, S. M. H. 2010. *Effectiveness of widely used critical velocity and bed shear stress equations for different types of sediment beds*: Master's Thesis, Washington State University [online], [cited 20 May 2014]. Available from Internet: http://www.dissertations.wsu.edu/Thesis/Fall2010/s_rashid_113010.pdf

Soulsby, R. L. 1997. *Dynamics of marine sands: a manual for practical application*. London: Thomas Telford.

Stratigaki, V.; Manca, E.; Prinos, P., *et al.* 2011. Large-scale experiments on wave propagation over *Posidonia oceanica*, *Journal of Hydraulic Research* 49: 31–43. http://dx.doi.org/10.1080/00221686.2011.583388

Suzuki, T.; Zijlema, M.; Burger, B.; Meijer, M.C.; Narayan, S. 2011. Wave dissipation by vegetation with layer schematization in SWAN, *Coastal Engineering* 59: 64–71. http://dx.doi.org/10.1016/j.coastaleng.2011.07.006

Thompson, C. E. L.; Amos, C. L.; Lecouturier, M.; Jones, T. E. R. 2004. Flow deceleration as a method of determining drag coefficient over roughened flat beds, *Journal of Geophysical Research* 109: 1–12. http://dx.doi.org/10.1029/2001JC001262

Suzuki, T.; Zijlema, M.; Burger, B.; Meijer, M. C.; Narayan, S. 2011. Wave dissipation by vegetation with layer schematization in SWAN, *Coastal Engineering* 59: 64–71.

Tschirky, P.; Hall, K.; Turcke, D. 2000. Wave attenuation by emergent wetland vegetation, in *Proceedings of the 27th International Conference on Coastal Engineering*, ASCE, 16–21 July 2000, Sydney, Australia, 865–877.

Yan, S.; Yu, H.; Zhang, L.; Xu, J.; Wang, Z. 2011. Water quantity and pollutant fluxes of inflow and outflow rivers of Lake Taihu, 2009, *Journal of Lake Science* 23(6): 855–862 (in Chinese). http://dx.doi.org/10.18307/2011.0605

Zhang, S. R. 2009. *Common wetland plants in China*. Beijing: Science Press.

Zhang, Y.; Xu, Y.; Xia, J. 2011. Analysis of dispersion and attenuation of surface waves in poroelastic media in the exploration-seismic frequency band, *Geophysical Journal International* 187: 871–888. http://dx.doi.org/10.1111/j.1365-246X.2011.05168.x

Li YIPING is an Associate professor at the College of Environment, Hohai University, China and the director of the Department of Environmental Engineering. He received his PhD in Environmental engineering from Hohai University, China in 2006. His research involves lake ecosystems and numerical models. He is especially interested in the whole process of hydrologic cycle and contaminant transport among land, surface water and groundwater and water environment protection. His present research is focused on the integrated watershed model linking hydrodynamic, water quality, optical, marine science and ecological models. He has published more than 50 peer-reviewed scientific articles in high impact journals and presented his work at international conferences.

Desmond Ofosu ANIM is a current postgraduate researcher (Environmental Engineering) at the College of Environment, Hohai University, China. Research interests include water environment protection, water pollution control, ecohydrology, storm water management. He is an author and co-author to over 15 scientific articles in high impact peer-reviewed journals and also presented his work at international conferences.

Ying WANG is a researcher at the Fujian Provincial Investigation, Design and Research Institute of Water Conservancy and Hydropower, China. She holds a Master's degree in Environmental engineering, Hohai University. Research interests include water resources management, environmental pollution control. She is an author and co-author to over 10 scientific articles in high impact peer-reviewed journals.

Chunyang TANG is a current PhD fellow (Environmental Science) at the College of Environment, Hohai University, China. Research interests include water environment protection, water pollution control, lake management, water resources hydrodynamic modeling. She is an author and co-author to over 20 scientific articles in high impact peer-reviewed journals. Tang has presented her works at various international conferences.

Wei DU is a current PhD fellow (Environmental Engineering) at the College of Environment, Hohai University, China. Research interests include water resources protection, water pollution control, river and lake hydrodynamic modeling. She is an author and co-author to over 7 scientific articles in high impact peer-reviewed journals.

Ni LIXIAO is currently an Associate professor at the College of Environment, Hohai University, China. She received her Ph.D degree from Nanjing University, China in 2004. Her research interests involve wastewater treatment and restoration of the aquatic ecosystems. She is especially interested in the algae control and wastewater treatment with constructed wetland. She is an author and co-author to over 30 scientific articles in high impact peer-reviewed journals.

Zhongbo YU, Professor, is currently the Dean of the College of Hydrology and Water Resources, Hohai University, China and also a professor of Hydrogeology and Hydrology at the Department of Geoscience, University of Nevada, Las Vegas. Zhongbo Yu has conducted research on development and application of distributed hydrologic models for hydrology and climate studies; characterization and evaluation of surface water, groundwater, and groundwater contamination; multiphase and multicomponent transport in saturated and unsaturated porous media; image processing and computing visualization. His current studies include the distributed modeling of hydrologic response to atmospheric forcing in the Susquehanna river basin, field monitoring and numerical modeling of dynamic hydrologic processes in small watersheds, and interactions among atmosphere, land surface, and groundwater. He is an author and co-author to over 200 scientific articles in high impact peer-reviewed journals and also serve as both editor and reviewer to various scientific journals. He has given presentation and lecture at various international conferences and workshops.

Kumud ACHARYA is a research professor of ecological engineering at the Division of Hydrologic Sciences, Desert Research Institute, Las Vegas. He hold PhD degree in Biology and Environmental Science, Saitama University, Saitama, Japan. His research interests include water quality, environmental restoration, invasive species, hydrology, eco-hydrology, nutrient cycling, wastewater treatment systems, groundwater management, and ecological modeling. His specific studies involve observational and experimental studies at various scales, including laboratory cultures (zooplankton, algal chemostats), short-term field experiments and sustained whole-ecosystem manipulations. He is an author and co-author to over 150 scientific articles in high impact peer-reviewed journals and also serve as both editor and reviewer to various scientific journals, such as Journal of Hydrologic Engineering, Journal of the American Water Resources Association and so on. He has given presentation and lecture at various international conferences and workshops.

Li CHEN is an Associate Research Professor at the Division of Hydrologic Sciences, Desert Research Institute, Las Vegas. He hold PhD degree in Fluid Mechanics, Chinese Academy of. Research interests include water movement and water-initiated material transport in surface and shallow ground environment that include dynamics and numerical modeling of rainfall-infiltration-runoff process; sediment transport monitoring and modeling in natural water body; hydrodynamic and contaminant transport modeling in near surface water system; soil erosion dynamics and geomorphologic evolution. He is an author and co-author to over 60 scientific articles in high impact peer-reviewed journals.

Permissions

All chapters in this book were first published in JEELM, by Vilnius Gediminas Technical University; hereby published with permission under the Creative Commons Attribution License or equivalent. Every chapter published in this book has been scrutinized by our experts. Their significance has been extensively debated. The topics covered herein carry significant findings which will fuel the growth of the discipline. They may even be implemented as practical applications or may be referred to as a beginning point for another development.

The contributors of this book come from diverse backgrounds, making this book a truly international effort. This book will bring forth new frontiers with its revolutionizing research information and detailed analysis of the nascent developments around the world.

We would like to thank all the contributing authors for lending their expertise to make the book truly unique. They have played a crucial role in the development of this book. Without their invaluable contributions this book wouldn't have been possible. They have made vital efforts to compile up to date information on the varied aspects of this subject to make this book a valuable addition to the collection of many professionals and students.

This book was conceptualized with the vision of imparting up-to-date information and advanced data in this field. To ensure the same, a matchless editorial board was set up. Every individual on the board went through rigorous rounds of assessment to prove their worth. After which they invested a large part of their time researching and compiling the most relevant data for our readers.

The editorial board has been involved in producing this book since its inception. They have spent rigorous hours researching and exploring the diverse topics which have resulted in the successful publishing of this book. They have passed on their knowledge of decades through this book. To expedite this challenging task, the publisher supported the team at every step. A small team of assistant editors was also appointed to further simplify the editing procedure and attain best results for the readers.

Apart from the editorial board, the designing team has also invested a significant amount of their time in understanding the subject and creating the most relevant covers. They scrutinized every image to scout for the most suitable representation of the subject and create an appropriate cover for the book.

The publishing team has been an ardent support to the editorial, designing and production team. Their endless efforts to recruit the best for this project, has resulted in the accomplishment of this book. They are a veteran in the field of academics and their pool of knowledge is as vast as their experience in printing. Their expertise and guidance has proved useful at every step. Their uncompromising quality standards have made this book an exceptional effort. Their encouragement from time to time has been an inspiration for everyone.

The publisher and the editorial board hope that this book will prove to be a valuable piece of knowledge for researchers, students, practitioners and scholars across the globe.

List of Contributors

Zongming Wang, Chunying Ren, Hao Yu, Zhangyu Dong, Chunyan Lu and Dehua Mao
Key Laboratory of Wetland Ecology and Environment, Northeast Institute of Geography and Agroecology, Chinese Academy of Sciences, 130102 Changchun, China

Yue Zhang
Key Laboratory of Wetland Ecology and Environment, Northeast Institute of Geography and Agroecology, Chinese Academy of Sciences, 130102 Changchun, China
College of Resources and Environment, Jilin Agricultural University, 130118 Changchun, China
University of Chinese Academy of Sciences, 100049 Beijing, China

Ivan Y. Tang and Irene M. C. Lo
Department of Civil and Environmental Engineering, The Hong Kong University of Science and Technology, Hong Kong, China

Dickson Y. S. Yan
Faculty of Science and Technology, Technological and Higher Education Institute of Hong Kong, Hong Kong, China

Tongzhou Liu
Harbin Institute of Technology Shenzhen Graduate School, Shenzhen Key Laboratory of Water Resource Utilization and Environmental Pollution Control, Shenzhen 518055, China

Antanas Juostas and Algirdas Janulevičius
Aleksandras Stulginskis University, Studentų g. 15, 53361 Akademija, Kauno r., Lithuania
Department of Environmental Engineering, Engineering Faculty, Cumhuriyet University, Sivas, Turkey

József Faitli and Imre Gombkötő
Institute of Raw Material Preparation and Environmental Processing, University of Miskolc, Miskolc - Egyetemváros H-3515, Hungary

Hongsheng Liu
Institute of Geography Sciences and Natural Resource Research, Chinese Academy of Sciences, 11A Datun Road, Beijing 100101, China
Department of Biology, University of Texas at Arlington, 701 Nedderman Dr. Arlington, Texas, 76013, USA

Zewei Miao
Nicholas School of the Environment, Duke University, 308 Research Drive, Durham, NC 27708, USA

Cengiz Kahraman, Sezi Cevik Onar and Basar Oztaysi
Department of Industrial Engineering, Istanbul Technical University, 34367 Macka, Istanbul, Turkey

Mehdi Keshavarz Ghorabaee
Department of Industrial Management, Faculty of Management and Accounting, Allameh Tabataba'i University, Tehran, Iran

Edmundas Kazimieras Zavadskas
Department of Construction Technology and Management, Faculty of Civil Engineering, Vilnius Gediminas Technical University, Lithuania

Morteza Yazdani
Department of Management Science, Universidad Europea de Madrid, Madrid, Spain

Gábor Mucsi, Ágnes Szenczi and Zoltán Molnár
Institute of Raw Material Preparation and Environmental Processing, University of Miskolc, 3515 Miskolc-Egyetemváros, Hungary

János Lakatos
Institute of Chemistry, University of Miskolc, 3515 Miskolc-Egyetemváros, Hungary

Jerzy Sęk, Olga S. Shtyka and Kamila Szymczak
Department of Chemical Engineering, Faculty of Process and Environmental Engineering, Lodz University of Technology, Wolczanska St. 213, 90-924 Lodz, Poland

Mariano Norzagaray-Campos, Leticia Espinosa-Carreón, Héctor González-Ocampo and Omar Llanes-Cárdenas
Instituto Politécnico Nacional-CIIDIR-Sinaloa, Blvd. Juan de Dios Bátiz Paredes No. 250, Colonia San Joachín, Guasave Sinaloa, México CP

Patricia Muñoz-Sevilla
Instituto Politécnico Nacional - CIIEMAD, Calle 30 de Junio de 1520 s/n, Barrio la Laguna Ticomán C.P. 07340 Del. Gustavo A. Madero, Ciudad de México, México

Rosario Ruíz-Guerrero
Instituto Politécnico Nacional- CIITEC, Cerrada de Cecati S/N. Col. Santa Catarina Azcapotzalco Ciudad de México, México

Debishree Khan and Sukha Ranjan Samadder
Department of Environmental Science & Engineering, Indian School of Mines, Dhanbad-826004, India

Renata Dagiliūtė and Gintarė Juozapaitienė
Department of Environmental Science, Vytautas Magnus University, Vileikos g. 8, LT-44404 Kaunas, Lithuania

Mohsen Mostafa and Shaban Shataee Jouibary
Department of Forestry, Faculty of Forest Sciences, Gorgan University of Agricultural Sciences and Natural Resources, Gorgan, Iran

Majid Lotfalian
Department of Forest Engineering, Faculty of Natural Resources, Sari University of Agricultural Sciences and Natural Resources, Sari, Iran

Amir Sadoddin
Department of Watershed Management, Faculty of Rangeland and Watershed, Gorgan University of Agricultural Sciences and Natural Resources, Gorgan, Iran

Ramunė Urbonienė
Landscape Architecture and Environmental Planning, Klaipėda University, Herkaus Manto g. 84, LT-92294 Klaipėda, Lithuania

Loreta Kelpšaitė
Geophysical Department, Klaipėda University, Herkaus Manto g. 84, LT-92294 Klaipėda, Lithuania

Ingrida Borisenko
Computer Science Department, Klaipėda University, Herkaus Manto g. 84, LT-92294 Klaipėda, Lithuania

Himadri Rajput, Anoop Verma and Manpreet Kaur
School of Energy and Environment, Thapar University, Thapar Technology Campus, Patiala 147004, India

Taranjeet Kaur and Amrit Pal Toor
Dr. S. S. Bhatnagar University Institute of Chemical Engineering and Technology, Panjab University, Sector 14, Chandigarh 160014, India

Deblina Maiti and Bably Prasad
Natural Resources and Environment Management, CSIR - Central Institute of Mining and Fuel Research, Dhanbad, India
Academy of Scientific and Innovative Research (AcSIR), CSIR - Central Institute of Mining and Fuel Research, Dhanbad, India

Yuvarajan D
Department of Mechanical Engineering, Jeppiaar Institute of Technology, Chennai, India

Venkata Ramanan M
Department of Mechanical Engineering, College of Engineering, Anna University, Chennai, India

Pradnya K. Ingle, Attarkar Karishma and Virendra K. Rathod
Department of Chemical Engineering, Institute of Chemical Technology, Matunga (E), Mumbai-400019, India

Saulius Vasarevičius and Vaidotas Danila
Department of Environmental Protection, Faculty of Environmental Engineering, Vilnius Gediminas Technical University, Saulėtekio al. 11, LT-10223 Vilnius, Lithuania

Gintautas Skripkiūnas
Department of Building Materials, Faculty of Civil Engineering, Vilnius Gediminas Technical University, Saulėtekio al. 11, LT-10223 Vilnius, Lithuania

Victoria O. Owolana
Department of Architecture and the Built Environment, Faculty of Environment and Technology, University of the West of England, Bristol BS16 1QY, UK

Colin A. Booth
Department of Architecture and the Built Environment, Faculty of Environment and Technology, University of the West of England, Bristol BS16 1QY, UK

Osman Çerezci, Baha Kanberoğlu and Şuayb Çağri Yener
Department of Electric-Electronic Engineering, Faculty of Engineering, Sakarya University, Esentepe Campus, 54055, Serdivan, Sakarya, Turkey

Index

www.ingramcontent.com/pod-product-compliance
Lightning Source LLC
LaVergne TN
LVHW060529210926
848485LV00005B/1172